THE SCATTERING OF ELECTROMAGNETIC WAVES FROM ROUGH SURFACES

The Artech House Radar Library

Radar System Analysis by David K. Barton
Introduction to Electronic Warfare by D. Curtis Schleher
Electronic Intelligence: The Analysis of Radar Signals by Richard G. Wiley
Electronic Intelligence: The Interception of Radar Signals by Richard G. Wiley
Principles of Secure Communication Systems by Don J. Torrieri
Multiple-Target Tracking with Radar Applications by Samuel S. Blackman
Solid-State Radar Transmitters by Edward D. Ostroff et al.
Logarithmic Amplification by Richard Smith Hughes
Radar Propagation at Low Altitudes by M.L. Meeks
Radar Cross Section by Eugene F. Knott, John F. Shaeffer, and Michael T. Tuley
Radar Anti-Jamming Techniques by M.V. Maksimov et al.
Introduction to Synthetic Array and Imaging Radars by S.A. Hovanessian
Radar Detection and Tracking Systems by S.A. Hovanessian
Radar System Design and Analysis by S.A. Hovanessian
Radar Signal Processing by Bernard Lewis, Frank Kretschmer, and Wesley Shelton
Radar Calculations Using the TI-59 Programmable Calculator by William A. Skillman
Radar Calculations Using Personal Computers by William A. Skillman
Techniques of Radar Reflectivity Measurement, Nicholas C. Currie, ed.
Monopulse Principles and Techniques by Samuel M. Sherman
Receiving Systems Design by Stephen J. Erst
Designing Control Systems by Olis Rubin
Advanced Mathematics for Practicing Engineers by Kurt Arbenz and Alfred Wohlhauser
Radar Reflectivity of Land and Sea by Maurice W. Long
High Resolution Radar Imaging by Dean L. Mensa
Introduction to Monopulse by Donald R. Rhodes
Probability and Information Theory, with Applications to Radar by P.M. Woodward
Radar Detection by J.V. DiFranco and W.L. Rubin
Synthetic Aperture Radar, John J. Kovaly, ed.
Infrared-to-Millimeter Wavelength Detectors, Frank R. Arams, ed.
Significant Phased Array Papers, R.C. Hansen, ed.
Handbook of Radar Measurement by David K. Barton and Harold R. Ward
Statistical Theory of Extended Radar Targets by R.V. Ostrovityanov and F.A. Basalov
Radar Technology, Eli Brookner, ed.
MTI Radar, D. Curtis Schleher, ed.

THE SCATTERING OF ELECTROMAGNETIC WAVES FROM ROUGH SURFACES

Petr Beckmann

André Spizzichino

ARTECH HOUSE, INC.
685 Canton Street
Norwood, MA 02062

International Standard Book Number: 0-89006-238-2
Library of Congress Catalog Card Number: 87-70042

10 9 8 7 6 5 4 3 2 1

CONTENTS

Part I — THEORY

By Petr Beckmann

Part II — APPLICATION

By ANDRÉ SPIZZICHINO

(*Translated by Petr Beckmann*)

Part I

THEORY

by

PETR BECKMANN

CHAPTER 1

INTRODUCTION

WHEN an electromagnetic wave is incident on a plane interface between two media, it is reflected according to well-known laws: the reflected field depends on the wavelength, the angle of incidence and the electrical properties (permittivity, permeability and conductivity) of the two adjoining media. The laws of reflection by a plane boundary are, in fact, so well understood that, conversely, the electrical properties of a material are often determined by measuring its reflection coefficient.

What happens, however, if the boundary is not plane, but irregular, with periodic or random variations of height measured from a certain mean plane, or simply "rough"?

The general and exact answer to this question is as yet unknown. The approximate solutions of the problem, solutions of specific types of rough surface, and their applications are the subject of this monograph.

Perhaps the most striking difference in the behaviour of a smooth and a rough surface is the fact that a smooth plane (of sufficiently large dimensions) will reflect the incident wave specularly in a single direction, whilst a rough surface will scatter it into various directions, though certain privileged directions may receive more energy than others. We shall use this fact to define a rough surface: it is a surface which will scatter the energy of an incident plane wave into various directions, whereas a surface that reflects specularly will be called smooth. According to this definition, the same surface may be rough for some wavelengths and smooth for others; or for the same wavelength it may be either rough or smooth for different angles of incidence. This is in harmony with common usage, for unpolished sheet metal is certainly considered smooth in the radio spectrum, though it reflects light only diffusely; and though the setting sun is reflected by an asphalt road in a phenomenon well-known to motorists as "road glare", it will not be mirrored by an asphalt surface when it stands high above the horizon. A smooth surface is thus the limiting case of a rough one. This limit depends, as we shall see, on the wavelength and angle of incidence of the incident radiation.

The problem of scattering from a rough surface, which we have as yet defined only very loosely, has aroused the interest of physicists and engineers for the last 60 years and has become of special interest in recent years, particularly in connection with the propagation of radio waves at frequencies above 30 Mc/s. The problem of scattering from rough surfaces arises, among others, in the following subjects:

In line-of-sight radio communication on very short waves the field at the receiving point may be broken up into that of the direct ray and the ray reflected from the surface of the earth. In the case of a smooth earth the field may be accurately computed; normally its phase is such that the total field is diminished by it. The terrain between the terminals of a communication path, however, is not usually smooth and in the absence of a reliable theory of rough-surface scatter the field strength at the receiving point cannot be accurately predicted. If the communication path lies above the surface of the sea, the scattering surface will not only be rough, but in addition time-varying; the received signal will therefore fade and the character of the fading will be affected by the form and movement of the water waves.

In tropospheric propagation, very short radio waves are scattered beyond the horizon by rough layers in the atmosphere; as they are in constant movement, the instantaneous field strength beyond the horizon will rapidly fluctuate. Similarly, short waves are often scattered by irregularities in the ionospheric F layer instead of being reflected by a smooth ionosphere.

In radar, rapidly fluctuating reflections from facets of the surface of the sea, known as "sea returns" or "sea clutter", may seriously impair effective signals from targets on or near the surface of the sea; in a rough sea the target may be obscured altogether. Effective measures against sea clutter depend on the knowledge of the scattering characteristics.

In radio astronomy, the surface roughness of the Moon is estimated by studying the distortion of the radar pulse back-scattered from the Moon to the Earth.

In acoustics, the sound engineer often endeavours to make the acoustic field diffuse by scattering sound waves from rough surfaces. Analytically, the case of sound waves is simpler than that of electromagnetic waves because the field of the latter is given by the solution of the vector wave equation, whilst the acoustic field is given by the solution of a scalar wave equation. Mathematically, the problem is to solve the wave equation

$$\nabla^2\psi + k^2\psi = 0 \tag{1}$$

where, in the acoustic case, the wave function ψ is the velocity potential, and in the electromegnetic case ψ stands for a component of any of the field vectors or their electromagnetic potentials; ψ (and all other time-dependent quantities in this monograph) are assumed harmonic in time and k is the propagation constant. Equation (1) must be solved for the boundary conditions imposed by the rough surface S. These boundary conditions are generally determined by the value of ψ and/or its normal derivative at that surface. There are two cases where the acoustic and electromagnetic boundary conditions are in exact analogy, namely

$$(\psi)_S = 0 \tag{2}$$

which in acoustics corresponds to the case of a completely free surface, and in electromagnetic problems to a perfectly conducting surface with ψ representing the tangential component of the electric field strength; and

$$\left(\frac{\partial \psi}{\partial n}\right)_S = 0 \tag{3}$$

which in acoustics corresponds to a rigid surface and in electromagnetic problems to a perfectly conducting surface with ψ representing the tangential component of the magnetic field strength. This monograph is primarily concerned with the scattering of electromagnetic waves and an attempt to solve the problem of scattering from a finitely conducting rough surface will be made. However, for the most part, we shall have to assume that the surface is perfectly conducting and in that case the boundary conditions (2) and (3) apply equally well to the scattering of sound waves from rigid or free rough surfaces. The restriction to perfect conductivity is actually not as serious as might appear at first sight and certainly not as serious as other assumptions that will be introduced at times to make the problem mathematically manageable; for it is a well-known fact that the roughness of a surface modifies the scattered field far more than its electrical properties.

A large number of papers have been published on the subject of scattering from rough surfaces, especially in the last 10 years. Many experimental data have been accumulated and many theories have been developed to explain and predict measured data. None of these theories is general and rigorous at the same time. In order to arrive at results that lend themselves to reasonably simple numerical calculation, or indeed, to arrive at any results at all, certain simplifying assumptions are introduced into these theories. Most theories of rough-surface scatter use one or more of the following assumptions or simplifying procedures:

1. The dimensions of the scattering elements of the rough surface are taken as either much smaller or much greater than the wavelength of the incident radiation;
2. The radius of curvature of the scattering elements is taken to be much greater than the wavelength of the incident radiation;
3. Shadowing effects are neglected;
4. Only the far field is calculated;

5. Multiple scattering is neglected;

6. The density of the irregularities (number of scatterers per unit length or area of the surface) is not considered;

7. The treatment is restricted to a particular model of surface roughness, e.g. sinusoidal or saw-tooth undulations, protrusions of definite shape in random positions, random variations in height given by their statistical distribution and correlation function, etc.

The theory developed in Chapter 3 and applied to both periodic and random surfaces makes the simplifying assumptions 2 to 5, in addition to assuming perfect conductivity. It would not be very honest to claim that any of these assumptions may be withdrawn at the cost of mathematical simplicity; for though such a claim would be theoretically quite true, in practice the cost, apart from a few exceptions, is such that it is hardly worth paying.

The results of the various treatments are limited to those conditions under which their initial assumptions may be considered valid. Most important among these conditions is the model of surface roughness for which the theory was developed. We shall therefore class the various theories from this point of view; in particular we shall distinguish surfaces with definite given profiles and surfaces with random profiles. The former are in principle more easily treated, but it is the latter that are encountered much more often in practical applications.

The first approach to any problem is almost invariably qualitative. We shall therefore start with the Rayleigh criterion of whether or not a surface is rough, and continue with a discussion of the relative importance of various areas of the scattering surface.

By far the largest number of rough-surface scatter theories is based on the Kirchhoff approximation of the boundary conditions which are required to evaluate the Helmholtz integral. Apart from the original Kirchhoff postulate, there is a number of methods to approximate these boundary conditions, e.g. estimating the surface current distribution from the incident magnetic field or expressing the total field by means of local reflection coefficients. In principle these approximations are but different versions of the same "Kirchhoff" method or "method of physical optics". Such a method is applied in Chapter 3 and results in a solution which is, within the validity of the basic postulate of the Method of Physical Optics, exact. It reduces to a simple formula when the surface is perfectly conducting and becomes more simple still if the surface is periodic.

Chapter 4 treats periodically rough surfaces by applying physical optics to periodic surfaces in general and to a number of particular profiles. The results confirm, generalize, and in one or two cases correct the results obtained by other authors by the more complicated methods of Fourier transformation, harmonic analysis and saddle-point integration.

Random rough surfaces are then treated in the following chapters. The most general and promising model is evidently a surface given by the two-dimensional statistical distribution of its height above a certain mean level, i.e. in the same way as a continuous random process. This kind of surface is treated in Chapter 5, again applying the general formula derived in Chapter 3. The general results are applied to the most important random surface—the normally distributed surface, and explicit expressions are obtained for the mean field and power scattered into the various directions for a given angle of incidence and any degree of roughness. Chapter 6 is devoted to a survey of rough surfaces formed by random planar arrays of objects including random bosses protruding from a conducting plane. Some of these models can be solved by non-Kirchhoff methods and a high degree of accuracy, though not of generality, can be obtained in special cases.

The total field scattered from a rough surface in a certain direction is the sum of elementary waves scattered in that direction by each elementary scatterer. As each elementary wave is given not only by its amplitude, but also by its phase, this sum will be a vector sum. If the scatterers are random, the total field will be the sum of random vectors and hence itself a random quantity. The next chapter, therefore, derives the general probability distribution of this sum of random vectors. The solution derived in Chapter 7 imposes practically no restrictions on the individual amplitude and phase distributions or the correlation between them. The resulting distribution is then expressed in terms of the mean scattered field and the mean scattered power, so that the results of Chapters 5 and 6 may be linked up with this distribution, giving the ultimate solution for the scattered field.

The scattered waves are described not only by their amplitudes and phases, but also by their polarization. One of the most difficult problems connected with rough-surface theory is the question of what happens to the original polarization of the incident wave after the wave has been scattered. This is a problem which, even in the case of single scatterers, has not yet been solved (perhaps not even seriously attacked). The reader, it is hoped, will hardly expect to find a general and rigorous solution in the case of a rough surface, which in essence is a certain configuration of a large number of small scatterers. Nevertheless, an attempt to attack the problem is included in Chapter 8.

In fairness to the reader, not only the subjects treated here are listed, but also those subjects which receive little or no attention in this monograph even though they are closely related to scattering from rough surfaces:

— Scattering from volumes (three-dimensional configurations) of scatterers;
— Scattering from gratings and lattices;
— Reflections from stratified media;†
— Propagation of surface waves over irregular or inhomogeneous media;
— Propagation in waveguides with corrugated walls.

At the end of the theoretical part of the book, a short mathematical appendix is included. It lists some formulae and definitions of probability theory, Bessel functions and other subjects. The purpose of the appendix is to remove explanatory notes from the main text and it is hoped that references to it will aid the newcomers to this subject without offending the more advanced readers.

† This subject is being treated in another monograph in this series [Wait, 1962].

CHAPTER 2

QUALITATIVE CONSIDERATIONS

2.1. The Rayleigh Criterion

Before attempting to determine the field scattered by a rough surface quantitatively, we shall first consider two more elementary questions. The first has already been indicated in the preceding pages. For what values of wavelength, surface roughness and angle of incidence does specular reflection change into diffuse scattering, or according to our previous definition, when does a smooth surface become rough?

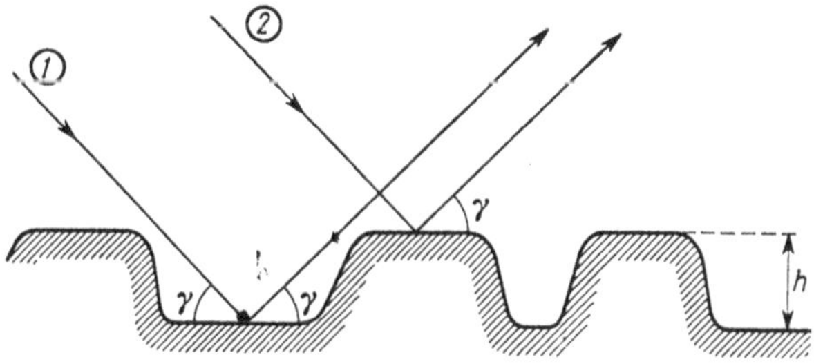

Fig. 2.1. Derivation of the Rayleigh criterion.

Rayleigh suggested a way of formulating the relation involving these parameters in the following way: Consider rays 1 and 2 (Fig. 2.1) incident on a surface with irregularities of height h at a grazing angle γ. The path difference between the two rays is

$$\Delta r = 2h \sin\gamma \tag{1}$$

and hence the phase difference is

$$\Delta\varphi = \frac{2\pi}{\lambda}\Delta r = \frac{4\pi h}{\lambda}\sin\gamma \tag{2}$$

If this phase difference is small, the two rays will be almost in phase as they are in the case of a perfectly smooth surface. If the phase difference increases, the two rays will interfere until for $\Delta\varphi = \pi$ they will be in phase opposition and cancel. If there is no energy flow in this direction, then it must have been redistributed in other directions, for it cannot have been lost. Thus for $\Delta\varphi = \pi$ the surface scatters and hence is rough, whilst for $\Delta\varphi = 0$ it reflects specularly and is smooth. We may now establish a value of phase difference between these two extremes to divide "rough" from "smooth". By simply arbitrarily choosing the value half-way between the two cases, i.e. $\Delta\varphi = \pi/2$ we obtain by substituting this value in (2) the relation usually known as the "Rayleigh criterion", namely that a surface is considered smooth for

$$h < \frac{\lambda}{8\sin\gamma} \tag{3}$$

Other values of $\Delta\varphi$, namely $\pi/4$ and $\pi/8$ have been called more realistic [Norton and Omberg, 1947; Kerr, 1951], thus replacing the factor 8 in (3) by 16 or 32; however, there seems to be little point in assessing an exact dividing line, which must in essence be more or less conventional. A safer way of expressing the basic idea of the Rayleigh criterion is to use the right side of (2),

$$\frac{4\pi h}{\lambda}\sin\gamma \tag{4}$$

as a measure of effective surface roughness; it may then be stated that a surface will tend to be effectively smooth only under two conditions:

$$\frac{h}{\lambda} \to 0 \quad \text{or} \quad \gamma \to 0 \tag{5}$$

The examples of road glare and reflection from unpolished sheet metal given in the introduction illustrate this point.

It is easy to point out the crudity of the Rayleigh criterion, which is based on simple ray theory and describes the irregularities of the rough surface by a single parameter; yet it must be said that to the present day it is, in engineering practice at least, still the most widely used approach to rough-surface scatter, evidently because it is simple and is expressed by the use of only three, easily measurable parameters (h, γ, λ). However, in Section 5.3 we shall obtain a *quantitative* expression for the mean power scattered into the direction of specular reflection in terms of practically the same three parameters. This expression includes the Rayleigh criterion as a simple limit and gives much additional information.

2.2. The Location and Size of the Active Scattering Region

The second elementary question is this: What regions of the scattering surface are the most important in contributing to the total field at a given receiving point when the surface is illuminated by a source at a given point?

The widely accepted answer to this question is "the first few Fresnel zones". We have no better answer to offer; however, for oblique incidence the above statement is not evident at all and has, to the present writer's knowledge, never been rigorously demonstrated. Closer investigation shows that the statement is, at best, highly probable, though for somewhat different reasons than those usually given in elementary textbooks on propagation theory.

We shall first sketch the derivation of the Fresnel zones and then discuss their relative importance.

Consider a smooth plane xy (Fig. 2.2) illuminated by a source at the point $A(0, 0, h_1)$. Physically, this source induces surface currents on the plane which radiate a secondary or reflected field. Let this reflected field

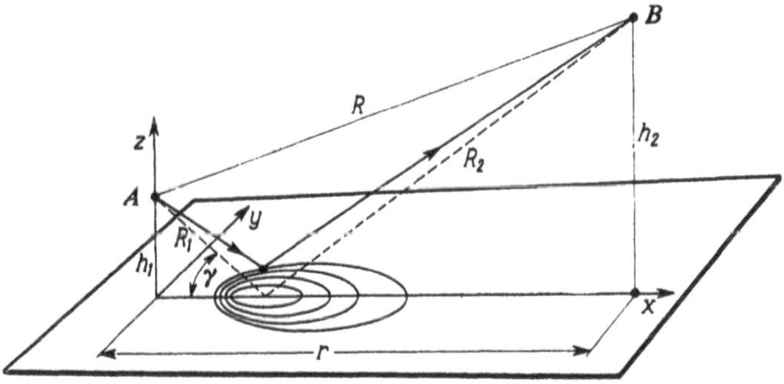

Fig. 2.2. The Fresnel zones on a reflecting plane.

be received at a point $B(r, 0, h_2)$. The locus of all points in the xy plane, from which the secondary radiation arrives at B with a constant phase difference δ with respect to the direct radiation along R is given by the relation

$$R_1+R_2-R = \delta \tag{1}$$

or, since R is constant, by

$$R_1+R_2 = \text{const} = R+\delta \tag{2}$$

which is the equation of ellipsoids of revolution (prolate spheroids) with foci at A and B. Hence the desired loci in the xy plane are ellipses determined by the intersection of that plane with the ellipsoids (2).

If now δ is increased in steps of $\lambda/2$, a family of ellipses in the xy plane will result; the average phase of the radiation from each zone bounded by two neighbouring ellipses will then differ from the adjacent zone by π, so that the elementary secondary waves radiated by successive zones will be in phase opposition.

The determination of the ellipses is a matter of straightforward, though somewhat tedious analytical geometry, which has been worked out in different terms by various authors [Vvedensky and Arenberg, 1948; Kerr, 1951; Alpert, Ginzburg and Feinberg, 1953].† Expressing the ellipses in terms of r, δ, h_1, h_2, one finds the centre of the ellipse corresponding to a given δ as

$$x_0 = \frac{r}{2}\left[1 - \frac{\left(\frac{h_2 - h_1}{r}\right)^2}{(\delta/r + \sec\theta)^2 - 1}\right]; \qquad y_0 = 0 \tag{3}$$

The semiminor axis is

$$y_1 = \frac{r}{2}\sqrt{\left\{\left[\left(\frac{\delta}{r}\right)^2 + \frac{2\delta}{r}\sec\theta\right]\left[1 - \frac{\tan^2\theta}{(\delta/r + \sec\theta)^2 - 1}\right]\right\}} \tag{4}$$

and the semimajor axis

$$x_1 = y_1\sqrt{\left\{1 + \frac{1}{(\delta/r + \sec\theta)^2 - 1}\right\}} \tag{5}$$

where θ is the slant angle of R, i.e.

$$\tan\theta = \frac{h_2 - h_1}{r} \tag{6}$$

These equations may be simplified if h_1, h_2 and δ are all much smaller than r. The approximate path difference δ_0 via the geometrical point of reflection s then

$$\delta_0 = \frac{2h_1 h_2}{r} \tag{7}$$

and the nth ellipse is determined by

$$\delta_n = \delta_0 + \frac{n\lambda}{2} \tag{8}$$

Using the above approximation and the transformation of δ we find the more useful formulae for the centre (x_{0n}) and the intercepts of the semimajor (x_{1n}) and semiminor axes (y_{1n}) of the nth ellipse:

$$x_{0n} \approx \frac{r}{2}\,\frac{1 + \frac{2h_1(h_1 + h_2)}{n\lambda r}}{1 + \frac{(h_1 + h_2)^2}{n\lambda r}} \tag{9}$$

$$x_{1n} \approx \frac{r}{2}\,\frac{\sqrt{\left(1 + \frac{2\delta_0}{n\lambda}\right)}}{1 + \frac{(h_1 + h_2)^2}{n\lambda r}} \tag{10}$$

† Cf. also Annex to Chapter 12 (p. 285).

$$y_{1n} \approx \frac{\sqrt{(n\lambda r)}}{2} \sqrt{\frac{\left(1+\frac{2\delta_0}{n\lambda}\right)}{1+\frac{(h_1+h_2)^2}{n\lambda r}}} \tag{11}$$

For small grazing angles the ellipses are very prolonged, getting longer and narrower with decreasing grazing angle γ. A numerical example, based on calculations by Kerr [1951], is given in Table 2.1.

TABLE 2.1—Dimensions of the First Fresnel Zone

λ	10 cm	1 m	10 cm
r	10 km		
h_1	100 m		
h_2	1000 m		2.5 m
γ	6.3°		0.59°
δ_0	20 m		5 cm
x_{01}	910 m	943 m	9350 m
x_{11}	83 m	262 m	615 m
y_{11}	9.1 m	29 m	6.6 m
Area	2370 m²	23,900 m²	128,000 m²

The argument, by which it is usually claimed that the first Fresnel zone makes the most important contribution to the total field received at B, runs along the following lines: the contribution of each zone is found by integrating the amplitude of excitation over the zone. Since successive zones are in phase opposition, the contributions of adjacent zones will tend to cancel, but not quite, as the amplitude of excitation decreases slowly from zone to zone, leaving the radiation of the area of approximately half the first Fresnel zone as the final total.

The original Fresnel zones, used in physical optics, are circular, concentric and their amplitudes of excitation are symmetrical about the direction of propagation, so that the amplitude of excitation is practically constant over the zone; in this case the above argument is unimpeachable. In the present case of oblique incidence, however, the amplitude of excitation will vary very markedly over a single zone, especially for the small grazing angles met in radio communications. As has been pointed out by Kerr [1951], it is not *a priori* evident that in this case the contribution of the first zone predominates. In an attempt to investigate the matter, Beckmann [1953] reached the conclusion that the first Fresnel zone is indeed likely to be the most important, but that further important contributions come from the areas in front of the transmitter and receiver; these areas are not identical with any Fresnel zones. In this paper [1953] the reflected field is, by applying the Helmholtz integral (cf. Appendix A) expressed as

$$E(r, 0, h_2) = iC \int_{-\infty}^{\infty} \int_{-\infty}^{\infty} \frac{h_2 R_1 + h_1 R_2}{R_1^2 R_2^2} e^{ik(R_1+R_2)} \mathrm{d}x \mathrm{d}y \tag{12}$$

where C is a constant and $k = 2\pi/\lambda$ is the propagation constant. It is also shown that this integral is not only equal to the reflected field as a result of integration, but also that it represents the actual physical summation in the process of integration, i.e. that the contribution of an elementary area $\mathrm{d}x\mathrm{d}y$ to the reflected field is proportional to

$$\mathrm{Re}\, \mathrm{d}E = C\frac{R_1+R_2}{R_1^2 R_2^2} \sin k(R_1+R_2) \tag{13}$$

where for simplicity it is assumed that $h_1 = h_2 = h$. From this it may be seen that the phase remains constant along the Fresnel zones (2) and that the field vanishes on the Fresnel ellipses, i.e. when R_1+R_2 equals an integral multiple of $\lambda/2$. But it can also be seen that the amplitude

$$A(x, y) = \frac{R_1+R_2}{R_1^2 R_2^2} \tag{14}$$

varies appreciably over the zone. The function $A(x, y)$ can be shown by elementary differential geometry to have the following properties:

(a) it is everywhere continuous;

(b) it has maxima at the points

$$P_1\big(r/2-\sqrt{[(r/2)^2-h^2]}, 0\big), \qquad P_2\big(r/2+\sqrt{[(r/2)^2-h^2]}, 0\big) \tag{15}$$

i.e. for $h \ll r$, very close to the transmitter and receiver site;

(c) it has a saddle-point at the geometrical point of reflection $P_0(r/2, 0)$ (however, for $h_1 \neq h_2$ the saddle point does not exactly coincide with the geometrical point of reflection);

(d) the curves of constant amplitude $A(x, y) = \text{const}$ are, for $h \ll r$, lemniscates with foci at the points $(0, 0)$ and $(0, r)$;

(e) $A(x, y) \to 0$ when $P(x, y) \to \infty$.

Figure 2.3 shows the surface $A(x, y)$ in a purely qualitative way. If this surface is taken as the "envelope" of the oscillating function

$$F(x, y) = \sin k(R_1+R_2)$$

i.e. a surface alternately positive and negative on the successive elliptical Fresnel zones, then the integral (12) gives the volume enclosed by this surface,

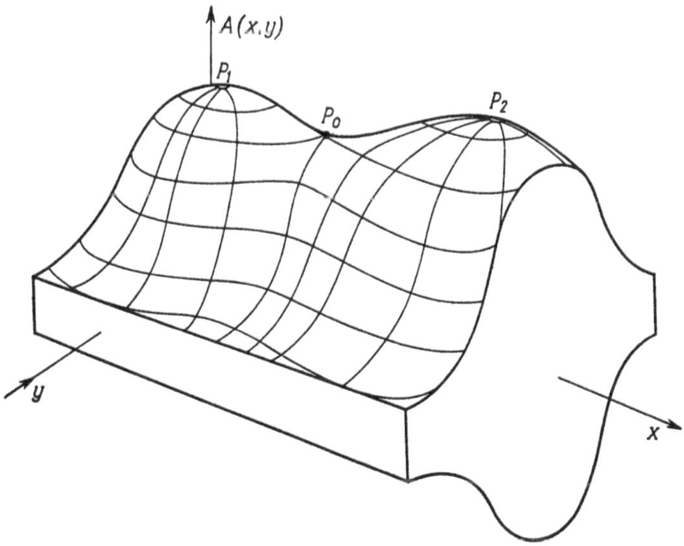

Fig. 2.3. A qualitative picture of the surface $A(x,y)$. The curves drawn on the surface correspond to $A = \text{const}$ and to the directions of grad A.

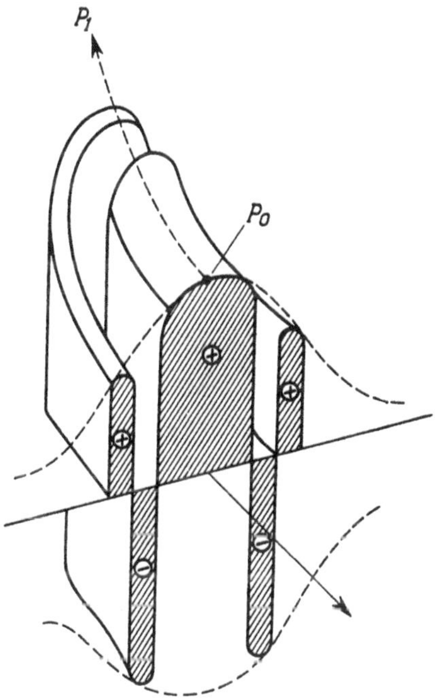

Fig. 2.4. An attempt to depict the volume enclosed by the surface $A(x,y)\,F(x,y)$ in the neighbourhood of the saddle-point P_0. The volume has been cut along the shaded plane.

the volume being considered negative below and positive above the xy plane. An attempt is made in Fig. 2.4 to depict the surface in the neighbourhood of the saddle point P_0. Evading rather intricate algebraic and numerical calculations, we may say that of all the Fresnel zones the first zone will make the largest contribution because it has the greatest area and thus generates the largest volume, and also because it is situated at the saddle point (though for $h_1 \neq h_2$ the saddle point does not coincide with the centre of the zone nor with the geometrical point of reflection). At the same time the regions where $A(x, y)$ attains its maxima should also be important.†

In those cases where the surface has finite dimensions and does not contain the whole of the first Fresnel zones, or if only parts of the surface are rough, the relative importance of different areas of the surface is still more difficult to estimate; in such cases there is apparently no other way but to enumerate the integral (12) with the corresponding integration limits.

† Cf. also Part II, pp. 262–263 and p. 266.

CHAPTER 3

THE GENERAL KIRCHHOFF SOLUTION FOR SCATTERING FROM ROUGH SURFACES

3.1. General Solution for a Surface Rough in One Dimension

The notation introduced in this section will be used throughout Part I of this book. We shall use Cartesian coordinates x, y, z with origin 0 and unit vectors $\mathbf{x}_0$, $\mathbf{y}_0$, $\mathbf{z}_0$. The rough surface will be given by the function

$$\zeta = \zeta(x, y) \tag{1}$$

The mean level of the surface is the plane

$$z = 0 \tag{2}$$

All quantities associated with the incident field will be denoted by the subscript 1 and those associated with the scattered field by the subscript 2. Thus the incident field is $\mathbf{E}_1$ and the scattered field $\mathbf{E}_2$. The medium in the space $z > \zeta$ is assumed to be free space. Until Chapter 8 we shall assume $\mathbf{E}_1$ to be linearly polarized; we shall consider either vertical or horizontal polarization (to be defined more exactly below) so that we shall deal only with the scalar value E_1 of the vector $\mathbf{E}_1$. We shall assume E_1 to be a harmonic plane wave of unit amplitude:

$$E_1 = e^{i\mathbf{k}_1 \cdot \mathbf{r} - i\omega t} \tag{3}$$

where

$$\mathbf{k}_1 = \frac{2\pi}{\lambda} \frac{\mathbf{k}_1}{k_1} \tag{4}$$

is the propagation vector, which will always lie in the xz plane (Fig. 3.1) and $\mathbf{r}$ is the radius vector:

$$\mathbf{r} = x\mathbf{x}_0 + y\mathbf{y}_0 + z\mathbf{z}_0 \tag{5}$$

In particular, for points on the surface S, we have

$$\mathbf{r} = x\mathbf{x}_0 + y\mathbf{y}_0 + \zeta(x, y)\mathbf{z}_0 \tag{6}$$

The angle of incidence, included between the direction of propagation of E_1 and the z axis, will be denoted by θ_1; the scattering angle, included between $\mathbf{z}_0$ and $\mathbf{k}_2$, where

$$k_2 = |\mathbf{k}_2| = |\mathbf{k}_1| = k = 2\pi/\lambda \tag{7}$$

will be denoted by θ_2, with θ_1 and θ_2 measured in opposite senses from the positive z axis (Fig. 3.1). For lateral scattering out of the plane of incidence $(\mathbf{k}_1, \mathbf{z}_0)$, a further angle θ_3 must be introduced (Fig. 3.2).

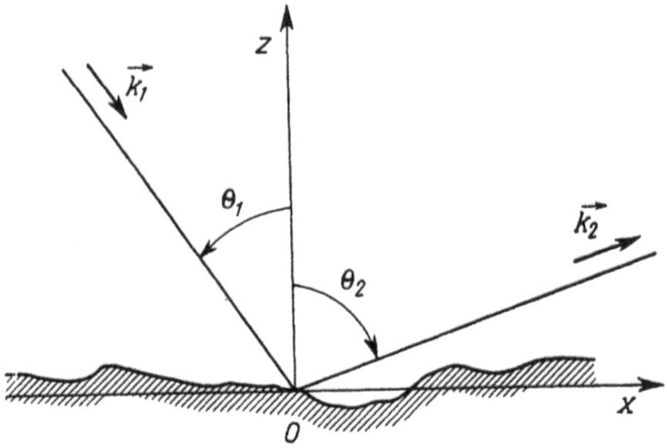

Fig. 3.1. Basic notation. Note that angle of incidence θ_1 and the scattering angle θ_2 are measured in opposite senses from the z axis.

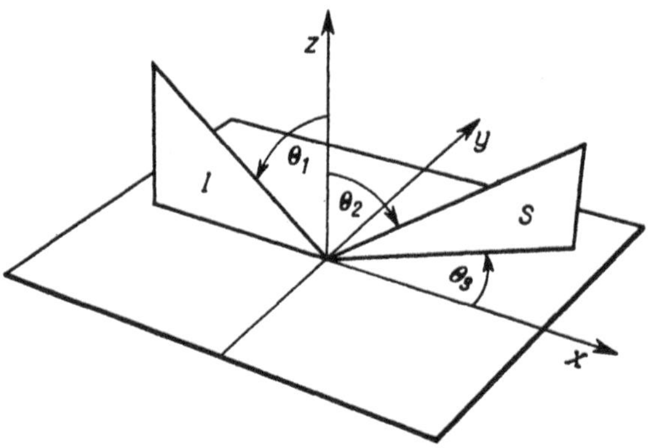

Fig. 3.2. The "scattering geometry" θ_1, θ_2, θ_3. I is the plane of incidence and S the scattering plane.

The time factor $\exp(-i\omega t)$ will henceforward be suppressed.

We next define the polarization of $\mathbf{E}_1$ and $\mathbf{E}_2$. We shall call the polarization of $\mathbf{E}_1$ vertical if $\mathbf{E}_1$ lies in the plane of incidence $\mathbf{k}_1$, $\mathbf{z}_0$ and similarly $\mathbf{E}_2$ will be called vertically polarized if it lies in the scattering plane $\mathbf{k}_2$, $\mathbf{z}_0$. Similarly, in horizontal polarization, $\mathbf{E}_1$ and $\mathbf{E}_2$ are normal to the plane of incidence and the scattering plane respectively. The quantities associated with vertical polarization will be denoted by the superscript "+" (e.g. E_1^+, E_2^+), and those associated with horizontal polarization by the superscript "$-$".† It should be understood that E is a scalar (unless printed in bold type as $\mathbf{E}$), so that

$$E = \mathbf{e} \cdot \mathbf{E}$$

where $\mathbf{e}$ is a unit vector in the direction of $\mathbf{E}$. In particular,

$$\mathbf{E}_1^- = E_1^- \mathbf{y}_0$$

Let P be the point of observation and let R' be the distance from P to a point $x, y, \zeta(x, y)$ on the surface S. Then as shown in Appendix A, the scattered field E_2 at P is given by the Helmholtz integral

$$E_2(P) = \frac{1}{4\pi} \iint_S \left(E \frac{\partial \psi}{\partial n} - \psi \frac{\partial E}{\partial n} \right) \mathrm{d}S \tag{8}$$

where

$$\psi = \frac{e^{ik_2 R'}}{R'}$$

In order to deal with plane scattered waves rather than spherical ones (i.e. with directions θ_2 rather than points P), we let $R' \to \infty$, i.e. we remove P to the Fraunhofer zone of diffraction; then as will be seen from Fig. 3.3,

$$k_2 R' = k_2 R_0 - \mathbf{k}_2 \cdot \mathbf{r} \tag{9}$$

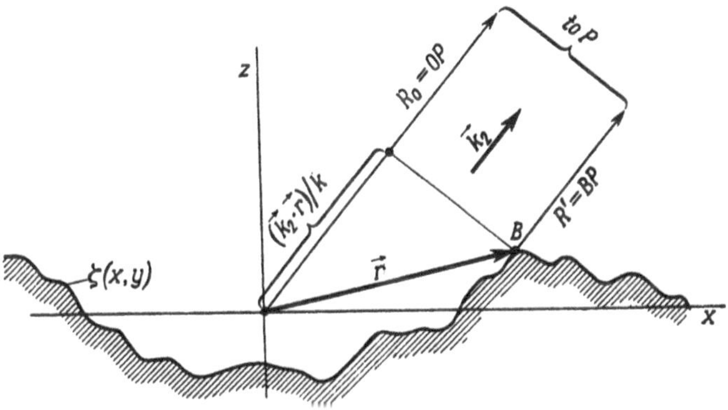

Fig. 3.3. Derivation of equation (9).

where R_0 is the distance of P from the origin, so that

$$\psi = \frac{e^{ik_2 R_0 - i\mathbf{k}_2 \cdot \mathbf{r}}}{R_0} \tag{10}$$

† For sound waves, the superscript "$+$" corresponds to a perfectly rigid, and the index "$-$" to a perfectly free (pressure release) surface.

E and $\partial E/\partial n$ are the field and its normal derivative on S. The exact value of these two quantities is in general unknown and the Kirchhoff or physical optics method consists essentially in approximating the values of E and $\partial E/\partial n$ on S and then evaluating the integral (8).†

In the present case we shall approximate the field at any point of the surface by the field that would be present on the tangent plane at that point (Fig. 3.4). The validity of this approximation will be discussed in Section 3.3; it will

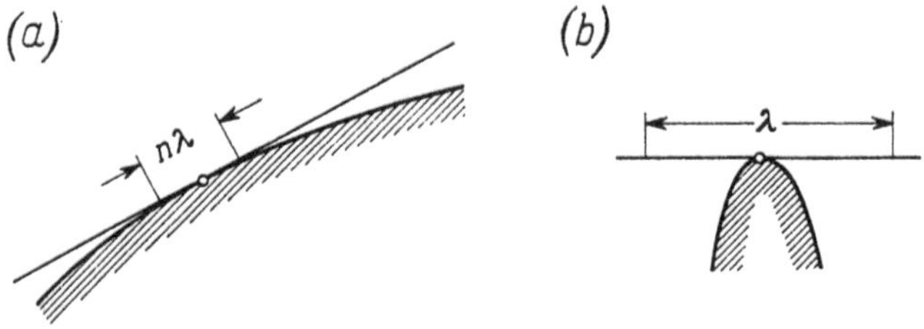

Fig. 3.4. The tangent plane at a general point of the rough surface. The radius of curvature is (a) large, (b) small in comparison with the wavelength.

obviously be very good when the radius of curvature of the irregularities is large compared with the wavelength (Fig. 3.4a), but will break down completely if the roughness includes sharp edges or sharp points (Fig. 3.4b).

Within this approximation the field on S will be

$$(E)_S = (1+R)E_1 \tag{11}$$

and

$$\left(\frac{\partial E}{\partial n}\right)_S = (1-R)E_1 \mathbf{k}_1 \cdot \mathbf{n} \tag{12}$$

where $\mathbf{n}$ is the normal to the surface at the considered point and R is the reflection coefficient of a smooth plane. The second relation follows from the first by differentiating the incident and reflected waves or by taking the magnetic field as

$$(H)_S = (1-R)H_1$$

and using the Maxwell equation

$$\nabla \times \mathbf{E} = -\varepsilon \frac{\partial \mathbf{H}}{\partial t}$$

The reflection coefficient R depends not only on the angle of incidence and the electrical properties of the reflecting material, but also on the pola-

† Originally, (8) was used in the special case of a plane conducting screen with an aperture; Kirchhoff postulated that E and $\partial E/\partial n$ are zero on the observer's side of the screen (opposite the source) and equal to the unperturbed incident field in the aperture. We are here using the name "Kirchhoff method" or "Physical Optics Method" to include more general approximations of the boundary conditions in (8), which seems to be consistent with modern usage.

rization of the incident wave. The Fresnel reflection coefficients for a smooth plane are

$$R^{+} = \frac{Y^2 \cos\vartheta - \sqrt{(Y^2 - \sin^2\vartheta)}}{Y^2 \cos\vartheta + \sqrt{(Y^2 - \sin^2\vartheta)}} \tag{13}$$

$$R^{-} = \frac{\cos\vartheta - \sqrt{(Y^2 - \sin^2\vartheta)}}{\cos\vartheta + \sqrt{(Y^2 - \sin^2\vartheta)}} \tag{14}$$

where ϑ is the "local" angle of incidence (Fig. 3.5), i.e. the angle concluded by $-\boldsymbol{k}_1$ and $\boldsymbol{n}$ at the considered point of the surface (as against the "overall" angle of incidence with respect to the xy plane $\theta_1 = \sphericalangle(-\boldsymbol{k}_1, z_0)$, which is

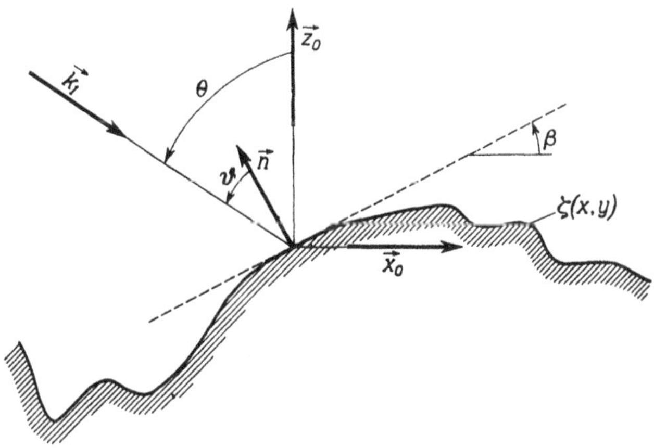

Fig. 3.5. The "local" scattering geometry. ϑ is the "local" angle of incidence with respect to the normal $\boldsymbol{n}$ at the considered point; θ_1 is the "overall" angle of incidence defined with respect to z_0 and constant for the whole surface.

constant for the whole surface). The quantity Y is the normalized admittance of the medium in the space $z < \zeta$ and is given by (15) [Beckmann 1956]

$$Y = \sqrt{\frac{\varepsilon_{rc}}{\mu_{rc}}} \tag{15}$$

where ε_{rc} is the relative complex permittivity

$$\varepsilon_{rc} = \frac{\varepsilon}{\varepsilon_0} + i60\lambda\sigma \tag{16}$$

with ε the dielectric constant, σ the conductivity† and $\mu_{rc} = \mu/\mu_0$ is the relative permeability (which may have an imaginary part if the medium has magnetic losses).

† MKS units are used throughout the book; ε_0 and μ_0 are the free-space values.

In order not to complicate matters by too many things at once, we temporarily limit our derivation to one-dimensionally rough surfaces

$$\zeta(x, y) = \zeta(x) \tag{17}$$

which are constant along the y coordinate, so that $\boldsymbol{n}$ always lies in the plane of incidence xz (Fig. 3.5). Then

$$\vartheta = \theta_1 - \beta = \theta_1 - \arctan \zeta'(x) \tag{18}$$

Substituting (10), (11) and (12) in (9) we find

$$E_2 = \frac{i e^{ikR_0}}{4\pi R_0} \int_S (R\mathbf{v} - \mathbf{p}) \cdot \mathbf{n} e^{i\mathbf{v}\cdot\mathbf{r}} \, dS \tag{18a}$$

where we have introduced the vectors

$$\mathbf{v} = \mathbf{k}_1 - \mathbf{k}_2 \tag{19}$$

$$\mathbf{p} = \mathbf{k}_1 + \mathbf{k}_2 \tag{20}$$

Resolving the vectors in (18) into their Cartesian components, we have

$$\mathbf{v} = k(\sin_1 - \sin\theta_2)\mathbf{x}_0 - k(\cos\theta_1 + \cos\theta_2)\mathbf{z}_0 \tag{21}$$

(This vector, or rather its components v_x and v_z, will accompany us all through Part I of this book; note that $v_x = 0$ for $\theta_1 = \theta_2$, i.e. for scattering into the direction of specular reflection, whereas v_z vanishes only for grazing incidence and reflection $\theta_1 = \theta_2 = \pi/2$.)

$$\mathbf{p} = k[(\sin\theta_1 + \sin\theta_2)\mathbf{x}_0 + (\cos\theta_2 - \cos\theta_1)\mathbf{z}_0] \tag{22}$$

$$\mathbf{n} = -\mathbf{x}_0 \sin\beta + \mathbf{z}_0 \cos\beta \tag{23}$$

$$\mathbf{r} = x\mathbf{x}_0 + \zeta(x)\mathbf{z}_0 \tag{24}$$

$$dS = \sec\beta \, dx, \quad \tan\beta = \zeta'(x) \tag{25}$$

For a surface extending from $x = -L$ to $x = L$ we may thus write (18) in the scalar form

$$E_2 = \frac{ike^{ikR_0}}{4\pi R_0} \int_{-L}^{L} (a\zeta' - b) e^{iv_x x + iv_z \zeta} \, dx \tag{26}$$

where

$$a = (1-R)\sin\theta_1 + (1+R)\sin\theta_2 \tag{27}$$

$$b = (1+R)\cos\theta_2 - (1-R)\cos\theta_1 \tag{28}$$

To get rid of the factor in front of the integral in (26) and also a number of difficulties associated with it, we normalize the expression (26) by introducing a scattering coefficient

$$\rho = \frac{E_2}{E_{20}} \tag{29}$$

where E_{20} is the field reflected in the direction of specular reflection ($\theta_2 = \theta_1$) by a smooth, perfectly conducting plane of the same dimensions under the same angle of incidence at the same distance, when the incident wave is horizontally polarized.† For $L \gg \lambda$, which we must assume for (11) and (12) to be valid, a smooth, perfectly conducting surface will concentrate practically all of the reflected energy into the direction $\theta_2 = \theta_1$, so that the law of the conservation of energy requires

$$|\rho| \leqslant 1 \tag{30}$$

Since by (21)

$$v_x = k(\sin\theta_1 - \sin\theta_2) \tag{31}$$

we have for the direction of specular reflection $v_x = 0$; for a smooth surface $\zeta = \zeta' = 0$, so that from (26) and (28) we find

$$E_{20} = \frac{ik\,\mathrm{e}^{ikR_0} L\cos\theta_1}{\pi R_0} \tag{32}$$

Hence from (26) and (29)

$$\rho = \frac{1}{4L\cos\theta_1} \int\limits_{-L}^{L} (a\zeta' - b)\,\mathrm{e}^{iv_x + iv_z\zeta}\,\mathrm{d}x \tag{33}$$

In a purely formal way this is the general solution for a one-dimensionally rough surface of finite conductivity, but as a and b are through (27), (28), (13), (14) and (18) (i.e. through R and ϑ) in general rather complicated functions of x, the integral (33) is useless in its present form except for the case of a smooth surface ($\zeta = 0$), when it yields

$$\rho = -\frac{(1+R)\cos\theta_2 - (1-R)\cos\theta_1}{2\cos\theta_1} \operatorname{sinc} v_x L \tag{34}$$

For horizontal polarization $R^- = -1$, so that (34) reduces to

$$\rho_0 = \operatorname{sinc} v_x L \tag{35}$$

In the last two formulae we have used the symbol

$$\operatorname{sinc} x = \frac{\sin x}{x} \tag{36}$$

which we shall use throughout this book.

Formula (35) gives the well-known scattering diagram of a conducting finite sheet; substituting (31) in this formula shows that as $\lambda/L \to 0$, the side

† The last condition is necessary to define the reference field. Although E_{20} is a scalar, its phase will differ for the two polarizations.

lobes will concentrate in the direction $\theta_2 = \theta_1$, making $\rho_0 = 1$ for $\theta_2 \approx \theta_1$ and $\rho_0 = 0$ in all other directions.

The integral (33) is, however, also easily integrated when a and b are constants. One way of achieving this is to average R over the surface, making a and b independent of x, which will be done in Section 5.4. The other, more important case is $Y \to \infty$ or a perfectly conducting surface. From (13) and (14) we then find

$$R^+ = 1, \quad R^- = -1 \tag{37}$$

so that a and b in (27) and (28) are independent of x. Removing the brackets in the integrand of (33), the first term is seen to be the product of $a\zeta' \exp(iv_z\zeta)$ and $\exp(iv_x x)$; integrating by parts, (33) then yields

$$\rho = \frac{1}{4L\cos\theta_1}\left[\left(b+\frac{av_x}{v_z}\right)\int_{-L}^{L} e^{i\mathbf{v}\cdot\mathbf{r}}dx - \frac{ia}{v_z}e^{i\mathbf{v}\cdot\mathbf{r}(x)}\Big|_{-L}^{L}\right] \tag{38}$$

Substituting (37) in (27) and (28) we obtain

$$\begin{aligned} a^+ &= 2\sin\theta_2, & b^+ &= 2\cos\theta_2 \\ a^- &= 2\sin\theta_1, & b^- &= -2\cos\theta_1 \end{aligned} \tag{39}$$

Substituting this in (34) and using (21), we finally obtain after elementary transformations

$$\rho^\pm(\theta_1, \theta_2) = \pm \sec\theta_1 \frac{1+\cos(\theta_1+\theta_2)}{\cos\theta_1+\cos\theta_2}\frac{1}{2L}\int_{-L}^{L} e^{i\mathbf{v}\cdot\mathbf{r}}dx + \frac{e^\pm(L)}{2L} \tag{40}$$

where

$$e^\pm(L) = \frac{i\sec\theta_1\sin\theta^\pm}{k(\cos\theta_1+\cos\theta_2)}e^{i\mathbf{v}\cdot\mathbf{r}(x)}\Big|_{-L}^{L} \tag{41}$$

with

$$\theta^+ = \theta_2, \quad \theta^- = \theta_1$$

The quantity $\mathbf{v}\cdot\mathbf{r}$ is given explicitly as

$$\mathbf{v}\cdot\mathbf{r} = \frac{2\pi}{\lambda}[(\sin\theta_1 - \sin\theta_2)x - (\cos\theta_1+\cos\theta_2)\zeta(x)] \tag{42}$$

Formula (40) is the general solution for a perfectly conducting, one-dimensionally rough surface, exact within the validity of (11). Its derivation is based on an approach of Isakovich [1952] who assumed perfect conductivity from the outset and used an approximate procedure for integration.

In subsequent applications we shall sometimes assume the surface S to be fixed at the "edges" with

$$\zeta(L) = \zeta(-L) = 0 \tag{43}$$

For $L \gg \lambda$, the second term of (40) is then easily shown negligible compared to the first, reducing (40) to the simple formula

$$\rho = \frac{F_2}{2L} \int_{-L}^{L} e^{i\mathbf{v}\cdot\mathbf{r}} dx \tag{44}$$

with

$$F_2(\theta_1, \theta_2) = \sec\theta_1 \frac{1+\cos(\theta_1+\theta_2)}{\cos\theta_1+\cos\theta_2} \tag{45}$$

The values of $F_2(\theta_1, \theta_2)$ are tabulated in Table 3.1.

TABLE 3.1—Values of $F_2(\theta_1, \theta_2)$

θ_2 \ θ_1	10°	20°	30°	40°	45°	50°	60°	70°	80°
80°	0.8765	0.7899	0.7308	0.6946	0.6847	0.6807	0.6946	0.7596	1.0000
70°	0.8982	0.8303	0.7899	0.7752	0.7783	0.7899	0.8485	1.0000	1.4961
60°	0.9178	0.8675	0.8453	0.8520	0.8683	0.8957	1.0000	1.2404	2.0000
50°	0.9358	0.9025	0.8982	0.9266	0.9563	1.0000	1.1515	1.4844	2.5196
40°	0.9528	0.9358	0.9495	1.0000	1.0437	1.1043	1.3054	1.7362	3.0642
30°	0.9689	0.9682	1.0000	1.0734	1.1317	1.2101	1.4641	2.0000	3.6446
20°	0.9846	1.0000	1.0505	1.1480	1.2217	1.3193	1.6304	2.2812	4.2743
10°	1.0000	1.0319	1.1018	1.2248	1.3153	1.4338	1.8077	2.5863	4.9710
0°	1.0154	1.0642	1.1547	1.3054	1.4142	1.5557	2.0000	2.9238	5.7588
−10°	1.0310	1.0975	1.2101	1.3913	1.5205	1.6879	2.2128	3.3054	6.6716
−20°	1.0472	1.1325	1.2692	1.4845	1.6371	1.8343	2.4532	3.7474	7.7588
−30°	1.0641	1.1698	1.3333	1.5876	1.7673	2.0000	2.7320	4.2743	9.1001
−40°	1.0822	1.2102	1.4043	1.7042	1.9164	2.1918	3.0642	4.9240	10.8237
−50°	1.1017	1.2548	1.4845	1.8392	2.0913	2.4202	3.4736	5.7587	13.1623
−60°	1.1234	1.3053	1.5773	2.0000	2.3032	2.7019	4.0000	6.8920	16.5830
−70°	1.1479	1.3640	1.6881	2.1984	2.5697	3.0641	4.7144	8.5492	22.1685
−80°	1.1763	1.4338	1.8247	2.4535	2.9212	3.5557	5.7592	11.2552	33.1730

Note: $F_2(0°, \theta_2) \equiv 1$.

3.2. General Solution for a Surface Rough in Both Dimensions

The results of the preceding section are easily extended to a surface $\zeta(x, y)$ rough in both dimensions. The principal difficulty is that R^+ and R^- as given by (13) and (14), Section 3.1, apply only to the "local" polarization, which is not necessarily identical with the polarization with respect to the xy plane, since $\mathbf{n}$ will generally not lie in the plane of incidence xz. This question will be investigated in more detail in Chapter 8; at present we may note that if the "local" plane of incidence $\mathbf{k}_1$, $\mathbf{n}$ is not identical with the main plane of incidence $\mathbf{k}_1$, $\mathbf{z}_0$, we must resolve $\mathbf{E}_1$ into a "locally horizontal" component (lying in the wave front and tangential to the surface element) and a "locally vertical" component (lying in the wave front and in the plane $\mathbf{k}_1$, $\mathbf{n}$). The two components will be reflected with reflection coefficients R^+ and R^-, and since in general these two (complex) coefficients are unequal, a linearly polarized wave will be depolarized into an elliptically polarized wave.

In order to avoid a tedious analysis of three-dimensional geometry, we note that a wave scattered into a direction θ_2, θ_3 (Fig. 3.2) is due to elementary waves reflected from tangent planes that are all mutually parallel. These planes will depolarize the incident wave in the same way. For finite conductivity the reflection coefficient R of the boundary conditions (11) and (12) of the preceding section will now be a rather complicated function of the electrical properties, the inclination, the angle of incidence and the polarization of the incident wave. We do not propose to calculate this function as we could not perform the ensuing integration in any case; but if the conductivity is perfect, then, as will be shown in Chapter 8, the reflecting planes leave the amplitude of the incident field unaltered, though they may change the direction of the $\mathbf{E}$ vector. This means that if we simply set the reflection coefficient equal to unity, this will give us the correct scalar value of $\mathbf{E}_2$, though it will tell us nothing as to how $\mathbf{E}_2$ is polarized. For the present we shall be satisfied with the scalar value of $\mathbf{E}_2$, leaving the question of its direction to Chapter 8.

Repeating the procedure of the preceding section we then find

$$\rho = \frac{1}{4XY\cos\theta_1}\int_{-X}^{X}\int_{-Y}^{Y}(a\zeta'_x+c\zeta'_y-b)\,e^{i\mathbf{v}\cdot\mathbf{r}}\,dx\,dy \tag{1}$$

where

$$\mathbf{v} = k[(\sin\theta_1-\sin\theta_2\cos\theta_3)\mathbf{x}_0-\sin\theta_2\sin\theta_3\mathbf{y}_0-(\cos\theta_1+\cos\theta_2)\mathbf{z}_0] \tag{2}$$

$$a = (1-R)\sin\theta_1+(1+R)\sin\theta_2\cos\theta_3 \tag{3}$$

$$b = (1+R)\cos\theta_2-(1-R)\cos\theta_1 \tag{4}$$

$$c = (1+R)\sin\theta_2\sin\theta_3 \tag{5}$$

where R is in general not equal to the Fresnel coefficient.†

† In acoustics, the troublesome polarization problems are not encountered and R in (3) to (5) is simply the acoustic reflection coefficient as before.

For a perfectly conducting surface, we set $R = 1$ with the reservation on polarization as above, and find

$$a = 2\sin\theta_2\cos\theta_3 \tag{6}$$

$$b = 2\cos\theta_2 \tag{7}$$

$$c = 2\sin\theta_2\sin\theta_3 \tag{8}$$

If a, b, c are due to infinite conductivity or for other reasons constant, (1) may be integrated by parts as in the preceding section, yielding

$$\rho = \frac{1}{4XY\cos\theta_1}\left[\left(b+\frac{av_x+cv_y}{v_z}\right)\int_{-X}^{X}\int_{-Y}^{Y} e^{i\mathbf{v}\cdot\mathbf{r}}\,dx\,dy - \right.$$

$$\left. -\frac{ic}{v_z}\int_{-X}^{X} e^{i\mathbf{v}\cdot\mathbf{r}}\Big|_{-Y}^{Y} dx - \frac{ia}{v_z}\int_{-Y}^{Y} e^{i\mathbf{v}\cdot\mathbf{r}}\Big|_{-X}^{X} dy\right] \tag{9}$$

Substituting (6) to (8) in (9), we find

$$\rho(\theta_1;\theta_2,\theta_3) = \frac{1+\cos\theta_1\cos\theta_2-\sin\theta_1\sin\theta_2\cos\theta_3}{\cos\theta_1(\cos\theta_1+\cos\theta_2)}\frac{1}{A}\int_{-X}^{X}\int_{-Y}^{Y} e^{i\mathbf{v}\cdot\mathbf{r}}\,dx\,dy +$$

$$+\frac{e(X,Y)}{A} \tag{10}$$

where $A = 4XY$ is the area of S projected onto the xy plane and $e(X, Y)$ is given by the last two terms of (9). The second term of (10) is an "edge effect" which will be shown negligible when compared to the first term for $A \gg \lambda^2$; in this case (10) reduces to the simple formula

$$\rho = \frac{F_3}{A}\iint_A e^{i\mathbf{v}\cdot\mathbf{r}}\,dx\,dy \tag{11}$$

with

$$F_3(\theta_1;\theta_2,\theta_3) = \frac{1+\cos\theta_1\cos\theta_2-\sin\theta_1\sin\theta_2\cos\theta_3}{\cos\theta_1(\cos\theta_1+\cos\theta_2)} \tag{12}$$

For longitudinal scattering ($\theta_3 = 0$), we have $F_3 = F_2$, so that (11) reduces to (44) of Section 3.1. Moreover, the longitudinally scattered waves (i.e. the waves scattered in the plane of incidence xz) are not depolarized if the incident wave is polarized either vertically or horizontally. This is best seen by substituting $\theta_3 = 0$ in (5) and the resulting $c = 0$ in (1); for longitudinally scattered waves the value of ζ_y' is irrelevant and hence no depolarization sets in, so that R will, as in the case of a one-dimensionally rough surface, be identical with the Fresnel coefficients R^+ or R^- depending on the polarization of $\mathbf{E}_1$. Physically, this is clear because longitudinally scattered waves originate only on tangent planes with $\mathbf{n}$ in the xz plane.

The first term of (10) was found by Isakovich [1952] by an approximate integration.

3.3. Properties of the General Solution

Formula (10) of the preceding section will be applied to the solution of both periodic and random surfaces. A number of effects, however, are common to the scattering of waves from any surface, and these may be found from the general solution.

Before we investigate these effects, we shall deal with the more important question of the conditions which ensure the validity of our solutions (40) of Section 3.1 and (10) of Section 3.2. We therefore summarize the assumptions under which they were derived:

A. The surface is perfectly conducting or, more generally, the quantities a, b, c in (3) to (5) of Section 3.2 are constant.

B. Mutual interaction of the irregularities (shadowing and multiple scattering) may be neglected.

C. The incident wave is plane and linearly polarized with the $\mathbf{E}$ vector either in the plane of incidence xz or perpendicular to it.

D. The point of observation is sufficiently far removed from the surface to regard the scattered waves as plane (but no assumption was made as to their polarization: any possible cross-polarized component has not been set equal to zero, but simply excluded from our present range of interest).

So far, these assumptions do not seriously impair the generality of our solution; besides, they can be withdrawn at the cost of exactness or simplicity: assumption A may be dropped by averaging a, b, c over the surface; B can be withdrawn by integrating over the illuminated parts of the surface† and introducing secondary reflection coefficients for multiple scattering; C will actually be withdrawn in Chapter 8 and D can be withdrawn by admitting a small error or sacrificing the simplicity of integration. But it is not these assumptions that curtail our solution to a certain class of surfaces.

The validity of our solution is essentially limited by the validity of the assumption

$$(E)_S = (1+R)\,E_1 \tag{1}$$

† More accurately, one replaces the function $\zeta(x)$ by $S(x)\,\zeta(x)$, where $S(x)$ is a "shadowing function"; to a first approximation $S(x) = 1$ on the illuminated parts and vanishes on the screened parts of the surface, but the solution of the wave equation requires a continuous function $S(x)$ which does not vanish over finite parts of the surface. For a more exact transition from light to shade cf. Fok [1945, 1946].

i.e. that the field at any point of the surface is expressed as the sum of the incident and reflected field with R the reflection coefficient of the plane which is tangent at the considered point. As (1) is fulfilled exactly in the case of an infinite plane, we conclude that the boundary condition (1) will be a very good approximation for "locally flat" surfaces composed of irregularities with small curvatures. Following Brekhovskikh [1952a], who used practically the same boundary condition, we require that at distances from the point of tangency, which are large compared with the wavelength of the incident radiation, the tangent plane shall not be appreciably distant from the rough surface (Fig. 3.4a). From elementary geometrical considerations, Brekhovskikh [1952a] finds the criterion

$$4\pi r_c \cos\vartheta \gg \lambda \tag{2}$$

where r_c is the radius of curvature and $\vartheta = \sphericalangle(-\mathbf{k}_1, \mathbf{n})$ the local angle of incidence. This is the principal limit of the validity of (1) and hence on the models of rough surfaces to which our solution will be applicable; in particular, the surface must not contain a large amount of sharp edges, sharp points or other irregularities with small radii of curvature.†

Another restriction on the use of the Fresnel coefficients is that $(kr_c/2)^{1/3}\cos\theta \gg 1$. This may be violated at low frequencies and near-grazing incidence [Wait and Conda, 1958].

Since a surface has two principal radii of curvature at every point, the smaller of the two must be considered when applying the criterion (2). For a short analysis of the adequacy of the criterion (2), the reader is referred to Brekhovskikh's paper [1952a].

Let us now investigate the factor

$$F_3(\theta_1;\theta_2,\theta_3) = \frac{1-\cos\theta_1\cos\theta_2-\sin\theta_1\sin\theta_2\cos\theta_3}{\cos\theta_1(\cos\theta_1+\cos\theta_2)} \tag{3}$$

which occurs in (10), Section 3.2, for any surface. We again note that in the plane of incidence, i.e. for $\theta_3 = 0$,

$$F_3(\theta_1;\theta_2,0) = \sec\theta_1 \frac{1+\cos(\theta_1+\theta_2)}{\cos\theta_1+\cos\theta_2} = F_2(\theta_1,\theta_2) \tag{4}$$

At first sight, two things seem wrong: for $\theta_1 = \pi/2$, this expression will become infinite; and F is not symmetrical in θ_1 and θ_2, so that the reciprocity theorem appears violated. Fortunately, neither is true. $\theta_1 = \pi/2$ corresponds

† In practical applications to irregular terrain, this concerns power lines, masts, roofs and most other man-made objects; Nature is usually more considerate.

to grazing incidence, when specular reflection will set in, as we shall show. For the direction of specular reflection ($\theta_2 = \theta_1$, $\theta_3 = 0$) we have, for any θ_1,

$$F_3(\theta_1; \theta_1, 0) = F_{\text{spec}} = \frac{1+\cos 2\theta_1}{2\cos^2\theta_1} \equiv 1 \tag{5}$$

so that (4) could become infinite only for non-specularly scattered waves at grazing incidence; such waves, as we shall show shortly, do not exist.

The reciprocity theorem, loosely formulated, states that the field at a point A due to a source at B equals the field at B when the same source is at A under equivalent conditions, provided the medium (or media) of propagation is linear. From this it follows that if a wave incident along the ray $A(\theta_1)$, Fig. 3.6, gives rise to a scattered wave, among others, along the ray $B(\theta_2)$, then a wave incident along $B(\theta_2)$ must give rise to a scattered wave, among others, along $A(\theta_1)$. Hence any formula giving the field as a function of θ_1 and θ_2 must yield the same value when θ_1 and θ_2 are interchanged.†

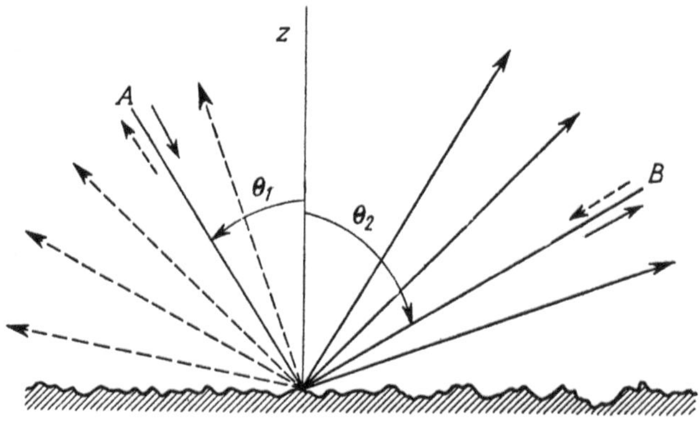

Fig. 3.6. Reciprocity in scattering.

The integral in (40), Section 3.1, is easily shown to be symmetrical in θ_1, θ_2, so that this will not compensate for the asymmetry of F_2. However, our solution gives the value of the scattering coefficient ρ, not the field E_2; returning to its definition in Section 3.1, we find from (29) and (32), Section 3.1, that the *field* is proportional to $\rho \cos\theta_1$, thus cancelling the factor $\sec\theta_1$, which disrupts the symmetry of our formula. Our one-dimensional formula is therefore consistent with the reciprocity theorem; in the two-dimensional case ($\theta_3 \neq 0$) the corresponding verification requires a transformation of coordinates and again confirms the reciprocity theorem.

† This suffices in our case. In general, other quantities may be affected, e.g. the input admittance at a boundary between two media, as this depends on which is its input (incident) side.

As this case shows, the reciprocity theorem should be used with caution when dealing with an incident wave which breaks up into several scattered waves. In particular, one should not use versions which apply only to undisrupted waves, as this may lead to paradoxes such as several incident waves uniting into a single scattered wave.

Besides scattering in the direction of specular reflection as in (5), there is one more case when F will equal unity: when $\theta_1 = 0$ (normal incidence), we find from (3) for any θ_2 or θ_3

$$F_3(0;\theta_2,\theta_3) \equiv 1$$

This does not, of course, mean that at normal incidence any surface will scatter isotropically into all directions θ_2, θ_3, for these two angles are yet contained in $\mathbf{v}$, appearing in the exponent of the integrand of the general solution.

We next consider the term $e(X, Y)/A$ in (10), Section 3.2, or $e(L)/2L$ in (36), Section 3.1. This term can be shown negligible when compared with the first term. This is particularly simple in the case of a periodic surface [Beckmann, 1961a], where $e(L) = 0$ if L equals an integral multiple of periods, so that $e(L)$ is the contribution of the fractional remainder, i.e. the incomplete periods at the edges of the surface. The same result may be obtained for a general surface from (10), Section 3.2, by showing the first term to be of the order F_3 and the second term of the order $1/XY$; however, instead of a mathematical proof (involving rigorous estimates of the integral), we shall make our point physically. Let S be an arbitrary surface of large dimensions ($X \gg \lambda$, $Y \gg \lambda$). Let S' be identical with S everywhere except near the edges $x = \pm X, y = \pm Y$, where it continuously approaches the value $\zeta = 0$. Then for S', we find from (9), Section 3.2,

$$\frac{e(X, Y)}{A} = -\frac{i(a+c)}{v_z}\,\frac{\sin v_x X \sin v_y Y}{XY} \tag{6}$$

The case $v_z = 0$ corresponds to grazing incidence and will be considered shortly; excluding this case for the present, we see that (6) vanishes for the direction of specular reflection ($v_x = v_y = 0$). For v_x and $v_y \neq 0$ we decrease X or Y, by "cutting off" strips from the surface S', until (6) again vanishes; since v_x and v_y contain the factor k, the necessary decrease will be at most $\lambda/2$. Thus the second term of our solution is an "edge effect", which represents the contribution of the narrow strips at the edges that have thus been "cut off" to make (6) vanish. Since S is identical with S' over an arbitrarily large area except for these narrow strips at the edges, it is physically clear that the two surfaces will also have practically the same scattered field and hence $e(X, Y)/A$ is always negligible for $A \gg \lambda^2$ or $L \gg \lambda$.

We now return to the excluded case $v_z = 0$. By (2), Section 3.2, this implies $\theta_1 = \theta_2 = \pi/2$, and this in turn implies $\theta_3 = 0$ (as the contrary case would require $\zeta_y' = \infty$ or vertical slopes, which are obviously not tolerated by the assumptions discussed above). Hence, again by (2), Section 3.2, $v = 0$. The last two terms of (9) then become indeterminate, but by repeated application of L'Hôpital's rule we find

$$\lim_{v \to 0} e(X, Y) = 0 \tag{7}$$

for any surface.

For the direction of specular reflection $\theta_2 = \theta_1$, $\theta_3 = 0$, which we shall briefly call the "specular direction", we have

$$F_2 = 1, \quad v_x = v_y = 0 \tag{8}$$

Neglecting the edge effect, we obtain for this case the simple formula

$$\rho_{\text{spec}} = \frac{1}{A} \iint_A e^{iv_z \zeta(x,y)} \mathrm{d}x \mathrm{d}y \tag{9}$$

with

$$v_z = -2k \cos\theta_1 \tag{10}$$

For $\theta_1 \to \pi/2$ (grazing incidence), $v_z \to 0$ and hence, by (8), $\mathbf{v} \to 0$, so that we need not, by (7), neglect the edge effect and from (9) we find

$$\lim_{\theta_1 \to \pi/2} \rho_{\text{spec}} = \frac{1}{A} \lim_{v_z \to 0} \iint_A e^{iv_z \zeta} \mathrm{d}x \mathrm{d}y = 1 \tag{11}$$

assuming the interchange of limiting and integration permissible.

Now, since $\rho \leqslant 1$, it follows that for $\theta_1 \to \pi/2$, (11) exhausts all possibilities, i.e. the specular direction $\theta_2 = \pi/2$ will receive all the available energy so that no other θ_2 will receive any scattered waves. We may therefore drop the index "spec" in (11) and write generally

$$\lim_{\theta_1 \to \pi/2} \rho = 1 \tag{12}$$

or more explicitly

$$\lim_{\theta_1 \to \pi/2} \rho(\theta_1; \theta_2, \theta_3) = \begin{cases} 1 & \text{for} \quad \theta_2 = \pi/2, \ \theta_3 = 0 \\ 0 & \text{for} \quad \theta_2 \neq \pi/2 \end{cases} \tag{13}$$

This result is not limited to perfectly conducting surfaces. The only reason why we assumed the surface to be perfectly conducting was to guarantee the constancy of *a*, *b*, *c* in (3) to (5) of Section 3.2, or *a*, *b* in (27) and (28) of Section 3.1, enabling us to perform the ensuing integration. But for $\theta_1 = \theta_2 = \pi/2$, $\theta_3 = 0$, these three quantities become independent of *R* and hence constant: $a = 2$, $b = c = 0$. The right side of (1), Section 3.2, then becomes indeterminate, but as before, setting $\theta_1 = \theta_2$, $\theta_3 = 0$ and then taking the limit $\theta_1 \to \pi/2$ again leads to (13).

Our derivation of (13) proves the empirically well-known result that any surface, regardless of its roughness or conductivity, will reflect specularly at grazing incidence.

There is an important corollary to this result. Let a wave be incident onto a rough surface along the ray *AO* (Fig. 3.7). It will be scattered into various directions θ_2, and we make the assumption, to be disproved by its failure, that one of these directions is that of a surface wave with $\theta_2 = \pi/2$

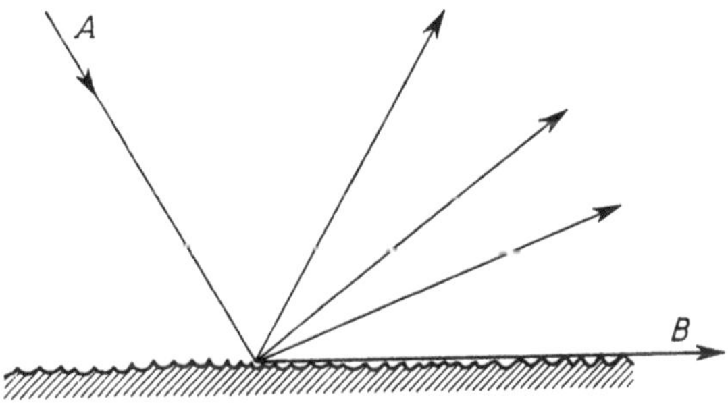

Fig. 3.7. The violation of the reciprocity theorem by a surface wave.

along the ray *OB*. We now interchange the transmitter and receiver, i.e. we let a wave be incident from *B* along the ray *BO* (grazing incidence). Invoking the reciprocity theorem, we must obtain (among others) a ray in the direction *OA*, which contradicts (13) unless *OAB* is a straight line.

The corollary of (13) is therefore: *A surface wave* ($\theta_2 = \pi/2$) *can exist at grazing incidence only.*

However, care should be taken not to generalize (13) or its corollary beyond the conditions under which they were derived; in particular, as our point of observation is in the Fraunhofer zone (or at infinity), our results do not contradict the existence of "leaky" or evanescent waves in the vicinity of the surface.† These waves are attenuated in the direction of their propagation; our results say nothing about them except for the trivial statement that there are no evanescent waves at infinity.

† Jordan, C. E. (Ed.): *Symposium on Electromagnetic Theory and Antennas* Copenhagen 1962. Pergamon Press 1963.

CHAPTER 4

PERIODICALLY ROUGH SURFACES

ALL possible models of a rough surface are conveniently divided into two classes: surfaces with exactly given profiles and surfaces with random irregularities. The two classes differ in their general treatment, since the first class does not involve any statistics; but they also differ in their applications, for the rough surfaces met in nature—terrain, sea, tropospheric and ionospheric layers, moon surface, etc.—are usually not known in exact detail, and even if the exact structure of these surfaces were known at every point and at any instant, any other but a statistical solution would be so specific as to be of little value, even if it were mathematically manageable. The same applies to most man-made surfaces met in practice.

There is, however, another class of rough surfaces, where a non-statistical approach is possible and desirable: rough surfaces with periodic irregularities such as sinusoidal undulations, saw-tooth profiles, regular corrugations, protrusions of equal shape spaced at regular intervals, etc. Historically, this class of rough surfaces was the first to be treated, evidently because of the greater simplicity of the non-statistical approach.† Though periodic surfaces are less often met in practice than randomly rough surfaces, there are at least two reasons why they merit close study. Firstly, their theory is sufficiently general to give some indication of the general behaviour of rough surfaces; secondly, if a surface is to be manufactured for the specific purpose of preventing specular reflection or even scatter in a certain direction, it is easier to make the roughness periodic than to make it of a random nature with a prescribed probability distribution.

The problem of scattering from a periodic surface was first investigated by Rayleigh [1895] in the case of normal incidence onto a sinusoidal surface. Rayleigh obtained an approximation of the first few scattered modes. His method is easily generalized for oblique incidence [LaCasce and Tamarkin, 1956] and the general Rayleigh approach was developed by Rice [1951] for slightly rough, random surfaces, by Deryugin [1952, 1953, 1960] and Myakishev [1958] for periodic rectangular corrugations and by Schouten

† The first treatment of the scattering of waves from a rough surface is evidently that of Rayleigh [1895, Section 272a], who considered normal incidence of sound waves onto a free or rigid sinusoidal surface.

and De Hoop [1957] for any analytically given rough surface. Kirchhoff methods for periodic surfaces were developed by Brekhovskikh [1951, 1952], (whose results were generalized for finite conductivity by Lysanov [1952]), Senior [1959], Beckmann [1961a] and by a number of authors who developed Kirchhoff methods for random surfaces, but whose results are easily adapted to periodic surfaces, e.g. Isakovich [1952], Eckart [1953] and others. Meecham [1956] and Heaps [1957] use variational methods to satisfy the boundary conditions, and non-Kirchhoff approaches limited to periodic profiles of a specific type have been developed by Twersky [1956a] and Biot [1957, 1958].

4.1. General Kirchhoff Solution of the Field Scattered by a Perfectly Conducting Periodic Surface

We first assume the surface S to be rough in one dimension (x) only. Let $\zeta(x)$ be periodic:

$$\zeta(x) = \zeta(x+\Lambda) \tag{1}$$

where Λ is the period (or "wavelength") of the surface.

Then the integrand $\exp(i\mathbf{v}\cdot\mathbf{r})$ of (40), Section 3.1, will be periodic with period Λ if

$$v_x \Lambda = 2\pi m \tag{2}$$

where m is any integer. Since by (21), Section 3.1,

$$v_x = \frac{2\pi}{\lambda}(\sin\theta_1 - \sin\theta_2) \tag{3}$$

we may write (2) as

$$\sin\theta_{2m} = \sin\theta_1 + m\frac{\lambda}{\Lambda} \qquad (m = 0, \pm 1, \pm 2, \ldots) \tag{4}$$

This equation is well known from grating theory and deserves some attention; we shall return to it presently.

Now let

$$\frac{L}{\Lambda} = n + n_1 \tag{5}$$

where n is an integer and n_1 is a pure fraction ($0 \leqslant n_1 < 1$). Then since $\exp(i\mathbf{v}\cdot\mathbf{r})$ is periodic with period Λ, we may write

$$\int_{-L}^{L} e^{i\mathbf{v}\cdot\mathbf{r}}\,dx = 2n\int_0^{\Lambda} + \int_0^{n_1\Lambda} + \int_{-n_1\Lambda}^{0} = \frac{2L}{\Lambda}\int_0^{\Lambda} + \int_{-n_1\Lambda}^{n_1\Lambda} - 2n_1\int_0^{\Lambda} \tag{6}$$

where we have omitted the integrands $\exp(i\mathbf{v}\cdot\mathbf{r})$.

Substituting (2) in (41), Section 3.1, yields

$$e(L) = e(2\pi m n_1/v_x) \tag{7}$$

if we assume that the origin is so chosen that $\zeta(x)$ is even.

Substituting (6) and (7) in the general solution (40) of Section 3.1, we obtain this basic formula for a perfectly conducting, one-dimensionally periodic surface:

$$\rho^{\pm}(\theta_1, \theta_2) = \pm \sec\theta_1 \frac{1+\cos(\theta_1+\theta_{2m})}{\cos\theta_1+\cos\theta_{2m}} \frac{1}{\Lambda} \int_0^{\Lambda} e^{i\mathbf{v}\cdot\mathbf{r}} dx + \frac{C(n_1)}{2L} \tag{8}$$

where

$$C(n_1) = F_2(\theta_1, \theta_2)\Big[\int_{-n_1\Lambda}^{n_1\Lambda} e^{i\mathbf{v}\cdot\mathbf{r}} dx - 2n_1 \int_0^{\Lambda} e^{i\mathbf{v}\cdot\mathbf{r}} dx\Big] + e(2\pi m n_1/v_x) \tag{9}$$

is a correction term or "edge effect" which vanishes for $n_1 = 0$, i.e. when L/Λ is integral; in any case, the second term in (8) is easily seen to be negligible when $L \gg \Lambda$.

Apart for the double signs, the first term of (8) can be shown equivalent to Brekhovskikh's result [1952b] for a finite surface, derived by Fourier transformation and harmonic analysis.†

The scattering angles θ_{2m} are determined by the grating equation (4). To each integer m there corresponds a scattered mode propagated in the direction θ_{2m}. The total number of possible modes is limited by the condition

$$|\sin\theta_{2m}| \leqslant 1 \tag{10}$$

The mode with $m = 0$ is seen from (4) to be the specular mode. The modes $m = \pm 1$ lie to either side of the specular direction as shown in Fig. 4.1. The modes continue thus to either side of the specular mode until the last modes that will still satisfy (10) are reached; they are propagated at angles near $\theta_{2m} = \pm \pi/2$.

If λ/Λ is small, it follows from (4) that m will run through a large number of integral values before (10) is violated, so that if the wavelength of the incident radiation is small compared to the period or "wavelength" of the surface, the incident wave will be broken up into many scattered waves. On the other hand one would conclude from (4) that for $\lambda/\Lambda \gg 1$ there will be only the mode $m = 0$, so that for $\lambda \gg \Lambda$ any periodic surface will reflect specularly regardless of its roughness. It so happens that this is quite true

† For an infinite surface, Brekhovskikh's result is multiplied by a factor $\sec\theta_2$. Senior [1959] obtains a similar result for the special case of normal incidence onto a sinusoidal surface. This result seems to be in error as it contradicts the reciprocity theorem ($\sec\theta_2$ spoils the symmetry of the formula for the *field*), cf. Section 3.3.

(cf. Section 4.2), but as yet we have no right to conclude this, as we are all the time working under the assumption $\lambda \ll \Lambda$, which obviously follows from (2) of Section 3.3.

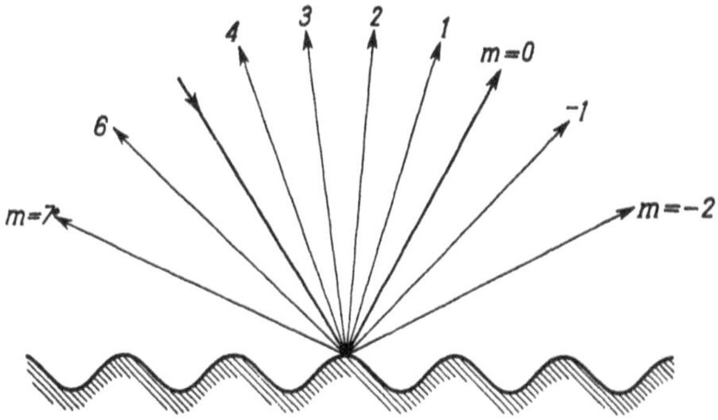

Fig. 4.1. The directions in which various modes m are scattered by a periodic surface, or directions of the side lobes of the scattering diagram.

Keeping m in (4) constant and increasing λ, we find that the direction in which this selected mode is propagated becomes more nearly horizontal until it becomes a surface wave and the limit of condition (10) is reached. It is then known as the "cut-off mode" for that particular frequency. The cut-off mode or surface wave was found by some investigators to have infinite amplitude; from the corollary of (13), Section 3.3, we find that its amplitude is zero.† In the present instance we may note that the Rayleigh method also gives zero amplitude for the cut-off mode and that the experimental measurements by LaCasce and Tamarkin [1956] show no increase of the amplitude as the mode approaches cut-off.

As can be seen from (4), there is no cut-off for the fundamental or specular mode $m = 0$.

The "grating equation" (4), which we have here obtained quite automatically, is obtained in grating theory by considering a grating of, say,

† A discrepancy of this kind may seem absurd, but it should be remembered that a surface wave of infinite amplitude is not easy to disprove: it does not violate the principle of the conservation of energy, as it does not carry away any energy from the (infinite) surface. The corollary of (13), Section 3.3, was therefore based on the reciprocity theorem rather than on considerations of the balance of energy. But our formulae have been derived under the assumption of an arbitrarily large, but finite surface so that they will not convincingly prove the non-existence of a surface wave in the case of an infinite surface. Difficulties of this kind are perhaps characteristic of the theory of surface wave propagation, e.g. the existence of a surface wave generated by a dipole on the surface of the earth as proposed by Zenneck and Sommerfeld in the early years of this century is still under dispute, though it now appears generally rejected. Cf. also last paragraph p. 33.

parallel wires (originally fine rulings on glass or molecules of crystals), spaced at intervals Λ (Fig. 4.2), excited by a wave incident at an angle θ_1. If the phase delay between the scattered waves A and B is calculated from the geometry of the figure, and all angles θ_2, for which this delay equals $2m\pi$ are determined (i.e. the angles for which the scattered waves will be in phase), then equation (4) is obtained. In our case the physical meaning of (4) is quite similar: θ_{2m} are the directions for which the waves scattered from individual periods of the surface will be in phase and will reinforce each other, giving rise to the modes or side-lobes of the scattering diagram.

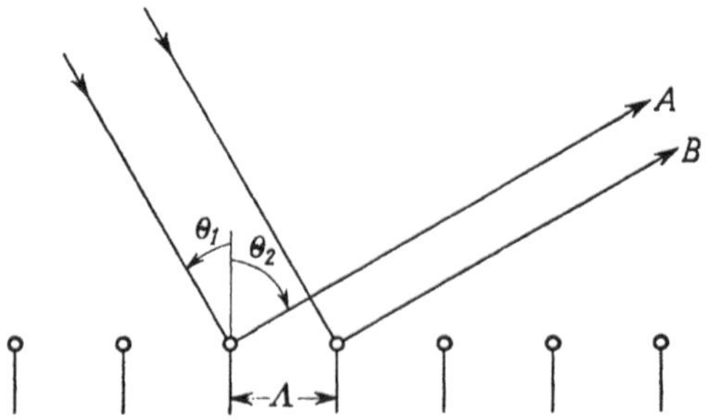

Fig. 4.2. Derivation of the grating equation by phase-delay.

If we want to calculate the field scattered in directions other than θ_{2m}, corresponding to the maxima of the side-lobes, we return to the general solution (40), Section 3.1. So as not to complicate the calculation by the irrelevant edge-effect term, we first assume $n_1 = 0$, so that $L/\Lambda = n$ is integral. We then divide the length $2L$ into $2n$ strips, each extending from $x = \mu\Lambda$ to $x = (\mu+1)\Lambda$, where μ is an integer $(-n \leqslant \mu \leqslant n-1)$. Then

$$\frac{1}{2L}\int_{-L}^{L} e^{i\mathbf{v}\cdot\mathbf{r}}\,dx = \frac{1}{2n\Lambda}\sum_{\mu=-n}^{n-1}\int_{\mu\Lambda}^{(\mu+1)\Lambda} e^{iv_x x+v_z\zeta(x)}\,dx$$

$$= \frac{1}{2n\Lambda}\sum_{\mu=-n}^{n-1} e^{iv_x\Lambda\mu}\int_0^{\Lambda} e^{iv_x x'+v_z\zeta(x')}\,dx' \qquad (11)$$

where the second equality follows from the substitution $x = x'+\mu\Lambda$. We now substitute

$$v_x\Lambda = 2\pi p \qquad (12)$$

or, on expressing v_x explicitly,

$$p = \frac{\Lambda}{\lambda}(\sin\theta_1 - \sin\theta_2) \qquad (13)$$

This equation is a generalized grating equation, formally identical with (4), except that p is not necessarily an integer. From (12) and (11) we then find

$$\frac{1}{2L}\int_{-L}^{L} e^{i\mathbf{v}\cdot\mathbf{r}}dx = \frac{1}{2n}\sum_{\mu=-n}^{n-1} e^{i2p\pi\mu}\cdot\frac{1}{\Lambda}\int_{0}^{\Lambda} e^{i\mathbf{v}\cdot\mathbf{r}}dx \tag{14}$$

which differs from our previous result by a factor

$$W = \frac{1}{2n}\sum_{\mu=-n}^{n-1} e^{2i\pi p\mu}, \qquad |W| \leqslant 1 \tag{15}$$

with $W = 1$ when p is integral. By treating (15) as a finite geometric series, we find

$$W = \frac{\sin 2np\pi}{2n\sin p\pi}e^{-ip\pi} \tag{16}$$

so that our general solution for a perfectly conducting periodic surface for any θ_2 (not necessarily equal to the angles θ_{2m} of the maxima of the side lobes) is from (40), Section 3.1,

$$\rho^{\pm}(\theta_1, \theta_2) = \pm W\sec\theta_1\frac{1+\cos(\theta_1+\theta_2)}{\cos\theta_1+\cos\theta_2}\frac{1}{\Lambda}\int_{0}^{\Lambda} e^{i\mathbf{v}\cdot\mathbf{r}}dx + O\left(\frac{n_1}{n}\right) \tag{17}$$

where $O(n_1/n)$ represents a correction term which vanishes when $L/\Lambda = n$ is integral; W is given by (16) and p by (13). The solution (17) reduces to (8) when p is integral.

For practical purposes it is unnecessary to calculate the scattering diagram point by point from (17); the directions of the maxima of the lobes are given by (4) and their values by (8), and the only other interesting quantity is usually the half-width of the lobes. This can be found from (16). $|W|$ attains

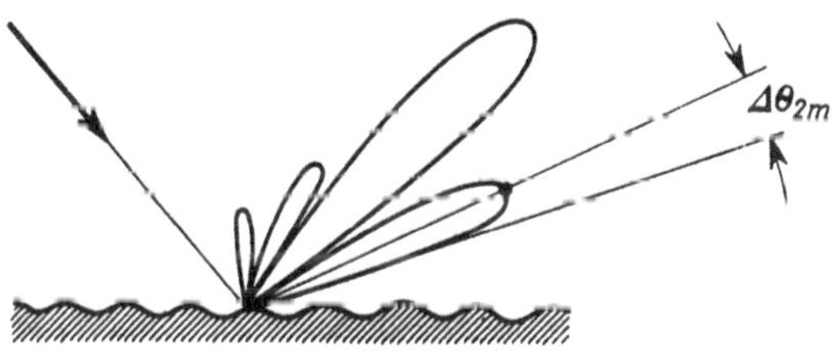

Fig. 4.3. The half-width of a side-lobe.

its maximum value unity for $p = m$ (integer) corresponding to the scattering angle θ_{2m}; the nearest zeros are $p = m \pm 1/2n$, corresponding to the angles

$\theta_{2m} \pm \Delta\theta_{2m}$, where $\Delta\theta_{2m}$ is the half-width of the lobe (Fig. 4.3). Substituting these values of p in (13) we have

$$\sin\theta_{2m} - \sin(\theta_{2m} \pm \Delta\theta_{2m}) = \frac{\lambda}{2n\Lambda} \tag{18}$$

from which the exact value of $\Delta\theta_{2m}$ may be found; for $\lambda \ll 2n\Lambda = 2L$ and θ_{2m} not too near $\pi/2$ we find

$$\Delta\theta_{2m} \approx \frac{\lambda}{2L}\sec\theta_{2m} \tag{19}$$

whilst for $\theta_{2m} \to \pi/2$, we find from (18) $\Delta\theta_{2m} \to 0$.

Formulae (18) and (19) have also been found by Brekhovskikh [1952b] by harmonic analysis and Fourier transformation.

The solution for any one-dimensionally periodic surface (8) is easily extended to two dimensions.

Let $\zeta(x, y)$ be periodic with periods Λ_x and Λ_y:

$$\zeta(x, y) = \zeta(x+\Lambda_x, y+\Lambda_y) \tag{20}$$

Then $\exp(i\mathbf{v}\cdot\mathbf{r})$ in the integrand of (10), Section 3.2, will be periodic with periods Λ_x, Λ_y, if

$$v_x\Lambda_x = 2m\pi, \quad v_y\Lambda_y = 2n\pi \tag{21}$$

where m and n are arbitrary integers, so that the grating equations are now, by expressing v_x and v_y explicitly from (2), Section 3.2,

$$\sin\theta_{2mn}\cos\theta_{3mn} = \sin\theta_1 + m\frac{\lambda}{\Lambda_x}, \quad \sin\theta_{2mn}\sin\theta_{3mn} = n\frac{\lambda}{\Lambda_y} \tag{22}$$

Then (10) of Section 3.2 may be written as

$$\rho = F_3(\theta_1; \theta_{2mn}, \theta_{3mn})\frac{1}{\Lambda_x\Lambda_y}\int_0^{\Lambda_x}\int_0^{\Lambda_y} e^{i\mathbf{v}\cdot\mathbf{r}}\,dx\,dy + O\left(\frac{\Lambda_x\Lambda_y}{A}\right) \tag{23}$$

where F_3 is given by (12) of Section 3.2, with the scattering angles θ_{2mn} and θ_{3mn} satisfying (22). To each couple of integers m, n there again corresponde a scattered mode; $m = n = 0$ is the fundamental or specular mode. The correction term in (23) will vanish if $A/\Lambda_x\Lambda_y$ is integral.

So far, all our considerations have been based on the assumption that the point of observation P is very far removed from the surface [cf. (8), Section 3.1], so that only a single plane wave is received at this point, i.e. in the direction θ_2 (or θ_2, θ_3). This is symbolically shown in Fig. 4.4a. For a periodic surface, the total field scattered into a certain direction is given by adding the contributions of each period to that direction, as may be seen from (5) and (6) if we set $n_1 = 0$. It is therefore plausible to assume that if

P is at a finite distance from S (Fig. 4.4b), the field at P can conversely be determined as the sum of the contributions, i.e. wave modes, scattered to P by different periods, now lying in different directions. (This result can in fact be derived by expressing the field at P as the sum of plane waves which are then found to be the above wave modes [Brekhovskikh, 1952a].)

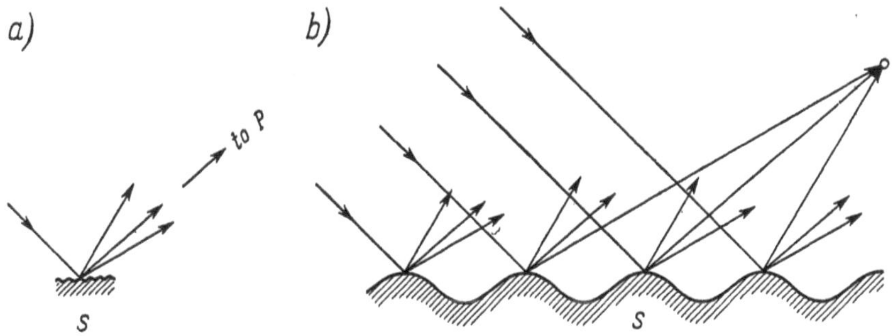

Fig. 4.4. The field at the point of observation consists (a) of a single mode in the Fraunhofer zone, (b) of the sum of many scattered modes in the Fresnel zone.

Thus

$$E_2(P) = \sum_m \sum_n E_2(\theta_{2mn}, \theta_{3mn}) \tag{24}$$

where the wave mode numbers m, n and the corresponding scattering angles are given by (22). Dividing (24) by the reference field E_{20} we have approximately (if $R' \approx R_0$ can still be considered valid)

$$\rho(P) = \sum_m \sum_n \rho(\theta_{2mn}, \theta_{3mn}) \tag{25}$$

where $\rho(\theta_{2mn}, \theta_{3mn})$ is given by (23), or in the case of a one-dimensionally periodic surface,

$$\rho(P) = \sum_m \rho(\theta_{2m}) \tag{26}$$

where $\rho(\theta_{2m})$ is given by (8).

4.2. The Rayleigh Method

The Krichhoff method, in the more general sense of Section 3.1, consists in postulating or approximating the boundary conditions on the surface S and then finding the corresponding field by using the Helmholtz integral. Rayleigh [1895], in what is evidently the first attempt to treat scattering from rough surfaces, proceeded in the opposite way: the field is postulated as the sum of plane waves of the type (24) of the preceding section, and the unknown terms of the sum are then found by satisfying the exact boundary conditions at the surface.

Rayleigh [1895] used this method in the special case of normal incidence onto a sinusoidal surface, but the method has been extended to oblique incidence [LaCasce and Tamarkin, 1956] and its principles applied to a different profile [Deryugin, 1952, 1953, 1960]. Its general formulation was given by Schouten and De Hoop [1957] and Rice [1951] based his treatment of a random, slightly rough surface on what is essentially a Rayleigh method. The variational method of Meecham [1956], used also by Heaps [1957], is also a Rayleigh rather than a Kirchhoff method.

For an infinitely conducting surface, the exact boundary condition is

$$(\mathbf{E}\times\mathbf{n})_s = 0 \tag{1}$$

and this should uniquely specify the solution—unfortunately only in theory, because Green's function of the surface must be known to eliminate the second term of the Helmholtz integral. For this second term, one therefore usually takes some equivalent of the boundary condition

$$\left(\frac{\partial}{\partial n}(\mathbf{E}\times\mathbf{n})\right)_s = 2\frac{\partial}{\partial n}(\mathbf{E}_1\times\mathbf{n}) \tag{2}$$

(as we have in principle done in Section 3.1), but this boundary condition (or its equivalents expressed in terms of $R^{\pm}$, the magnetic field $\mathbf{H}$ or surface current density $\mathbf{J}$) is exact only in the case of a plane.

The Rayleigh method, on the other hand, uses the exact boundary condition (1) only, and one might conclude that it is therefore preferable. However, except for certain cases where the Kirchhoff method becomes very inexact (presence of edges, etc.), or where certain special techniques can be applied, this is not so. The reason is that satisfying the boundary conditions leads to an infinite set of equations and their solution is usually too cumbersome to treat anything but very slight roughness or other particular cases, so that (with some noteworthy exceptions) the final result is usually a worse approximation than the Kirchhoff solution, which can be expressed in a closed form or, if need be, found more quickly by numerical integration.

A typical example of this is the case treated by Rayleigh [1895, vol. II, Section 272a], namely

$$\zeta(x) = h\cos Kx \tag{3}$$

where

$$K = \frac{2\pi}{\Lambda} \tag{4}$$

We shall obtain a solution of this case in a closed form in a few lines, in Section 4.3A by applying the general Kirchhoff solution (8) of Section 4.1, whereas the Rayleigh solution leads to an infinite series for each mode.

The general procedure of the Rayleigh method is the following: We assume the field to consist of the sum of plane waves propagated in the directions θ_{2m} given by the grating equation (4), Section 4.1:

$$\sin\theta_{2m} = \sin\theta_1 - \frac{mK}{k} \tag{5}$$

This equation may also be obtained independently either by grating theory as explained in Section 4.1, or by substituting the terms of the proposed solution in the wave equation.

The total field in the space $z \geqslant \zeta$ will then be

$$E = E_1 + E_2 = \exp(i\mathbf{k}_1 \cdot \mathbf{r}) - \sum_{m=-\infty}^{\infty} A_m \exp(i\mathbf{k}_2^{(m)} \cdot \mathbf{r}) \tag{6}$$

with

$$\mathbf{k}_2^{(m)} = k(\mathbf{x}_0 \sin\theta_{2m} + \mathbf{z}_0 \cos\theta_{2m}) \tag{7}$$

and

$$\mathbf{r} = x\mathbf{x}_0 + y\mathbf{y}_0 + z\mathbf{z}_0 \tag{8}$$

The minus sign in (6) is taken for convenience, as the coefficients A_m giving the amplitude of the mth mode, are as yet unknown.

On the surface S we must have, for the vertically polarized component

$$\left(\frac{\partial E^+}{\partial n}\right)_{z=\zeta} = 0 \tag{9}$$

and for the horizontally polarized component

$$(E^-)_{z=\zeta} = 0 \tag{10}$$

Limiting ourselves to the latter case ($\mathbf{E}_1 \parallel \mathbf{y}_0$), we find from (10) and (5) to (8)

$$e^{-ik\zeta(x)\cos\theta_1} = \sum_{m=-\infty}^{\infty} A_m e^{-imKx + ik\zeta(x)\cos\theta_{2m}} \tag{11}$$

Both sides of this equation are now expanded in a Fourier series with respect to x (which will in general result in a double series on the right side) and the resulting Fourier coefficients are equated. This results in an infinite set of linear equations, each of which contains the unknown coefficient A_m. By progressive solution (or successive approximation) the coefficients A_m are then approximated.

For example, if $\zeta(x)$ is given by (3), the Fourier expansion of the left side of (11) is by (7), Appendix B,

$$e^{-ikh\cos\theta_1\cos Kx} = \sum_{m=-\infty}^{\infty} (-i)^m J_m(kh\cos\theta_1)e^{-imKx} \tag{12}$$

The right side of (11) may by use of the same formula be written as

$$\sum_{m=-\infty}^{\infty} \sum_{n=-\infty}^{\infty} (-i)^n A_m J_n(kh\cos\theta_{2m})e^{-i(n-m)Kx} \tag{13}$$

The corresponding terms in (12) and (13) are those for which $imKx = i(n-m)Kx$. Taking the terms for which m and $n-m$ equal a certain integer q, we find

$$J_q(kh\cos\theta_1) = \sum_{m=-\infty}^{\infty} (-i)^m A_m J_{q+m}(kh\cos\theta_{2m}) \tag{14}$$

For each q we gain an equation of the type (14). The system can then be solved for the coefficients A_m (in principle equivalent to the scattering coefficients ρ_m), e.g. by taking a limited number of terms, starting with $m = q = 0$, and solving (14) as a finite system of linear equations; this will be a good approximation when the series (14) converges quickly (slightly rough surface, $kh \ll 1$). The reader is invited to turn to p. 48, where formula (8) gives the Kirchhoff solution for the same surface for any kh; the comparison leaves little doubt that in this instance the Kirchhoff solution is superior.

However, this does not by any means imply that the Rayleigh method is useless.

Firstly, it is not restricted, like the Kirchhoff method, by the condition $\Lambda \gg \lambda$, and this can be used to demonstrate a result which we only suspected in Section 4.1. From the grating equation (5) we find that for $|m|$ greater than corresponds to the cut-off modes, i.e. for

$$m > \frac{\Lambda}{\lambda}(1-\sin\theta_1) \tag{15}$$

and

$$m < \frac{\Lambda}{\lambda}(\sin\theta_1 - 1) \tag{16}$$

θ_{2m}, and hence $k_2^{(m)}$ in (7) will be complex, which by (6) corresponds to rapidly attenuated standing waves that can be important only in the immediate neighbourhood of the surface. For $\Lambda \ll \lambda$ it therefore follows from (15) and (16) that the only propagated mode will be $m = 0$: any periodic surface will therefore reflect specularly if its period is smaller than the wavelength of the incident radiation.

Secondly, the Rayleigh method does not fail if edges (or small radii of curvature) are present in the surface. An example of this is Deryugin's method [1952] of solving a surface consisting of rectangular corrugations (Section 4.3D).

Thirdly, the Rayleigh approach is the basis of a variational method due to Meecham [1956], which we shall now briefly sketch.

We return to equation (11), which is the exact boundary condition with the scattered field on the right side of the equation. For briefness, we write (11) as

$$\varphi(x) - \sum_{m=-\infty}^{\infty} A_m f_m(x) = 0 \tag{17}$$

where $\varphi(x)$ is the left side of (11) and $f_m(x)$ are the exponential functions on the right side of (11). So as to start the summation from zero, we renumber the modes $m = -\infty$ to $m = \infty$ as follows:

$$\left.\begin{array}{ll} F_1(x) = f_0(x) & \bar{A}_1 = A_0 \\ F_2(x) = f_1(x) & \bar{A}_2 = A_1 \\ F_3(x) = f_{-1}(x) & \bar{A}_3 = A_{-1} \\ F_4(x) = f_2(x) & \bar{A}_4 = A_2 \\ \cdots\cdots & \cdots\cdots \end{array}\right\} \tag{18}$$

Then (11) may be written as

$$\varphi(x) - \sum_{m=0}^{\infty} \bar{A}_m F_m(x) = 0 \tag{19}$$

It is at this point that Meecham's method breaks away from the usual Rayleigh method in that it attempts to determine the coefficients $\bar{A}_m$ more exactly. If the series (19) is broken off after the Nth term, as must be done in practice, the left side of (19) will in general not equal zero. We shall therefore choose the constants $\bar{A}_m$ in such a way that the resulting error in the boundary condition (10), or its equivalents, (11) and (19), is minimized. This will evidently be the case when the integral of the absolute square of the left side of (19) (broken off after N terms) over the surface $\zeta(x)$ is minimum. Since the surface is periodic with period Λ, it is sufficient to minimize the integral from $x = 0$ to $x = \Lambda$. We thus have to determine the coefficients $\bar{A}_m$, for which

$$\int_0^{\Lambda} \left| \varphi(x) - \sum_{m=0}^{N} \bar{A}_m F_m(x) \right|^2 dx = \text{minimum} \tag{20}$$

By setting the N derivatives with respect to the $\bar{A}_m$'s equal to zero, we find that the minimum of (20) will be achieved by choosing the coefficients $\bar{A}_m$ so that they satisfy the set of equations

$$\sum_{m=1}^{N} \bar{A}_m (F_j \cdot F_m) = (F_j \cdot \varphi) \quad (j = 1, 2, \ldots, N) \tag{21}$$

where the symbol $(g \cdot h)$ denotes the "scalar product" of two functions g, h, defined as

$$(g \cdot h) = \frac{1}{\Lambda} \int_0^{\Lambda} g^* h \, \mathrm{d}x \tag{22}$$

with g^* the complex conjugate of g.

Instead of solving the system (21) directly, one may also return to (20) and make use of a theorem from the theory of orthogonal functions, according to which (20) will be achieved if the $\bar{A}_m$'s are the coefficients of the expansion of $\varphi(x)$ in a series of orthogonal functions $F_m(x)$.† In our case the functions $F_m(x)$ will of course in general not be mutually orthogonal; however, an orthonormal set of functions $G_m(x)$ can always be constructed from linear combinations of functions $F_m(x)$; expanding $\varphi(x)$ in a series of $G_m(x)$ then immediately yields the desired coefficients $\bar{A}_m$. Referring the reader to the work of Courant and Hilbert for details of the method,‡ we shall here only give the final results for the computation of the coefficients $\bar{A}_m$ as given by Meecham [1956]:

$$\bar{A}_m = \sum_{k=m}^{N} B_k \Gamma_m^{(k)} \tag{23}$$

where

$$B_k = \sum_{m=1}^{k} \Gamma_m^{(k)} (F_m \cdot \varphi) \tag{24}$$

and the quantities $\Gamma_m^{(k)}$ are defined by

$$\Gamma_m^{(k)} = \frac{\gamma_m^{(k)}}{\sqrt{M_k}} \tag{25}$$

where

$$\gamma_m^{(k)} = \left\{ \begin{array}{ll} 1 & \text{for } m = k \\ -\sum_{\beta=1}^{k-1} \sum_{\alpha=1}^{\beta} \Gamma_\alpha^{(\beta)*} \Gamma_m^{(\beta)} (F_\alpha \cdot F_\beta) & \end{array} \right\} \tag{26}$$

† Courant-Hilbert: *Methods of Mathematical Physics*, Vol. I, Chapter II, Section 1.3, Interscience, New York, 1953.

‡ *Loc. cit.* Section 1.2.

and

$$M_k = \sum_{\mu=1}^{k} \sum_{\nu=1}^{k} \gamma_\mu^{(k)*} \gamma_\nu^{(k)} (F_\mu \cdot F_\nu) \tag{27}$$

Formulae (25) to (27) have to be computed cyclically starting with $m = k = 1$; it will be seen that the formulae are generally manageable only by computing machines.

Whilst there can be little doubt that Meecham's method gives more exact results for the total field at any point $P(x, z)$ (Fig. 4.4b), it is open to some criticism in the physical interpretation of the coefficients $\bar{A}_m$. When computed by the above calculation, these are, as we have seen, coefficients in the expansion by some orthogonal functions G_m, which are not individual wave modes, but linear combinations of several such wave modes F_m; hence these coefficients can hardly be interpreted as reflection coefficients or amplitudes of the individual modes of which the total field is made up. It is therefore not surprising to find discrepancies between the numerical values of Meecham's minimizing coefficients and the amplitudes of wave modes derived by Kirchhoff methods, and from this point of view it is quite natural that Meecham's low-order orthogonal functions should carry a greater fraction of the total energy than low-order wave modes. For a given number of terms in the series expansion Meecham's method will therefore give better results for the total field at a given point (calculated as the sum of all wave modes), whereas the Kirchhoff method should yield better results for the individual wave modes propagated in a given direction.

4.3. Application to Specific Profiles

We shall now apply the general formulae of the preceding two sections for the scattering from a periodic surface to specific profiles.

A. *Sinusoidal Surfaces*

Let the surface S be given by

$$\zeta(x) = h \cos Kx \tag{1}$$

where

$$K = \frac{2\pi}{\Lambda} \tag{2}$$

is the "phase constant" of the surface, and Λ, as before, is its period or "wavelength".

We apply formula (8) of Section 4.1 and find

$$\frac{1}{\Lambda}\int_0^\Lambda e^{i\mathbf{v}\cdot\mathbf{r}}dx = \frac{1}{\Lambda}\int_0^\Lambda e^{iv_x x + iv_z h\cos Kx}dx \tag{3}$$

where by (21), Section 3.1,

$$v_x = k(\sin\theta_1 - \sin\theta_2) \tag{4}$$

$$v_z = -k(\cos\theta_1 + \cos\theta_2) \tag{5}$$

with $k = 2\pi/\lambda$.

Substituting $Kx = t$ and using the grating equation (4), Section 4.1, this becomes

$$\frac{1}{2\pi}\int_0^{2\pi} e^{imt + iv_z h\cos t}dt = i^m J_m(v_z h) \tag{6}$$

where we have used formula (10) of Appendix **B**.

Setting

$$s = -v_z h = kh(\cos\theta_1 + \cos\theta_2) \tag{7}$$

and using (3) of Appendix **B**, we then find from (8), Section 4.1, the solution

$$\rho^{\pm}(\theta_1, \theta_{2m}) = \pm\sec\theta_1 \frac{1+\cos(\theta_1+\theta_{2m})}{\cos\theta_1+\cos\theta_{2m}}(-i)^m J_m(s) + \frac{C(n_1)}{2L} \tag{8}$$

where the angles θ_{2m} are given by the grating equation (4), Section 4.1, and the "edge effect" $C/2L$ which vanishes for integral L/λ and in any case goes to zero for $L \gg \Lambda$, is given by (9), Section 4.1. We shall discuss this result in a little more detail.

The behaviour of ρ for grazing incidence ($\theta_1 = \theta_2 \to \pi/2$) and slight roughness ($h \to 0$) can again be verified (more easily than in the general case of Section 3.3), remembering that $J_0(0) = 1$ and $J_m(0) = 0$ for $m \neq 0$. Note that (8) always stays finite.

The limits of the validity of (8) are given by the conditions under which the Kirchhoff approximation may be considered valid (Section 3.3). In particular, the condition of small radii of curvature (2), Section 3.3, requires

$$\frac{\lambda^2 h}{\Lambda^3} \ll \cos\vartheta \tag{9}$$

where ϑ is the "local" angle of incidence (Section 3.1). This condition will be most critical at the points of inflection $x = \pm\pi/2 + 2n\pi$, where by (18), Section 3.1,

$$\vartheta = \theta_1 \pm \arctan Kh \tag{10}$$

so that (8) will be exact for

$$\lambda^2 \ll \frac{\Lambda^3}{h} \cos\left(\theta_1 + \arctan 2\pi \frac{h}{\Lambda}\right) \tag{11}$$

In particular, for normal incidence ($\theta_1 = 0$) this becomes

$$\lambda^2 \ll \frac{\Lambda^4}{h\sqrt{[(2\pi h)^2 + \Lambda^2]}}$$

To prevent shadowing (screening) we must always have $\vartheta < \pi/2$, i.e.

$$\theta_1 + \arctan Kh < \pi/2 \tag{12}$$

and similarly,

$$\theta_{2m} + \arctan Kh < \pi/2 \tag{13}$$

Condition (12) is less stringent than (11). Condition (13) shows that apart from all other arguments, a surface wave ($\theta_{2m} = \pi/2$) can be considered only if a shadowing function is introduced, or if the surface is smooth ($h = 0$).†

Formula (8) gives the scattering coefficient only for the directions θ_{2m} as given by the grating equation (4), Section 4.1, i.e. in the direction of the maxima of the side lobes. The value of ρ for any θ_2 (not necessarily equal to θ_{2m}) may be found by applying formula (17), Section 4.1. Using (12) of Appendix B, we have

$$\rho^{\pm}(\theta_1, \theta_2) = \pm \frac{\sin 2np\pi}{2n \sin p\pi} \sec\theta_1 \frac{1+\cos(\theta_1+\theta_2)}{\cos\theta_1 + \cos\theta_2} e^{-ip\pi} \times$$

$$\times \left[J_{-p}(s) + \frac{\sin p\pi}{\pi} \int_0^{\infty} e^{pt - s\sinh t} dt\right] + O\left(\frac{n_1}{n}\right) \tag{14}$$

where p is given by (13) of Section 4.1 and n, n_1 by (5), Section 4.1. When $p = m$ (integer), the factor before the integral sign equals zero and (14) reduces to (8) as may easily be verified by applying (3) of Appendix B. The half-widths of the side lobes are, as always, given by (19), Section 4.1.

Brekhovskikh [1952b], using a formula equivalent to the first term of (8), has calculated the scattering diagrams of sinusoidal surfaces for a constant ratio $\Lambda/\lambda = 10$, two angles of incidence ($\theta_1 = 45°$ and $\theta_1 = 80°$) and various values of kh. The results (after slight modifications necessitated by

† Our arguments against the existence of an infinite surface wave in Section 3.3, in particular (13) and its corollary, apply to any $\zeta(x)$ and therefore include the surface $\zeta(x) = S(x)\, h \cos x$, where $S(x)$ is a shadowing function as indicated in the footnote on p. 28.

our definition of the scattering coefficient) are shown in Figs. 4.5–4.17 and their parameters are summarized in Table 4.1. The width of the side lobes as drawn in these figures is more or less symbolic, because it will depend on the length of the surface (the number of periods in the active scattering region) and is given by (18) or (19) of Section 4.1.

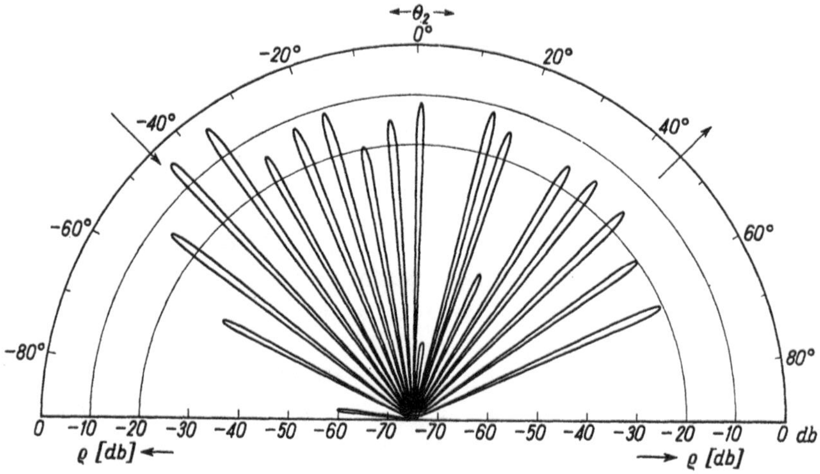

Fig. 4.5. Scattering by a sinusoidal surface. $\Lambda = 10\lambda$, $\theta_1 = 45°$, $kh = 10$.†

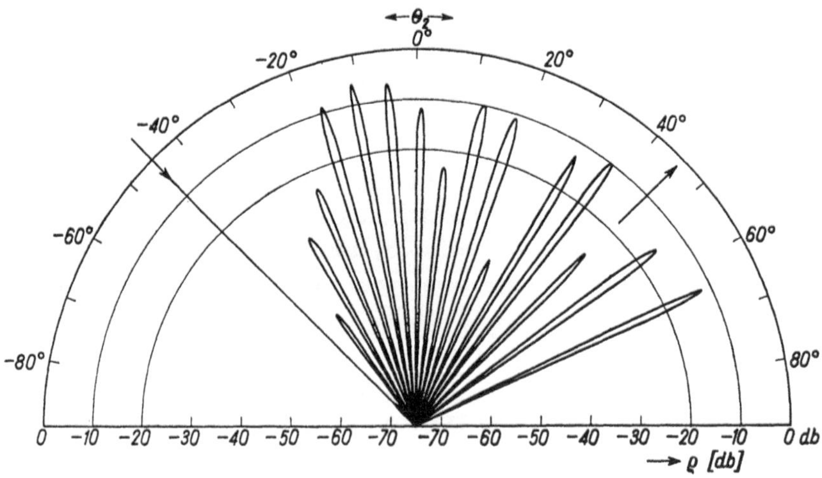

Fig. 4.6. Scattering by a sinusoidal surface. $\Lambda = 10\lambda$, $\theta_1 = 45°$, $kh = 6$.†

† Adapted from [Brekhovskikh, 1952b].

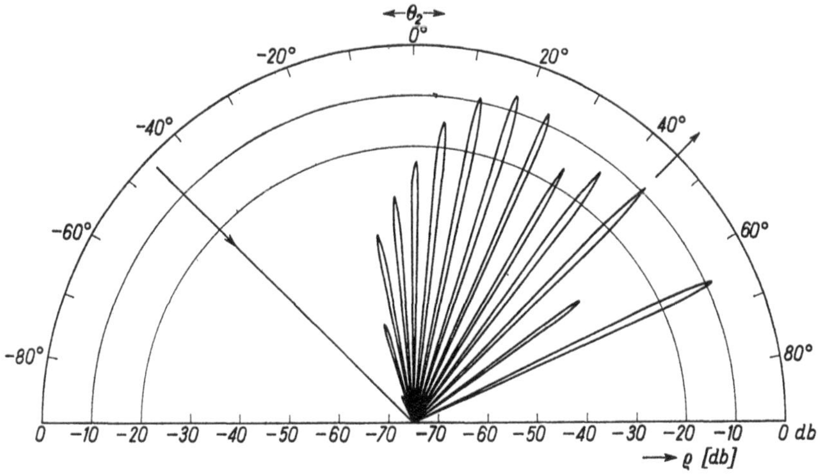

Fig. 4.7. Scattering by a sinusoidal surface. $\Lambda = 10\lambda$, $\theta_1 = 45°$, $kh = 3$.†

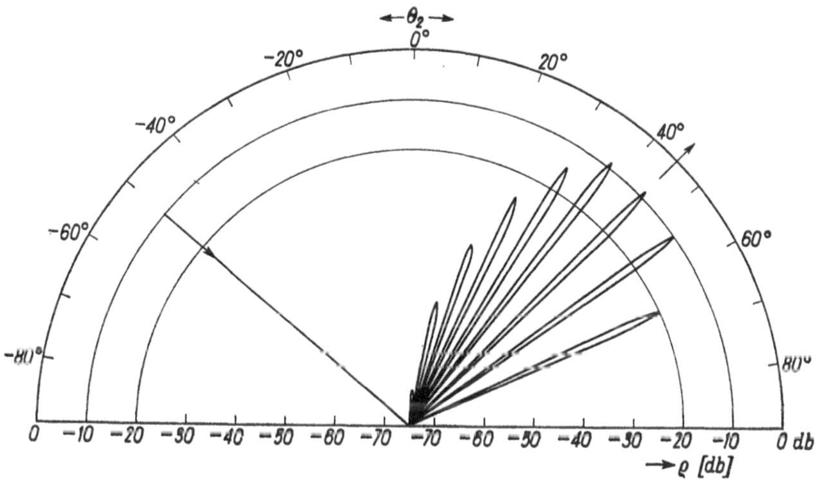

Fig. 4.8. Scattering by a sinusoidal surface. $\Lambda = 10\lambda$, $\theta_1 = 45°$, $kh = 1$.†

† Adapted from [Brekhovskikh, 1952b].

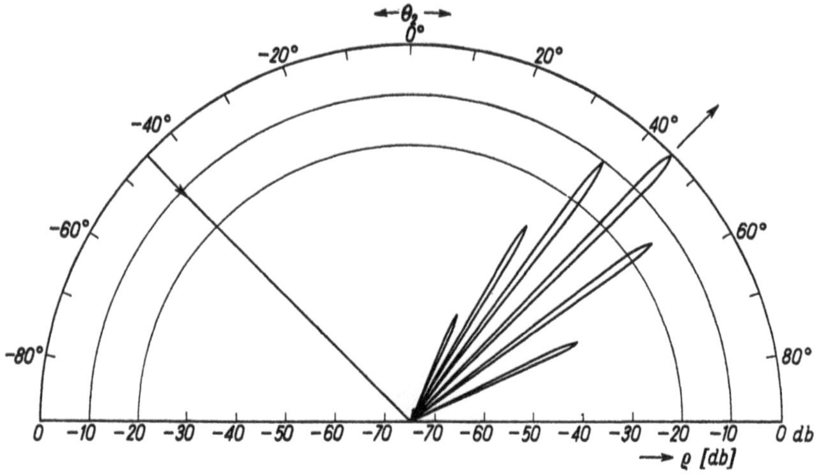

Fig. 4.9. Scattering by a sinusoidal surface. $\Lambda = 10\lambda$, $\theta_1 = 45°$, $kh = 0.3$.†

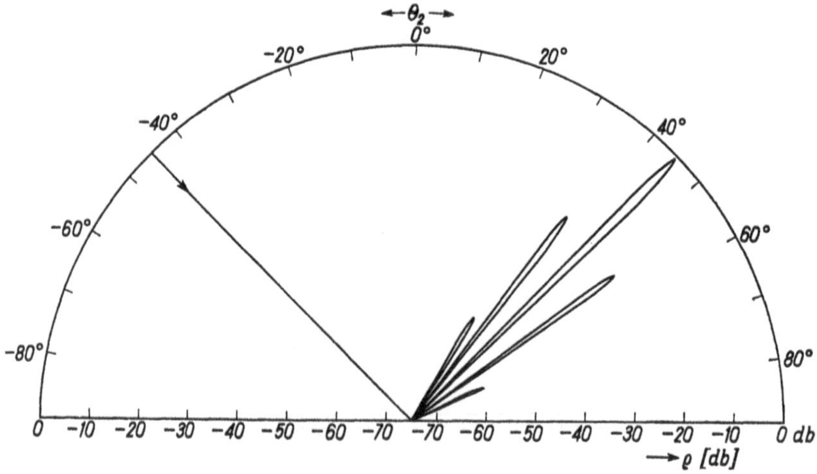

Fig. 4.10. Scattering by a sinusoidal surface. $\Lambda = 10\lambda$, $\theta_1 = 45°$, $kh = 0.1$.†

† Adapted from [Brekhovskikh, 1952b].

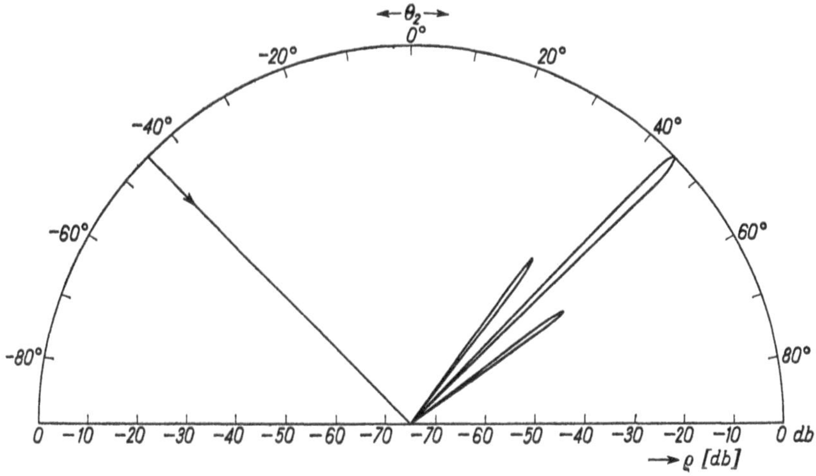

Fig. 4.11. Scattering by a sinusoidal surface. $\Lambda = 10\lambda$, $\theta_1 = 45°$, $kh = 0.03$.†

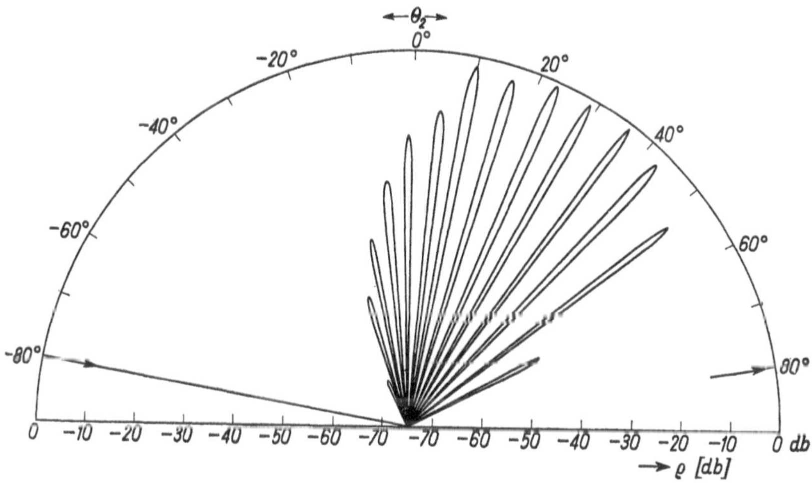

Fig. 4.12. Scattering by a sinusoidal surface. $\Lambda = 10\lambda$, $\theta_1 = 80°$, $kh = 6$.†

† Adapted from [Brekhovskikh, 1952b].

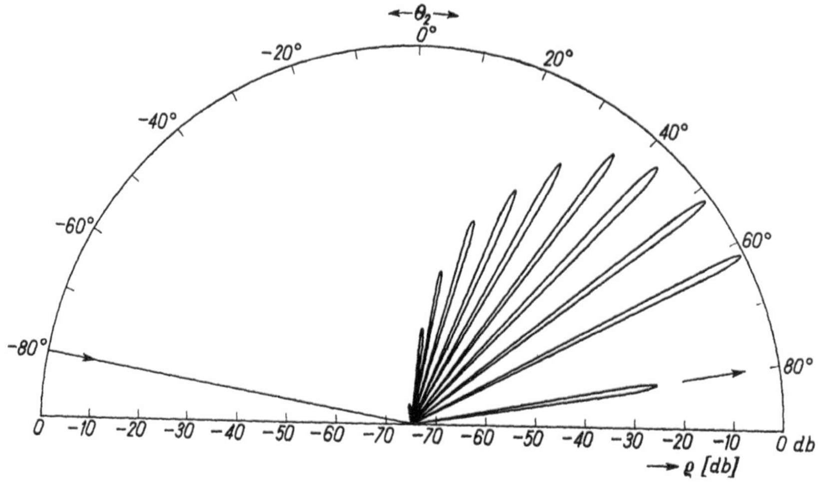

Fig. 4.13. Scattering by a sinusoidal surface. $\Lambda = 10\lambda$, $\theta_1 = 80°$, $kh = 3$.†

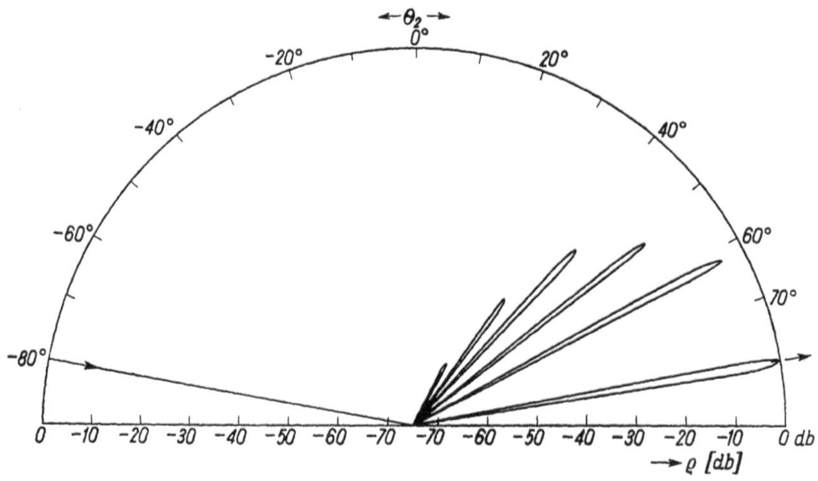

Fig. 4.14. Scattering by a sinusoidal surface. $\Lambda = 10\lambda$, $\theta_1 = 80°$, $kh = 1$.†

† Adapted from [Brekhovskikh, 1952b].

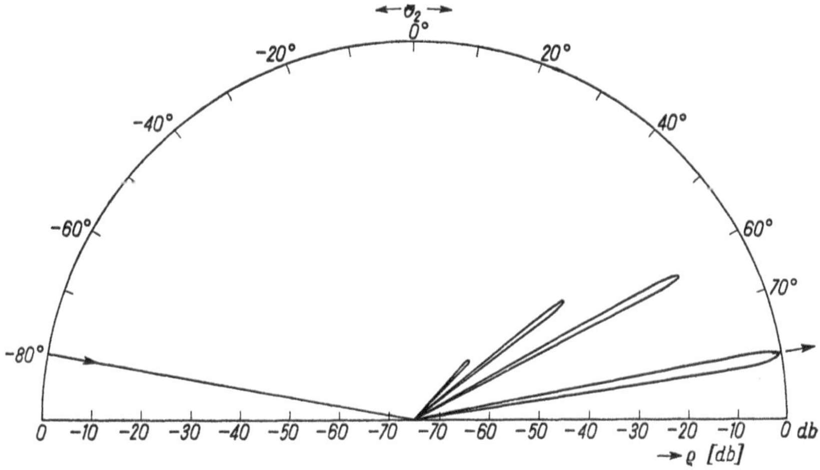

Fig. 4.15. Scattering by a sinusoidal surface. $\Lambda = 10\lambda$, $\theta_1 = 80°$, $kh = 0.3$.†

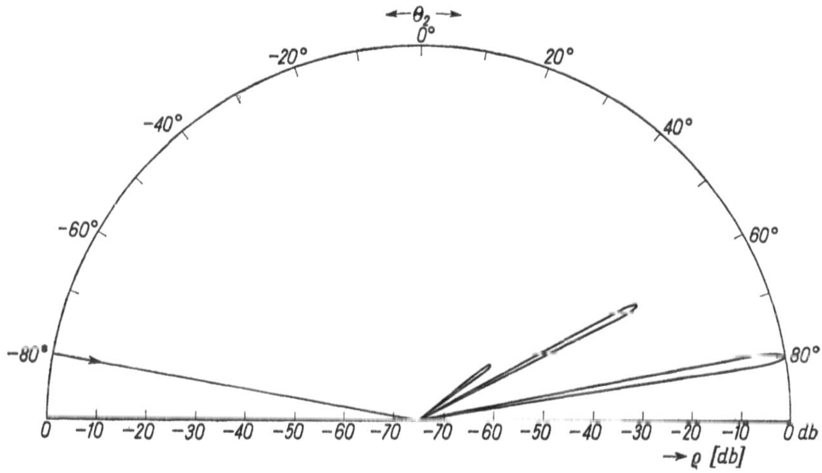

Fig. 4.16. Scattering by a sinusoidal surface. $\Lambda = 10\lambda$, $\theta_1 = 80°$, $kh = 0.1$.†

† Adapted from [Brekhovskikh, 1952b].

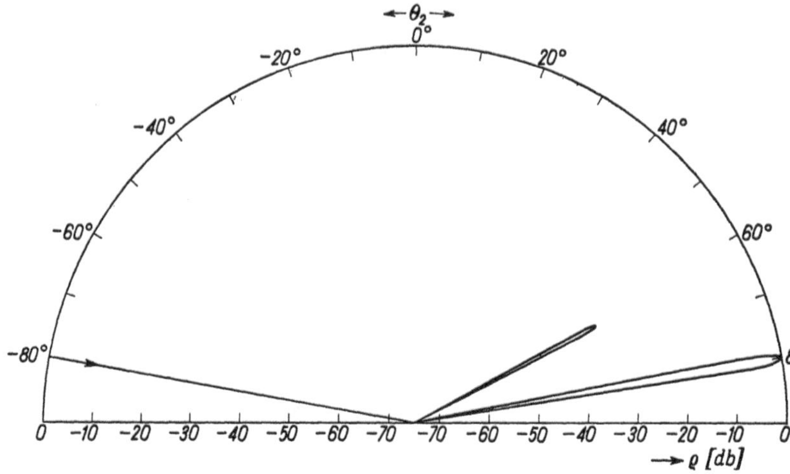

Fig. 4.17. Scattering by a sinusoidal surface. $\Lambda = 10\lambda$, $\theta_1 = 80°$, $kh = 0.03$.†

TABLE 4.1—Key to Figs. 4.5–4.17 [Brekhovskikh, 1952b]

Fig.	$kh = 2\pi h/\lambda$	Λ/h	θ_1	Λ/λ
4.5	10	6.28		
4.6	6	10.5		
4.7	3	20.9		
4.8	1	62.8	45°	
4.9	0.3	209		
4.10	0.1	628		
4.11	0.03	2090		
				10
4.12	6	10.5		
4.13	3	20.9		
4.14	1	62.8	80°	
4.15	0.3	209		
4.16	0.1	628		
4.17	0.03	2090		

As may be seen from these figures, diffuse scattering takes place for the relatively slight roughness $kh = 10$ (i.e. $h/\lambda = 1.6$) of Fig. 4.5 and even for $ka = 3$ (i.e. $h/\lambda = 0.48$), at an angle of incidence of 45°, the maximum lobe does not lie in the direction of specular reflection (Fig. 4.7).

If we use these figures to check the Rayleigh criterion (3) of Section 2.1, considering the reflection specular if the first side lobe is at least 30 db below the main lobe, we must confirm previous suggestions [Kerr, 1951]

† Adapted from [Brekhovskikh, 1952b].

that the factor 8 in the Rayleigh criterion is too low and is more realistically replaced by 32.

The scattering patterns shown in Figs. 4.5–4.17 were calculated by Brekhovskikh [1952b] by using his formula for a *finite* surface, which is equivalent to our formula (8). Brekhovskikh's formula for an *infinite* surface differs by an additional factor $\sec\theta_2$, which we have not found correct (cf. footnote on p. 36). The latter formula was experimentally checked by LaCasce and Tamarkin [1956] by means of underwater sound waves scattered from a sinusoidal surface floating on the water. Fairly good agreement was found except for large scattering angles (when the factor $\sec\theta_2$ will be important).†

Doubly-sinusoidal (and also doubly-trochoidal) surfaces are considered by Hoffman [1955], but only in the direction of back-scatter.

B. *Surfaces Generated by Superposition*

Let the profile $\zeta(x)$ consist of a fundamental undulation and a higher harmonic:

$$\zeta(x) = h_1 \cos Kx + h_2 \cos MKx \tag{15}$$

where M is an integer indicating the order of the harmonic. Then in analogy to (3) and (6)

$$\frac{1}{\Lambda}\int_0^\Lambda e^{i\mathbf{v}\cdot\mathbf{r}}\,dx = \frac{1}{2\pi}\int_0^{2\pi} e^{iv_x x - s_1\cos Kx - s_2 \cos MKx}\,dx \tag{16}$$

where s_1 and s_2 are given by (7) for $h = h_1$ and $h = h_2$ respectively. Expanding $\exp(is_2 \cos MKx)$ in Bessel functions by (7), Appendix B, changing the order of integration and summation, and substituting in (8), Section 4.1, we find

$$\rho^\pm(\theta_1, \theta_{2m}) = \pm\sec\theta_1 \frac{1+\cos(\theta_1+\theta_{2m})}{\cos\theta_1+\cos\theta_{2m}} \sum_{\nu=-\infty}^{\infty} (-i)^{m-\nu M+\nu} \times$$

$$\times J_{m-\nu M}(s_1) J_\nu(s_2) + O\left(\frac{n_1}{n}\right) \tag{17}$$

An equivalent result has been obtained by Brekhovskikh [1952b] by harmonic analysis. Figure 4.18 shows a surface with $kh_1 = 3$, $kh_2 = 2$ and

† The same paper also checks the Eckart theory [1953] adapted to periodic surfaces Eckart replaces $\partial/\partial n$ by $\partial/\partial z$ in the Helmholtz integral, an approximation permissible only for rough surfaces with small slopes, which in the case of a sinusoidal surface means small K or small h, the latter corresponding to a slightly rough surface. This is in fact borne out by the experimental results of LaCasce and Tamarkin [1956].

$M = 4$, and the corresponding scattering pattern for $\theta_1 = 45°$, $\Lambda/\lambda = 10$ [Brekhovskikh, 1952b].

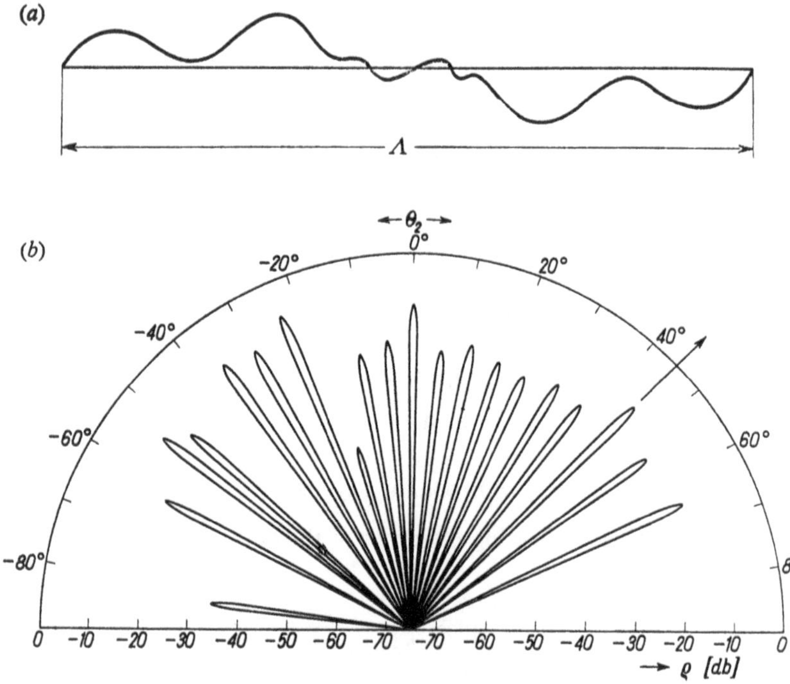

Fig. 4.18. (a) Surface generated by a fundamental and higher harmonic as given by equation (15). (b) Scattering diagram of this surface for $kh_1 = 3$, $kh_2 = 2$, $M = 4$, $\Lambda = 10\lambda$ [Adapted from Brekhovskikh, 1952b].

C. *Saw-tooth Profiles*

A saw-tooth profile (Fig. 4.19) contains two edges in each period and the Kirchhoff method can therefore not be expected to yield exact results. However, it is hard to resist the temptation of the particularly easy integration of $\exp(i\boldsymbol{v} \cdot \boldsymbol{r})$ corresponding to this case; for $w = \Lambda/2$ (Fig. 4.19) we find from (8)

$$\rho^{\pm}(\theta_1, \theta_{2m}) = \pm F_2(\theta_1, \theta_{2m}) \frac{s \sin(m\pi/2 + s)}{(m\pi/2)^2 - s^2} \tag{18}$$

where s is given by (7) and F_2 by (41), Section 3.1.

For more exact results, a Rayleigh method must be used in this case. Meecham [1956] applied his variational method (Section 4.2) to this case, and graphical results are given in his paper. However, as explained in Section

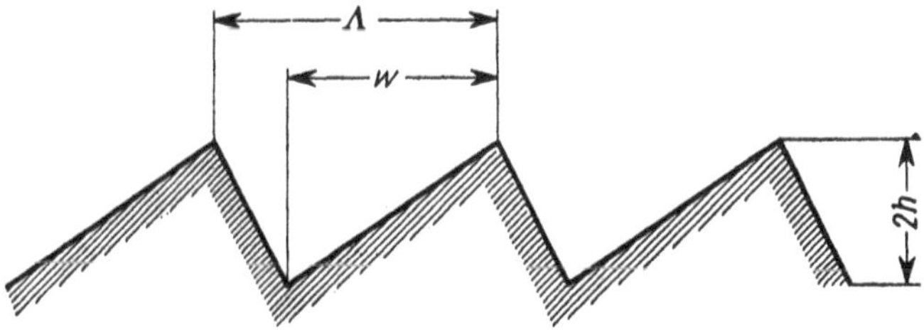

Fig. 4.19. Quantities defining a saw-tooth surface.

4.2, the total field, rather than the individual wave modes of which it is made up, is found in this way.

D. *Rectangular Corrugations*

Surfaces consisting of rectangular corrugations (Fig. 4.20) are met in waveguide theory [Elliott, 1954; Semenov *et al.*, 1958]. For scattering in free-space propagation, the Kirchhoff solution will be inexact, as there are now 4 edges in each period.

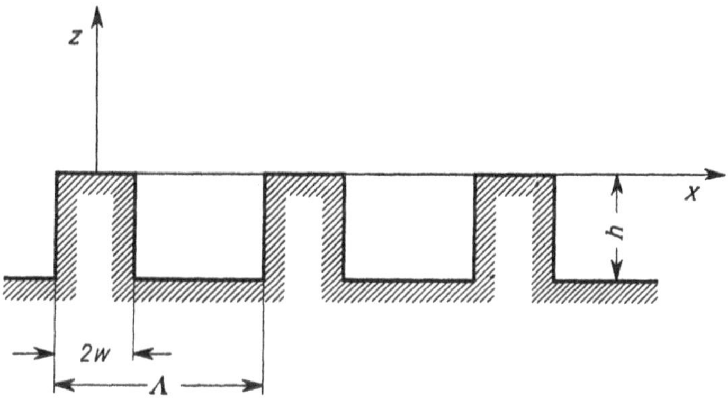

Fig. 4.20. Quantities defining rectangular corrugations.

An exact theory for this profile has been developed by Deryugin [1952, 1953, 1960a, 1960b] and Myakishev [1958], though the final numerical computations have to be performed by solving an infinite set of equations by successive approximation, but any desired degree of accuracy can be obtained. The results have been applied to the design of a diffractional antenna [Deryugin and Myakishev, 1956].

Only an outline of the principle and the results of the theory will be given here. The principle of the method is in essence a Rayleigh method, as the field is expressed as the sum of plane waves and the boundary conditions are then satisfied by determining the appropriate coefficients in the corresponding expansions. However, Deryugin's method differs from the original Rayleigh method in that it first determines the exact structure of the field in the slots, not neglecting the waves reflected downwards by the side-walls into the slots, and then the field in the space above the corrugations is linked up with the field in the slots by using the fact that the tangential components of **E** and **H** must be continuous across the plane $z = 0$ (Fig. 4.20).

We limit the following summary to normal incidence with **E** polarized parallel to the slots [Deryugin, 1952], though we shall also give the results of Deryugin's more recent work [1960]. The wave equation for the magnetic field is first solved in Cartesian coordinates. The solution has five denumerable sets of unknown constants. Considering the periodicity and symmetry of the surface, the solution in the space $z > 0$ (Fig. 4.20) is again found to be

$$H = e^{ik_0 z} + \sum_{m=-\infty}^{\infty} A_m e^{-ik_m z} \cos mKx \tag{19}$$

with $K = 2\pi/\Lambda$ and

$$k_m = \begin{cases} \sqrt{(k^2-m^2)} & \text{for} \quad k > m \\ -i\sqrt{(m^2-k^2)} & \text{for} \quad k < m \end{cases} \tag{20}$$

The field in the slots ($z < 0$) is also represented by a solution of the wave equation in rectangular coordinates. Imposing the boundary condition

$$\frac{\partial H}{\partial n} = 0 \tag{21}$$

for $z = -h$, $x = \pm w$ (i.e. at the bottom and the side walls of the slots, Fig. 4.20) on this solution, the field in the slots can be expressed by a solution with one denumerable set of unknown constants. From H one finds E and then satisfies the additional conditions that H and E must be continuous across the plane $z = 0$. The set of constants A_m, representing by (19) the amplitudes of the wave modes or reflection coefficients, is then found by solving a system of equations involving the unknown constants. The computation leads to the following results:

$$A_0 = \frac{i\pi k_0 Kw + Q_0}{i\pi k_0 Kw - Q_0} + \sum_{n=0}^{\infty} \frac{Q_{0n}}{ik_0 Kw - Q_0} A_n \tag{22}$$

$$A_m = \frac{Q_{0m}}{i\pi k_m Kw - 2Q_m} + \sum_{n=0}^{\infty} \frac{Q_{nm}}{i\pi k_m Kw - 2Q_m} A_m \tag{23}$$

$$A_{-m} = A_m \tag{24}$$

In these equations

$$Q_{mn} = \begin{cases} nm \sin nKw \sin mKw \sum_{j=0}^{\infty} \varepsilon_j \dfrac{q_j \tan q_j Kh}{(n^2-p_j^2)(m^2-p_j^2)} & \text{for } n \neq m \\ 0 & \text{for } m = n \end{cases} \tag{25}$$

$$Q_m = m^2 \sin mKw \sum_{j=0}^{\infty} \varepsilon_j \frac{q_j \tan q_j h}{(m^2-p_j^2)^2} \tag{26}$$

$$Q_{m0} = Q_{0m} = \frac{Kwk_0}{m} \sin mKw \tan k_0 Kh \tag{27}$$

$$Q_0 = K^2 w^2 k_0 \tan k_0 Kh \tag{28}$$

$$p_j = j\frac{\Lambda}{2w}; \quad q_j = \sqrt{(k_0^2-p_j^2)}; \quad \varepsilon_j = \begin{cases} 1 & \text{for } j = 0 \\ 2 & \text{for } j \neq 0 \end{cases} \tag{29}$$

Equations (22) and (23) can be used to determine A_0 and A_m by successive approximation as follows: As a first approximation, the first terms of (22) and (23) are calculated; the results are then substituted in the sums and thus a second approximation is obtained; continuing this process, A_0 and A_m are determined to any desired degree of accuracy.

The above summary is based on Deryugin's original paper [1952]. Deryugin subsequently extended his method to oblique incidence for horizontal polarization (1960a) and vertical polarization (1960b).† Numerical calculations have been carried out by Deryugin [1960] for normal incidence and Myakishev [1958] for oblique incidence. Deryugin's results are shown in Figs. 4.20–4.23 for longitudinal polarization, and in Figs. 4.24–4.28 for lateral polarization. The broken curves show the results obtained by a phase-delay method [Deryugin, 1954], which is in essence equivalent to a Kirchhoff method.

Since Deryugin's method is exact, it provides an important clue to the validity of the Kirchhoff method. Considering that there are 4 edges in each period and that the condition $\Lambda \gg \lambda$ is not satisfied in any of the cases calculated by Deryugin, one would have expected the Kirchhoff method to break down completely, yet Deryugin's curves show that the discrepancy,

† In the case of any one-dimensionally rough surface (unlike the case of a smooth plane), the two polarizations remain distinguishable even at normal incidence. In the present case the **E** vector will lie either across the slots or along them. For roughness in one dimension only, the two polarizations are therefore more aptly called "longitudinal" ($\mathbf{E} \perp \mathbf{y}_0$) and "lateral" ($\mathbf{E} \parallel \mathbf{y}_0$).

especially for lateral polarization, is in view of these conditions surprisingly small. In particular, it may be verified from Deryugin's curves that the condition

$$\Lambda \gg \lambda, \quad w \gg \lambda \tag{30}$$

is more important than the presence of edges. It may be seen that in all cases the agreement improves as the ratio Λ/λ and w/λ is increased; for $\Lambda/w = 4$

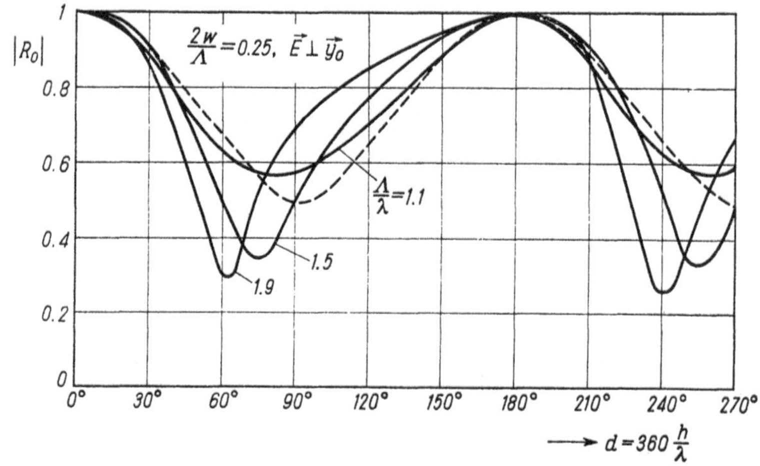

Fig. 4.21. Scattering by rectangular corrugations. Lateral polarization, normal incidence, $2w/\Lambda = 0.25$. Reflection coefficient of the specular mode $|R_0|$ as a function of the "electrical depth" $d = (360h/\lambda)^\circ$ for various ratios Λ/λ. The broken curve was calculated by a Kirchhoff method [Deryugin, 1960a].

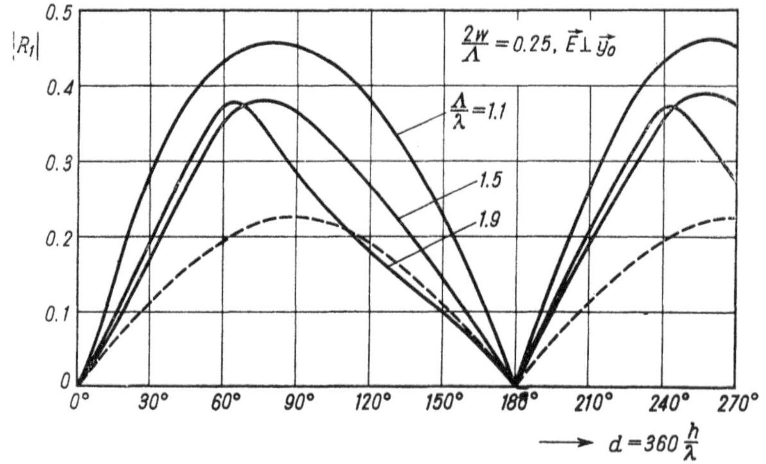

Fig. 4.22. Scattering by rectangular corrugations. Reflection coefficient $|R_1| = |R_{-1}|$ (modes $m = \pm 1$); other conditions as in Fig. 4.21 [Deryugin, 1960a].

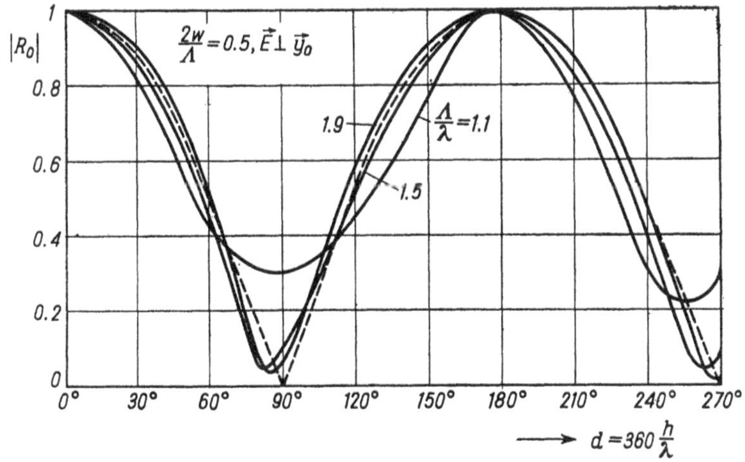

Fig. 4.23. Scattering by rectangular corrugations. $2w/\Lambda = 0.5$, other conditions as in Fig. 4.21 [Deryugin, 1960a].

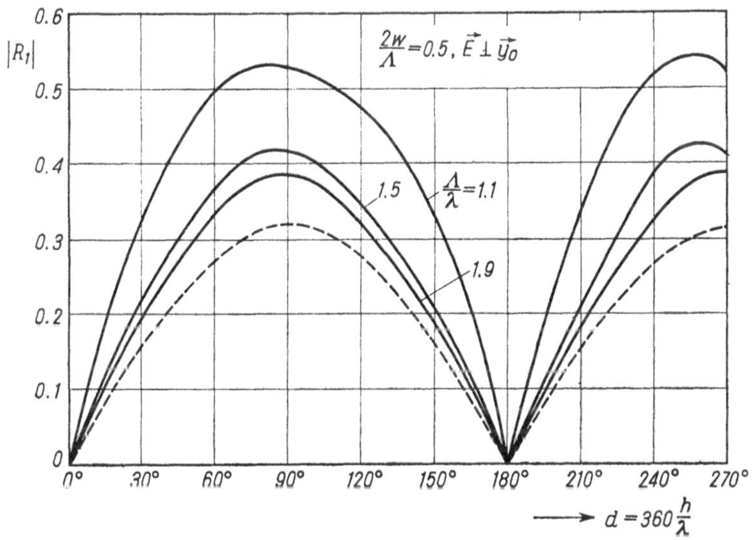

Fig. 4.24. Scattering by rectangular corrugations. $2w/\Lambda = 0.5$, reflection coefficient $|R_1| = |R_{-1}|$ (modes $m = \pm 1$); other conditions as in Fig. 4.21 [Deryugin, 1960a].

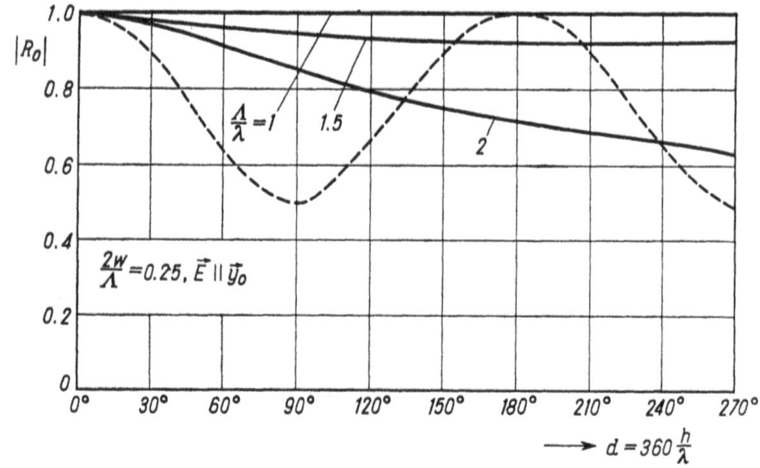

Fig. 4.25. Scattering by rectangular corrugations. Longitudinal polarization, normal incidence, $2w/\Lambda = 0.25$. Reflection coefficient $|R_0|$ of the specular mode ($m = 0$) as a function of the "electrical depth" $d = (360h/\lambda)°$ for various ratios Λ/λ. The broken curve was calculated by a Kirchhoff method [Deryugin, 1960b].

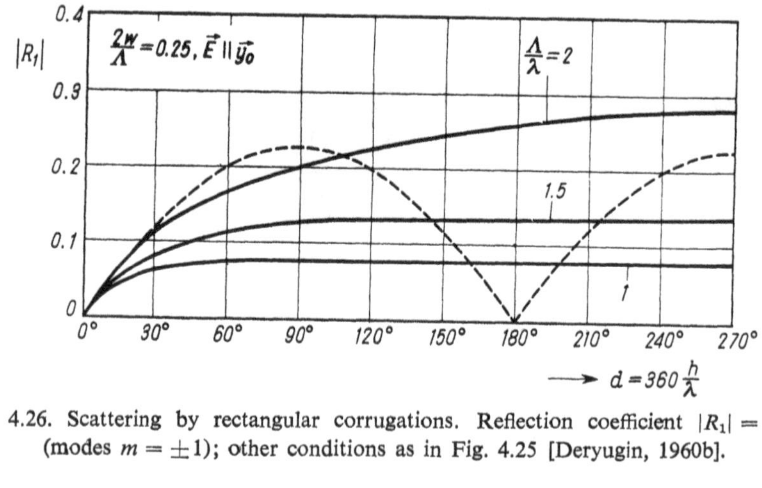

Fig. 4.26. Scattering by rectangular corrugations. Reflection coefficient $|R_1| = |R_{-1}|$ (modes $m = \pm 1$); other conditions as in Fig. 4.25 [Deryugin, 1960b].

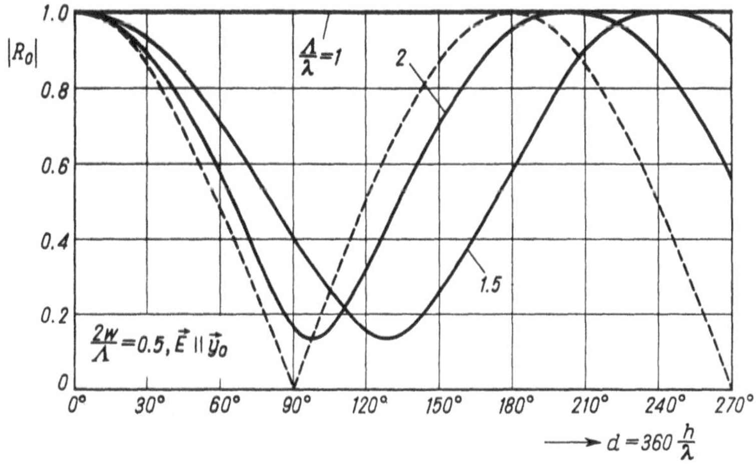

Fig. 4.27. Scattering by rectangular corrugations. Reflection coefficient $|R_0|$ of the specular mode, $2w/\Lambda = 0.5$, other conditions as in Fig. 4.25 [Deryugin, 1960b].

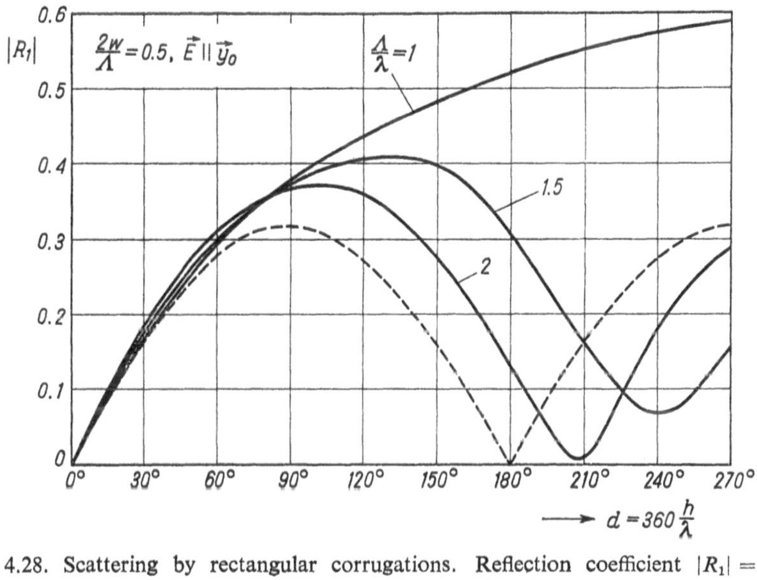

Fig. 4.28. Scattering by rectangular corrugations. Reflection coefficient $|R_1| = |R_{-1}|$ (modes $m = \pm 1$), $2w/\Lambda = 0.5$; other conditions as in Fig. 4.25 [Deryugin, 1960b].

and $\Lambda/\lambda = 1.9$ (or 2), which is still far removed from (30), the Kirchhoff approximation is quite good (very good for lateral polarization and R_0, Fig. 4.23), in spite of the presence of edges.

This comparison shows that the Kirchhoff method may still be a good approximation when edges are present and even when the basic condition $\Lambda \gg \lambda$ is not adequately satisfied (we shall find some more evidence for the latter in Sections 5.3 and 6.1); this leads one to believe that for profiles where no other method is available, one does not risk too much by using the Kirchhoff method to "trespass in forbidden territory".

E. *Regular Distribution of Bosses on a Plane*

Rayleigh [1907] showed that the field scattered by a circular semicylinder protruding from a smooth conducting plane is the same as that scattered by a complete circular cylinder in free space under the incident wave and its image with respect to the plane. Twersky [1950, 1951, 1953, 1957] extended this principle to semicylinders of arbitrary cross-section. The same image technique may be used in the case of a hemispherical boss on a smooth plane by constructing the scattered field from the field scattered by a complete sphere under the incident plane wave and the image of that wave with respect to the smooth plane. Again, Twersky has shown that the principle holds for a boss (protuberance) of arbitrary shape, and by introducting a neat notation he treats the two-dimensional case (cylindrical boss of arbitrary cross-section), the three-dimensional case (boss of arbitrary shape), horizontal polarization and vertical polarization all in parallel [Twersky, 1957a, 1957b]. If the far-field response to an incident wave of an individual scatterer, consisting of a single boss and its image in the plane, is known as a function of the scattering angle, then the field scattered by a distribution of bosses on a plane may in principle be calculated by summing the contributions of all individual scatterers. This is the Twersky method, which uses a distribution of bosses on a plane as a model of a rough surface.

The Twersky theory will be met again in the treatment of random rough surfaces (Section 6.1), though even from the above brief sketch one can see that it may readily be adapted to periodic surfaces, i.e. to surfaces with regular spacing between the bosses: Knowing the far-field response of a body symmetrical about a plane and using the image technique, the scattering coefficient of a boss consisting of one symmetrical half of that body protruding from a plane can be found; the field scattered by a regular array of such bosses can then be found in almost the same way as will be done for a random array of bosses [cf. equation (14), Section 6.1].

The scattering coefficient of a single body (consisting of the boss and its image) may be determined by measurement; analytical solutions are known only for certain geometrical shapes (permitting separation of variables) and of these only the cylinder and the sphere are mathematically reasonably simple, though even in these two cases numerical calculations are cumbersome unless the radii of the spheres or cylinders are either much larger or much smaller than the wavelength of the incident radiation. The latter case, small hemispherical bosses on a conducting plane, has been treated by Biot [1957, 1958], who also uses the image technique for very small hemispherical bosses replacing the spheres by equivalent electric and magnetic dipoles.† Biot then finds the boundary conditions for the electric and magnetic Hertz potentials and hence the reflection coefficient as a function of the angle of incidence. However, though the method is exact in that it respects the mutual interaction (coupling) among the bosses, it can be applied only if the hemispherical bosses are so small that no real scattering sets in: the absolute value of the reflection coefficient remains unchanged (unity) and only its phase is affected by the roughness.

4.4. Finite Conductivity

If the conductivity of the surface is finite, no solution which is general, exact and explicit at the same time is at present known. If the Kirchhoff approximation is valid, the exact solution is given by (33) of Section 3.1, or its equivalent for a two-dimensionally rough surface (1), Section 3.2. The integration can now no longer be performed as in (38), Section 3.1, because a and b are not constant, the reflection coefficient R being a function of the local slope of the surface.

The exact solution may therefore be obtained from (33), Section 3.1, if one is willing to undertake the arduous task of integrating this expression when a and b are functions of x given through R and ϑ by (27), (28), (13), (14) and (18) of Section 3.1.

Similarly, Brekhovskikh's method of harmonic analysis has been extended to finite conductivity by Lysanov [1952], but the integration necessary to find the corresponding Fourier coefficients is equally laborious. Lysanov has calculated the field scattered by a sinusoidal and trochoidal surface of finite conductivity, as has also Aksenov [1958].‡

† Stratton, J. A.: *Electromagnetic Theory*, Section 9.27, McGraw-Hill, New York, 1941.

‡ Since the foregoing was written, Aksenov [1961] has published a more general theory of scattering by a periodic surface of finite conductivity.

There is, however, a number of special cases where the integration can be evaded with varying degrees of accuracy. In the general case a number of approximations is possible; these become coarser and coarser as the range of variation of $a(x)$ and $b(x)$ in the interval $x(0, \Lambda)$ increases.

The simplest (and a most important) case is

$$|\sqrt{Y}| \gg 1 \tag{1}$$

when the incident radiation is horizontally polarized. The admittance Y (Section 3.1) satisfies this condition for practically all kinds of terrain (except very dry soil at very short waves) and from (14), Section 3.1, we find, when (1) is valid,

$$R^- = -1 \tag{2}$$

so that a^- and b^- are constant and all our derivations and results obtained for a perfectly conducting surface remain valid.

It will be seen from (13), Section 3.1, that (1) is not sufficient to guarantee $R^+ = 1$, so that for vertical polarization and finite conductivity our previous results will, when (1) is valid, be a good approximation only if ϑ remains small or fairly constant over the whole surface (e.g. slight roughness near normal incidence).

For the direction of specular reflection ($\theta_2 = \theta_1$) we find from (27) and (28), Section 3.1,

$$a_0 = 2\sin\theta_1 \tag{3}$$

$$b_0 = 2R\cos\theta_2 \tag{4}$$

so that a is now constant. Since for $\theta_1 = \theta_2$, we have $v_x = 0$, the integration (33), Section 3.1, simplifies to

$$\rho = \frac{1}{2L}\int_{-L}^{L} R(\vartheta)\,e^{2ik\cos\theta_1\zeta(x)}\,dx \tag{5}$$

where we have neglected the edge-effect term.

In the general case one may always make a and b constant by taking their averages over the surface (or period). If $\zeta(x)$ is symmetrical (i.e. if positive slopes are compensated by equally large negative slopes) one may take R for $\vartheta = \theta_1$ as such an average.

The simplest approximation is to multiply the field scattered from a perfectly conducting rough surface by $R(\theta_1)$, which will by (18), Section 3.1, be a good approximation for

$$\max|\arctan\zeta'(x)| \ll \theta_1 \tag{6}$$

This will apply to a surface with very small slopes or for any slightly rough surface (with large radii of curvature).

The general conclusion is that finite conductivity can affect the character of the scattering from a rough surface only if the local reflection coefficients are determined by the local angle of incidence rather than by the electrical properties of the scattering material; this will be the case when the admittance at the surface is near unity (e.g. in the case of tropospheric layers).

In most cases, therefore, one would expect the scattered field to be determined by the roughness of the surface rather than by its electrical properties; this is also borne out by experimental measurements.

Note that in this brief consideration we have nowhere used the periodicity of $\zeta(x)$, so that it will, as far as it goes, apply to any rough surface of finite conductivity. The actual averaging will be briefly indicated in Section 5.4.

CHAPTER 5

RANDOM ROUGH SURFACES: I. SURFACES GENERATED BY RANDOM PROCESSES

5.1. Classification of Approaches

The surfaces where the need to predict the scattered field is most urgent—sea, terrain, atmospheric layers, etc.—are neither periodic nor explicitly given; in practice they can only be described by their statistical properties.

A systematic classification of existing theories developed for random rough surfaces is not an easy task and we shall try to accomplish it by grouping the theories according to the models of rough surfaces to which they apply.

The two main groups of random rough surfaces are those generated by a stochastic process and those formed by a random planar array of objects.

A stochastic process (such as thermal noise) is usually a random function of time; in our case it is a function of space coordinates, in other words, the graph of a noise voltage as a function of time, as recorded by some recording instrument, could serve as the profile of a rough surface. The theory of stochastic processes has in recent years made much progress in connection with noise theory and information theory and some of its results are readily applied to random rough surfaces. This approach has been used, among others, by Isakovich [1952], Ament [1953], Eckart [1953], Feinstein [1954], Davies [1954] (continuous Gaussian process) and Beckmann [1957] (Markov chain). In the case of a Gaussian process, the surface is described by its statistical distribution and correlation function; in the case of a Markov chain, it is described by its stochastic matrix.

If a rough surface consists of a planar array of objects, the size, shape, position and/or separation of the objects may be random quantities. Examples of such surfaces are bosses on a plane [Twersky, 1957], arrays of inclined half-planes [Ament, 1960] and arrays of individual unspecified point scatterers [Spetner, 1958].

The two main groups do not exhaust all models. A rough surface may be given by random Fourier coefficients of its profile [Rice, 1951], or by the statistical distribution of the slopes of its plane facets [Ornstein and van der Burg, 1937] or radii of curvature [Du Castel and Spizzichino, 1959];

one may also directly attack the scattered field by considering the distribution of the phase delay caused by the rough surface among the individual rays scattered in the specular or any other direction [Bullington, 1954; Beckmann, 1957; Du Castel and Spizzichino, 1959].

In the following we shall simply divide the models of a rough surface into those generated by a continuous stationary random process and other models. The former type is doubtlessly the most realistic and general and the results gained with the aid of this very flexible model include practically all the results gained by the use of more specific models. In this chapter we therefore investigate the field scattered by a rough surface generated by a continuous stationary random process and find the mean field and mean power scattered into an arbitrary direction for any given angle of incidence. In the next chapter we briefly survey the mean values of the field and power scattered by other models of rough surfaces; but mean values by themselves are incomplete information, the final solution must yield the entire probability distribution about these mean values. The statistical distribution of the scattered field will therefore be calculated in Chapter 7.

The question of mean values and the probability distributions about them brings us to a point on the general use of statistical methods, on which we should like to digress briefly. Even after half a century of quantum theory and under the present impact of information theory and cybernetics, one often finds the rather outdated opinion that the application of statistical methods to a problem is dictated by our inability to solve it "exactly". Yet unless statistics are understood as a mere averaging process, the opposite is in fact true. A random quantity can be considered as found when not only its mean value, but the entire probability distribution about this mean value is known. In certain rare and usually idealized cases, this probability distribution is such that the mean value is, in the limit, the only value the quantity may assume and in this case the quantity is no longer random. In this special case of a single value instead of an entire distribution, we have a (so-called) "exact" solution; this is of course much easier to find than the statistical solution, which includes the "exact" solution as a special case. From this point of view the two opposites are not "statistical" and "exact solution", but "general, statistical solution" and "idealized limiting case".†

† Besides the question of generality, there is of course also the question of economy. From the "exact" or "deterministic" point of view, there is nothing random about tossing a coin: given its mass, its dimensions, the forces acting on it, the air resistance, the elasticity of the material it bounces on, etc., etc., there must be a unique solution indicating whether heads or tails will turn up when it comes to rest. But this is a case not only of our ability, but also of our willingness to find such a solution.

In the present instance, the Kirchhoff solution for the scattering from a rough surface is given by (10), Section 3.2, and the scattering coefficient ρ can be determined by this formula if $\zeta(x, y)$ is given. We have found such a solution in the idealized cases that $\zeta(x, y)$ is periodic or constant (smooth surface). The more realistic and also more difficult problem is to determine the probability distribution of ρ when ζ is a random function of x and y; both idealized cases will then be included in the general solution: a smooth surface is the limiting case when the variance of ζ goes to zero, and a periodic surface is the case when the correlation function of the process is periodic.

We shall attempt to find this general statistical solution in the ensuing chapters. The reader unacquainted with statistical methods will find a very brief explanation of probability theory terms and formulae in Appendix C, to which we shall constantly refer in the remaining chapters.

5.2. Rough Surfaces as Random Processes

The rough surfaces met in nature (terrain, sea, atmospheric layers) are best described by the statistical distribution of their deviation from a certain mean level. This does not yet, however, describe the surface completely; this distribution does not tell us whether the hills and valleys of the surface are crowded close together or whether they are far apart. A second function, the correlation function or autocorrelation coefficient, describes this aspect of the surface. It turns out that the statistical distribution suffices for determining the mean value of the field, but to calculate its variance (or power) the general two-dimensional distribution is needed.

In recent years, electrical engineers, physicists and applied mathematicians have devoted much attention to the theory of random processes. A typical example of a random process is the thermal noise voltage $e(t)$ generated by a resistance. This voltage is a function of time, but its values are random. Suppose now that we record a graph of the noise voltage $e(t)$; this graph may be taken as one of many models of the profile of a rough surface. Instead of being a random function of time, the height coordinate ζ of the surface will be a random function of the coordinate x or of the coordinates x and y, i.e. $\zeta(x)$ or $\zeta(x, y)$, which from the point of view of the mathematical apparatus is immaterial and hence we may use the methods and results developed in noise theory and other branches of the theory of random processes.†

† Cf. Davenport, W. N. and Root, W. L.: *Introduction to Random Signals and Noise*. McGraw-Hill, New York, 1958; Doob, J. L.: *Stochastic Processes*. Chapman and Hall, London; Wiley and Sons, New York, 1953; Blanc-Lapierre, A. and Fortet, R.:

A number of authors have used this model of a rough surface, notably Isakovich [1952], Ament [1953], Eckart [1953], Davies [1954] and Feinstein [1954]. A different model having, however, much in common with the above approach, was used by Booker, Ratcliffe and Shinn [1950].

In this section we shall apply the general solution (10), Section 3.2, to the case when $\zeta(x, y)$ is a random function. Some of our results will be similar to those of Isakovich [1952], which is not surprising, since the equivalent of the first term of (10), Section 3.2, was derived by Isakovich by approximate methods. We shall not, however, limit our investigation to very rough surfaces as Isakovich has done in most cases.

We again start with the one-dimensional case when the surface is given by a random stationary process $\zeta(x)$ and is constant in the y direction.

Let $\zeta(x)$ be a random variable assuming values z with a probability density $w(z)$; let the mean value, which throughout this monograph will be denoted by angular brackets $\langle\ \rangle$, be

$$\langle\zeta\rangle = 0 \tag{1}$$

and consider the mean value of the integral

$$\langle \int_{-L}^{L} e^{i\mathbf{v}\cdot\mathbf{r}} dx \rangle = \int_{-L}^{L} e^{iv_x x} \langle e^{iv_z\zeta} \rangle dx = \langle e^{iv_z\zeta} \rangle \int_{-L}^{L} e^{iv_x x} dx \tag{2}$$

But since

$$\langle e^{iv_z\zeta} \rangle = \int_{-\infty}^{\infty} w(z) e^{iv_z z} dz = \chi(v_z) \tag{3}$$

is by (31), Appendix C, the definition of the characteristic function $\chi(v_z)$ associated with the distribution $w(z)$,† we have

$$\langle \int_{-L}^{L} e^{i\mathbf{v}\cdot\mathbf{r}} dx \rangle = \chi(v_z) \int_{-L}^{L} e^{iv_x x} dx \tag{4}$$

Théorie des Fonctions Aléatoires. Masson, Paris, 1953; to those reading Russian, the present writer would recommend LEVIN, B. R.: *The Theory of Random Processes and its Application to Radio Engineering*, 2nd ed. Sovyetskoye Radio, Moscow, 1960, for its clarity and wealth of useful material.

† Except for the unimportant factor $1/\sqrt{(2\pi)}$, the characteristic function $\chi(v_z)$ is thus the Fourier transform of $w(z)$ and communication engineers may find it useful to look on $\chi(v_z)$ as the spectral density of a signal $w(z)$ with z representing time and v_z angular frequency. However, it should be remembered that $w(z) \geqslant 0$, so that in this analogy $w(z)$ is not an alternating signal, but a d.c. pulse of the shape given by the probability distribution.

Substituting (3) and (4) in (40), Section 3.1, we find the important relation

$$\langle\rho\rangle = \chi(v_z)\rho_0 \tag{5}$$

where

$$\rho_0 = \operatorname{sinc} v_x L \tag{6}$$

is, as in (35) and (36), Section 3.1, the scattering coefficient of a smooth surface.

Note that when $L \gg \lambda$ (as we must assume according to Section 3.3 all the time for our solution to be exact), ρ_0 will equal unity for the specular direction ($\theta_2 = \theta_1$, $v_x = 0$), but will tend rapidly to zero as θ_2 leaves the specular direction. Hence $\langle\rho\rangle$ will equal zero for any direction of scattering except a narrow wedge about the specular direction; if $\chi(v_z)$ equals zero in the specular direction, $\langle\rho\rangle$ will equal zero for all directions. However, it should be remembered that ρ is a complex quantity and that we may not infer $\langle|\rho|\rangle = 0$ from $\langle\rho\rangle = 0$ (unless ρ is real and non-negative).

Note also that the quantity $\langle\rho\rangle$ will be real if $\chi(v_z)$ is real; this will be the case if $w(z)$ is symmetrical about zero, for which (1) is necessary, but not sufficient. Most distributions that will be met in practice are symmetrical about their mean value (e.g. the normal, uniform and Simpson distributions). An example of a distribution that is not symmetrical about its mean value is the Rayleigh distribution, which we shall meet frequently in Chapter 7 as a distribution of the scattered field, but it is doubtful whether any practical *surface* is Rayleigh-distributed. On the other hand it may well be that the surface of the sea is at certain times distributed asymmetrically about its mean value and we shall therefore regard $\chi(v_z)$ generally as a complex quantity, though in most practical cases it will be real. The mean value zero as in (1) is of course only a matter of normalization, which can always be achieved by defining ζ as the deviation from the mean height.

A rough surface $\zeta(x, y)$ is often generated by superposition of several random variables with different distributions:

$$\zeta = \zeta_1 + \zeta_2 + \ldots \tag{7}$$

Thus ζ_1 might be the surface of the earth (terrain) and ζ_2 the vegetation growing on it; or ζ might be the surface of the sea consisting of "swell", "sea" and "ripple". If $\chi(v_z)$ is the characteristic function of ζ, and $\chi_1, \chi_2, \ldots$ are the characteristic functions of $\zeta_1, \zeta_2, \ldots$ respectively, then using the basic property of characteristic functions (36), Appendix C, we find from (5)

$$\langle\rho\rangle = \chi_1\chi_2 \cdots \rho_0 \tag{8}$$

This is one of the great advantages of expressing the average field (or scattering coefficient) by means of the characteristic functions associated with the surface and has been made possible by expressing the general solution of Sections 3.1 and 3.2 in such a way that $\zeta(x, y)$ is contained in the exponent only (in $\mathbf{r}$), and its derivatives are not contained in the expression at all.†

Since ρ is a complex quantity [owing to the possible phase shift between E_2 and E_{20}, cf. (29), Section 3.1], its mean value is of little use except as a stepping stone to determine the mean value of

$$|\rho| = \sqrt{\rho\rho^*} \tag{9}$$

where the asterisk denotes the complex conjugate quantity.

Note that the mean square of (9),

$$\langle\rho\rho^*\rangle = \langle|\rho|^2\rangle = \left\langle\left|\frac{E_2^2}{E_{20}^2}\right|\right\rangle \tag{10}$$

is proportional to the mean scattered power; it is also related to the variance of ρ, denoted by $D\{\rho\}$, and to the variance of the scattered field $D\{E_2\}$ by

$$\langle\rho\rho^*\rangle = \langle\rho\rangle\langle\rho\rangle^* + D\{\rho\} = \langle\rho\rangle\langle\rho^*\rangle + \frac{1}{|E_{20}|^2} D\{E_2\} \tag{11}$$

as follows from (29) and (17) of Appendix C.

The root-mean-square value of ρ is

$$\rho_{\text{RMS}} = \sqrt{\langle\rho\rho^*\rangle} \tag{12}$$

and the mean scattered power is given by

$$\langle P_2\rangle = \tfrac{1}{2}Y_0\langle E_2E_2^*\rangle = \tfrac{1}{2}Y_0|E_{20}|^2\langle\rho\rho^*\rangle \tag{13}$$

where $Y_0 = 1/120\pi$ is the admittance of free space.

Thus we can find all the quantities of interest by determining the value $\langle\rho\rho^*\rangle$.

To find $\langle\rho\rho^*\rangle$, we shall assume the surface S (or its active region) large enough to rid us of the edge-effect terms (Section 3.3), and again start with the case of one-dimensional roughness. From (44), Section 3.1, we have

$$\langle\rho\rho^*\rangle = \frac{F_2^2}{4L^2}\int_{-L}^{L}\int_{-L}^{L} e^{iv_x(x_1-x_2)}\langle e^{iv_z(\zeta_1-\zeta_2)}\rangle\,dx_1\,dx_2 \tag{14}$$

where we have for briefness set

$$\zeta_1 = \zeta(x_1), \quad \zeta_2 = \zeta(x_2) \tag{15}$$

† This is also true of the solutions found by Eckart [1953], Feinstein [1954] and Davies [1954], but Eckart replaces $\partial/\partial n$ by $\partial/\partial z$, whilst Feinstein and Davies use the Helmholtz integral in a form which is valid only for a plane (though possibly perforated) surface, so that in all three cases the solution can be expected to be valid only in the case of slightly rough surfaces or surfaces with very gentle slopes.

But by (37) of Appendix C,

$$\langle e^{iv_z(\zeta_1-\zeta_2)}\rangle = \int_{-\infty}^{\infty}\int_{-\infty}^{\infty} W(z_1, z_2)e^{iv_z(\zeta_1-\zeta_2)}\,dz_1dz_2 = \chi_2(v_z, -v_z) \qquad (16)$$

is the two-dimensional characteristic function of the distribution $W(z_1, z_2)$, which will equal the product of the distributions $w(z_1)$ and $w(z_2)$ only if ζ_1 and ζ_2 are independent. Now we assume that $\zeta(x)$ is a "purely random" process, that does not contain any non-random periodic component (in the terminology of noise theory, ζ corresponds to pure "noise", not to "signal plus noise"). Then ζ_1 and ζ_2 will obviously be idependent if they are far apart, i.e. for large values of the "separation parameter", defined by

$$\tau = x_1 - x_2 \qquad (17)$$

But when τ is small, i.e. when x_1 and x_2 are taken near each other, then ζ_1 and ζ_2 will be correlated; when $\tau = 0$, they will be identical. Thus the distribution of ζ, $w(z)$, is insufficient to determine the two-dimensional characteristic function (16); in addition to the statistical distribution of ζ we must therefore know its correlation function

$$B(\tau) = \lim_{L\to\infty} \frac{1}{2L}\int_{-L}^{L} \zeta(x)\zeta(x+\tau)\,dx \qquad (18a)$$

or its normailzed correlation function (better known as the "autocorrelation coefficient"), which is related to the correlation function $B(\tau)$ by

$$C(\tau) = \frac{B(\tau)-\langle\zeta\rangle^2}{D\{\zeta\}} = \frac{\langle\zeta_1\zeta_2\rangle-\langle\zeta_1\rangle\langle\zeta_2\rangle}{\langle\zeta_1^2\rangle-\langle\zeta_1\rangle^2} \qquad (18b)$$

In the following, we shall always use $C(\tau)$, which owing to (1) reduces to

$$C(\tau) = \frac{\langle\zeta_1\zeta_2\rangle}{\langle\zeta_1^2\rangle} \qquad (18c)$$

It follows from (18c) that for a purely random surface

$$\lim_{\tau\to 0} C(\tau) = 1 \qquad (19)$$

(full correlation, linear dependence), and

$$\lim_{\tau\to\infty} C(\tau) = 0 \qquad (20)$$

(independence).†

† If the surface contains a non-random periodic component, (20) will not hold; in particular, if the surface is periodic with period Λ, then $C(m\Lambda) = 1$ for any integral m, as follows from (18c). We also exclude, at present, a surface corresponding to "white

For a purely random surface, $C(\tau)$ will decrease monotonously from its maximum value $C(0) = 1$ to $C(\infty) = 0$; let the distance in which $C(\tau)$ drops to the value e^{-1} be T. This distance, which we shall call the "correlation distance", will of course be quite small compared with L, for in the contrary case we should not have a rough surface but just one or two irregularities. We thus have

$$T \ll L \tag{21}$$

From (11), (14), (5) and (3) we find

$$D\{\rho\} = \langle\rho\rho^*\rangle - \langle\rho\rangle\langle\rho^*\rangle$$

$$= \frac{F^2}{2L^2}\int_{-L}^{L}\int_{-L}^{L} e^{iv_x(x_1-x_2)}[\chi_2(v_z, -v_z) - \chi(v_z)\chi^*(v_z)]\,dx_1\,dx_2 \tag{22}$$

When ζ_1 and ζ_2 are independent, we find from (38) and (33) of Appendix C

$$\chi_2(v_z, -v_z) = \chi(v_z)\chi^*(v_z) \tag{23}$$

But ζ_1 and ζ_2 are independent for all but very small τ; hence the square bracket in (22) will vanish for all but very small τ. Substituting (17) in (22) and using (21), we therefore have after elementary manipulations

$$D\{\rho\} = \frac{F^2}{2L}\int_{-L}^{L} e^{iv_x\tau}[\chi_2(v_z, -v_z) - \chi(v_z)\chi^*(v_z)]\,d\tau \tag{24}$$

with the only significant contributions to this integral coming from the region near $\tau = 0$.

Note that χ_2, but not χ, is a function of τ, and therefore we need not have taken $\chi\chi^*$ into the integrand; we have nevertheless done so, as this will keep us out of trouble with diverging integrals later on.

Equation (24) shows that to determine the variance $D\{\rho\}$ we must know the two-dimensional distribution $W(z_1, z_2; \tau)$ of the surface, i.e. the two-dimensional distribution of ζ at two points x_1 and x_2 separated by any distance τ.

The next step is to perform the integration (24); however, because this is not simply done in very general terms, we shall delay this step to the next section, where we shall adopt a specific kind of distribution for ζ and a correlation coefficient $C(\tau)$. The most typical case is of course the normal

---

noise", for which $C(\tau) = 0$ for $\tau \neq 0$ and $C(0) = 1$, as (19) will obviously not hold for such a surface. A "white noise" surface, as follows from its correlation function, is discontinuous everywhere and therefore does not lend itself to treatment by Kirhchhoff methods (cf. Section 5.3).

distribution, which will be substituted in the above general formula; other distributions can then be treated quite analogically.

Meanwhile we shall apply the same method to a surface rough in both dimensions, i.e. to (9) and (10) of Section 3.2.

We assume that the surface is isotropically rough, i.e. that ζ is distributed with the same statistical distribution $w(z)$ in all directions over the surface. In the contrary case, when $w(z)$ is different for the x and y directions, the calculation becomes very involved and we shall not attempt a solution of the general "anisotropic" case. In practical applications, anisotropic roughness may occur for the surface of the sea, which may have a different distribution along and across the direction of the wind; in this case one may, in the absence of a more general "anisotropic" solution, use the above formulae for one-dimensional roughness (which is but a simple special case of anisotropic roughness), regarding the surface of the sea as constant in the cross-wind direction.

As before, we find (without neglecting the "edge effect")

$$\langle\rho\rangle = \frac{1}{A}\int_{-X}^{X}\int_{-Y}^{Y} e^{iv_x x + iv_y y}\langle e^{iv_z\zeta}\rangle \,dx\,dy = \chi(v_z)\rho_0 \tag{25}$$

where ρ_0 is the scattering coefficient of a plane surface of area $A = XY$, i.e.

$$\rho_0 = \operatorname{sinc} v_x X \operatorname{sinc} v_y Y \tag{26}$$

where we use the symbol sinc as in (36), Section 3.1.

Since $X \gg \lambda$, $Y \gg \lambda$, (26) will equal unity for $v_x = v_y = 0$, i.e. for the specular direction $\theta_2 = \theta_1$, $\theta_3 = 0$, but will tend to zero rapidly as θ_2 leaves the specular direction, so that $\langle\rho\rangle$ will equal zero everywhere except (possibly, if $\chi(v_z) \neq 0$) inside a narrow cone about the specular direction. Therefore (25) and (26) is equivalent to (5) and (6), so that, as far as $\langle\rho\rangle$ is concerned, there is practically no difference between a surface rough in one dimension or (isotropically) in two dimensions.

The formula for the variance $D\{\rho\}$, however, is different. From (11), Section 3.2, we find

$$\langle\rho\rho^*\rangle = \frac{F_3^2}{A^2}\int_{-X}^{X}\int_{-X}^{X}\int_{-Y}^{Y}\int_{-Y}^{Y} e^{iv_x(x_1-x_2)+iv_y(y_1-y_2)}\,dx_1\,dx_2\,dy_1\,dy_2 \times \times \langle e^{iv_z(\zeta_1-\zeta_2)}\rangle \tag{27}$$

where we have again for briefness set

$$\zeta_1 = \zeta(x_1, y_1), \qquad \zeta_2 = \zeta(x_2, y_2) \tag{28}$$

The separation parameter τ, giving the distance between the variable points (x_1, y_1) and (x_2, y_2) must now be expressed in polar coordinates:

$$\begin{aligned} x_1 - x_2 &= \tau \cos\varphi \\ y_1 - y_2 &= \tau \sin\varphi \end{aligned} \tag{29}$$

Then using (11) and remembering that for the same reasons as above, only the regions near $\tau = 0$ will contribute to the integral, we find after elementary manipulations

$$D\{\rho\} = \frac{F_3^2}{A^2} \int_0^{\infty} \int_0^{2\pi} e^{iv_x \tau \cos\varphi + iv_y \tau \sin\varphi} [\chi_2(v_z, -v_z) - \chi(v_z)\chi^*(v_z)] \tau \, d\tau \, d\varphi \tag{30}$$

The integral over φ leads to the Bessel function of order zero as in (13) of Appendix B; (36) then becomes

$$D\{\rho\} = \frac{2\pi F_3^2}{A^2} \int_0^{\infty} J_0\left(\tau \sqrt{(v_x^2 + v_y^2)}\right) [\chi_2(v_z, -v_z) - \chi(v_z)\chi^*(v_z)] \tau \, d\tau \tag{31}$$

or introducing

$$v_{xy} = \sqrt{(v_x^2 + v_y^2)} \tag{32}$$

our result may be written as

$$D\{\rho\} = \frac{2\pi F_3^2}{A^2} \int_0^{\infty} J_0(v_{xy}\tau) [\chi_2(v_z, -v_z) - \chi(v_z)\chi^*(v_z)] \tau \, d\tau \tag{33}$$

As before, only χ_2, not χ, is a function of τ, so that the second term in the square bracket might be integrated as in (25); nevertheless, we shall find the present form of (31) and (24) more useful for evaluating the integrals, as otherwise we might run into difficulties in the integration where we should obtain expressions which will diverge when the roughness (i.e. $D\{\zeta\}$) tends to zero.

At this point we again leave the general treatment and rather adopt a specific kind of distribution $w(z)$ and a specific autocorrelation coefficient $C(\tau)$ to show how (24) and (33) are integrated.

The most typical and important distribution is of course the normal distribution. As we may choose any variance to represent any degree of roughness, and any correlation distance to represent any density of the irregularities, this will not mean a great loss of generality, but in any case, the same procedure may be adopted with any other kind of distribution. There will generally be only one slight point of difference, namely that if the distribution is unsymmetrical, its characteristic function will be complex instead of real, which leads to a small, but easily surmountable difficulty in the calculation of the distribution of the scattered field in Chapter 7.

5.3. The Normally Distributed Surface

We take the normal distribution as the most important and typical of a rough surface for substitution in the formulae of the preceding section; in fact, the results obtained in this section, when tied up with the entire statistical distribution of the scattered field (to be derived in Chapter 7), represent about the most general solution of the rough-surface scatter problem known at the present time.

Let ζ be distributed normally with mean value

$$\langle\zeta\rangle = 0 \tag{1}$$

and standard deviation σ, i.e. let the distribution of ζ be given by

$$w(z) = \frac{1}{\sigma\sqrt{(2\pi)}}\, e^{-z^2/2\sigma^2} \tag{2}$$

The standard deviation σ, which in this case, due to (1) and to (15) of Appendix C, is also the root-mean-square value of ζ, describes the roughness of the surface and corresponds to the amplitude h of the periodic roughness considered in Chapter 4; σ^2 is the variance of ζ and large σ correspond to large roughness. When $\sigma \to 0$, (2) will tend to the Dirac distribution (infinite for $z = 0$, vanishing for all other values of z) and ζ will cease to be a random variable because ζ will now assume a single value 0, i.e. correspond to a smooth surface.†

The characteristic function associated with the normal distribution (2) is by (3), Section 5.2,

$$\chi(v) = \exp(-\tfrac{1}{2}\sigma^2 v_z^2) \tag{3}$$

The rough surface is not uniquely described by the statistical distribution of ζ, as this tells us nothing about the distances between the hills and valleys of the surface, i.e. about the density of the irregularities. This is described by the correlation function $B(\tau)$ or its normalized equivalent, the autocorrelation coefficient as given by the formulae (18) of the preceding section. We shall use the autocorrelation coefficient $C(\tau)$ as given by (18c), Section 5.2, which, as we have seen, gives the correlation (or lack of independence)

† This is one of the idealized cases easily solved by deterministic methods as mentioned in Section 5.1. By finding the statistical solution for any σ, this "exact" or "deterministic" case will of course be included in the solution

between the random values assumed by ζ at two points (x_1, y_1) and (x_2, y_2) separated by a distance τ. As shown by formula (54) of Appendix C, $C(\tau)$ must be an even function of τ, and to satisfy (19) and (20) of the preceding section, we take as a sufficiently general autocorrelation coefficient the function

$$C(\tau) = \mathrm{e}^{-\tau^2/T^2} \tag{4}$$

where T is the "correlation distance", for which $C(\tau)$ will drop to the value e^{-1}.

Our surface $\zeta(x)$ or $\zeta(x, y)$ is now described by (2) and (4). By varying the constants σ and T we may vary our model to conform to almost any rough surface met in practice, and if this should not suffice, we may use another distribution instead of the normal distribution (2) or another correlation function instead of the autocorrelation coefficient (4), as indicated at the end of this section.

Substituting (3) in (5) or (25) of the preceding section, and expressing v_z explicitly from (6), Section 3.2, we find

$$\langle \rho \rangle = \rho_0 \exp\left[-\frac{2\pi\sigma^2}{\lambda^2}(\cos\theta_1 + \cos\theta_2)^2\right] \tag{5}$$

or, since by (6) and (26), Section 5.2, ρ_0 vanishes everywhere except near the direction of specular reflection $(v_x = v_y = 0)$,

$$\langle \rho_{\text{spec}} \rangle = \exp\left[-\frac{1}{2}\left(\frac{4\pi\sigma\cos\theta_1}{\lambda}\right)^2\right] \tag{6}$$

a result that has been obtained in various equivalent forms by most of the authors mentioned in Section 5.2.† We must be careful, however, in the physical interpretation of this important result, as ρ is a complex quantity, e.g. it does not follow from $\langle \rho \rangle = 0$ that $\langle |\rho| \rangle = 0$, and we shall therefore rather delay any comment until we find $\langle \rho\rho^* \rangle = \langle |\rho|^2 \rangle$.

The two-dimensional normal distribution of two random variables ζ_1 and ζ_2, with mean values zero and variances σ^2 and correlated by a correlation coefficient C, is

$$w(z_1, z_2) = \frac{1}{2\pi\sigma^2\sqrt{(1-C^2)}} \exp\left[-\frac{z_1^2 - 2Cz_1z_2 + z_2^2}{2\sigma^2(1-C^2)}\right] \tag{7}$$

so that from (14), Section 5.2, the characteristic function of this distribution is

$$\chi_2(v_z, -v_z) = \exp\left[-v_z^2\sigma^2(1-C)\right] \tag{8}$$

† Ament [1953] states that this result was derived by Pekeris and, independently, by MacFarlane during World War II.

Substituting (4) in this expression and expanding (8) in an exponential series we have

$$\chi_2(v_z, -v_z) = e^{-v_z^2\sigma^2} \sum_{m=0}^{\infty} \frac{v_z^{2m}\sigma^{2m}}{m!} e^{-m\tau^2/T^2} \tag{9}$$

We shall work with the quantity $v_z^2\sigma^2$ for some time; for briefness we therefore introduce a new symbol g, where

$$\sqrt{g} = v_z\sigma = 2\pi\frac{\sigma}{\lambda}(\cos\theta_1+\cos\theta_2) \tag{10}$$

Using (3) we then find

$$[\chi_2(v_z, -v_z)-\chi(v_z)\chi^*(v_z)] = e^{-g}\sum_{m=1}^{\infty}\frac{g^m}{m!}e^{-m\tau^2/T^2} \tag{11}$$

and this expression may be substituted in either (24) or (31) of Section 5.2.

We first take the case of a surface rough in one dimension only. Substituting (11) in (24), Section 5.2, we obtain

$$D\{\rho\} = \frac{F^2}{2L}\int_{-\infty}^{\infty} e^{iv_x-g}\sum_{m=1}^{\infty}\frac{g^m}{m!}e^{-m\tau^2/T^2}d\tau$$

where we have replaced the integration limits $\pm L$ by $\pm\infty$; this is permissible, since from Section 5.2 we know that the integral receives significant contributions only from the region near $\tau = 0$.

Using the integral†

$$\int_{-\infty}^{\infty} e^{-at^2}\cos bt\,dt = \sqrt{\left(\frac{\pi}{a}\right)}e^{-b^2/4a} \qquad (a > 0) \tag{12}$$

we then find

$$D\{\rho\} = \frac{\sqrt{\pi}\cdot F^2T}{2L}e^{-g}\sum_{m=1}^{\infty}\frac{g^m}{m!\sqrt{m}}e^{-v_x^2T^2/4m} \tag{13}$$

or by (5) and (10) of Section 5.2,

$$\langle\rho\rho^*\rangle = D\{\rho\}+\langle\rho\rangle\langle\rho\rangle^*$$

$$= e^{-g}\left(\operatorname{sinc} v_xL+\frac{\sqrt{\pi}\cdot F^2T}{2L}\sum_{m=1}^{\infty}\frac{g^m}{m!\sqrt{m}}e^{-v_x^2T^2/4m}\right) \tag{14}$$

---

† Ryzhik, I. M. and Gradstein, I. S.: *Tables of Series, Products and Integrals*, Verl. d. Wissenschaften, Berlin, 1957, No. 3.564.5. Derivation: Integrate $\exp(az^2)$ in the complex z-plane from $-\infty$ to ∞ along the real axis and back along $\operatorname{Im} z = b/2a$; since no residua are enclosed, equate to zero and compare the real parts.

The dependence of (13) and (14) on L is correct, but somewhat misleading unless the correctly normalized quantity is considered. In Chapter 7 we shall find that the most useful normalization is to divide by the standard deviation; in the meantime we may note that if we had worked with the scattered field E_2, the sum in the expression for $D\{E_2\}$ and $\langle E_2 E_2^* \rangle$ would have been proportional to L instead of L^{-1}.

For $v_x \neq 0$, we have from (13) and (14)

$$D\{\rho\} = \langle \rho\rho^* \rangle \tag{15}$$

for any g, whilst for $v_x = 0$, i.e. $\theta_1 = \theta_2$, we obtain from (13) and (14)

$$\frac{\langle \rho\rho^* \rangle}{D\{\rho\}} = 1 + \left[\frac{\sqrt{\pi} \cdot F^2 T}{2L} \sum_{m=1}^{\infty} \frac{g^m}{m! \sqrt{m}} \right]^{-1} \tag{16}$$

For sufficiently large g ($g \gg 1$), the second term will be negligible compared with unity, so that for $g \gg 1$, (15) will hold also for $v_x = 0$. Since

$$\langle \rho\rho^* \rangle = \frac{\langle E_2 E_2^* \rangle}{|E_{20}|^2}, \quad D\{\rho\} = \frac{D\{E_2\}}{|E_{20}|^2} \tag{17}$$

we find from (15) and the definition of the root-mean-square value (12), Section 5.2,

$$D\left\{\frac{E_2}{E_{2\,\mathrm{RMS}}}\right\} = 1 \begin{cases} \text{for any } g, \ \theta_1 \neq \theta_2 \\ \text{for } g \gg 1, \ \theta_1 = \theta_2 \end{cases} \tag{18}$$

Relation (15), and its normalized equivalent (18), is typical for the Rayleigh distribution.

From (14) or (13) we may obtain the mean scattered power and the variance and RMS value of the scattered field (Section 5.2).

The series (13) converges more rapidly than the exponential series, but for numerical calculations this is not rapid enough when g is large.† To analyse the dependence of $D\{\rho\}$ on g, v_x and T, we therefore consider three cases: (A) $g \ll 1$, (B) $g \approx 1$, (C) $g \gg 1$. Since by (10) $\sqrt{g}$ is proportional to σ/λ, these three cases correspond to a slightly, moderately and very rough surface.

---

† Most textbooks on higher mathematics ignore the question of numerical computation and give the impression that the exponential series always converges quickly; writing down the first few terms of the series for, say, e^{-10} will cure one of this delusion. In the case of lower transcendental functions, series are of course used only for small arguments, and functions of large arguments are then found by the simple addition formulae that make these functions "lower" transcendental. For higher transcendental functions such as (13), there are no simple addition theorems and infinite series, however well they may pass a convergence test, are usually useless for large arguments unless one is willing to add some thousand terms.

(A) $g \ll 1$. This is the case of a slightly rough surface. The series (13) will converge so quickly that we may take only its first term and find

$$D\{\rho\} \approx \frac{\sqrt{\pi} \cdot gF^2T}{2L} e^{-v_x^2T^2/4} \qquad (g \ll 1) \tag{19}$$

In particular, for the specular direction ($v_x = 0$), we have on substituting for g and F

$$D\{\rho\} \approx 2\pi^{5/2} \frac{\sigma^2}{\lambda^2} \frac{T}{L} [1 + \cos(\theta_1 + \theta_2)]^2 \tag{20}$$

As a check on our calculations, we set $g = 0$ (smooth surface) and obtain from (14) the correct result $\langle \rho\rho^* \rangle = \text{sinc}^2 v_x L$.

(B) $g \approx 1$. In this case, corresponding to a moderately rough surface, the series (13) is still useful and the exact formula (13) may still be employed with advantage. As g grows larger, an increasing number of terms of the series must be taken into consideration. In this case we may also use the estimate (which is valid always, but trivial in the other two cases)

$$g e^{-v_x^2T^2/4} \leqslant \sum_{m=1}^{\infty} \frac{g^m}{m! \sqrt{m}} e^{-v_x^2T^2/4m^2} < e^g \tag{21}$$

The first inequality follows from comparison with the first term of the series, and the second inequality from comparison with the exponential series; the equality sign on the left corresponds to $g = 0$.

From (21) and (13) we then find

$$\frac{\sqrt{\pi} \cdot F^2Tg}{2L} e^{-g - v_x^2T^2/4} \leqslant D\{\rho\} < \frac{\sqrt{\pi} \cdot F^2T}{2L} \tag{22}$$

As a particular example we have for $g = 1$, $v_x = 0$

$$D\{\rho\} = \frac{\sqrt{\pi} \cdot F^2T}{2L} e^{-1} \sum_{m=1}^{\infty} \frac{1}{m! \sqrt{m}} = 0.43 \frac{F^2T}{L} \tag{23}$$

since

$$\sum_{m=1}^{\infty} \frac{1}{m! \sqrt{m}} = 1.308\ldots \tag{24}$$

(C) $g \gg 1$. This is the most interesting case, as it corresponds by (10) to a very rough surface. In this case the series (13) converges too slowly to be of any practical use and we therefore return to the integral (24), Section 5.2. First we note that for $g = v_z^2\sigma^2 \gg 1$,

$$D\{\rho\} = \langle \rho\rho^* \rangle - \langle \rho \rangle \langle \rho \rangle^* \approx \langle \rho\rho^* \rangle \tag{25}$$

since $\langle\rho\rangle$ will by (3) be negligible for $g \gg 1$. (It may easily be shown from the properties of characteristic functions that quite generally $\langle\rho\rangle \to 0$ for $D\{\zeta\} \gg 1$, i.e. for any very rough, not necessarily normally distributed surface.) Substituting (8) in (24), Section 5.2, we then obtain

$$D\{\rho\} = \langle\rho\rho^*\rangle = \frac{F^2}{L}\int\limits_0^L e^{-v_z^2\sigma^2(1-C)}\cos v_x\tau\,d\tau \tag{26}$$

or using (4) and (10),

$$D\{\rho\} = \frac{F^2}{L}\int\limits_0^L e^{-g[1-\exp(\tau^2/T^2)]}\cos v_x\tau\,d\tau \tag{27}$$

It may again easily be verified that for $g \gg 1$ the only significant contribution to this integral comes from the region near $\tau = 0$; we may therefore set $\exp(\tau^2/T^2) \approx 1-\tau^2/T^2$ and replace the upper limit of integration by ∞, obtaining

$$D\{\rho\} = \frac{F^2}{L}\int\limits_0^\infty e^{-g\tau^2/T^2}\cos v_x\tau\,d\tau \tag{28}$$

Applying (12), this becomes

$$D\{\rho\} = \frac{TF^2}{L}\sqrt{\left(\frac{\pi}{g}\right)}\exp\left(-\frac{v_x^2T^2}{4g}\right) \tag{29}$$

or using (10) to discard the symbol g,

$$D\{\rho\} = \frac{TF^2\sqrt{\pi}}{Lv_z\sigma}\exp\left(-\frac{v_x^2T^2}{4v_z\sigma^2}\right) \tag{30}$$

In particular, for the direction of specular reflection ($v_x = 0$, $F = 1$), this becomes

$$D\{\rho\} = \langle\rho\rho^*\rangle \approx \frac{T\sqrt{\pi}}{Lv_z\sigma} = \frac{\lambda T}{4\sqrt{\pi}\cdot\sigma L\cos\theta_1} \tag{31}$$

Before we discuss these results, we shall determine the corresponding quantities for a surface rough in two dimensions. Substituting (11) in (33), Section 5.2, we now have

$$D\{\rho\} = \frac{2\pi F^2 e^{-g}}{A}\sum_{m=1}^{\infty}\frac{g^m}{m!}\int\limits_0^\infty J_0(v_{xy}\tau)e^{-m\tau^2/T^2}\tau\,d\tau \tag{32}$$

where we have interchanged summation and integration, since the convergence of (11) is uniform. As before,

$$v_{xy} = \sqrt{(v_x^2+v_y^2)} = k\sqrt{(\sin^2\theta_1-2\sin\theta_1\sin\theta_2\cos\theta_3+\sin^2\theta_2)} \tag{33}$$

Using the integral (14) of Appendix B, we have, since $m > 0$,

$$\int_0^\infty J_0(v_{xy}\tau)\,e^{-m\tau^2/T^2}\tau\,d\tau = \frac{T^2}{2m}e^{-v_{xy}^2T^2/4m} \tag{34}$$

Substituting this in (32), our solution is now

$$D\{\rho\} = \frac{\pi T^2F^2e^{-g}}{A}\sum_{m=1}^{\infty}\frac{g^m}{m!m}e^{-v_{xy}^2T^2/4m} \tag{35}$$

As before, we distinguish three cases: (A) $g \ll 1$, (B) $g \approx 1$, (C) $g \gg 1$.

(A) $g \ll 1$ (slightly rough surface). Then

$$D\{\rho\} = \frac{\pi gT^2F^2}{A}\exp\left[-\left(g+\frac{v_{xy}^2T^2}{4}\right)\right] \quad (g \ll 1) \tag{36}$$

where we have only taken the first term of the series.

(B) $g \approx 1$ (moderately rough surface). In this case one may still use (35) with a reasonable number of terms. For an estimate of the series we have in analogy to (21)

$$ge^{-v_{xy}^2T^2/4} \leqslant \sum_{m=1}^{\infty}\frac{g^m}{m!m}e^{-v_{xy}^2T^2/4m} < e^g \tag{37}$$

so that

$$\frac{\pi gT^2F^2e^{-g-v_{xy}^2T^2/4}}{A} \leqslant D\{\rho\} < \frac{\pi T^2F}{A} \tag{38}$$

In particular, for $v_{xy} = 0$, g = 1, we obtain from (35)

$$D\{\rho\} = \frac{\pi T^2}{Ae}\sum_{m=1}^{\infty}\frac{1}{m!m} = 0.95\frac{T^2}{A} \tag{39}$$

(C) $g \gg 1$ (very rough surface). In this case we may again use a saddle-point integration, but for the direction of specular reflection ($v_{xy} = 0$), the result may be obtained directly from (35) as follows:

Since

$$m!m = \left(1-\frac{1}{m+1}\right)(m+1)! \tag{40}$$

and for $g \gg 1$ no significant contributions are made to the sum (35) by the terms with small m, we may put $m!m \approx (m+1)!$ in (35); since for the specular direction $F = 1$, we find

$$D\{\rho\} \approx \frac{\pi T^2e^{-g}}{Ag}\sum_{m=1}^{\infty}\frac{g^{m+1}}{(m+1)!} = \frac{\pi T^2e^{-g}}{Ag}(e^g-1-g) \tag{41}$$

Again using the fact that $g \gg 1$, we find $1+g$ negligible when compared with e^g, so that for the specular direction

$$D\{\rho\} = \frac{\pi T^2}{Ag} \qquad (g \gg 1,\ v_{xy} = 0) \tag{42}$$

or more explicitly,

$$D\{\rho\} = \frac{\lambda^2 T^2}{16\pi A\sigma^2 \cos^2\theta_1} \tag{43}$$

Note that here, as in (21), $g \gg 1$ and $v_{xy} = 0$ precludes $\theta_1 \to \pi/2$, as will be seen from (10) for $\theta_1 = \theta_2$.

For $v_{xy} \neq 0$, we cannot use the above artifice and have to resort to saddle-point integration. Substituting (10) and (8) in (33), Section 5.2, we now have

$$\langle\rho\rho^*\rangle = \frac{2\pi F^2}{A} \int_0^\infty J_0(v_{xy}\tau)\, e^{-g(1-C)}\tau\, d\tau \tag{44}$$

or using (4),

$$\langle\rho\rho^*\rangle = \frac{2\pi F^2}{A} \int_0^\infty J_0(v_{xy}\tau)\, e^{-g[1-\exp(\tau^2/T)]}\ \tau\, d\tau \tag{45}$$

It is again easily shown that for $g \gg 1$ this integral receives significant contributions only from the neighbourhood of $\tau = 0$.†

Hence we may use only the first two terms of the expansion of $\exp(\tau^2/T^2) \approx 1-\tau^2/T^2$, so that (45) may be written as

$$\langle\rho\rho^*\rangle = \frac{2\pi F^2}{A} \int_0^\infty J_0(v_{xy}\tau) \exp\left(-\frac{g\tau^2}{T^2}\right)\tau\, d\tau \tag{46}$$

This integral may be evaluated by using (14) of Appendix B, p. 183 yielding

$$\langle\rho\rho^*\rangle = \frac{\pi F^2 T^2}{Ag} \exp\left(-\frac{v_{xy}^2 T^2}{4g}\right) \tag{47}$$

or using (10),

$$\langle\rho\rho^*\rangle = \frac{\pi F^2 T^2}{A v_z^2 \sigma^2} \exp\left(-\frac{v_{xy}^2 T^2}{4 v_z^2 \sigma^2}\right) \tag{48}$$

Since for $g \gg 1$

$$D\{\rho\} = \langle\rho\rho^*\rangle - e^{-g}\rho_0^2 \approx \langle\rho\rho^*\rangle \tag{49}$$

the expression (48) holds also for $D\{\rho\}$.

† E.g. by writing $J_0(v_{xy}\tau)$ as $\exp[\ln J_0(v_{xy}\tau)]$ and differentiating the resulting exponent of the integrand twice with respect to τ. Setting the first derivative equal to zero yields the maximum of the integrand for $\tau = 0$; the large negative second derivative obtained for $g \gg 1$ shows that this maximum is very sharp.

For the specular direction we have $v_x = v_y = 0$, $F = 1$, $v_z = 2k\cos\theta_1$, so that (48) reduces to (43); this provides a check on our approximate evaluation of the integral.

These results are more easily evaluated if we consider the quantity

$$\langle\rho\rho^*\rangle = \langle|\rho|^2\rangle = D\{\rho\} + \langle\rho\rangle\langle\rho\rangle^* \tag{50}$$

which is proportional to the mean scattered power.

From (5) we obtain

$$\langle\rho\rangle = \rho_0 e^{-g/2} \tag{51}$$

where $\rho_0 = \operatorname{sinc} v_x L$ for one dimension, and $\rho_0 = \operatorname{sinc}(v_x X)\operatorname{sinc}(v_y Y)$ for two dimensions, so that practically

$$\langle\rho\rangle\langle\rho\rangle^* = \begin{cases} e^{-g} & \text{for} \quad v_{xy} = 0 \\ 0 & \text{for} \quad v_{xy} \neq 0 \end{cases} \tag{52}$$

where $v_{xy} = \sqrt{(v_x^2 + v_y^2)} = 0$ implies the specular direction.

From (52) and (50) we obtain

$$\langle\rho\rho^*\rangle = D\{\rho\} + e^{-g}\rho_0^2 \tag{53}$$

Thus the mean scattered power in the case of one-dimensional roughness is determined by the following expressions:

$$\langle\rho\rho^*\rangle = \left(\rho_0^2 + \frac{\sqrt{\pi}\cdot F^2 T}{2L}\sum_{m=1}^{\infty}\frac{g^m}{m!\sqrt{m}}\, e^{-v_x^2 T^2/4m}\right)e^{-g} \tag{54}$$

as follows from (13), with g given by (10). Special cases of (54) are from (19) and (30)

$$\langle\rho\rho^*\rangle = e^{-g}\rho_0^2 + \frac{\sqrt{\pi}\cdot F^2 T g}{2L}\, e^{-v_x^2 T^2/4m} \qquad (g \ll 1) \tag{55}$$

$$\langle\rho\rho^*\rangle = \frac{TF^2\sqrt{\pi}}{L v_z \sigma}\exp\left(-\frac{v_x^2 T^2}{4 v_z^2 \sigma^2}\right) \qquad (g \gg 1) \tag{56}$$

For a surface rough in both dimensions we have from (35) and (53)

$$\langle\rho\rho^*\rangle = e^{-g}\left(\rho_0^2 + \frac{\pi T^2 F^2}{A}\sum_{m=1}^{\infty}\frac{g^m}{m!\,m}\, e^{-v_{xy}^2 T^2/4m}\right) \tag{57}$$

with the special cases given by (36) and (48):

$$\langle\rho\rho^*\rangle = e^{-g}\left(\rho_0^2 + \frac{\pi T^2 F^2}{A}\, e^{-v_{xy}^2 T^2/4}\right) \qquad (g \ll 1) \tag{58}$$

$$\langle\rho\rho^*\rangle = \frac{\pi F^2 T^2}{A v_z^2 \sigma^2}\exp\left(-\frac{v_{xy}^2 T^2}{4 v_z^2 \sigma^2}\right) \qquad (g \gg 1) \tag{59}$$

Equation (59) is extensively used in Part II for further derivations. The author of Part II introduces the quantities β_0 and β, defined by

$$\tan\beta_0 = \frac{2\sigma}{T} \tag{60}$$

and

$$\tan\beta = \frac{v_{xy}}{v_z} \tag{61}$$

Since σ is the standard deviation of the heights of the points of the rough surface and T the correlation distance, β_0 is a measure of the mean (absolute) slope of the irregularities forming the rough surface (cf. Appendix D for a more exact statement); β is the angle made by the bisector of the incident and scattered rays with the z axis. Thus β_0 describes the rough surface and β the scattering geometry; both are independent of the wavelength. Using (61) and (62) to express (59), we find

$$\langle\rho\rho^*\rangle = \frac{F^2\lambda^2\cot^2\beta_0}{A\pi(\cos\theta_1+\cos\theta_2)^2}\exp\left(-\frac{\tan^2\beta}{\tan^2\beta_0}\right) \tag{62}$$

The mean power reflection coefficient of a surface of area A is then from (29) and (32) of Section 3.1 found to be

$$\langle|R_1|^2\rangle = \frac{\langle|E_2|^2\rangle}{E_0^2} = \langle\rho\rho^*\rangle\frac{A^2\cos^2\theta_1}{\lambda^2 r^2}$$

where r is the distance from the surface to the point of observation; substituting from (62) and expressing F, $\cos\theta_1$ and $\cos\theta_2$ in terms of the vector $\boldsymbol{v}$, we finally find

$$\langle|R_1|^2\rangle = \frac{A}{\pi r^2}\frac{v^4}{v_z^4}\cot^2\beta_0\exp\left(-\frac{\tan^2\beta}{\tan^2\beta_0}\right) \tag{63}$$

a result equivalent to that obtained by Isakovich [1952] for a very rough surface. It forms the starting point of certain derivations in Part II.

The results for a one-dimensional and two-dimensional surface differ, since a one-dimensionally rough surface will scatter cylindrical and not spherical waves†; the power scattered into a certain direction will therefore be naturally smaller in the latter case. Otherwise the character of the scattered power shows much the same kind of dependence on F, T/L (or T^2/A) and g.

The general solutions (54) and (57) consist of two terms: the "specular" term $e^{-g}\rho_0^2$, corresponding to the direction of specular reflection only, and the term corresponding to the diffusely scattered field. It is interesting to note the general behaviour of the scattered field as g increases from

† Cf. also the remarks following equation (4a), Section 6.1.

$g = 0$ (smooth surface) to $g \gg 1$ (very rough surface). When $g = 0$ (smooth surface or grazing incidence), there is a single reflected wave in the specular direction (plus small side lobes for small L, cf. Section 3.1); this is the special case obtainable by non-statistical methods, cf. Section 5.1. For $0 < g \ll 1$ (slightly rough surface), we can see from (59) and (62) that the specular term is the dominating term, since $T \ll L$ or $T^2 \ll A$; but when $g \gg 1$, (60) and (63) show that the specular term has lost its privileged position altogether, the dependence on the scattering angle θ_2, θ_3 being reduced to that contained in F, v_{xy} and g. This dependence of the scattering pattern on g, i.e. on the surface roughness σ/λ is symbolically shown in Fig. 5.1. This behaviour of the scattering pattern has been derived by several authors using other methods, e.g. Davies [1954] and DuCastel and Spizzichino [1959].

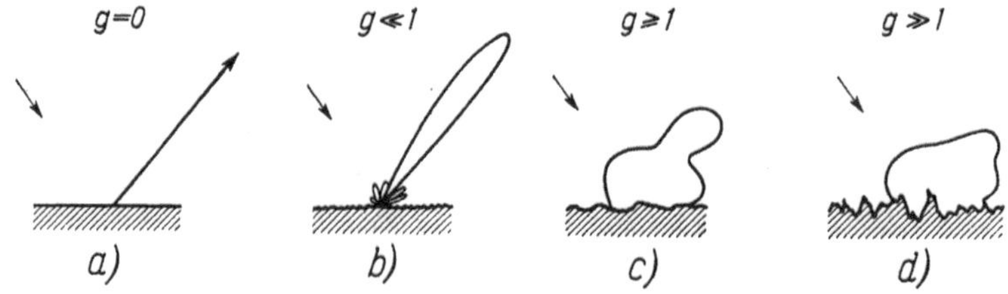

Fig. 5.1. Transition from specular reflection to diffuse scattering. The surfaces are: (a) smooth, (b) slightly rough, (c) moderately rough, (d) very rough.

Before discussing the dependence of the scattered power on the correlation distance T, we recall its physical significance. From Section 5.2 and (4) we know that T is the distance for which the correlation between ζ_1 and ζ_2 will drop to e^{-1}, where ζ_1 and ζ_2 are the values of ζ (deviation from the mean level) at two different points. Thus $T = 0$ implies that there will be no correlation between the values assumed by ζ at two different points, however close to each other they may be; hence $T = 0$ implies a ("white noise") surface that tends to be everywhere discontinuous. Similarly, $T \to 0$ implies a surface with very densely packed irregularities such as that formed by, say, the tree-tops of a dense forest. If T is large, points on the surface will be correlated even if they are far apart, hence large T imply a gently rolling surface, i.e. large distances between its hills and valleys. A surface with large T is shown in Fig. 5.2a, whereas Fig. 5.2b shows the profile of a surface with small T, for the same standard deviation σ. (The drawings are symbolic, as we have not taken the trouble to make the profiles exactly normally distributed.) Thus the correlation distance T is to a certain extent an analogy of the period Λ of a periodic surface, just as the standard deviation σ is an analogy to the amplitude h. Since the Kirchhoff approximation will only be accurate if the radii of curvature of the irregularities are large when compared with the wavelength (Section 3.3), we must require

$$T \gg \lambda \tag{64}$$

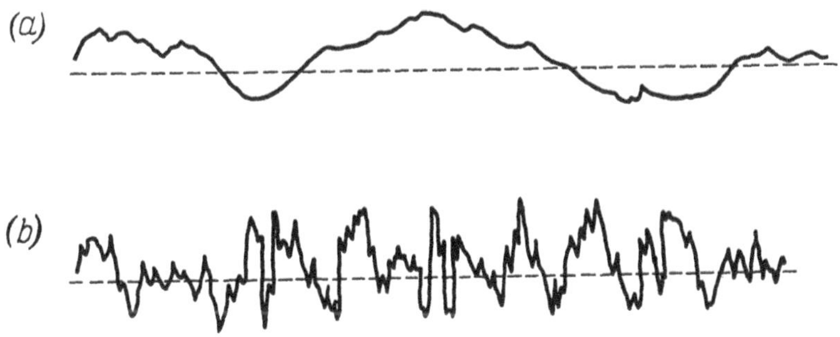

Fig. 5.2. A random surface with (a) large, (b) small correlation distance.

Also, for the reasons given in Section 5.2, we have

$$T \ll L, \qquad T^2 \ll A \tag{65}$$

Our results will therefore be accurate only if (64) and (65) hold. In particular, the Kirchhoff approximation will break down completely if $T \to 0$.

It should also be noted that we have derived our results from (40), Section 3.1, and (10), Section 3.2, which are valid only if the local reflection coefficient of the surface is independent of the local angle of incidence, e.g. in the case of a perfect conductor. Therefore the total scattered power (scattered into all directions) must be constant and varying T (or any other quantity) in our solution may redistribute the power scattered into various directions, but must not alter the total power scattered into all directions.

The dependence of (61) or (58) on T is not monotonous. Differentiating (61) with respect to T, we obtain

$$\frac{\partial\langle\rho\rho^*\rangle}{\partial T} = \frac{\pi F^2 e^{-g}}{A} \sum_{m=1}^{\infty} \frac{g^m}{m!\,m}\left(T - \frac{v_{xy}^2 T^3}{4m}\right) e^{-v_{xy}^2 T^2/4m} \tag{66}$$

with a very similar result if (58) is differentiated with respect to T. For $v_{xy} = 0$, this is positive, so that as the correlation distance is increased, the specular direction will receive more power at the expense of other directions. *Of two equally rough surfaces the one with the more gentle slopes of hills and valleys should therefore scatter less diffusely.* This result holds equally well for one-dimensionally or two-dimensionally rough surfaces. This effect will not, however, be very pronounced in the case of a slightly rough surface, since the specular trem will, owing to (65), predominate, and the specular term, as may be seen from (62) or (59), is independent of the correlation distance.

Other effects of varying T are interlinked with the wavelength dependence, which we shall discuss next.

The wavelength λ occurs in our solutions (61) and (58) in g and v_x or v_{xy}. In all cases it is contained either as σ/λ or T/λ. Therefore, as was to be expected, the scattering pattern will be independent of wavelength if the roughness and correlation distance is measured in wavelengths of the incident radiation; in other words, increasing the wavelength has the same effect as decreasing the roughness parameter (standard deviation) σ and the correlation distance T.

We next consider the wavelength dependence of the specularly and diffusely scattered power in two limiting cases. When λ is large ($g \ll 1$), it follows from (62) that the mean square value of the scattering coefficient is independent of the wavelength for the specular direction (for which the first term dominates), but for the diffusely scattered waves it varies as λ^{-2}; when λ is small ($g \gg 1$), the mean square value of the scattering coefficient is proportional to λ^2 for all directions (for a one-dimensionally rough surface it is proportional to λ).

This result which gives the wavelength dependence caused by the roughness (not the dimensions) of the surface, as our reference field (Section 3.1) is the field reflected by a smooth surface of equal dimensions. If we now consider the wavelength dependence of the scattered field E_2, this will include not only the effects of the roughness, but also the effects of the dimensions of the surface, as even a smooth surface of constant dimensions will scatter more power into the specular direction when the wavelength is smaller. Since by Section 3.1 the scattered field E_2 is proportional to $kA\rho \cos \theta_1$, or simply to $\lambda^{-1}\rho$, we find from (62) and (63) the following result: When the wavelength is large ($g \ll 1$), the mean scattered power of the specularly reflected wave is proportional to λ^{-2} and the diffusely scattered power is proportional to λ^{-4}; when the wavelength is small ($g \gg 1$), the mean scattered power is independent of the wavelength.

Slightly different results may be obtained from (59) and (60) for a one-dimensionally rough surface. If one takes A to be the active area of the surface and sets it equal to the first Fresnel zone (for finite distances of source and point of observation), A will itself be a function of λ, and the wavelength dependence is further complicated.

The independence of the mean scattered power on the wavelength when the wavelength is small or the surface very rough ($g \gg 1$) has also been obtained by Isakovich [1952], Du Castel and Spizzichino [1959] and Spizzichino [1959]. This result may also be obtained from (60) or (63) by taking L/λ or A/λ^2 constant. It is then easily seen that $\langle\rho\rho^*\rangle$ will depend only on T/σ.

Numerical calculations show that for $T/\sigma = 20$ the scattering diagram has a (not very sharp) maximum in the specular direction; when the ratio T/σ is decreased to 10, the maximum recedes into the diffuse scattering diagram and when decreased still further, the maximum will shift, tending toward $\theta_2 = 0$. If we now "trespass in forbidden territory" and let $T \to 0$, thus violating (64), we find from (58) and (63) that only the specular term remains, making it plausible that a "white noise" surface will reflect specularly.

Owing to (65), we find that for the specular direction both (54) and (57) reduce to

$$\langle\rho\rho^*\rangle = e^{-g} \tag{67}$$

since $\rho_0 = 1$ for $v_{xy} = 0$. Using (10), we obtain

$$\langle\rho\rho^*\rangle = \exp\left[-\left(\frac{4\pi\sigma\cos\theta_1}{\lambda}\right)^2\right] \tag{68}$$

and with it a quantitative formulation of the Rayleigh criterion. Although many authors have derived formula (6), this point does not seem to have been sufficiently appreciated. For 80 years, the Rayleigh criterion has been in wide use, although it is only qualitative and unreliable; its popularity is evidently due to the fact that it is simple and contains only three parameters (Section 2.1): the height of the roughness h, the grazing angle γ and the wavelength λ. As we have seen, the only form in which the Rayleigh criterion agrees with practice, is the following: a surface will reflect specularly, if

$$\frac{h\sin\gamma}{\lambda} \to 0 \tag{69}$$

By the definition of ρ (Section 3.1), the quantity

$$\langle\rho\rho^*\rangle = \langle|\rho|^2\rangle = \frac{\langle|E_2|^2\rangle}{|E_{20}^2|} \tag{70}$$

is the mean square of the scattering coefficient, or the power reflection coefficient; σ is the standard deviation of the surface, i.e. the root-mean-square deviation from its mean level, and is therefore equivalent to h; and since $\sin\gamma = \cos\theta_1$, the Rayleigh criterion in terms of our parameters becomes

$$\langle\rho\rho^*\rangle \to 1 \quad \text{for} \quad \frac{\sigma\cos\theta_1}{\lambda} \to 0 \tag{71}$$

Formula (68) gives the same, and much additional information for as little trouble. Although it applies only to a normally distributed surface, there can perhaps be little doubt that it is superior to the Rayleigh criterion.

A useful alternative of (68) is

$$\left.\begin{matrix}\langle \rho\rho^* \rangle \\ \rho_{\mathrm{RMS}}\end{matrix}\right\} = -670\left(\frac{\sigma \sin\gamma}{\lambda}\right)^2 \text{ db} \tag{72}$$

where $\gamma = 90° - \theta_1$ is the grazing angle.

Comparing our results with other authors who investigated the normally distributed surface, we find our results nearest to those of Isakovich [1952], who investigated the case of a very rough surface, but also used the method we have mostly applied in Section 5.2. Our formulae (31) and (42) are in agreement with those of Isakovich; otherwise Isakovich uses a more refined saddle-point method, leading in principle to the same results for $g \gg 1$. Figures 5.3–5.5 show the scattering patterns of normally distributed, perfectly conducting, very rough surfaces ($g \gg 1$) for various angles of incidence and various T/σ as calculated by Isakovich [1952]. The curves have been taken from Isakovich's paper [1952] with the scale of the radial axis modified to conform to our definition of the scattering coefficient. All the figures give $\langle \rho\rho^* \rangle$, i.e. the ratio of the mean scattered power to the power reflected by a smooth surface in the specular direction; this ratio is expressed in decibels. (In these figures $\theta_3 = 0$.)

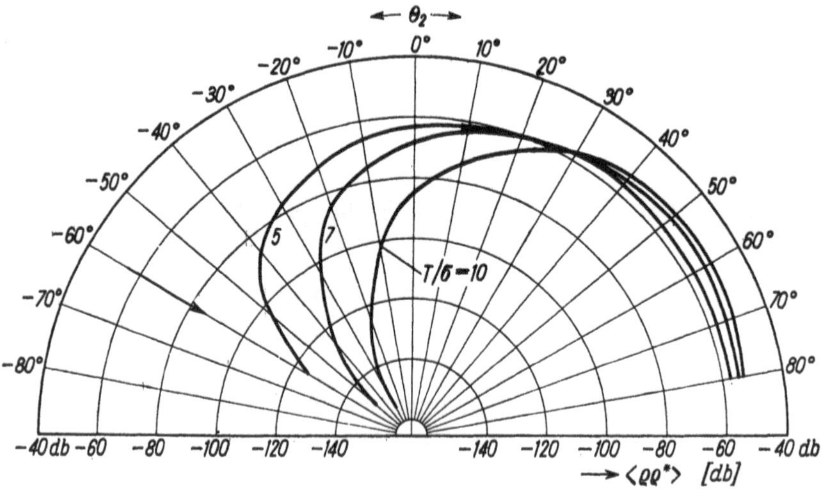

Fig. 5.3. Scattering diagram of a one-dimensional, very rough, normally distributed surface; $L = 1000\lambda$, $\theta_1 = 60°$, $T/\sigma = 5$, 7, 10 [Adapted from Isakovich, 1952].

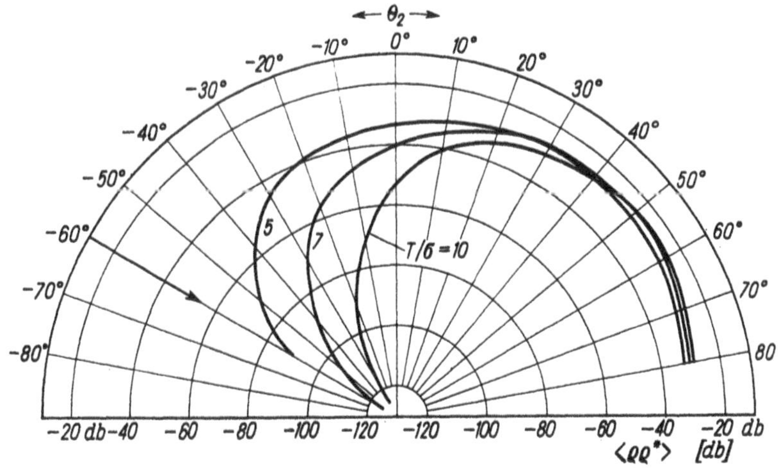

Fig. 5.4. Scattering diagram of a two-dimensional, very rough, normally distributed surface; $A = 10^6\lambda^2$, $\theta_1 = 60°$, $T/\sigma = 5, 7, 10$ [Adapted from Isakovich, 1952].

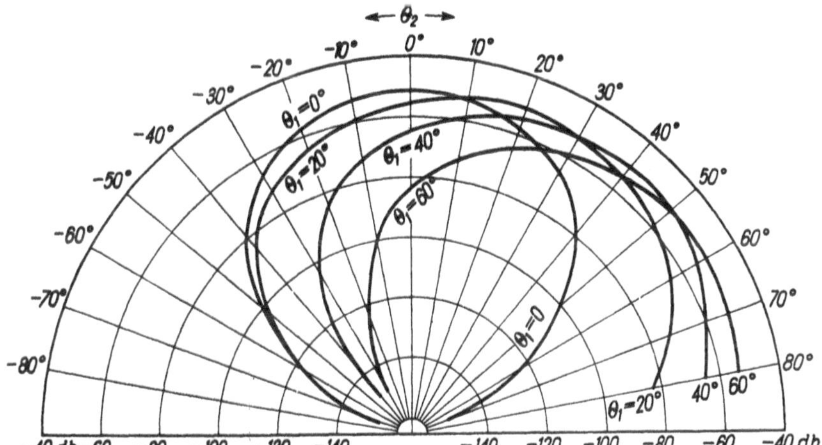

Fig. 5.5. Scattering diagram of a two-dimensional, very rough, normally distributed surface; $A = 10^6\lambda^2$, $T/\sigma = 10$, $\theta_1 = 0°, 20°, 40°, 60°$ [Adapted from Isakovich, 1952].

The figures apply to surfaces with $L = \sqrt{A} = 1000\,\lambda$; for one-dimensional roughness, $\langle\rho\rho^*\rangle$ is proportional to λ/L, and for two-dimensional roughness to λ^2/A, whilst $\langle E_2 E_2^*\rangle$ is independent of the wavelength.

It is hoped that the above analysis of the field scattered by a normally distributed surface will sufficiently illustrate the application of the general formulae of Section 5.2 to any other kind of distribution.

In the general case of a surface generated by a continuous, stationary random process (not necessarily normal), the procedure is quite analogical. We assume the two-dimensional distribution $W(z_1, z_2; \tau)$ to be given, i.e. the two-dimensional distribution of ζ at any two points x_1, y_1 and x_2, y_2, separated by a distance τ. If this is not so, some equivalent function, from which we may find the characteristic function $\chi_2(v_z, -v_z; \tau)$ must be given, e.g. the set of moments of the distribution. The procedure is then the following:

1. Find the one-dimensional distribution $w(z)$ from

$$w(z) = \sqrt{\left(\left[\lim_{\tau\to\infty} W(z_1, z_2; \tau)\right]_{z_1=z_2=z}\right)} \tag{73}$$

where instead of $\tau \to \infty$, we may also write $C(\tau) \to 0$. As a rule, $w(z)$ will already be given (as it was in this Section for a normally distributed surface), but in can always be obtained from $W(z_1, z_2)$ by use of (73) or an equivalent formula.

2. Find the characteristic function of $w(z)$

$$\chi(v_z) = \int_{-\infty}^{\infty} w(z) e^{iv_z z} dz \tag{74}$$

Substitute this in (5), Section 5.2, to obtain the mean value $\langle\rho\rangle$.

3. Determine the two-dimensional characteristic function

$$\chi_2(v_z, -v_z; \tau) = \int_{-\infty}^{\infty}\int_{-\infty}^{\infty} W(z_1, z_2; \tau) e^{iv_z(z_1-z_2)} dz_1 dz_2 \tag{75}$$

4. Substitute (74) and (75) in (24), Section 5.2, for a one-dimensionally rough surface, and in (33), Section 5.2, for a two-dimensionally rough surface. As a rule, the integration will not be straightforward; use series integration for slightly and moderately rough surfaces, and saddle-point integration for very rough surfaces.

The correlation function $B(\tau)$ is found from the two-dimensional distribution $W(z, z; \tau)$ from

$$B(\tau) = \int_{-\infty}^{\infty}\int_{-\infty}^{\infty} z_1 z_2 W(z_1, z_2; \tau) dz_1 dz_2 \tag{76}$$

or from the two-dimensional characteristic function $\chi_2(v_z, -v_z; \tau)$ from

$$B(\tau) = \left[\frac{\partial^2 \chi_2(v_z, -v_z; \tau)}{\partial v_z^2}\right]_{v_z=0} \tag{77}$$

The derivation of these formulae will be found in the literature on random processes; not all of them are general, but they hold under the assumptions we have made, namely that the process generating the rough surface is stationary, ergodic and "purely" random.

The above procedure yields the mean value and variance of ρ, from which other quantities, such as the mean field and mean power may be obtained as a function of $\theta_1, \theta_2, \theta_3$, $D\{\zeta\}$ and T; before we link up these quantities with the entire probability distribution of the scattered field (Chapter 7), we shall, after briefly considering finite conductivity, consider other models of random rough surfaces.

5.4. Finite Conductivity

We shall now make the general remarks of Section 4.4 on finite conductivity a little more concrete in the application to random rough surfaces generated by a stationary random process. For a random rough surface, R is a random function of ζ', since the local angle of incidence ϑ is by (18), Section 3.1, a function of $\zeta'(x)$. In a stationary random process, $\zeta(x)$ and $\zeta'(x)$ are everywhere uncorrelated.

We shall use the approximation

$$\langle a\zeta'\rangle \approx \langle a\rangle\langle\zeta'\rangle \tag{1}$$

where a is given by (3), Section 3.2, thus ignoring the covariance of a and ζ' caused by $R(\zeta')$. Then from (33) and (28), Section 3.1, we find

$$\langle\rho\rangle = -\frac{\langle b\rangle}{4L\cos\theta_1}\int_{-L}^{L} e^{iv_x x}\langle e^{iv_z\zeta}\rangle\, dx \tag{2}$$

Substituting for b from (28), Section 3.1, this yields

$$\langle\rho\rangle = \frac{(1+\langle R\rangle)\cos\theta_2-(1-\langle R\rangle)\cos\theta_1}{2\cos\theta_1}\operatorname{sinc} v_x L\cdot\chi(v_z) \tag{3}$$

In this expression, $\langle R\rangle$ is the mean value of the reflection coefficient taken over the surface, i.e. over the possible values of ζ', for R is by (12) and (14), Section 3.1, a function of the local angle of incidence ϑ, and this is in turn by (18), Section 3.1, a function of ζ'. Thus if the distribution of ζ' is given by $p(\zeta')$, then

$$\langle R\rangle = \int_{-\infty}^{\infty} p(\zeta')R(\zeta')\, d\zeta' \tag{4}$$

In practice this formula will of course be of little use and we may approximate $\langle R\rangle$ by

$$\langle R(\zeta')\rangle \approx R(\zeta' = \langle\zeta'\rangle) = R(\vartheta = \theta_1) \tag{5}$$

where the last equality follows from (18), Section 3.1, for $\langle\zeta'\rangle = 0$. The approximation of the first equality will improve with decreasing $D\{\zeta\}$, i.e. with decreasing roughness.

For $2L \gg \lambda$, (3) vanishes everywhere except in the specular direction $\theta_2 = \theta_1$, $\theta_3 = 0$, when it yields

$$\langle\rho\rangle = \langle R\rangle\chi(v_z) \tag{6}$$

a result which we might have obtained directly from (5), Section 4.4.

Thus by averaging over R, we convert the functions $a(x)$ and $b(x)$ in (32) to constants and may then apply the method used for perfectly conducting surfaces. The averages $\langle a\rangle$ and $\langle b\rangle$ will be a function of $\langle R\rangle$, which is given exactly by (4) and can be approximated by $R(\vartheta = \theta_1)$, i.e. by the Fresnel coefficient for a smooth surface of the same conductivity and under the same angle of incidence.

In general, we may by using the approximation (1) calculate $\langle\rho\rangle$ and $\langle\rho\rho^*\rangle$ for a surface of finite conductivity by applying the general formula (9), Section 3.2, and replacing a, b, c by $\langle a\rangle$, $\langle b\rangle$, $\langle c\rangle$, i.e. replacing R by $\langle R\rangle$ in (3)–(5) of Section 3.2. This leads, as is easily verified, only to a multiplication of $F_3(\theta_1; \theta_2, \theta_3)$ by $\langle R\rangle$; hence under the approximation (1)

$$\langle\rho\rangle_f = \langle R\rangle\langle\rho\rangle_\infty \tag{7}$$

$$\langle\rho\rho^*\rangle_f = \langle RR^*\rangle\langle\rho\rho^*\rangle_\infty \tag{8}$$

$$D\{\rho\}_f = \langle RR^*\rangle D\{\rho\}_\infty \tag{9}$$

where the indices f and ∞ denote finite and infinite conductivity respectively, and $\langle R\rangle$ is given approximately by (5).

The relation

$$\rho_f = \langle R\rangle\rho_\infty \tag{10}$$

does not follow immediately from (7); we shall, however, find that the distribution of $|\rho|$ is determined uniquely by $\langle\rho\rangle$, $D\{\rho\}$ and the distribution of ζ. In the application to rough surfaces generated by a stationary random process we may therefore use (10) within the validity of (1) and (5).

An exact calculation based on (33), Section 3.1 (or its equivalent for a surface rough in both dimensions) would be very troublesome and it is doubtful whether the results would justify such a complicated procedure.†

† Since the above was written, Kovalev and Pozdnyak [1961] investigated the scattering by slightly rough surfaces of finite conductivity. They assume the roughness to be sufficiently slight for the second and higher powers of n_x (projection of the normal onto the x axis) to be neglected in the expansion of the reflected field (given by the Green–Kirchhoff formula) in a power series. For a one-dimensional rough, normally distributed surface the authors derive relatively simple formulae for the mean scattered field for both polarizations. The results justify most of the above qualitative remarks.

CHAPTER 6

RANDOM ROUGH SURFACES: II. OTHER MODELS

In the preceding chapter we have found the mean value and variance of the scattering coefficient and hence the mean field and mean power scattered by a random surface generated by a stochastic process and given by its two-dimensional statistical distribution. In particular, we have investigated the normally distributed surface given by its (normal) distribution and correlation function. This is probably the most general and realistic type of a rough surface.

However, a statistical problem is completely solved only when the entire probability distribution (not just the mean value and variance) of the random quantity is found. Before we go on to find this distribution, we shall in this chapter digress to other models of random rough surfaces that, though less general, have other advantages, such as not being restricted to the Kirchhoff approximation, enabling multiple scattering to be respected, or being more easily calculable. The theories developed in connection with these models generally yield the mean scattering coefficient (or its equivalent) as a function of the surface parameters and the scattering geometry θ_1, θ_2, θ_3.

However, as the aim of Part I of this monograph is to give a general and integral theory of rough-surface scatter rather than a comprehensive and detailed survey of past work in this field, in this chapter we shall only give a brief outline of the theories developed by Twersky, Ament, Rice, Beckmann, Du Castel, Spizzichino and others, restricting our account to main principles and results, which will enable us to link up these theories with the main thread of our development.

6.1. Twersky: Bosses on a Conducting Plane

One of the most successful theories of rough-surface scatter, though limited to a not very general profile, has been developed by Twersky. The Twersky model of a rough surface consists of regularly or irregularly spaced bosses (protrusions) of almost arbitrary (but usually identical) shape on a perfectly conducting plane. The distance between neighbouring bosses should be large compared to their size. Although such a model is seen to be less general than a random surface given by its statistical distribution and correlation function, the Twersky method has several advantages over the former, especially the following two: (1) it does not completely neglect multiple scattering, i.e. the mutual interaction among the bosses, (2) the scattered field is easily calculable (and not restricted to the validity of the Kirchhoff approximation) if the scattering function of the body consisting of a single boss and image with respect to the plane is known.

Twersky's method is based on an image method of Rayleigh [1907], who showed that the field scattered by a circular semicylinder protruding from a conducting plane under an incident plane wave equals the field scattered by an entire cylinder in free space illuminated by two incident waves: the originally incident wave and its image with respect to the plane (Fig. 6.1).

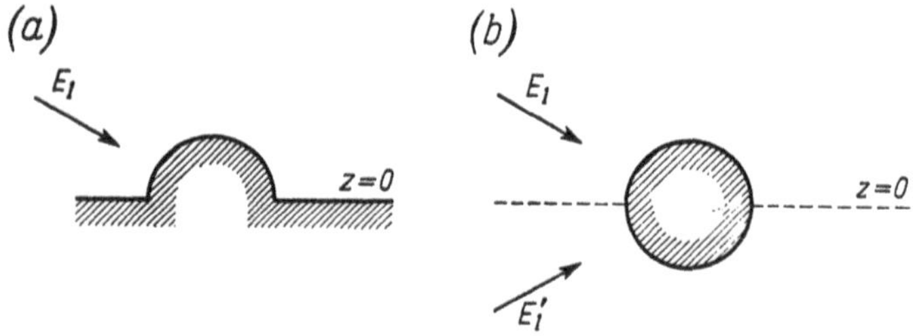

Fig. 6.1. The Rayleigh image method.

This result may be generalized to bodies symmetrical with respect to the plane $z = 0$ of otherwise almost arbitrary shape. Without going into a detailed proof of this statement, its validity may be seen by considering the boundary conditions, which must uniquely determine any electromagnetic problem. On the surface of the body in Fig. 6.2b the total incident field will be the sum of $\mathbf{E}_1$ and $\mathbf{E}_1'$ exactly as it would be if $\mathbf{E}_1'$ were due to the reflection of $\mathbf{E}_1$ by a perfectly conducting plane along $z = 0$. On the (fictitious) plane $z = 0$, the total field will equal $\mathbf{E}_1 + \mathbf{E}_1'$. If $\mathbf{E}_1$ and $\mathbf{E}_1'$ are vertically polarized (i.e. lying in the plane of incidence xz), we let E_1 be in phase with E_2, so that $\mathbf{E}_1 + \mathbf{E}_1' = 2(\mathbf{E}_1 \cdot \mathbf{z}_0)\mathbf{z}_0$ for $z = 0$; if $\mathbf{E}_1$ and $\mathbf{E}_1'$ are horizontally polarized, we let E_1 and E_1' be in antiphase, so that $\mathbf{E}_1 + \mathbf{E}_1' = 0$ for $z = 0$, exactly as if a conducting plane were present at $z = 0$. Hence Figs. 6.2a and 6.2b are equivalent. The same result is obtained when one considers also the waves scattered by the boss back onto the plane.

Twersky defines the "scattering amplitude" $g^{\pm}(\theta_1, \theta_2)$ as the far field response in the direction θ_2 when the scatterer is excited by a plane wave

$$E_1^{\pm} = e^{i\mathbf{k}_1 \cdot \mathbf{r}} \tag{1}$$

where we use the symbols of Section 3.1. The scattering amplitude is thus in principle equivalent to our scattering coefficient $\rho^{\pm}(\theta_1, \theta_2)$.

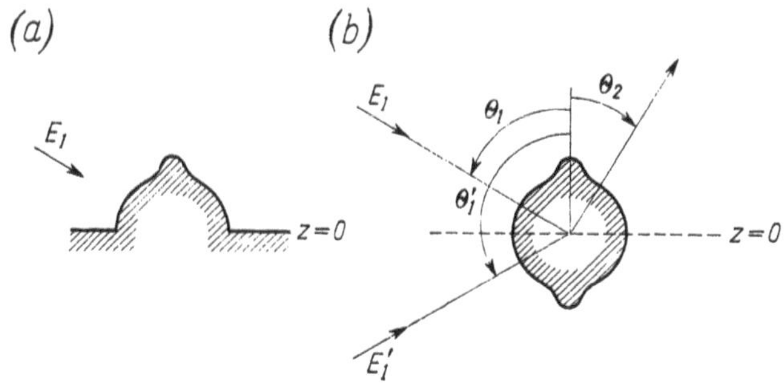

Fig. 6.2. The image method.

For a finite ("three-dimensional") scatterer, the far field has the form

$$E_2^{\pm} = \frac{e^{ikR_0}}{ikR_0} g^{\pm}(\theta_1, \theta_2)\,|E_1| \tag{2}$$

and for an infinite cylindrical scatterer of finite cross-section ("two-dimensional")

$$E_2^{\pm} = \sqrt{\left(\frac{2}{i\pi kR_0}\right)} e^{ikR_0} g^{\pm}(\theta_1, \theta_2)\,|E_1| \tag{3}$$

Equations (2) and (3) define the scattering amplitude $g^{\pm}(\theta_1, \theta_2)$, which we have here slightly modified. To convert $g^{\pm}(\theta_1, \theta_2)$ to $\rho^{\pm}(\theta_1, \theta_2)$, we extend the definition of ρ from a surface to a body by defining it as in (29), Section 3.1, with the reference field E_{20} defined as the field reflected by a perfectly conducting plane surface of unit area ($A = 2L = 1$) into the direction of specular reflection when the incident wave is vertically polarized. Comparing (2) with (32) of Section 3.1, we then find that Twersky's scattering amplitude $g(\theta_1, \theta_2)$ is related to our scattering coefficient $\rho(\theta_1, \theta_2)$ in the case of a three-dimensional scatterer by

$$\rho^{+}(\theta_1, \theta_2) = -\frac{\pi}{k^2 \cos\theta_1} g^{+}(\theta_1, \theta_2) \tag{4a}$$

In the case of a two-dimensional scatterer, i.e. one that is infinite and constant in one dimension (cylinder, strip, etc.), the scattered wave (3) represents a cylindrical wave, which may be recognized from the distance dependence $R_0^{-1/2}$. In Section 3.1 we obtained for a one-dimensionally rough surface the dependence R_0^{-1}, which is typical of a spherical wave. The reason for this is to be found in (10) of Section 3.1, where we considered R' in the denominator constant and equal to R_0 and removed it outside the sign of integration. This corresponds to a surface finite in the y direction and the point of observation is so far removed from the surface that the scattered wave will be spherical rather than cylindrical. If the surface is infinite in the y direction, R' can of course not be considered constant and must be kept under the integral sign. This exact integration is unnecessary owing to our subsequent normalization, where the distance dependence and the constant factors drop out in any case; however, to find the relation of ρ and g in the case of cylindrical waves, this integration must be performed. One then obtains instead of (32), Section 3.1, for $2L = 1$,

$$E_2 = \rho E_{20} = -\rho \sqrt{\left(\frac{2}{i\pi k R_0}\right)} \frac{k}{2} \cos\theta_1$$

and comparing this with (3), the required relation for a cylindrical scatterer is

$$\rho^{\pm}(\theta_1, \theta_2) = -\frac{2}{k\cos\theta_1} g^{\pm}(\theta_1, \theta_2) \tag{4b}$$

Having thus obtained the relation between our scattering coefficient and Twersky's scattering amplitude through (4a) and (4b), we shall use the latter in this section.

Let $f^{\pm}(\theta_1, \theta_2)$ be the scattering amplitude of an individual boss (e.g. a hemisphere or semicylinder) and let $g^{\pm}(\theta_1, \theta_2)$ be the scattering amplitude of the body consisting of the boss plus its image in the plane (e.g. a sphere or a cylinder); then the image theorem tells us that

$$f^{\pm}(\theta_1, \theta_2) = g^{\pm}(\theta_1, \theta_2) \pm g^{\pm}(\theta_1', \theta_2) = g^{\pm}(\theta_1, \theta_2) \pm g^{\pm}(\pi-\theta_1, \theta_2) \tag{5}$$

where θ_1' is shown in Fig. 6.2.

Equation (5) is the basic equation on which the Twersky theory is founded; the $g^{\pm}$'s are known exactly for several types of bodies (although the corresponding expressions are reasonably simple only in the case of the sphere and the circular cylinder) and may always be obtained either approximately† or by measurement.

The scattering amplitude $f(\theta_1, \theta_2)$ as given by (5) is that of a single boss on a conducting plane. When there are many bosses on the plane, this will

† A recent and powerful method of calculating scattering diagrams approximately is that of KELLER, J. B.: Diffraction by an aperture, *J. Appl. Phys.* **28**, 426–444 [1957]; LEVY, B. R. and KELLER, J. B.: Diffraction by a smooth object, *Comm. Pure Appl. Math.* **12**, 159–209 [1959]. Cf. also Keller [1958].

change, as the boss is excited not only by the incident wave, but also by the waves scattered from all the other bosses by multiple scattering. Integrating this contribution from the other bosses over the plane, using the reciprocity theorem and assuming the separation between neighbouring bosses to be large compared to their sizes, Twersky then derives the "reflection amplitude" defined as

$$A = \frac{\langle E_2(\theta_1, \theta_2)\rangle}{E_1} \tag{6}$$

and analogous to our $\langle \varrho_{\text{spec}}\rangle$, obtaining

$$A = \frac{1+Z^{\pm}}{1-Z^{\pm}} \tag{7}$$

where, if N is the average number of bosses per unit area (or length),

$$Z^{\pm} = -\frac{\pi N}{k^2 \cos\theta_1} f^{\pm}(\theta_1, \theta_1) = N\rho_f^{\pm} \tag{8}$$

for finite bosses, and

$$Z^{\pm} = -\frac{N}{k \cos\theta_1} f^{\pm}(\theta_1, \theta_1) = 2N\rho_c^{\pm} \tag{9}$$

for cylindrical bosses. In these expressions, ρ_f and ρ_c are the scattering coefficients for individual finite and cylindrical bosses as given by (4a) and (4b) respectively.

For the "differential scattering cross-section" $\sigma(\theta_1, \theta_2)$, defined as the incoherent power scattered by unit area into unit solid angle (or unit angle wedge) around θ_2 per unit incident power density, and thus analogous to our quantity $\langle\rho\rho^*\rangle$, Twersky derives the expression

$$\sigma(\theta_1, \theta_2) = \frac{N}{k^2}\left|\frac{f(\theta_1, \theta_2)}{1-Z}\right|^2 \tag{10}$$

for finite bosses, to be divided by $\pi/2k$ for the case of cylindrical bosses.

For the derivation of these results, the reader is referred to Twersky's papers [1957a] for a heuristic treatment applied to electromagnetic waves, and to [1957b] for an analytic treatment applied to the acoustic analogy. The method can also be applied to a number of related problems, e.g. to scattering volumes [Twersky, 1959], which, as stated in the introduction, is outside the scope of this monograph. (For an extensive bibliography of this and other allied subjects, the reader is referred to Twersky [1960].)

Comparing the Twersky theory with the general method of Chapter 3, which may also be applied to the Twersky model of a rough surface, but which neglects multiple scattering, it is easily shown that as far as multiple scattering is concerned, the two solutions for the field scattered into the specular direction differ by a factor of the order

$$\frac{1}{1+NA_b} \tag{11}$$

where A_b is the area occupied by one boss. Since the assumption of small bosses separated by large distances implies

$$NA_b \ll 1$$

it follows from (11) that the discrepancy caused by neglecting multiple scattering is very slight. However, for bosses on a plane, the Twersky solution is far superior to the general Kirchhoff method of Chapter 3 for another reason, namely that g in (5), and hence f may be substituted without having recourse to the Kirchhoff approximation.

As an example, consider a surface of length $2L$ with $2NL$ semicylindrical bosses of circular cross-section with radius a and axes perpendicular to the plane of incidence xz. Since (10) is derived neglecting multiple scattering, whereas (9) takes multiple scattering into account, we shall check our previous method only for the specular direction. Applying (44), Section 3.1, to find ρ_{spec}, we have

$$\rho_{spec} = \frac{1}{2L}\left[\sum_{j=1}^{2NL}\int_{x_{1j}}^{x_{2j}} dx + 2NL\int_{-a}^{a} \exp\left(iv_z\sqrt{(a^2-x^2)}\right) dx\right]$$

The first term in the square bracket is simply the total length of the surface unoccupied by bosses $(2L-4aNL)$; the integral is evaluated by substituting $x = a\cos\varphi$ and using (6) of Appendix B. After some manipulation with the resulting series, this yields

$$\rho_{spec} = 1-2aN+2aN\sum_{m=0}^{\infty}\varepsilon_m\frac{J_{2m}(2ka\cos\theta_1)}{1-4m^2} \tag{12}$$

where

$$\varepsilon_m = \begin{cases} 1 & \text{for} \quad m = 0 \\ 2 & \text{for} \quad m \neq 0 \end{cases} \tag{13}$$

Note that we have put ρ, not $\langle\rho\rangle$, in (12), for the mean value is automatically contained in the average number of bosses per unit length N; in fact (12) will apply to any spacing of the $2NL$ bosses on the length $2L$, so that (12) will equally well apply to a periodic surface (regularly spaced bosses), in which case we just set

$$N = \frac{1}{\Lambda} \tag{14}$$

where Λ is the period of the surface, i.e. the distance between neighbouring bosses. (For an application of the Twersky theory to a periodic surface consisting of bosses on a plane, cf. Twersky [1956a].) For the same reason, the field scattered by the Twersky model in the specular direction will not fluctuate when a constant number of bosses moves at random over the plane. It will, however, fluctuate, if N is time-varying, i.e. if bosses appear and disappear at random.

Returning to (12), we note that our solution (unlike Twersky's) neglects not only multiple scattering, but also shadowing effects, unless $\theta_1 = 0$.†

The second term in (12) represents the area occupied by the bosses and the third term is simply N times the Kirchhoff approximation for the scattering coefficient of a completely illuminated semicylindrical boss. Comparing (12) with (7), we find (after a conversion from ρ to A) that the two solutions differ by a factor of approximately $1+Na \approx 1$, if Z is found by the Kirchhoff approximation. Thus as far as multiple scattering is concerned, the method of Chapter 3 is not in appreciable error when applied to the Twersky model of a rough surface; the superiority of Twersky's method for bosses on a plane lies rather in the fact that by using the image method for scatterers with known $g^{\pm}$, it derives exact values for Z.

This is especially important for problems of polarization, where the Kirchhoff method yields only a change in sign for the two basic polarizations, whereas the exact values of ρ^+ and ρ^- may differ appreciably.

Thus although the Twersky method is limited to bosses protruding from a plane with some additional restrictions (large separation of bosses, analytical treatment practically restricted to hemispheres and circular semicylinders), it is, in addition to having the advantages mentioned at the beginning of this Section, at present also the only model of a rough surface permitting exact theoretical investigation of polarization problems.

6.2. Ament: Randomly Spaced Half-Planes

The Twersky model of a rough surface was evidently selected not because of its practical interest as such, but because its effect on electromagnetic waves is easily calculated. Also, the random element of this surface is the spacing between the objects of the surface. By working with the parameter N,

† There is little point in improving the approximation by respecting shadowing effects, e.g. integrating in (12) over φ from 0 to $\pi-\theta_1$ (instead of from 0 to π) and replacing $2aN$ in (12) by $2aN \sec\theta_1$, as shadowing effects are already included in the exact value of $g^{\pm}$, on which Twersky's solution is based.

the average number of objects per unit area or length, the "statistics" of the surface are described from the outset.

The same two simplifying elements are included in the model investigated by Ament [1954, 1956, 1960], whose work on random-process surfaces [1953] we have already noted in Chapter 5. Ament's model of a rough surface consists of perfectly conducting infinite half-planes in the space $z < 0$ with their edges parallel to the y axis and lying in the plane $z = 0$, and tilted at an angle β with respect to the z axis (Fig. 6.3). There are N half-planes per unit length (average).

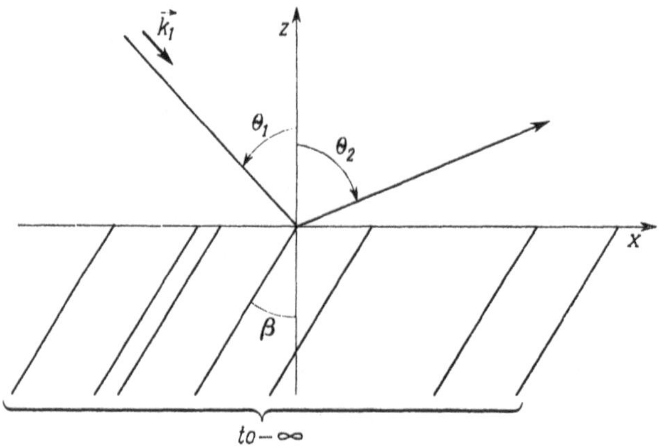

Fig. 6.3. Ament's model of a rough surface consisting of tilted half-planes.

Let this "rough surface" be illuminated by a plane wave of unit amplitude

$$E_1^{\pm} = e^{i\mathbf{k}_1 \cdot \mathbf{r}} \tag{1}$$

where we continue to use the notation introduced in Section 3.1. Let R be the (unknown) reflection coefficient of the surface, defined by

$$R = \frac{E_{2\,\mathrm{spec}}}{E_1} \tag{2}$$

In analogy to the Kirchhoff postulate of the unperturbed field, it is assumed that the field incident onto any one plane equals the field that would be present at the same point with that particular plane removed (i.e. the originally incident field plus the field scattered by all other planes); to determine this field and the currents induced by it, a reciprocity theorem is derived and applied, and by using the free-space Green's function $\exp(ikr)/r$ and integrating over the plane $z = 0$, Ament [1960] derives the following expressions for the reflection coefficient in the specular direction:

$$R^{-} = \frac{k\cos\theta_1 - \mu}{k\cos\theta_1 + \mu} \tag{3}$$

$$R^{+} = \frac{k\cos\theta_1 - \mu\sin(\theta_1 + \beta)}{k\cos\theta_1 + \mu\sin(\theta_1 - \beta)} \tag{4}$$

where μ is a properly selected root of

$$k^2 - \mu^2 - k^2\sin^2\theta_1 + 2Ni\sec\beta\,\sqrt{[k^2 - (k\sin\theta_1\sin\beta + \mu\cos\beta)^2]} = 0 \tag{5}$$

For the differential scattering cross-section, as defined by Twersky (Section 6.1), the resulting expressions are fairly complicated, but for θ_1, $\theta_2 \to \pi/2$ (grazing incidence), $\beta = 0$ and an average spacing of many wavelengths between the half-planes, Ament [1960] finds σ independent of polarization and equal to

$$\sigma = \frac{\gamma_1^2\gamma_2^2 k^2}{8N^2} \tag{6}$$

where $\gamma_{1,2} = \pi/2 - \theta_{1,2}$ are the grazing angles of incidence and scattering.

As with other theories mentioned in this chapter, we have given only a very brief sketch of the principle and results of Ament's theory. Ament's results [1960] are actually much more general than given in this brief mention, although they have been evaluated only in the case of rather artificial surfaces.

6.3. Rice: A Randomized Rayleigh Method

We have seen in Section 4.2 that the Rayleigh method is in theory more exact than the Kirchhoff method, but that in practice the infinites series representing the solution converges reasonably quickly only for very slightly rough surfaces. The same is true if the Rayleigh method is applied to a random surface. As the equation of the surface $\zeta = \zeta(x, y)$ has to be expanded in a Fourier series for the Rayleigh method to be applied, the surface is most easily "randomized" by using random Fourier coefficients of this expansion. This has been done by Rice [1951].

Let the rough surface $\zeta(x, y)$ be given by

$$\zeta(x, y) = \sum_{m,n} P(m, n)\,e^{-iK(mx+ny)} \tag{1}$$

where, as in Chapter 4,

$$K = \frac{2\pi}{\Lambda} \tag{2}$$

but Λ is now arbitrarily large. The coefficients $P(m, n)$ are taken as random quantities, subject to the condition

$$P(-m, -n) = P^*(m, n) \tag{3}$$

where the asterisk denotes the complex conjugate. This condition has to be imposed to make $\zeta(x, y)$ real. It is further assumed that the coefficients $P(m, n)$ are distributed normally about zero and that the four independent random variables formed by the real and imaginary parts of $P(m, n)$ and $P(m, -n)$ all have the same variance. This variance is a function of m and n, or rather of the quantities

$$p = Km, \qquad q = Kn \tag{4}$$

Since

$$\langle P(m, n)\rangle = 0 \tag{5}$$

the variance is equal to the mean square value [cf. Appendix C, (15)]; we set it equal to

$$\langle P(m, n)\, P^*(m, n)\rangle = \tfrac{1}{4}K^2 W(p, q) \tag{6}$$

where $W(p, q)$ describes the statistical properties of the surface. The reason why the variance is chosen in this way is that, as may be seen from the Fourier transform of (1) (by letting $\Lambda \to \infty$ and changing the sum to an integral), $W(p, q)$ represents the relative strength of the components between p and $p+\mathrm{d}p$ (for the x direction) and q and $q+\mathrm{d}q$ (for the y direction), their strength being measured by their contribution to the mean square value of $\zeta(x, y)$. This is of course an analogy of a power-frequency spectrum, the frequency in this case being expressed in radians per unit length rather than in cycles per second.

Since $P(m, n)$ and $P(u, v)$ are independent, we have

$$\langle P(m, n)\, P(u, v)\rangle = 0 \quad \text{for} \quad m, n \neq u, v \tag{7}$$

After this description of Rice's model of a rough surface we shall again only give a brief outline of his method and results. For a horizontally polarized incident wave, let the total electric field strength in the space $z > \zeta(x, y)$ be

$$\begin{aligned} E_x &= \sum_{m,n} A_{mn} E(m, n, z) \\ E_y &= 2i \sin k\theta_1 z\, \mathrm{e}^{-iK\nu x} + \sum_{m,n} B_{mn} E(m, n, z) \\ E_z &= \sum_{m,n} C_{mn} E(m, n, z) \end{aligned} \tag{8}$$

where

$$E(m, n, z) = \mathrm{e}^{-iK(mx+ny)-ib(b(m,n)z} \tag{9}$$

with

$$b(m, n) = \sqrt{[k^2 - K^2(m^2+n^2)]} \tag{10}$$

In (8), ν is an integer, making the field periodic with period Λ; the scattered field is determined through the reflection coefficients A_{mn}, B_{mn}, C_{mn}, which are found by satisfying (approximately) the boundary condition that the tangential component of **E** must vanish at $z = \zeta(x, y)$, and from the relation $\nabla \cdot \mathbf{E} = 0$, which gives

$$KmA_{mn}+KnB_{mn}+b(m, n)C_{mn} = 0 \tag{11}$$

The method so far is thus seen to be in essence similar to Rayleigh's method of Section 4.2. The same approach has also been used in a quite rigorous manner by Schouten and Hoop [1957], but as might be expected, the mathematical difficulties become such that it is barely possible to determine the first reflection coefficient (wave mode).

Rice's treatment is more approximate (higher order terms are neglected in the boundary conditions) and is further simplified by introducing the statistics of the coefficients $P(m, n)$ and taking an average. Only one component (E_y) has a non-vanishing average, and one of the double sums in the general expression for it vanishes owing to (7). The average reflection coefficient is then found to be

$$1-\frac{k}{2}\int_{-\infty}^{\infty}\int_{-\infty}^{\infty}\left(\frac{\theta_1 q^2}{b}+\theta_1 b\right)W(p, q)\,dp\,dq \tag{12}$$

This integral is evaluated by Rice [1951] in special cases, e.g. if $W(p, q)$ is zero except for a small region around $p = 0$, $q = 0$, which corresponds to a large average distance between the hills of the surface compared to the wavelength of the incident radiation, then the average reflection coefficient for horizontal polarization is

$$1-2k^2\gamma^2\langle\zeta^2(x, y)\rangle \tag{13}$$

For details, the reader is referred to Rice's thorough account of the treatment [1957]. The mathematical difficulties in Rice's theory can be surmounted only in the case of slightly rough surfaces, as is typical of the Rayleigh method. It seems that Rice's theory is better suited to surface wave (ground wave) propagation, which, as stated in the introduction, is outside the scope of this book.

6.4. Beckmann: Facet Surface Generated by a Markov Chain

In any application of rough-surface scatter, the ultimate object of interest is not the rough surface itself, but the field scattered by it; moreover, in many applications such as rough layers in the troposphere and ionosphere, we have no direct access to the surface and cannot measure the statistics of its geometry. Under these circumstances it seems more logical to adopt a model of a rough surface that will scatter a realistic field rather than attempt to calculate the field scattered by a realistic model. The field scattered by a rough surface (or by any configuration of scatterers) must always consist of elementary waves in mutual phase interference. A simple model of a rough surface that will have this effect and easily lends itself to calculations is one consisting of plane elements (facets) of random slope if the following two assumptions are made:

A. Every element reflects the beam incident upon it with the Fresnel reflection coefficient in the locally specular direction only (Fig. 6.4), which will be a good approximation if the facets are large when compared with the wavelength.

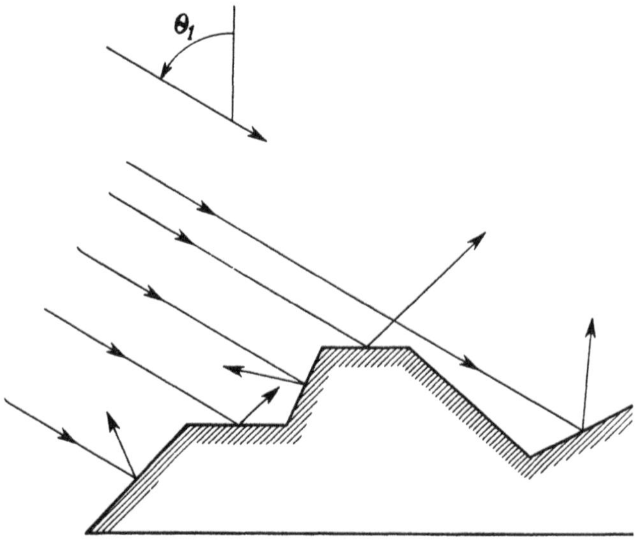

Fig. 6.4. Scattering by a surface of large plane facets.

B. The elementary waves thus reflected in the same direction (by facets of the same slope) are summed with respect to their phases to form one resultant wave in that direction.

A surface consisting of random plane facets is best described by a Markov chain as follows:

Consider a random function of time $Z(t)$, which may assume s discrete values, changing its value at discrete moments $t_1, t_2, t_3, \ldots$ The intervals $t_1, t_2-t_1, t_3-t_2, \ldots$ will be assumed to have a constant length τ, but Z need not necessarily at every point $n\tau$ (where n is an integer) change its value, but may preserve its previous value (remain in the same state). Thus Z may (but need not necessarily) change its value at the points τ, 2τ, ..., $n\tau$. Let Z have the value Z_i ("be in state i") at the point $n\tau$; let the probability that at the next step, i.e. at the point $(n+1)\tau$, Z will assume the value Z_j ("pass into state j") be

$$p_{ij} = P\{Z_i, n\tau; Z_j, (n+1)\tau\} \tag{1}$$

and let this transition probability be independent of n; then the entire process will be described by the ("stochastic") matrix

$$\Pi = \begin{Vmatrix} p_{11} & p_{12} & p_{13} \cdots p_{1s} \\ p_{21} & p_{22} & p_{23} \cdots p_{2s} \\ p_{31} & p_{32} & p_{33} \cdots p_{3s} \\ \cdot & \cdot & \cdot \quad \cdot \quad \cdot \\ p_{s1} & p_{s2} & p_{s3} \cdots p_{ss} \end{Vmatrix} = \|p_{ij}\| \tag{2}$$

The ith row of this matrix gives the conditional probabilities that if Z is in state i, it will pass at the next step into state j.

The random process briefly sketched here is a special case of a Markov process, namely a Markov chain with a finite number of states and discrete moments of transition.†

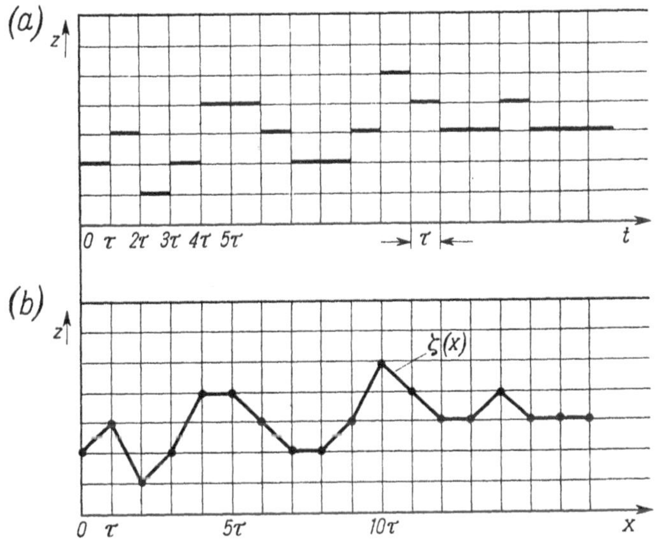

Fig. 6.5. Generation of a facet-surface by a Markov chain.

† Doob, J. L.: *Stochastic Processes.* Chapman and Hall, London, Wiley, New York 1953; Blanc-Lapierre, A. and Fortet, R.: *Théorie des Fonctions Aléatoires.* Masson, Paris 1953; Sarymsakov, T. A.: *Foundations of the Theory of Markov Processes* (in Russian). Gostekhizdat, Moscow 1954.

A realization of a Markov chain is shown in Fig. 6.5a. Now consider only the instantaneous values of Z at the moments τ, 2τ, ..., $n\tau$, and connect the resulting points by straight segments as in Fig. 6.5b; when the time axis t is replaced by the length-coordinate axis x, the resulting broken line of Fig. 6.5b is the profile $\zeta(x)$ of our model of a rough surface. This model is thus seen to be generated by a random process, but was not included in the preceding chapter, as the approach to surfaces generated by a continuous process is quite different than in the case of a Markov chain. By a suitable choice of the matrix Π a very wide class of rough surfaces may be constructed; furthermore to each matrix Π there corresponds an infinite number of different realizations of the same Markov chain, i.e. an infinite number of rough surfaces with the same statistical properties (the same is of course true of the realizations of any random process). As a numerical example, consider the matrix

$$\Pi = \frac{1}{16}\left\|\begin{array}{l} 11\ 4\ 1\ 0\ 0\ \ldots \qquad\qquad\qquad 0 \\ \ 5\ 6\ 4\ 1\ 0\ 0\ \ldots \qquad\qquad 0 \\ \ 1\ 4\ 6\ 4\ 1\ 0\ 0\ \ldots \qquad 0 \\ \ 0\ 1\ 4\ 6\ 4\ 1\ 0\ 0\ \ldots \quad 0 \\ \ 0\ 0\ 1\ 4\ 6\ 4\ 1\ 0\ 0\ \ldots\ 0 \\ \ \cdot\ \cdot\ \cdot\ \cdot\ \cdot\ \cdot\ \cdot\ \cdot\ \cdot\ \cdot\ \cdot\ \cdot\ \cdot\ \cdot \\ \ldots \qquad \ldots\ 0\ 0\ 1\ 4\ 6\ 4\ 1\ 0\ 0\ 0 \\ \ldots \qquad\quad \ldots\ 0\ 0\ 1\ 4\ 6\ 4\ 1\ 0\ 0 \\ \ldots \qquad\qquad \ldots\ 0\ 0\ 1\ 4\ 6\ 4\ 1\ 0 \\ \ldots \qquad\qquad\quad \ldots\ 0\ 0\ 1\ 4\ 6\ 4\ 1 \\ \ldots \qquad\qquad\qquad \ldots\ 0\ 0\ 1\ 4\ 6\ 5 \\ \ldots \qquad\qquad\qquad\quad \ldots\ 0\ 0\ 1\ 4\ 11 \end{array}\right\| \tag{3}$$

The irregularities of the first and last two rows are due to the fact that the surface may not pass below the first or above the sth level; the transition probabilities of the other lines are binomially distributed. Realizations of the surface given by (3) are shown in Fig. 6.6. They were actually constructed by tossing 4 coins at each step: for 4 heads the surface rose by two levels, for 3 heads by one level, for 2 heads and 2 tails it remained at the same level, for 3 tails it dropped by one level and for 4 tails it dropped by two levels. The probabilites of these events are 1/16, 4/16, 6/16, 4/16 and 1/16 respectively, as prescribed by (3). Other examples will be found in [Beckmann, 1957].

Working on the assumptions A and B stated above, the field is then calculated as the vector sum of n unit vectors with random phases. Using the properties of Markov chains, it is shown that in certain cases the phases

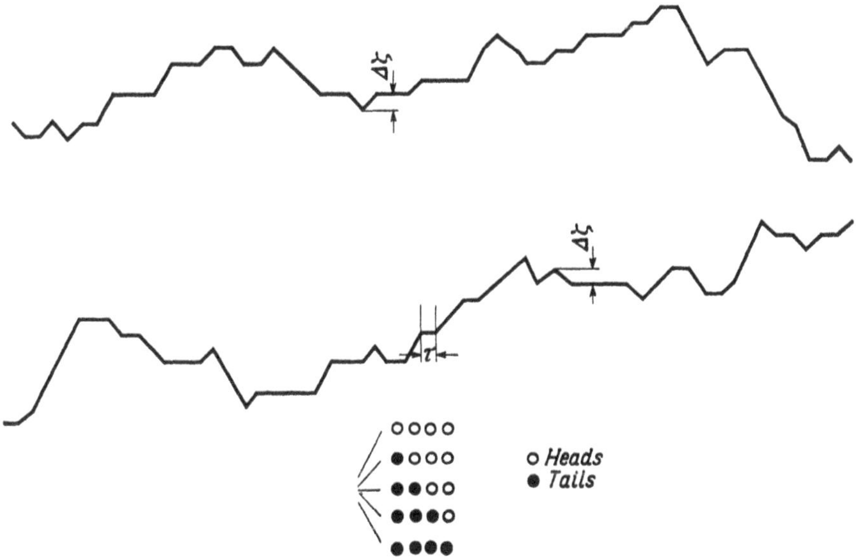

Fig. 6.6. Model of a rough surface realized by tossing four coins.

of the individual vectors (elementary waves) are distributed uniformly from $-a$ to a, where a need not necessarily (as in a Rayleigh-distributed vector sum) equal an integral multiple of π, but is given by

$$a = \frac{2\pi h \cos\theta_1}{\lambda} \tag{4}$$

where h is the range of the roughness, i.e. $h = \zeta_{max} - \zeta_{min}$.

At this point we shall only derive a result that was subsequently used by Du Castel and Spizzichino (Section 6.5). If the n elementary waves scattered into the specular direction have unit amplitude, then the total scattered field is

$$E_2 = \sum_{j=1}^{n} e^{i\varphi_j} \tag{5}$$

and hence

$$\langle E_2 E_2^* \rangle = n + n(n-1)\langle e^{i\varphi_i - i\varphi_j} \rangle \tag{6}$$

Since in the above case the φ_i and φ_j are independent and uniformly distributed from $-a$ to a, we have (for $\varphi_i \neq \varphi_j$)

$$\langle e^{i\varphi_i - i\varphi_j} \rangle = \left(\frac{1}{2a}\int_{-a}^{a} e^{i\varphi}\, d\varphi\right)\left(\frac{1}{2a}\int_{-a}^{a} e^{-i\varphi_j}\, d\varphi\right) = \mathrm{sinc}^2 a \tag{7}$$

where we again use the symbol $\mathrm{sinc}\, a = (\sin a)/a$.

From (7) and (6) we find

$$\langle |E_2|^2 \rangle = n + n(n-1)\,\text{sinc}^2 a \tag{8}$$

which is the result used by Du Castel and Spizzichino [1959], who generalized and refined the method developed by Beckmann [1957]. Note that when $a = 0$ (all elementary waves in phase), the scattered power is proportional to n^2; when $a = \pi$, or $a \gg \pi$ (uniform phase distribution over the phase cycle), the mean scattered power is proportional to n. This point will be important when we consider the coherence and incoherence of the scattered waves.

However, having described the model of a surface generated by a Markov chain, we shall not continue with the above method of solving it, for in the original paper [Beckmann, 1957] only an approximation of the distribution of the scattered field was obtained; subsequently the exact solution was found [Beckmann, 1959], and in the next chapter the exact and general distribution of any two-dimensional random vector sum will be derived. This distribution may be linked up not only with the above simplified model, but with any other model of a rough surface, including those generated by a continuous random process. Thus except for the actual model of the surface which we have here described, we consider the original theory [Beckmann, 1957] as made obsolete by Chapter 7 of this monograph, and will therefore not dwell on it any further.

6.5. Du Castel and Spizzichino: Surface of "Brilliant Points"

The theory developed by Beckmann [1957] was modified, generalized and made into a more useful theory by Du Castel and Spizzichino [1959] and Spizzichino [1959]. The model was generalized from an infinite, one-dimensional surface with plane facets to a finite, two-dimensional surface with continuous radii of curvature, and the number of elementary waves n propagated into a certain direction was eliminated in a more natural manner.

Consider a surface element ΔS with principal radii of curvature r_{c1} and r_{c2} (Fig. 6.7), which will scatter the incident energy into a solid angle

$$\Delta\Omega = \frac{4\Delta S \cos\theta_1}{r_{c1} r_{c2}} \tag{1}$$

The power scattered by the element ΔS is, at a distance R_0 (per unit incident power density)

$$|\Delta E|^2 = \frac{\cos\theta_1 \Delta S}{R_0^2 \Delta\Omega} \tag{2}$$

where for simplicity we assume the surface to be perfectly conducting, though this is not necessary either here or in the theory of Section 6.4.

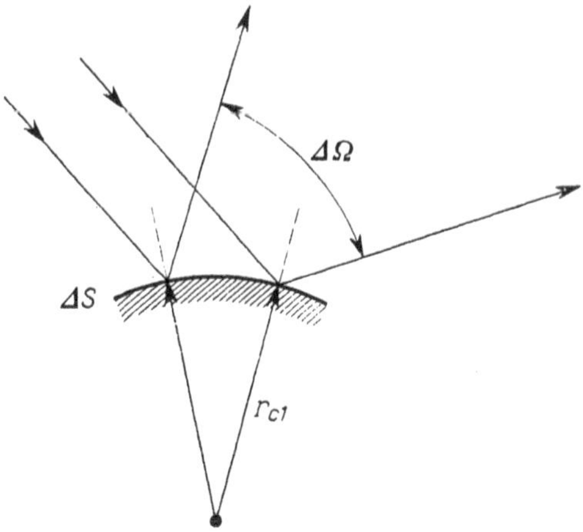

Fig. 6.7. The energy scattered into the solid angle $\Delta\Omega$ by an element of surface ΔS with radius of curvature r_{c1}.

Only certain elements ΔS will reflect the incident energy in a given direction. If we let $\Delta S \to 0$, then also $\Delta\Omega \to 0$, but ΔE_2 will stay finite. Thus in the limit, the elementary waves ΔE_2 scattered into a certain direction may be considered to originate from a number of "brilliant points" reminiscent, for example, of the glistening facets of a sunlit sea. For the mean value of $|\Delta E_2|^2$, Du Castel and Spizzichino [1959] find

$$\langle|\Delta E_2|^2\rangle = \frac{r_c^2}{4R_0^2} \tag{3}$$

and for a one-dimensional surface Spizzichino [1959] finds

$$\langle|\Delta E_2|^2\rangle = \frac{r_c}{R_0}\sqrt{\cos\theta_1} \tag{4}$$

where r_c is a mean value of the radius of curvature. Assuming the surface to be uniformly distributed in a height interval $h = \zeta_{max} - \zeta_{min}$, and using (4) and (8) of Section 6.4, we have for the direction of specular reflection

$$E_{2\,RMS} = \Delta E_2 \sqrt{(n+n(n-1)\,\mathrm{sinc}^2 a)} \tag{5}$$

Assuming the slopes of the surface to be uniformly distributed with a maximum value

$$\beta_{max} = \max(\arctan\zeta'_x, \arctan\zeta'_y),$$

Du Castel and Spizzichino find the number of "brilliant points" per unit area as

$$n = \frac{1}{r_c^2 \beta_{\max}^2} \tag{6}$$

The phase range of the elementary waves in the specular direction is given by (4), Section 6.4; if, however, the surface is finite (length L in the plane of incidence), this range can be shown to equal

$$a' = a + \frac{L}{\lambda} \left| \frac{\theta_1 - \theta_2}{2} \right| \cos\theta_1 \tag{7}$$

reducing to a for $\theta_1 = \theta_2$.

Substituting (7) and (6) in (5) we obtain the RMS value of the scattered field in any direction: if h is large ($a \gg \pi$), then $E_{2\,\mathrm{RMS}}$ is proportional to $\sqrt{n}$ and independent of $(\theta_1 - \theta_2)$; if $a < \pi$ (slightly rough surface or grazing incidence), $E_{2\,\mathrm{RMS}}$ is proportional to n in a narrow cone about the specular direction $\theta_2 = \theta_1$ and proportional to $\sqrt{n}$ in all other directions. We are thus once more led to the behaviour of the field scattered by a rough surface shown in Fig. 5.1 (p. 90); in the next chapter the exact relations governing the specular (coherent) and diffuse (incoherent) field for any rough surface (or other configuration of scatterers) will be derived.

Du Castel and Spizzichino [1959] also calculate the scattered field in which the transmitter and receiver are at finite distances R_1 and R_2 from the surface. For the mean power scattered diffusely, they find

$$P_d = P_0 \frac{1}{16\beta_{\max}^2} \left(\frac{1}{R_1} + \frac{1}{R_2} \right)^2 S_d \cos\theta_2 \tag{8}$$

where S_d is the area active in the diffuse scattering and P_0 is the free-space power at a distance $R_1 + R_2$. For the specularly reflected power Du Castel and Spizzichino find

$$P_s = P_0 \frac{1}{4 r_c^2 \beta_{\max}^2} \frac{\lambda^2}{\cos\theta_1} \left(1 - \frac{a}{\pi} \right) \operatorname{sinc} a \tag{9}$$

Formulae (8) and (9) again apply to a uniformly distributed surface.

We have again only given a very brief outline of the theory, as the dependence of the scattered field and power on the angle of incidence and wavelength derived by Du Castel and Spizzichino has already been obtained more exactly in Chapter 5, and the exact statistical distribution of the specular and diffuse field will be obtained in the next chapter; for further details of the theory we therefore again refer the reader to Du Castel and Spizzichino [1959] and Spizzichino [1959].

6.6. Other Models—A Guide to the Literature

In Chapter 5 we have considered rough surfaces generated by a random process, paying particular attention to the normally distributed surface. In the preceding sections we have briefly surveyed other models used by Twersky, Ament, Rice, Beckmann, Du Castel and Spizzichino. But these do not yet, by a long way, exhaust the models and methods developed for solving the problem of the scattering of electromagnetic waves from a rough surface. However, many (perhaps even the greater part) of the remaining theories are either closely associated with a particular application (to be treated in Part II), or covered by other material in Part I of this monograph. Other theories use a very general mathematical approach which does not lead to very explicit results and may rather be classified as a rigorous formulation of the problem. Still further theories are less concerned with the actual scattering of waves from rough surfaces than with subjects outside the scope of this book such as surface wave propagation, wave guide theory and scattering from volumes of scatterers. Rather than attempt to survey the remaining theories, we shall therefore restrict ourselves to a brief guide to the literature on the subject.

The first attempt to treat scattering by a random rough surface is apparently due to Andronov and Leontovich [1926]. In one of the earlier attempts to derive the energy scattered by a rough surface, Ornstein and van der Burg [1937] consider a surface of plane facets and regard the normals as points on a sphere. Solving the diffusion equation for points on a sphere originating from the pole, they find an approximate expression for the mean value of the scattered power (light intensity) in terms of Legendre polynomials.

General mathematical approaches of solving the wave equation were used by Hufford [1952] and Hoffman [1955].

Of the applications that have given rise to theories, radar sea clutter (cf. Chapter 1) seems to have been the most prolific. We have already noted the work of Davies [1954] and Ament [1953, 1956, 1960], which was written with an eye to the scattering properties of the rough surface of the sea. This problem was also treated by Goldstein [1946a, 1946b]. A notable theory of radar sea clutter including the effect of wind speed was developed by Katzin [1957] by considering the slope and size distribution of the facets of the sea surface. A facet model of the sea surface has also been worked out by Spetner and Katz [1960] and Braude [1958]. The theory of Beard Katz and Spetner [1956], Spetner [1958] and Braude, Komarov and Ostrovsky [1958] considers the field scattered by the surface of the sea as a vector sum of constant and random components, and the distribution of the scattered field was considered in a more elementary way by Blake [1950]; this problem will be solved exactly in the next chapter. There is of course a large number of papers on experimental measurements of scattering from the surface of the sea and their interpretation.† It is probably futile to hope that all of them have been included in the bibliography (p. 476) but perhaps the majority is included.

Scattering by irregular terrain is another application that has given rise to a number of theories, such as the one by Bullington [1954], who uses a phase method to approximate the RMS value of the reflection coefficient. It may in fact be shown from the general distribution of Chapter 7 that Bullington's method gives the RMS value of the constant component for a symmetrically distributed surface and is thus a good approximation.‡ There is of course a voluminous literature on surface wave propagation of radio waves over rough ground, which is outside the scope of this monograph, but some theories work with sufficiently elevated antennae to make them applicable to the problem under investigation. Typical examples are the theories by Bass and Bocharov [1958a, 1958b], Braude [1958], Men', Braude and Gorbach [1959], and many others that will be found in the bibliography. Important theories already outside the scope of this book were also developed by Wait [1959a, 1959b, 1962].

A theory closely associated with beyond-the-horizon troposheric propagation by scattering from atmospheric sheets (*feuillets*) was developed by Voge [1960], which does not differ very radically from that of Du Castel and Spizzichino [1959] and a comparison of the two theories is included in [Du Castel, 1960]. Another theory with the same application is that by Friis, Crawford and Hogg [1957]. Here again we touch on the subject of scattering by volumes, which is extensively reviewed in [Twersky, 1960].

Hiatt, Senior and Weston [1960], using an approximate boundary condition by Senior, have investigated the scattering by a sphere with a rough surface and the results were used by Senior and Siegel [1960] to develop a theory of radar scattering by the moon.

Finally, this brief guide to the literature on rough-surface scatter should include an admirable survey of the field by Bachynski [1959].

---

† Note added in proof: Cf. also Braude, S. Ya. (Ed.): *Radio-oceanographic Investigations of Sea Waves* (in Russian). Ukr. Acad. Sci. Kiev. 1962.

‡ A theory of reflection by rough terrain was published by Grosskopf [1961] as the manuscript was going to press.

CHAPTER 7

THE STATISTICAL DISTRIBUTION OF THE SCATTERED FIELD

THE total field scattered by a rough surface (or, for that matter, by any configuration of scatterers) at a given point or in a given direction is necessarily always the sum of elementary waves in mutual phase interference. If we represent the elementary waves by complex numbers $Ae^{i\varphi}$, suppressing, as usual, the time factor $\exp(i\omega t)$, each wave will be represented by a vector in the complex plane and hence the resultant field will be given as a vector sum in the complex plane (Fig. 7.1). If the surface (or array of scatterers) is random, this vector sum will also be random. In Section 1 of this chapter we

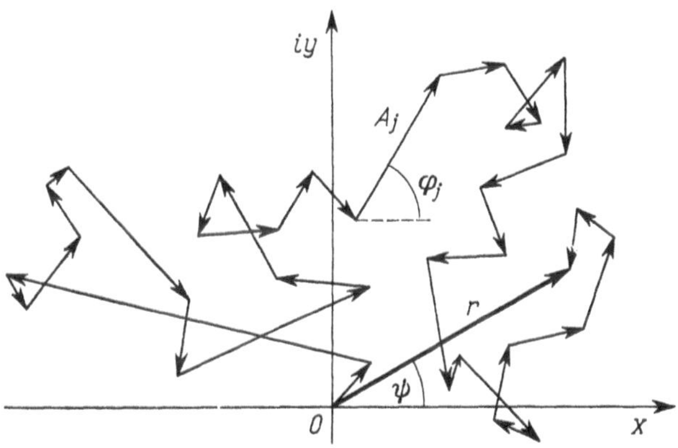

Fig. 7.1. The vector sum of n random vectors or the sum of n random complex numbers.

therefore derive the probability distribution of the sum of arbitrary coplanar random vectors. In Section 2 we link up this distribution with the mean field and power scattered by a rough surface as obtained by our general method of Chapter 5 or by the more special methods applicable to the models of Chapter 6. This approach will provide us with the general solution of the rough-surface scatter problem, not restricted to mean values, but giving the entire distribution about these mean values; it will be exact to the extent that these mean values are exact.

7.1. The Probability Distribution of the Sum of Arbitrary Random Coplanar Vectors

Consider the sum

$$U = r e^{i\psi} = \sum_{j=1}^{n} A_j e^{i\varphi_j} \tag{1}$$

where the A_j are real. The sum (1) may be interpreted as a vector sum in the complex plane as shown in Fig. 7.1. In our case the A_j correspond to the amplitudes of the elementary scattered waves and the φ_j to their phases; r is the resultant amplitude and ψ the resultant phase of the total scattered field.

Let the A_j and φ_j be random quantities distributed with probability densities $w_j(A)$ and $w_j(\varphi)$ respectively.†

Our task is to find the distribution of the random quantities r and ψ when n is large. We shall impose almost no restrictions on the distributions of the A_j and φ_j (requiring only that they satisfy the Central Limit Theorem) and we do not even require that A_j and φ_j be independent.

Rayleigh [1896, Section 42a] solved this problem in its most elementary form, namely when all A_j equal unity and all φ_j are distributed uniformly between $-\pi$ and π, i.e.

$$w(\varphi) = \begin{cases} \dfrac{1}{2\pi} & \text{for} \quad -\pi \leqslant \varphi \leqslant \pi \\ 0 & \text{for} \quad |\varphi| > \pi \end{cases} \tag{2}$$

The required probability density $p(r)$ is then given by the well-known Rayleigh distribution‡

† In order not to run out of symbols or to overburden them with indices, we economize by using w_j as an operator rather than a symbol for a function: $w_j(A)$ and $w_j(\varphi)$ are in general given by different functions, but we shall distinguish them by writing $w_{jA}(A)$ and $w_{j\varphi}(\varphi)$ only if there is any danger of confusion. We shall also economize by using the same symbol for a random variable and the values it assumes ("x is distributed with a probability density $w(x)$") unless there is danger of confusion, and some symbols, such as x, y, R and k, do not have the same meaning as in Section 3.1, the notation of which does not apply to this section.

‡ Rayleigh, who was not familiar with today's routine methods of probability theory, used an ingenious method of adding one more vector to the sum and transforming the resulting difference equation to a partial differential equation, which turns out to be the equation of heat conduction; on solving it, (3) is obtained [Rayleigh, 1896, Section 42a]. Note also that the Rayleigh distribution is closely related to the "χ_k^2-distribution" (as it is known in probability theory), which is the distribution of

$$\chi_k^2 = \xi_1^2 + \xi_2^2 + \dots + \xi_k^2$$

where the ξ_i are distributed normally with mean value zero and variance σ^2. The Rayleigh distribution (3) is the distribution of χ_k for $k = 2$ and $\sigma^2 = n/2$.

$$p(r) = \frac{2r}{n} e^{-r^2/n} \tag{3}$$

and ψ is again uniformly distributed between $-\pi$ and π. We shall obtain (3) as a simple special case of our general distribution.

A number of phase distributions similar to (2), e.g. φ distributed uniformly in any interval equal to an integral multiple of 2π, also leads to the Rayleigh distribution; so does any phase distribution with variance much larger than π.

However, if a surface is only slightly rough, or if the grazing angle is sufficiently small, the phases of the elementary waves will not be distributed uniformly over an interval 2π or more; a small band of phases will be privileged (in the limit, for a smooth surface, all elementary waves will have the same single phase) and neither (2) nor its equivalents will apply, and we therefore have to find a more general distribution.

The problem of finding the probability distribution of r and ψ in (1) when all A_j equal unity, but the φ_j are distributed with almost arbitrary, not necessarily identical symmetrical phase distributions $w_j(\varphi)$ was solved by Beckmann [1959]. In the present section we shall find the distributions of r without any restrictions other than that n is large and that the distributions of $A_j \cos \varphi_j$ and $A_j \sin \varphi_j$ satisfy the Central Limit Theorem (cf. Appendix C). Thus the A_j may be random and even correlated with the corresponding φ_j and the distribution of φ_j need not be symmetrical.

We shall first derive the distribution of r and ψ in the special case that all A_j are equal to unity and the $w_j(\varphi)$ symmetrical about zero. The solution can be expressed in terms of three parameters:

$$\alpha = \langle \operatorname{Re} U \rangle, \qquad s_1 = D\{\operatorname{Re} U\}, \qquad s_2 = D\{\operatorname{Im} U\} \tag{4}$$

which are easily obtainable from the given $w_j(A)$ and $w_j(\varphi)$. As before, $\langle\,\rangle$ denotes the mean value and D the variance.

It turns out that the solution remains formally the same when the above restrictions are withdrawn and only the value of the parameters (4) changes (if necessary, after a rotation of the real and imaginary axes in the complex plane).

The principle by which the distribution of r and ψ is calculated, is the following:

We first calculate the parameters (4) by elementary theorems of probability theory; since n is large, $\operatorname{Re} U = x$ is distributed normally with mean value α and variance s_1, and $\operatorname{Im} U = y$ is distributed normally with mean value zero (if the coordinate axes have been correctly chosen) and variance s_2. This approach gives us the two-dimensional normal distribution of $U(x, y)$;

transforming to polar coordinates, we obtain the required distribution of $U(r, \psi)$ and from it the distributions $p(r)$ and $p(\psi)$.

We therefore consider first the case when all $A_j = 1$. Let

$$x = \sum_{j=1}^{n} x_j = \sum_{j=1}^{n} \cos\varphi_j \tag{5}$$

$$y = \sum_{j=1}^{n} y_j = \sum_{j=1}^{n} \sin\varphi_j \tag{6}$$

so that

$$r^2 = x^2+y^2; \quad x = r\cos\psi; \quad y = r\sin\psi \tag{7}$$

The mean values of x and y are from (5) and (6)

$$\alpha = \langle x\rangle = \sum_{j=1}^{n} \alpha_j, \quad \beta = \langle y\rangle = \sum_{j=1}^{n} \beta_j \tag{8}$$

where

$$\alpha_j = \langle x_j\rangle = \int_{-\infty}^{\infty} w_j(\varphi)\cos\varphi\, d\varphi$$
$$\beta_j = \langle y_j\rangle = \int_{-\infty}^{\infty} w_j(\varphi)\sin\varphi\, d\varphi \tag{9}$$

The variance of x is found from (15) and (18) of Appepnix C:

$$s_1 = D\{x\} = \sum_{j=1}^{n} (\langle x_j^2\rangle - \alpha_j^2) = \sum_{j=1}^{n} \Big(\int_{-\infty}^{\infty} w_j(\varphi)\cos^2\varphi\, d\varphi - \alpha_j^2\Big) \tag{10}$$

and similarly

$$s_2 = D\{y\} = \sum_{j=1}^{n} \Big(\int_{-\infty}^{\infty} w_j(\varphi)\sin^2\varphi\, d\varphi - \beta_j^2\Big) \tag{11}$$

In the special case that all the $w_j(\varphi)$ are mutually equal and that $w(\varphi)$ is symmetrical about its mean value zero (or other integral multiple of π), making $\beta = 0$ and eliminating the sums, the above formulae reduce to

$$\alpha = n \int_{-\infty}^{\infty} w(\varphi)\cos\varphi\, d\varphi \tag{12}$$

$$\beta = 0 \tag{13}$$

$$s_1 = n \int_{-\infty}^{\infty} w(\varphi)\cos^2\varphi\, d\varphi - \frac{\alpha^2}{n} \tag{14}$$

$$s_2 = n \int_{-\infty}^{\infty} w(\varphi)\sin^2\varphi\, d\varphi \tag{15}$$

TABLE 7.1

$w(\varphi)$	α	s_1	s_2
(16)	$n e^{-\frac{1}{2}\sigma^2}$	$\frac{n}{2}(1-e^{-\sigma^2})^2$	$\frac{n}{2}(1-e^{-2\sigma^2})$
(17)	$n \operatorname{sinc} a$	$\frac{n}{2}(1+\operatorname{sinc} 2a-2\operatorname{sinc}^2 a)$	$\frac{n}{2}(1-\operatorname{sinc} 2a)$
(18)	$n \operatorname{sinc}^2 a$	$\frac{n}{2}(1+\operatorname{sinc}^2 2a-2\operatorname{sinc}^4 a)$	$\frac{n}{2}(1-\operatorname{sinc}^2 2a)$

Table 7.1 lists the values of α, s_1 and s_2, computed from (12) to (15) for the following three distributions:

(a) normal with standard deviation σ:

$$w(\varphi) = \frac{1}{\sigma\sqrt{2\pi}} e^{-\varphi^2/2\sigma^2} \tag{16}$$

(b) uniform from $-a$ to $+a$:

$$w(\varphi) = \begin{cases} \dfrac{1}{2a} & \text{for} \quad -a \leqslant \varphi \leqslant a \\ 0 & \text{for} \quad |\varphi| > a \end{cases} \tag{17}$$

(c) Simpson-distributed from $-2a$ to $+2a$:

$$w(\varphi) = \begin{cases} \dfrac{1}{2a}\left(1+\dfrac{\varphi}{2a}\right) & \text{for} \quad -2a \leqslant \varphi \leqslant 0 \\ \dfrac{1}{2a}\left(1-\dfrac{\varphi}{2a}\right) & \text{for} \quad 0 \leqslant \varphi \leqslant 2a \\ 0 & \text{for} \quad |\varphi| \geqslant 2a \end{cases} \tag{18}$$

As before, we use the symbol

$$\operatorname{sinc} a = \frac{\sin a}{a} \tag{19}$$

Now if n, the number of elementary waves is large, and the distributions $w_j(\varphi)$ are such that the quantities x_j and y_j satisfy the Central Limit Theorem (Appendix C), which in practice will mostly be the case, then the x and y components of r will be normally distributed with mean values α and β and variances s_1 and s_2 respectively. Moreover, if the distributions $w_j(\varphi)$ are symmetrical about zero, as we temporarily assume, the two quantities will be uncorrelated. (This follows from the symmetry of the two-dimensional distribution with respect to the coordinate axes, so that the covariance (22) of Appendix C vanishes; note also that for normally distributed quantities non-correlation implies independence.) Thus the two-dimensional distribution of x and y is the product of the distributions of x and y, i.e.

$$W(x, y) = \frac{1}{2\pi \sqrt{(s_1 s_2)}} \exp\left[-\frac{(x-\alpha)^2}{2s_1} - \frac{y^2}{2s_2}\right] \tag{20}$$

Transforming to polar coordinates as in (7), the required probability density of r is

$$p(r) = \frac{r}{2\pi \sqrt{(s_1 s_2)}} \int_0^{2\pi} \exp\left[-\frac{(r\cos\psi-\alpha)^2}{2s_1} - \frac{r^2\sin^2\psi}{2s_2}\right] d\psi \tag{21}$$

After some algebraic manipulation, the integral may be transformed to

$$e^{-D} \int_0^{2\pi} e^{-P\cos^2\psi+Q\cos\psi}\, d\psi \tag{22}$$

where

$$D = \frac{s_2\alpha^2+s_1 r^2}{2s_1 s_2}$$

$$P = \frac{s_2-s_1}{2s_1 s_2} r^2$$

$$Q = \frac{\alpha}{s_1} r$$

As shown in Appendix E, (5),

$$\int_0^{2\pi} e^{-P\cos^2\psi-Q\cos\psi}\, d\psi = 2\pi e^{-P/2} \sum_{m=0}^{\infty} (-1)^m \varepsilon_m I_m(P/2) I_{2m}(Q) \tag{23}$$

where I_m is the modified Bessel function of order m and

$$\varepsilon_m = \begin{cases} 1 & \text{for} \quad m=0 \\ 2 & \text{for} \quad m \neq 0 \end{cases} \tag{24}$$

Using (23) and (22) we find from (21)

$$p(r) = \frac{r}{\sqrt{(s_1 s_2)}} \exp\left[-\frac{\alpha^2}{2s_1} - \frac{(s_1+s_2)r^2}{4s_1 s_2}\right] \times$$

$$\times \sum_{m=0}^{\infty} (-1)^m \varepsilon_m I_m\left(\frac{s_2-s_1}{4s_1 s_2} r^2\right) I_{2m}\left(\frac{\alpha r}{s_1}\right) \tag{25}$$

This general result includes as special cases three distributions that are well known and that may be used to check the validity of (25).

(a) The Rayleigh distribution must be obtained if $w(\varphi)$ is given by (17) with $a = \pi$; in this case we find from (12) to (15), or directly from Table 7.1, $\alpha = 0$, $s_1 = s_2 = n/2$. Since

$$I_m(0) = \begin{cases} 1 & \text{for} \quad m = 0 \\ 0 & \text{for} \quad m \neq 0 \end{cases} \tag{26}$$

we find on substituting these values in (25)

$$p(r) = \frac{2r}{n} e^{-r^2/n} \tag{27}$$

which is identical with (3).

(b) The distribution of the amplitude of a constant vector plus a Rayleigh vector was found by Rice [1944]; it is discussed in detail and tabulated in [Norton, Vogler *et al.*, 1955]. If the constant vector is **C**, we direct it along the x axis, and obviously have $\alpha = C$, $s_1 = s_2 = n/2$; substituting this in (25), we obtain

$$p(r) = \frac{2}{n} r e^{-(r^2+C^2)/n} I_0\left(\frac{2Cr}{n}\right) \tag{28}$$

which is the Rice distribution [1944].

(c) Hoyt [1947] considered the distribution of the amplitude of a vector whose x and y components are distributed normally with mean value zero and unequal variances s_1 and s_2. Here we have $\alpha = 0$, which substituted in (25) yields

$$p(r) = \frac{r}{\sqrt{(s_1 s_2)}} \exp\left[-\frac{(s_1+s_2)r^2}{4s_1 s_2}\right] I_0\left(\frac{s_2-s_1}{4s_1 s_2} r^2\right) \tag{29}$$

in agreement with Hoyt [1947].

Let us now calculate $\langle r^2 \rangle$ (which is proportional to the mean power of the resultant scattered field). Again using the theorem

$$D\{z\} = \langle z^2 \rangle - \langle z \rangle^2 \tag{30}$$

we have

$$\langle r^2 \rangle = \langle x^2 \rangle + \langle y^2 \rangle = s_1 + \langle x \rangle^2 + s_2 = s_1 + s_2 + \alpha^2 \tag{31}$$

Now the resulting vector can obviously be represented as the sum of a constant vector $\vec{\alpha}$ (directed along the x axis) and a Hoyt vector described under (c) above (Fig. 7.2). Equation (31) simply tells us that the mean power of the resulting field is the sum of the power of the constant component (α^2) and the power of the fluctuating or Hoyt component. The ratio of the two components (unlike the number of elementary waves n) can easily be measured and is also easily calculable from the results of Chapter 5. Let this ratio be

$$\frac{\alpha^2}{s_1+s_2} = B^2 \tag{32}$$

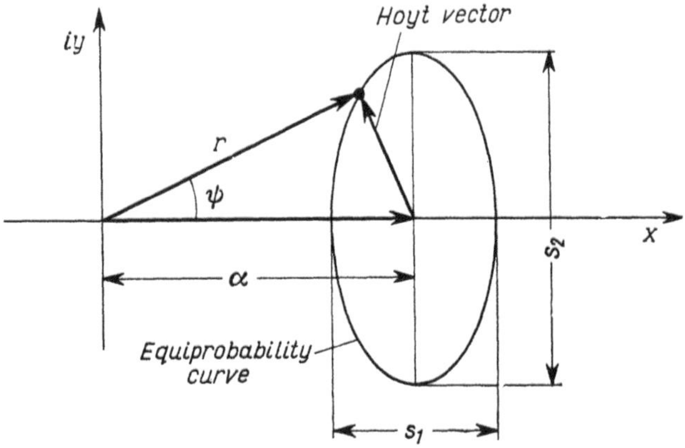

Fig. 7.2. A random vector sum may be represented as the sum of a constant vector $(\vec{\alpha})$ and a Hoyt vector. The x and y components of the latter are normally distributed with mean values zero, but generally unequal variances.

so that B is the ratio of the RMS values of the constant and random components. If we introduce the "normalized amplitude"

$$u = \frac{r}{\sqrt{(s_1+s_2)}} \tag{33}$$

and the "asymmetry factor"

$$K = \sqrt{\frac{s_1}{s_2}} \tag{34}$$

we finally find the general solution of our problem in the form†

$$p(u) = \frac{K^2+1}{K} u \exp\left[-\frac{1+K^2}{2}\left(B^2+\frac{1+K^2}{2K^2}u^2\right)\right] \times$$

$$\times \sum_{m=0}^{\infty} (-1)^m \varepsilon_m I_m\left(\frac{K^4-1}{4K^2}u^2\right) I_{2m}[B(1+K^2)u] \tag{35}$$

Written in this form, we find the special cases of the Rayleigh distribution ($K=1$, $B=0$)

$$p(u) = 2u\mathrm{e}^{-k^2} \tag{36}$$

† The normalization described by equations (32) to (34) represents an improvement on the normalization chosen in [Beckmann, 1959].

the Rice distribution ($K = 1$)

$$p(u) = 2u\,\mathrm{e}^{-(B^2+u^2)}\,I_0(2Bu) \tag{37}$$

and the Hoyt distribution ($B = 0$)

$$p(u) = \frac{K^2+1}{K}\,u\exp\left[-\left(\frac{1+K^2}{2K}\right)^2 u^2\right] I_0\left(\frac{K^4-1}{4K^2}u^2\right) \tag{38}$$

A discussion and tables of the latter will be found in Beckmann [1959]; tables of the Rice distribution will be found in Norton, Vogler *et al.* [1955]. Curves of the general distribution (35) are given in Appendix F (p. 199).

We now turn to the phase distribution $p(\psi)$ of the resulting vector sum (1). This is found from (20) by transforming to polar coordinates and integrating over r from 0 to ∞:

$$p(\psi) = \frac{1}{2\pi\sqrt{(s_1 s_2)}}\int_0^\infty r\exp\left[-\frac{(r\cos\psi-\alpha)^2}{2s_1}-\frac{r^2\sin^2\psi}{2s_2}\right]\mathrm{d}r \tag{39}$$

Introducing the symbol

$$G = BK\cos\psi\sqrt{\left(\frac{1+K^2}{2(K^2\cos^2\psi+\sin^2\psi)}\right)} \tag{40}$$

(39) yields after somewhat tedious calculations

$$p(\psi) = \frac{K\mathrm{e}^{-\frac{1}{2}B^2(1+K^2)}}{2\pi(K^2\cos^2\psi+\sin^2\psi)}\,[1+G\sqrt{\pi}\cdot\mathrm{e}^{G^2}(1+\operatorname{erf}G)] \tag{41}$$

where

$$\operatorname{erf}G = \frac{2}{\sqrt{\pi}}\int_0^G \mathrm{e}^{-t^2}\,\mathrm{d}t \tag{42}$$

The next step is to generalize our result for vectors with random amplitudes, which may be correlated with the phase.

Let the jth vector be

$$A_j\mathrm{e}^{i\varphi_j} \tag{43}$$

where A_j and φ_j are both random and governed by the two-dimensional probability density $w_j(A, \varphi)$. Then

$$x = \sum_{j=1}^{n} A_j\cos\varphi_j \tag{44}$$

and

$$\langle x\rangle = \sum_{j=1}^{n}\int_{-\infty}^{\infty}\int_{-\infty}^{\infty} w_j(A,\varphi)A\cos\varphi\,\mathrm{d}A\mathrm{d}\varphi \tag{45}$$

We first assume that A_j and φ_j are independent; then

$$w_j(A, \varphi) = w_{jA}(A) w_{j\varphi}(\varphi) \tag{46}$$

which substituted in (45) yields

$$\langle x \rangle = \alpha = \sum_{j=1}^{n} \alpha_j = \sum_{j=1}^{n} \int_{-\infty}^{\infty} A w_{jA}(A) \mathrm{d}A \int_{-\infty}^{\infty} w_{j\varphi}(\varphi) \cos\varphi \, \mathrm{d}\varphi \tag{47}$$

Again assuming the distributions $w_{j\varphi}(\varphi)$ to be symmetrical (i.e. even functions), we have $\langle y \rangle = 0$.

The variance of x is

$$D\{x\} = \langle x^2 \rangle - \alpha^2 = \sum_{j=1}^{n} \left(\int_{-\infty}^{\infty} A^2 w_{jA}(A) \mathrm{d}A \int_{-\infty}^{\infty} w_{j\varphi}(\varphi) \cos^2\varphi \, \mathrm{d}\varphi - \alpha_j^2 \right) = s_1 \tag{48}$$

and similarly

$$D\{y\} = \sum_{j=1}^{n} \int_{-\infty}^{\infty} A^2 w_{jA}(A) \mathrm{d}A \int_{-\infty}^{\infty} w_{j\varphi}(\varphi) \sin^2\varphi \, \mathrm{d}\varphi = s_2 \tag{49}$$

Substituting for α, s_1 and s_2 from (47) to (49) in (20), we continue from there to the solution (35) as before; hence (47) to (49) solve our problem.

In the special case that all $w_{jA}(A)$ are the same and equal to $w_A(A)$, and that all $w_{j\varphi}(\varphi)$ are equal to $w_\varphi(\varphi)$, (47) to (49) simplify to

$$\alpha = n\langle A \rangle \int_{-\infty}^{\infty} w_\varphi(\varphi) \cos\varphi \, \mathrm{d}\varphi \tag{50}$$

$$s_1 = n\langle A^2 \rangle \int_{-\infty}^{\infty} w_\varphi(\varphi) \cos^2\varphi \, \mathrm{d}\varphi - \frac{\alpha^2}{n} \tag{51}$$

$$s_2 = n\langle A^2 \rangle \int_{-\infty}^{\infty} w_\varphi(\varphi) \sin^2\varphi \, \mathrm{d}\varphi \tag{52}$$

For example, if $w_\varphi(\varphi)$ is uniform from $-\pi$ to π, we find

$$\alpha = 0, \qquad s_1 = s_2 = \tfrac{1}{2} n\langle A^2 \rangle \tag{53}$$

which substituted in (25) yields the distribution of the amplitude of the vector sum of vectors with uniformly distributed phase and random amplitude A, i.e. the amplitude of a Rayleigh vector composed not of unit vectors, but of vectors with random amplitudes:

$$p(r) = \frac{2r}{n\langle A^2 \rangle} \mathrm{e}^{-r^2/n\langle A^2 \rangle} \tag{54}$$

This is seen to be equivalent with the Rayleigh distribution (3), provided the number of vectors n is multiplied by the mean square value (mean power) of the actual components.

For a generalized Rayleigh vector composed of vectors with constant, but different amplitudes A_j, we find from (47) to (49)

$$\alpha = 0, \quad s_1 = s_2 = \tfrac{1}{2}\sum_{j=1}^{n} A_j^2 \tag{55}$$

For a vector of this kind Norton, Vogler *et al.* [1955] indicated that (1) the amplitude r of the resulting vector is distributed so that

$$P\{r > z\} = \mathrm{e}^{-z^2/k^2} \tag{56}$$

where k^2 is constant and equal to $\langle r^2 \rangle$, i.e. $k^2 = 2s_1$; (2) that the resulting phase is uniformly distributed between $-\pi$ and π, i.e.

$$P\{\psi > \psi_0\} = 1 - \frac{\psi_0}{2\pi} \qquad (|\psi_0| < \pi) \tag{57}$$

(3) that the random variables r and ψ are independent.

Using the above formulae, it is easily shown that all three assumptions are correct. (The Central Limit Theorem also requires $\langle A_j^2 \rangle \ll k^2$ for all j.)

Substituting (55), i.e. $\alpha = 0$, $s_1 = s_2 = k^2/2$ in (25), we have

$$p(r) = \frac{2r}{k^2}\mathrm{e}^{-r^2/k^2}$$

hence, by (5) and (7) of Appendix C,

$$P\{r > z\} = 1 - \frac{2}{k^2}\int_0^z r\mathrm{e}^{-r^2/k^2}\mathrm{d}r = \mathrm{e}^{-z^2/k^2} \tag{58}$$

which proves (56). Substituting (55) in (32) and (34) gives $B = 0$, $K = 1$, whence by (40) and (41)

$$p(\psi) = \frac{1}{2\pi}$$

from which the complement of the distribution function (57) follows. Finally, two random variables r and ψ are independent if and only if

$$p(r, \psi) = p_r(r)p_\psi(\psi) = \int_{-\infty}^{\infty} p(r, \psi)\mathrm{d}\psi \int_{-\infty}^{\infty} p(r, \psi)\mathrm{d}r \tag{59}$$

Substituting $\alpha = 0$, $s_1 = s_2$ in (20) and transforming to polar coordinates it is immediately seen that (59) holds; hence ψ and r are independent. At the same time, we note that $\alpha = 0$, $s_1 = s_2$ is the only condition under which (59) will hold; hence the phase and amplitude of the resulting vector sum is independent only in the case of a Rayleigh vector (components with constant amplitudes) or a generalized Rayleigh vector (components with random amplitudes), but not in the case of components with non-uniform phase distributions over the basic phase cycle.

We now further generalize the distribution of the sum of random vectors $A_j \exp(i\varphi_j)$ by dropping the assumption that A_j and φ_j are independent or even uncorrelated. (This point arises in the case of a rough surface of finite conductivity, where the slope of an element of the surface determines the local reflection coefficient, which in turn affects both the amplitude and the phase of the elementary wave scattered by the element of surface.) Then (46) no longer holds and instead we have

$$\alpha = \langle x \rangle = \sum_{j=1}^{n} \int_{-\infty}^{\infty} \int_{-\infty}^{\infty} w_j(A, \varphi) A \cos\varphi \, dA \, d\varphi \tag{60}$$

$$s_1 = D\{x\} = \sum_{j=1}^{n} \Big[\int_{-\infty}^{\infty} \int_{-\infty}^{\infty} w_j(A, \varphi) A^2 \cos^2\varphi \, dA \, d\varphi - \alpha_j^2 \Big] \tag{61}$$

$$s_2 = D\{y\} = \sum_{j=1}^{n} \int_{-\infty}^{\infty} \int_{-\infty}^{\infty} w_j(A, \varphi) A^2 \sin^2\varphi \, dA \, d\varphi \tag{62}$$

This includes the special case when all w_j are equal to $w(A, \varphi)$:

$$\alpha = n \int_{-\infty}^{\infty} \int_{-\infty}^{\infty} w(A, \varphi) A \cos\varphi \, d\varphi \tag{63}$$

$$s_1 = n \int_{-\infty}^{\infty} \int_{-\infty}^{\infty} w(A, \varphi) A^2 \cos^2\varphi \, dA \, d\varphi - \frac{\alpha^2}{n} \tag{64}$$

$$s_2 = n \int_{-\infty}^{\infty} \int_{-\infty}^{\infty} w(A, \varphi) A^2 \sin^2\varphi \, dA \, d\varphi \tag{65}$$

Thus, if the two-dimensional distribution $w(A, \varphi)$ is given then α, s_1 and s_2 may be calculated immediately and the problem is solved by substituting in (63) to (65) and then in (32) to (35). If other quantities are given, we must first make sure that the problem is uniquely soluble, e.g. the distributions $w_A(A)$, $w_\varphi(\varphi)$ and the correlation coefficient $C(A, \varphi)$ are not enough, for s_1 and s_2 cannot be uniquely determined. However, it may still be shown that the mean square value (average power) of the fluctuating component always equals the sum of the mean square values of the individual components, for from (64) and (65) we have

$$s_1+s_2 = n \int_{-\infty}^{\infty} A^2\left[\int_{-\infty}^{\infty} w(A,\varphi)\,d\varphi\right]dA - \frac{\alpha^2}{n}$$
$$= n \int_{-\infty}^{\infty} A^2 w_A(A)\,dA - \frac{\alpha^2}{n}$$
$$= n\langle A^2\rangle - \frac{\alpha^2}{n} \tag{66}$$

From (31) we therefore find for the mean total power

$$\langle r^2\rangle = n\langle A^2\rangle + \alpha^2\left(1-\frac{1}{n}\right) \tag{67}$$

As n is by hypothesis large,† the second term is just α^2, or the mean power of the steady component; the first term must therefore be the mean power of the fluctuating component, which is seen to equal the sum of the mean power contributed by each individual vector.

Two assumptions remain that we have not yet dropped, namely that $w(\varphi)$ is symmetrical about its mean value (normalized to zero), and that the individual vector components of the sum are mutually independent.

If the $w_j(\varphi)$ are not symmetrical about zero, β in (9) does not vanish; moreover, x and y will be correlated. To investigate this case, we assume for simplicity that all the $w_j(\varphi)$ are equal to the same phase distribution $w(\varphi)$, and that all $A_j = 1$. (Both assumptions are easily withdrawn as in the other cases above and are immaterial with respect to the problem we wish to investigate.)

The mean values are, as in (9),

$$\left.\begin{aligned} \alpha &= \langle x\rangle = n\int_{-\infty}^{\infty} w(\varphi)\cos\varphi\,d\varphi \\ \beta &= \langle y\rangle = n\int_{-\infty}^{\infty} w(\varphi)\sin\varphi\,d\varphi \end{aligned}\right\} \tag{68}$$

but since $w(\varphi)$ is now asymmetrical about $\langle\varphi\rangle = 0$, i.e. not an even function of φ, β does not vanish.

† Actually in (67) we need n larger than is usually necessary for the Central Limit Theorem. If the ξ_i's of (55), Appendix C, are distributed normally, Lyapunov's theorem holds immediately, i.e. for $n \geqslant 1$; if they are distributed uniformly, ξ is for all practical purposes normally distributed for $n \geqslant 4$; if, as in the Rayleigh distribution, the ξ_i's are of the kind $\sin\varphi_i$ or $\cos\varphi_i$ with φ_i distributed uniformly over an interval 2π, then ξ is for all practical purposes distributed normally as soon as $n \gtrsim 6$ (cf. [Greenwood and Durand, 1955]); in most other cases of practical significance the convergence to the normal distribution is also very rapid.

The variances s_1 and s_2 are given by (14) and (15). But our derivation must now break away from the previous method as (20), which applies to the case when x and y are independent, is no longer valid; the two-dimensional distribution will now be given by the two-dimensional distribution of two normally distributed random variables x and y, correlated by a correlation coefficient C, i.e.

$$W(x, y) = \frac{1}{2\pi\sqrt{[s_1 s_2(1-C^2)]}} \times$$

$$\times \exp\left\{-\frac{1}{2(1-C^2)}\left[\frac{(x-\alpha)^2}{s_1} - 2C\frac{(x-\alpha)(y-\beta)}{s_1 s_2} + \frac{(y-\beta)^2}{s_2}\right]\right\} \tag{69}$$

To calculate the correlation coefficient, we use its definition

$$C = \frac{K_{xy}}{\sqrt{(s_1 s_2)}} \tag{70}$$

where K_{xy} is the covariance

$$K_{xy} = \langle xy\rangle - \langle x\rangle\langle y\rangle \tag{71}$$

To determine K_{xy}, we must in turn find $\langle xy\rangle$, which may be done as follows:

$$\langle xy\rangle = \left\langle\left(\sum_{j=1}^{n}\cos\varphi_j\right)\left(\sum_{j=1}^{n}\sin\varphi_j\right)\right\rangle = \left\langle\tfrac{1}{2}\sum_{j=1}^{n}\sin 2\varphi_j + \sum_{j\neq i}\cos\varphi_j\sin\varphi_i\right\rangle \tag{72}$$

where the second sum is extended over the remaining $n(n-1)$ terms. Since φ_i and φ_j are for $i\neq j$ independent, we find

$$\langle xy\rangle = \tfrac{1}{2}n\langle\sin 2\varphi\rangle + n(n-1)\langle\cos\varphi\rangle\langle\sin\varphi\rangle$$

$$= \tfrac{1}{2}n\int_{-\infty}^{\infty} w(\varphi)\sin 2\varphi\, d\varphi + n(n-1)\int_{-\infty}^{\infty} w(\varphi)\cos\varphi\, d\varphi \int_{-\infty}^{\infty} w(\varphi)\sin\varphi\, d\varphi \tag{73}$$

From this expression we can now clearly see that if $w(\varphi)$ is symmetrical about zero, i.e. if $w(\varphi)$ is an even function, $\langle xy\rangle$ will vanish, as the integrands are odd functions and the integration limits symmetrical; hence in this case the covariance (71) will vanish (since $\langle y\rangle = 0$), so that x and y are by (70) uncorrelated, and since they are normally distrbuted, they are also independent.

Returning to the general case of an asymmetrical distribution $w(\varphi)$, we find from, (68), (70), (71) and (73)

$$C = \frac{\tfrac{1}{2}\int w(\varphi)\sin 2\varphi\, d\varphi - \int w(\varphi)\cos\varphi\, d\varphi \int w(\varphi)\sin\varphi\, d\varphi}{\sqrt{\left[\int w(\varphi)\sin^2\varphi\, d\varphi \int w(\varphi)\cos^2\varphi\, d\varphi\right]}} \tag{75}$$

where we have omitted the integration limits $-\infty$, ∞.

We now introduce new coordinate axes x', y', turned through an angle φ_0 with respect to the axes x, y, choosing the angle φ_0 in such a way that the quantities $x' = r \cos \psi'$ and $y' = r \sin \psi'$ are independent; ψ' is given by

$$\psi' = \psi - \varphi_0 \tag{76}$$

This follows almost immediately from (22), Appendix C, according to which two random variables are uncorrelated (and hence, if they are normally distributed, independent), if their two-dimensional distribution $W(x, y)$ in the xy plane is symmetrical with respect to a line parallel to one of the coordinate axes. The curves $W(x, y) = \text{const}$ in (69) are concentric

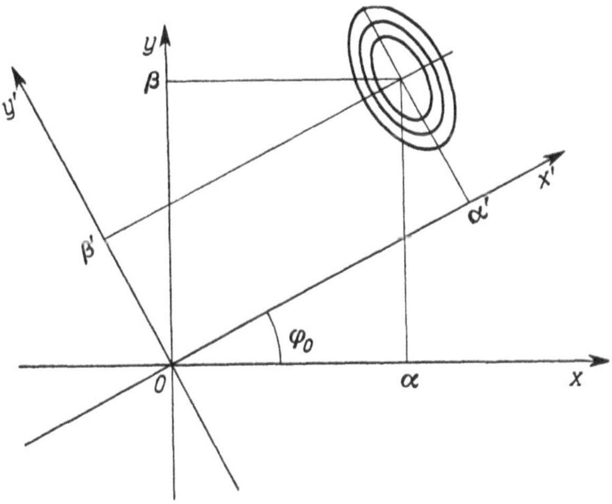

Fig. 7.3. Choosing the coordinate axes so as to make x' and y' independent. This will be the case when the coordinate axes are parallel to the axes of the equiprobability ellipses.

ellipses with centres α, β (Fig. 7.3), so that all we have to do to make x' and y' independent is to rotate the coordinate system in such a way that the new coordinate axes x' and y' are parallel to the axes of the family of ellipses. This is a well-known transformation in probability theory; the required angle φ_0 is found from†

$$\tan 2\varphi_0 = \frac{2C\sqrt{(s_1 s_2)}}{s_1 - s_2} \tag{77}$$

where s_1, s_2 are given by (14) and (15), and C is given by (75).

In the new coordinate system,

$$\varphi' = \varphi - \varphi_0 \tag{78}$$

† Hristow, W. K.: *Grundlagen der Wahrscheinlichkeitsrechnung*. Verl. Bauwesen, Berlin 1961, p. 125.

Hence

$$w'(\varphi') = w(\varphi+\varphi_0) \tag{79}$$

$$\alpha' = n\int_{-\infty}^{\infty} w'(\varphi')\cos\varphi'\,d\varphi'; \quad \beta' = n\int_{-\infty}^{\infty} w'(\varphi')\sin\varphi'\,d\varphi'\,d\psi \tag{80}$$

$$s_1' = n\int_{-\infty}^{\infty} w'(\varphi')\cos^2\varphi'\,d\varphi' - \frac{\alpha'^2}{n}; \quad s_2' = n\int_{-\infty}^{\infty} w'(\varphi')\sin^2\varphi'\,d\varphi' - \frac{\beta'^2}{n} \tag{81}$$

$$s_1'+s_2' = s_1+s_2 \tag{82}$$

As x' and y' are now independent, we may proceed as in (20) and (21), but as now $\beta \neq 0$, this results in

$$p(r) = \frac{r}{2\pi\sqrt{(s_1 s_2)}}\int_0^{2\pi}\exp\left[-\frac{(r\cos\psi-\alpha)^2}{2s_1} - \frac{(r\sin\psi-\beta)^2}{2s_2}\right]d\psi \tag{83}$$

where we have omitted the primes at s_1, s_2, α, β and ψ. The integral (83) is not as easily integrated as (21), and though it is possible to evaluate it by series expansion [cf. (19), Appendix E], the numerical computation of the series is not as simple as direct numerical integration of (83) and we shall therefore content ourselves with (83) as the solution of our problem without attempting to express it more explicitly.

We may note that the rotation of the coordinate system through an angle φ_0 is, from the physical point of view, completely immaterial, as we are only multiplying (1) by a factor $\exp(i\varphi_0)$, which is equivalent to shifting all phases by φ_0. However, the initial phase (or initial time in $\exp(i\omega t)$) is in all practical electromagnetic problems left to the choice of convenience. Another way to express this is to say that since the choice of the x and y axes in the complex plane is left to our convenience, we always choose them in such a way that x and y will be independent.

The last assumption concerns the mutual independence of the individual vectors (which is not to be confused with the independence of the amplitude and phase of every single vector as above). If the individual vector terms are not independent, the general solution for the amplitude and phase distribution of the sum (1) becomes a difficult problem. In principle, the problem may still be solved by working with an n-dimensional distribution of r, transforming to a new random quantity r' with uncorrelated components,† solving the problem for r' as above, and then

† Such a transformation is always possible by constructing the new components as linear combinations of the original components, using the elements of the correlation matrix of the original distribution, cf. PUGACHEV, V. S.: *Theory of Random Functions*, 2nd ed., Fizmatgizdat, Moscow 1960, Chapt. 3, Sec. 21 (in Russian).

retransforming back to r, but the labour involved in such a treatment becomes prohibitive unless the correlation matrix is particularly simple. Fortunately, in practical applications this difficulty which concerns only a few of the large number of interdependent vector terms can usually be evaded. In our derivation we have only once made use of the independence of the vector terms. Namely it has been assumed that the variance of the sum is the sum of variances in (10) and (11) or in (48) and (49). Furthermore, if the variance of the vector sum is determined by other methods with due respect to the correlation among some of the terms (as has been the case in determining the variance of the scattering coefficient of a rough surface), the results of this section will hold even for interdependent vectors.

We may now summarize our results for the most general case which has been developed by successively dropping the restrictions on the sum (1). This case occurs when the A_j and φ_j are both random and given by two-dimensional distributions $w_j(A, \varphi)$, which may be all different, asymmetrical and such that A_j and φ_j are correlated. The only restrictions that are now left are that the individual vectors $A_j \exp \varphi_j$ are mutually independent and that the quantities $A_j \cos\varphi_j$, $A_j \sin\varphi_j$ satisfy the conditions of the Central Limit Theorem. The first restriction may be withdrawn if $D\{\mathrm{Re}\, U\}$ and $D\{\mathrm{Im}\, U\}$ are known (as they are in applications to rough-surface scatter), whilst the second restriction is hardly likely to be of importance in practical applications.

Under these conditions, we determine the quantities α, s_1, s_2 from (60) to (62) and

$$\beta = \langle y \rangle = \sum_{j=1}^{n} \int_{-\infty}^{\infty} \int_{-\infty}^{\infty} w(A, \varphi) A \cos\varphi \, \mathrm{d}A \mathrm{d}\varphi \tag{84}$$

We then calculate the covariance of x and y as

$$K_{xy} = \tfrac{1}{2} \sum_{j=1}^{n} \int w_j(A, \varphi) A \sin 2\varphi \, \mathrm{d}A \mathrm{d}\varphi +$$

$$+ \sum_{i=1}^{n} \sum_{j=1}^{n}{}' \int_{-\infty}^{\infty} \int_{-\infty}^{\infty} w_i(A, \varphi) A \cos\varphi \, \mathrm{d}A \mathrm{d}\varphi \int_{-\infty}^{\infty} \int_{-\infty}^{\infty} w_j(A, \varphi) A \sin\varphi \, \mathrm{d}A \mathrm{d}\varphi - \alpha\beta \tag{85}$$

where the apostrophe at the double sum indicates that the terms $i = j$ are to be omitted. The correlation coefficient C is then calculated from (70) and if it does not vanish, the coordinate axes are turned through an angle φ_0 given by (77). These new coordinates x and y, in which the prime is omitted, are independent. We now calculate α, β, s_1 and s_2 with respect to the new coordinates as before; the required distribution $p(r)$ is given by (83) and the phase distribution $p(\psi)$ is given by the same expression integrated over r from 0 to ∞ instead of over ψ from 0 to 2π. If $\beta = 0$, the integral simplifies and may be evaluated by series expansion.

We now return once more to the special case $\beta = 0$ to discuss the numerical computation of $p(r)$. In its normalized form, $p(u)$ is given by (35), which converges quickly only for small arguments of the Bessel functions; if it does not converge quickly, tabulated values of modified Bessel functions† run out and there is no direct possibility of using their asymptotic expressions.

For individual numerical calculations it is therefore usually more advantageous to use (8) of Appendix E instead of (23) of the present section. Using the normalization (32) to (34) we then find

$$p(u) = \frac{1+K^2}{K} \exp\left[-\frac{1+K^2}{2}(B^2+u^2)\right] \times \\ \times \sum_{m=0}^{\infty} A_m \left(\frac{K^4-1}{K^2}\right)^m u^{2m} i_m[Bu(1+K^2)] \tag{86}$$

where

$$i_m(x) = x^{-m} I_m(x) \tag{87}$$

and the coefficients

$$A_m = \frac{1\cdot3\cdot5\ldots(2m-1)}{2\cdot4\cdot6\ldots2m} \tag{88}$$

are tabulated in Appendix E, Table E1.

For large B, we find from (15), Appendix E,

$$p(u) \approx \frac{\sqrt{(1+K^2)}}{K} \sqrt{u} \frac{e^{-\frac{1+K^2}{2}(B-u)^2}}{(\sqrt{2\pi}B)} \sum_{m=0}^{\infty} A_m \left(\frac{K^2-1}{BuK^2}\right)^m \tag{89}$$

For computations on a larger scale it is evidently quickest (if the necessary computor is available) to integrate (21) directly; in the case of (83) there is no other choice.

We may note in passing that the results of the present section are not limited to an electromagnetic (or acoustic) field scattered from a rough surface or to other scattering configurations. They are applicable to any random coplanar vector sum, e.g. to the Brownian movement of molecules and the kinetic theory of gases. Among others, the present section solves

† To the author's knowledge the most extensive tables of modified Bessel functions are the *Tables of Bessel Functions*, Vol. II, Comm. Math. Tables, Cambridge 1952, containing $i_m(x) = x^{-m}I_m(x)$ for $0 \leqslant m \leqslant 20$ and $0 \leqslant x \leqslant 20$.

the two-dimensional "random walk" problem (as it is known in probability theory) under general conditions. A similar technique has been applied to a three-dimensional vector sum in connection with satellite-induced ionization of the atmosphere.†‡

7.2. The Statistical Distribution of the Amplitude of the Scattered Field

We now return to the notation introduced in Section 3.1.

From (11), Section 3.2, we know that the scattering coefficient of a two-dimensionally rough surface is given by

$$\rho = \frac{F_3}{A} \iint_A e^{i\mathbf{v}\cdot\mathbf{r}} dx dy \tag{1}$$

and for a one-dimensionally rough surface by (44), Section 3.2,

$$\rho = \frac{F_2}{L} \int_{-L}^{L} e^{vi\cdot r} dx = \frac{F_2}{L} \lim_{n\to\infty} \sum_{j=1}^{n} e^{iv_x x_j + iv_z \xi} \Delta x_j \tag{2}$$

Thus ρ is the sum of vectors with constant amplitude ($F_2 \Delta x_j / L$ or $F_3 \Delta x_j \Delta y_i / A$) and, since ζ is random, with random phase

$$\varphi = \mathbf{v}\cdot\mathbf{r} = v_x x + v_y y + v_z \zeta(x, y) \tag{3}$$

These vectors are not all independent, for the phases of vectors which correspond to elementary waves radiated by neighbouring elements of the surface will be correlated. However, we have already calculated the variance of ρ (Sections 5.2, 5.3) with due respect to the correlation function (or, more generally, to the two-dimensional distribution of the surface), and as this is the only point of our derivation in the preceding section which assumes the independence of the vector terms, we may use all other results of the preceding section.

Using (4), Section 7.1, and the definition of the mean value and variance of a complex random quantity (27) and (28), Appendix C, we immediately see that for a symmetrically distributed surface

$$\alpha = \langle \rho \rangle \tag{4}$$

$$s_1 + s_2 = D\{\rho\} \tag{5}$$

$$s_1 = D\{\mathrm{Re}\,\rho\}; \quad s_2 = D\{\mathrm{Im}\,\rho\} \tag{6}$$

† Beckmann, P.: The probability distribution of the velocities of gas molecules after collision with a rapidly moving rigid body. *Czechosl. J. Phys.* **9**, 557–561 (1959).

‡ Cf. also footnote on p. 198.

Using the results of Section 5.2, we may write (4) as

$$\alpha = \langle \rho \rangle = \chi(v_z)\rho_0 \tag{7}$$

where $\chi(v_z)$ is the characteristic function of the distribution of the surface i.e. of ζ, and

$$\rho_0 = \begin{cases} \operatorname{sinc} v_x X \operatorname{sinc} v_y Y & \text{for two-dimensional roughness} \\ \operatorname{sinc} v_x L & \text{for one-dimensional roughness} \end{cases} \tag{8}$$

If ζ is not distributed symmetrically about the mean level zero (e.g. long valleys and short high peaks), $\chi(v_z)$ will be complex, so that

$$\alpha + i\beta = \langle \rho \rangle = \chi(v_z)\rho_0 \tag{9}$$

In this case we turn the coordinate axes through an angle φ_0 given by (77), Section 7.1, and consider the quantity $\rho \exp(i\varphi_0)$ which differs from ρ only by a phase shift φ_0.

Normalizing as in (32) to (34) of Section 7.1, we find on using (4), (5) and (7)

$$B = \frac{\langle \rho \rangle}{\sqrt{(D\{\rho\})}}; \qquad B^2 = \frac{\rho_0^2 \chi(v_z)}{\langle \rho\rho^* \rangle - \rho_0^2 \chi^2(v_z)} \tag{10}$$

$$u = \frac{|\rho|}{\sqrt{(D\{\rho\})}}; \qquad u^2 = \frac{\rho\rho^*}{D\{\rho\}} \tag{11}$$

The quantities $\langle \rho \rangle$, $\langle \rho\rho^* \rangle$ and $D\{\rho\}$ have been found in terms of the properties of the rough surface and the scattering geometry in Section 5.2, and in the special case of a normally distributed surface in Section 5.3.

Since by (29), Section 3.1, $\rho = E_2/E_{20}$ where E_{20} is constant, we may also write (6) and (7) as

$$B = \frac{\langle |E_2| \rangle}{\sigma\{E_2\}}; \qquad B^2 = \frac{\langle E_2 \rangle \langle E_2 \rangle^*}{D\{E_2\}} \tag{12}$$

where $\sigma\{E_2\}$ denotes the standard deviation of the amplitude of the scattered field; it equals the square root of the variance and is measurable as the RMS value of the fluctuations about the mean value. Also,

$$u = \frac{|E_2|}{\sigma\{E_2\}}; \qquad u^2 = \frac{E_2 E_2^*}{D\{E_2\}} \tag{13}$$

To determine the "asymmetry factor" K of (34), Section 7.1, which in terms of ρ is given by

$$K^2 = \frac{D\{\operatorname{Im} \rho\}}{D\{\operatorname{Re} \rho\}} \tag{14}$$

we could calculate $D\{\operatorname{Im} \rho\}$ and $D\{\operatorname{Re} \rho\}$ in close analogy to $D\{\rho\}$ as in Section 5.2; however, for a random rough surface the correlation distance

T must always be much smaller than L (or $\sqrt{A}$) and we incur only a negligible error by neglecting the interdependence of (a very few of) the vector terms; using (10) and (11) of Section 7.1, we find

$$K^2 = \frac{\iint\left[\int w(\varphi)\sin^2\varphi\, d\varphi - \left(\int w(\varphi)\sin\varphi\, d\varphi\right)^2\right] dx\, dy}{\iint\left[\int w(\varphi)\cos^2 d\varphi - \left(\int w(\varphi)\cos\varphi\, d\varphi\right)^2\right] dx\, dy} \tag{15}$$

where the integrations extend over the entire area of the surface for x and y, and from $-\infty$ to ∞ for φ.

Substituting for φ from (3) and expressing $\sin^2\varphi$ and $\cos^2\varphi$ in terms of $\cos^2\varphi$ we find after some manipulation (in analogy to the calculation of $\langle\rho\rangle$ in Section 5.2)

$$K^2 = \frac{1-\rho_0^{(2)}\operatorname{Re}\chi(2v_z)-2[\operatorname{Im}\chi(v_z)]^2\rho_0^2}{1+\rho_0^{(2)}\operatorname{Re}\chi(2v_z)-2[\operatorname{Re}\chi(v_z)]^2\rho_0^2} \tag{16}$$

where

$$\rho_0^{(2)} = \begin{cases} \operatorname{sinc} 2v_x X \operatorname{sinc} 2v_y Y & \text{for two-dimensional roughness} \\ \operatorname{sinc} 2v_x L & \text{for one-dimensional roughness} \end{cases} \tag{17}$$

If the distribution of the surface $w(z)$ is symmetrical about its mean value (zero), the characteristic function χ is real, so that in this more usual case (16) reduces to

$$K^2 = \frac{1-\rho_0^{(2)}\chi(2v_z)}{1+\rho_0^{(2)}\chi(2v_z)-2\rho_0^2\chi^2(v_z)} \tag{18}$$

Now it follows from (8) and (17) that for X, $Y \gg \lambda$, both ρ_0 and $\rho_0^{(2)}$ will vanish unless $v_x = v_y = 0$ (implying the direction of specular reflection). Hence, as already noted in Chapter 5, ρ_0 and $\rho_0^{(2)}$ will vanish outside a narrow cone about the direction of specular reflection. Substituting $\rho_0 = 0$ in (16) and (10), we find $K = 1$, $B = 0$; comparing this with (36) of Section 7.1, we see that outside a narrow cone (or wedge, for one-dimensional roughness) about the specular direction, the amplitude of the field scattered by any rough surface is Rayleigh-distributed. As $A = XY \to \infty$, the cone reduces to a dimensionless ray.

Now let the surface be very rough:,

$$D\{\zeta\} \gg \lambda^2 \tag{19}$$

and let

$$v_z = \frac{2\pi}{\lambda}(\cos\theta_1 + \cos\theta_2) \tag{20}$$

differ appreciably from zero (precluding grazing incidence or large wavelengths). Then it is easily shown from the properties of characteristic functions that $|\chi(v_z)| \to 0$ and hence $|\chi(2v_z)| \to 0$; substituting this in (16) and (10) we again find $B \to 0$, $K \to 1$, so that n this case we are led to the Rayleigh distribution even for $\rho_0 = 1$ (specular direction).

We have thus found the following result:

Outside a narrow cone (or wedge) about the direction of specular reflection, the amplitude of the field scattered by a rough surface is always Rayleigh-distributed; if the surface is very rough [cf. (19)], *and grazing incidence is excluded, the amplitude of the scattered field is Rayleigh-distributed everywhere.*

This result has also been derived by heuristic methods under less general conditions [Beckmann, 1957; Du Castel and Spizzichino, 1959].

The Rayleigh distribution is an important limiting distribution of the general probability distribution (35) or (86), Section 7.1, of the amplitude of the scattered field, and is included in these two general formulae for $K = 1$, $B = 0$. Thus, for instance, we note from (10) and (16) that the conditions that make B vanish, make K equal to unity, i.e. $B \to 0$ implies $K \to 1$, hence we find from (36), Section 7.1, that the amplitude of a field scattered from a rough surface that has no constant (non-random, coherent) component, must be Rayleigh-distributed.

Conversely, we find from (16) and (10) that $K = 1$ implies (with some very theoretical exceptions) $B = 0$. Hence, if the amplitude of the field scattered by a rough surface is not Rayleigh-distributed, it is never distributed with the Rice distribution (37), Section 7.1 ($K = 1$, $B > 0$, i.e. constant vector plus Rayleigh vector), but with the distribution (35), Section 7.1, for $K \neq 1$, $B > 0$ (constant vector plus Hoyt vector).

From (4) and (32), Section 7.1, we have

$$D\{\rho\} = \langle\rho\rho^*\rangle - \alpha^2 = \langle\rho\rho^*\rangle - B^2(s_1 + s_2) \tag{21}$$

and since $D\{\rho\} \geqslant 0$, the variance will attain its maximum value for $B = 0$, which implies $K = 1$.† Hence *of all the possible distributions of the amplitude of a field scattered by a symmetrically distributed rough surface, included in* (35), *p.* 126, *the Rayleigh distribution is the one with the greatest variance.* This result may also be obtained by physical reasoning [Beckmann, 1957].

We have now in (10), (11) and (16) linked up the quantities $\langle\rho\rangle$ and $D\{\rho\}$, obtained in Chapter 5 from the properties of the rough surface and the scattering geometry, with the entire statistical distribution of the amplitude of the scattered field (35) or (86) of Section 7.1; and except for the

† There are some exceptional cases when $B = 0$ and K nevertheless differs from unity, e.g. if $w(\varphi)$ has two identical isolated maxima, one at $\varphi = 0$ and the other at $\varphi = \pm\pi$; then $\langle \varphi \rangle = 0$, $w(-\varphi) = w(\varphi)$ as we require, but $\alpha = \beta = B = 0$ and $K \neq 1$, i.e. one obtains the Hoyt distribution alone (Hoyt vector without constant vector). The above distribution of D would correspond to a rather unrealistic surface which fluctuates alternately about two constant levels. We exclude such surfaces by the requirement that $w(z)$ has no periodic maxima.

polarization of the scattered field, which we have so far ignored, this completes our solution of the electromagnetic field scattered by a random rough surface.

To summarize our results, we have a random rough surface given in general by its two-dimensional distribution $w(z_1, z_2)$, or in the case of a normally distributed surface, by its distribution $w(z)$ and autocorrelation-coefficient $C(\tau)$; from this, by use of the characteristic function χ of the roughness distribution, we obtained the mean value and the variance of the scattering coefficient ρ in Section 5.2. In particular, for a normally distributed surface, $\langle\rho\rangle$, $\langle\rho\rho^*\rangle$ and $D\{\rho\}$ are derived in Section 5.3, and the same quantities may also be derived for special models of rough surfaces by other theories, e.g. the Twersky theory of Section 6.1. From (10), (11) and (16) of the present section we then find u, B and K; the distribution of u, which is related to the amplitude of the scattered field by (13), is then for the given values of B and K given by (35) or (86) of Section 7.1, and this completes the (scalar) solution, as it yields the probability distribution of the amplitude of the scattered field in various directions for a given angle of incidence. The distribution of the phase of the scattered field (which is usually of less interest than the amplitude) is given by (41), Section 7.1. The effect of varying the roughness and the correlation distance of the surface, and the angle of incidence and wavelength of the incident radiation on the mean field and mean power scattered in a given direction has already been discussed in Section 5.3.

Our solution is thus complete as far as the scalar value of the scattered field is concerned; but, as yet we have no clue as to what happens to its polarization.

As a numerical example, consider a normally distributed two-dimensional surface with standard deviation (RMS roughness) (a) $\sigma = 0.1\lambda$, (b) $\sigma = 10\lambda$, and correlation distance $T = 200\lambda$. Let the angle of incidence be $\theta_1 = 45°$ and let the active area of the surface be $A = 4\times 10^6\lambda^2$. We wish to find the mean field and mean power scattered in the specular direction ($\theta_2 = 45°$, $\theta_3 = 0$) or any other direction, and the probability that the amplitude of the scattered field in these directions will exceed a given value z or a given level Z decibels below $|E_{20}|$, the amplitude of the field that would be reflected by a smooth plane of the same dimensions.

We first take Case (a), $\sigma = 0.1\lambda$. From (6), Section 5.3, we find $\langle\rho\rangle = 0.674$ for the specular direction $\theta_2 = \theta_1 = 45°$ and, as always, zero for the non-specular directions. The parameter g of (10), Section 5.3, equals 0.7896; we therefore use (35), Section 5.3, to calculate the variance. For the specular direction we obtain (using two terms of the series) $D\{\rho\}$

$= 0.0138$, and hence the mean power coefficient $\langle\rho\rho^*\rangle = D\{\rho\} + \langle\rho\rangle\langle\rho\rangle^* = 0.468$. In comparison with this value the variance and mean power for the other directions are negligible as may be seen from (35), Section 5.3; owing to the large correlation distance, the factor $\exp(-v_{xy}^2 T^2/m)$ will make the field practically vanish outside a small cone about the specular direction ($v_{xy} = 0$).

To find the distribution of $|\rho|$, we determine B, u and K. From (10) of this section we have $B = 5.737$, and from (11)

$$|\rho| = 0.1175u \tag{22}$$

Using (3) and (10) of Section 5.3, we find that for the normal distribution, formula (18) of this section reduces after elementary manipulations to

$$K^2 = \frac{1+e^{-g}}{1-e^{-g}} = \coth\frac{g}{2} \tag{23}$$

so that in our case $K^2 = 2.661$.

Substituting these values for B and K in the asymptotic formula (89), Section 7.1, we find

$$p(u) = 1.173\sqrt{u}\cdot e^{-1.83(5.737-u)^2}\left(1+\frac{0.109}{u}+\dots\right) \tag{24}$$

This probability density obviously equals practically zero everywhere except in the neighbourhood of $u = 5.737$, where it is almost identical

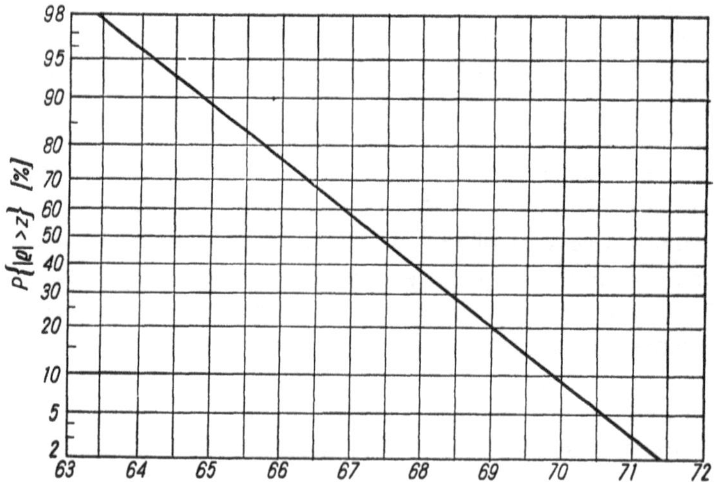

Fig. 7.4. The distribution of the amplitude of the field scattered into the specular direction by a normally distributed surface of active area $A = 4\times 10^6\lambda^2$, RMS roughness $\sigma = 0.1\lambda$ and correlation distance $T = 200\lambda$ for an angle of incidence $\theta_1 = 45°$.

with the normal distribution, since the factor $\sqrt{u}$ is almost constant in the small interval where $p(u)$ attains significant values. (For small u the asymptotic expansion is not valid as may be seen from its derivation in Appendix E.) We may thus replace $p(u)$ in (24) by the normal distribution with mean value $\langle u\rangle = 5.737$ and variance $D\{u\} = 0.2733$. Transforming back to $|\rho|$ by (22) and using (17), Appendix C, we find $|\rho|$ normally distributed with mean value $\langle|\rho|\rangle = 0.674$ and variance $D\{|\rho|\} = 3.774\times10^{-3}$,† or standard deviation $\sigma\{|\rho|\} = 0.01943$. Hence $P\{|\rho| > z\}$ is found, according to (5) and (7) of Appendix C, by integrating the normal distribution ($a = 0.674$, $\sigma = 0.01943$) from $-\infty$ to z; the result is shown in Fig. 7.4, where the ordinate gives the probability that $|\rho|$ will exceed the value indicated on the abscissa.‡

As was to be expected from the slight roughness ($\sigma = 0.1\lambda$) and the long correlation distance ($T = 200\lambda$), the fluctuations of the amplitude scattered by this surface are very small ($\sigma\{|\rho|\} < 2\%$). Nevertheless, only 46.8% ($= \langle\rho\rho^*\rangle$) of the incident energy is reflected in the specular direction; the remaining energy is scattered diffusely in the other directions, though each individual direction receives only a negligible amount of energy (an average of 0.3% per degree, including the immediate neighbourhood of the specular direction, which receives most of the remaining energy).

We now turn to Case (b), which differs from (a) only in the roughness, $\sigma = 10\lambda$. From (10), Section 5.3, we now find $g = 7896$; hence $\langle\rho\rangle = e^{-3948} \approx 0$. Since $g \gg 1$, we may use (59), Section 5.3, to obtain $D\{\rho\} = \langle\rho\rho\rangle^*$. For the plane of incidence ($\theta_3 = v_y = 0$) the results of this calculation are shown in Fig. 7.5, where $\rho_{RMS} = \sqrt{\langle\rho\rho^*\rangle}$ is plotted as a function of θ_2. The reader should not be misled by this figure, in which ρ_{RMS} is plotted along a linear scale; the diffuse character of the scattering characteristic becomes apparent when ρ_{RMS} is plotted in decibels (cf. below) as has been done in previous figures.

† It is easily verified that if $B > 0$, $K > 1$, then $D\{\rho\} > D\{|\rho|\}$.

‡ There is no need for calculations if one plots $P\{|\rho| > z\}$ on probability paper as in Fig. 7.4, in which the normal distribution function

$$L(z) = \frac{1}{\sigma\sqrt{2\pi}} \int_{-\infty}^{z} \exp\left[-\frac{(t-\langle t\rangle)^2}{2\sigma^2}\right] dt$$

or its complement

$$1 - L(z)$$

appears as a straight line. This is determined by two points, e.g. $P = 50\%$ for $z = \langle t\rangle$, and $P = 2.275\%$ for $z = \langle t\rangle \pm 2\sigma$ (lower sign for the distribution function, upper sign for its complement).

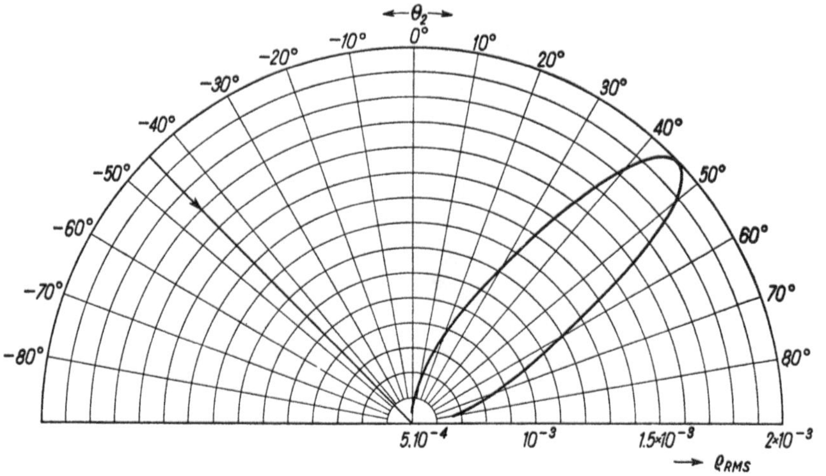

Fig. 7.5. Scattering diagram of a normally distributed surface with active area $A = 4\times 10^6\lambda^2$, RMS roughness $\sigma = 10\lambda$ and correlation distance $T = 200\lambda$ for an angle of incidence $\theta_1 = 45°$ (Linear power scale.)

The distribution of $|\rho|$ is the Rayleigh distribution, as $g = 7896 \gg 1$ (very rough surface). The required probability is therefore

$$P\left\{\frac{|\rho|}{\rho_{\mathrm{RMS}}} > z\right\} = P\left\{\frac{|\rho|^2}{\langle\rho\rho^*\rangle} > z^2\right\} = e^{-z^2} \tag{25}$$

which is determined by a single constant $\langle\rho\rho^*\rangle$. This is found from (59), Section 5.3, for various scattering angles. The probability (25) for various scattering angles θ_2 is shown in Fig. 7.6.

If (25) is plotted on Rayleigh paper as in Fig. 7.6, there is again no need for any calculations (other than the basic calculation of the coordinates of Rayleigh paper), for the relation (25) appears on it as a straight line with 45° slope and is thus determined by a single point, e.g. $P = e^{-1} \approx 37\%$ for $z = 1$.

Figure 7.7 shows the statistics of the scattering diagram of this surface. The mean power coefficient $\langle\rho\rho^*\rangle$ is calculated as a function of θ_2 from (59), Section 5.3; since $|\rho|$ is Rayleigh-distributed, the mean power is exceeded with a probability of 37%. Other levels and the probabilities with which they are exceeded are found from (25) or directly from Fig. 7.6. Thus Fig. 7.7 gives the scattering diagram with various levels and the corresponding probabilities with which they are exceeded. The levels are given in decibels below the power (or field) reflected by a perfectly conducting smooth plane of equal dimensions ($4\times 10^6\lambda^2$) under the same angle of incidence (45°).

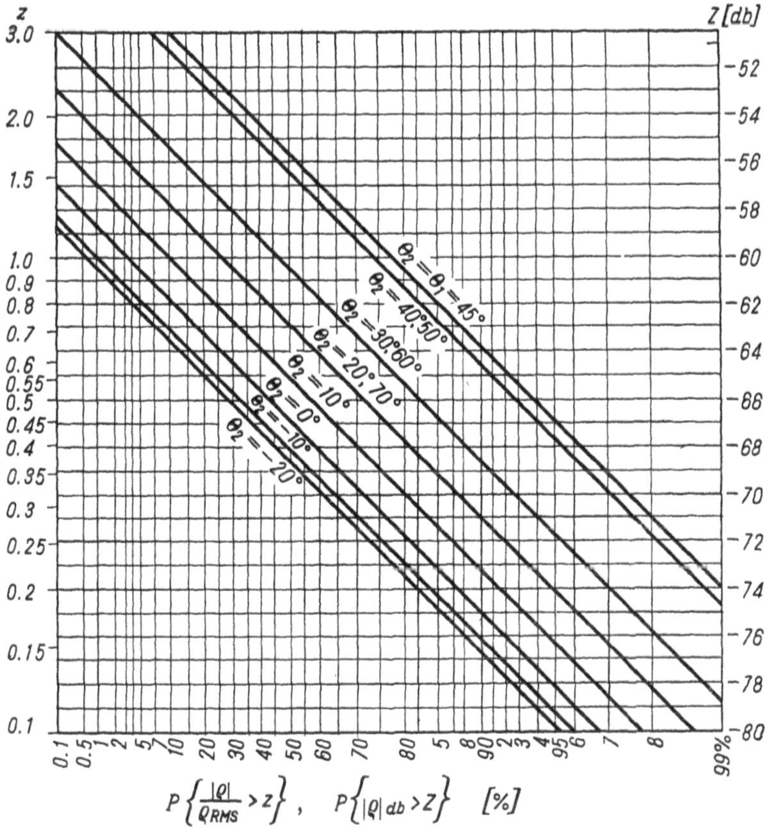

Fig. 7.6. The distribution of $|\rho|$ for various scattering angles. Example: The probability that in the direction $\theta_2 = 70°$ the field (or power) will exceed a level of 60 db below the field (or power) reflected by a smooth plane of the same dimensions, is 25%. To simplify the figure, the distribution curves have been very slightly displaced to unite the curves for $\theta_2 = 50°$ and $\theta_2 = 40°$ and for other pairs of θ_2 symmetrical to the specular direction $\theta_2 = 45°$.

Figure 7.7 shows that this very rough surface ($\sigma = 10\lambda$) scatters the incident energy diffusely. Nevertheless, there is a clear maximum in the specular direction, which is conspicuous not only when the figure is plotted linearly instead of logarithmically (Fig. 7.5), but also when Fig. 7.7 is compared with Isakovich's figures (Figs. 5.3 to 5.5, p. 497, 95). The reason is that in our example we have chosen a large ratio $T/\sigma = 20$, which will favour the specular direction, as may be seen by writing (53), Section 5.3, as

$$\langle \rho\rho^* \rangle = \frac{\lambda^2}{A} \frac{F^2(\theta_1, \theta_2)\exp(-v_{xy}^2 T^2/4g)}{(\sin\theta_1 - \sin\theta_2)^2 + \left(\frac{\sigma}{T}\right)^2 (\cos\theta_1 + \cos\theta_2)^2} \tag{26}$$

The maximum for $\theta_1 = \theta_2$ will therefore increase with decreasing σ/T; this may be seen from Isakovich's calculations for $T/\sigma = 5$, 7, 10 and the present calculation for $T/\sigma = 20$. This confirms our conclusion that for

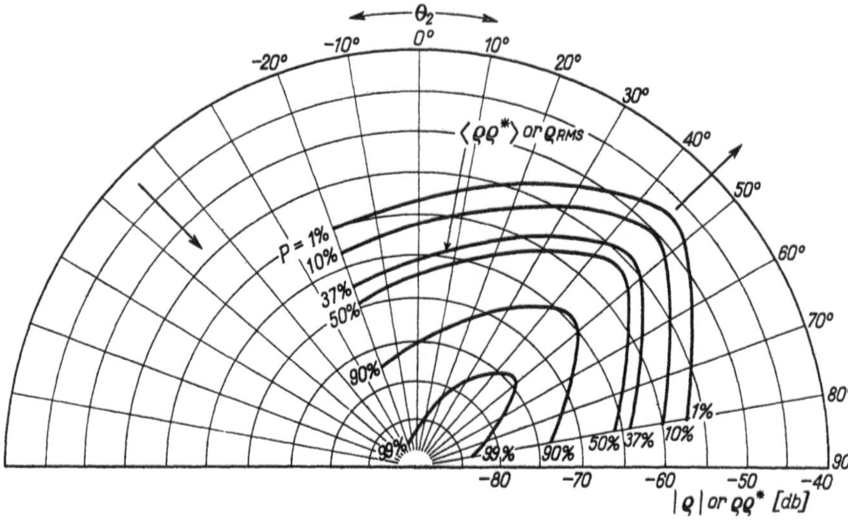

Fig. 7.7. The complete scattering diagram of a normally distributed surface as in Fig. 7.5. Examples: The scattered field (or power) will exceed the curve marked "90%" with a probability of 90%, i.e. in 90% of a large number of samples, or if the surface is time-varying, in 90% of the time. In the direction $\theta_2 = 60°$, the field (or power) will in 1% of all cases exceed a level 68 db below the field (or power) reflected by a smooth plane of the same area ($4 \times 10^6 \lambda^2$) under the same angle of incidence (45°). (Note added in proof: An unfortunate error in the numerical calculations distorts the curves which should go to zero more rapidly as θ_2 leaves the specular direction 45°. The curves are correct near the specular direction and the error does not affect the conclusions drawn in the main text.)

large T (gently rolling surfaces with long distances between hills and valleys) the specular direction remains privileged even if the surface is very rough.

7.3. Coherence and Incoherence of the Scattered Field

We have so far avoided the terms "coherent" and "incoherent" scattering, as they are sometimes used in a confusing manner. Originally, a coherent wave means a wave of the type $\exp(i\omega t + \varphi)$ where φ is constant as in most man-made radio waves; and an incoherent wave means a wave in which the phase φ fluctuates at random as in light waves. We shall presently give a more exact definition of coherence and incoherence.

The essential difference between the two types lies in the way in which the power of the sum several waves depends on the power of the individual summed waves. If two radio waves of equal amplitude interfere, the total power density may be anything from zero (if the waves are in phase opposition) to four times the power density of either wave (if they are in phase), but two light sources of equal power will always give double the luminous intensity of either source. If n independent waves of unit amplitude interfere at a point in space, the total power denisty is

$$W = \frac{1}{\eta}\left|\sum_{j=1}^{n} E_j\right|^2 = \frac{1}{\eta}\left|\sum_{j=1}^{n} e^{i\varphi_j}\right|^2 \tag{1}$$

where η is the impedance of the (lossless) medium (120π Ω for free space). To get rid of the constant η which is immaterial for our present purpose, we may either consider waves of amplitude η instead of unit

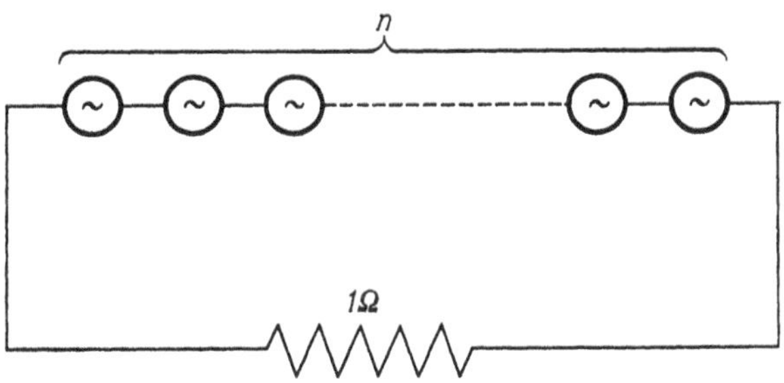

Fig. 7.8. Circuit analogy for wave interference. If the voltage of each generator is 1 V, the power expended in the 1Ω resistor will vary between 0 and n^2 depending on the phases of the n randomly phased generators.

amplitude, or else we can consider the analogy of Fig. 7.8, where n independent generators of 1 V each work into a resistance of 1 Ω, so that the total expended power is

$$W = \left|\sum_{j=1}^{n} e^{i\varphi_j}\right|^2 = \left(\sum_{j=1}^{n} \cos\varphi_j\right)^2 + \left(\sum_{j=1}^{n} \sin\varphi_j\right)^2 \tag{2}$$

where φ_j is the phase of the jth generator relative to some reference phase. On performing the squares, we may write (2) as

$$W = n + \sum_{i\neq j} \cos(\varphi_i - \varphi_j) \tag{3}$$

where the sum is taken over the $n(n-1)$ terms with $i \neq j$.

Let

$$\varphi_i - \varphi_j = \varepsilon_j \tag{4}$$

Then

$$W = n + \sum_{j=1}^{n(n-1)} \cos\varepsilon_j \tag{5}$$

If all the φ_j are equal (i.e. all the generators in phase), then from (4) all $\varepsilon_j = 0$, so that

$$W = n+n(n-1) = n^2 \tag{6}$$

a result which we could of course have obtained directly from (2).

New let the φ_j be random, independent and distributed uniformly between $-\pi$ and π. Then ε_j will, in effect, also be distributed uniformly from $-\pi$ to π.

This point needs some clarification. Strictly speaking, if the φ_j are distributed uniformly between $-\pi$ and π, then according to (4), ε_j will be Simpson-distributed from -2π to 2π [cf. (3), Appendix C], but owing to the periodicity of $\cos\varepsilon_j$, this is completely equivalent to a uniform distribution from $-\pi$ to π, as the probability densities in any other interval from $m\pi$ to $(m+2)\pi$, where m is an integer, will be added to those in the basic interval $(-\pi, \pi)$ or any other chosen basic interval of length 2π. If the variance $D\{\varphi_j\}$ is large ($\gg \pi^2$), the same result is obtained even for a non-uniform distribution of φ_j, unless the function $w(\varphi)$ happens to contain a function with period $2m\pi$. (This would correspond to the rather artificial case of a surface that fluctuates about certain regularly spaced height-levels, and we exclude this kind of a surface.) From another point of view, it is immaterial for the interference of waves whether the phase of a wave is $\pi/2$ or $72\pi+\pi/2$; selecting any interval of length 2π and superposing consecutively the probability densities of all the remaining intervals of length 2π in this basic interval, we must obtain a uniform distribution, if the original distribution extends over a sufficiently large number of intervals 2π (and if certain exceptions, revealed by a rigorous analysis, but very artificial from a physical point of view, are disregarded). Thus if $D\{\varphi\} \gg \pi^2$, then the distribution of φ, whatever it is, will be equivalent to a uniform distribution over the basic phase cycle and we shall say that φ is "in effect distributed uniformly from $-\pi$ to π".

If now the ε_j are in effect distributed uniformly from $-\pi$ to π, then the mean value of the power is from (5)

$$\langle W \rangle = n + \frac{n(n-1)}{2\pi}\int_0^{2\pi} \cos\varepsilon \, d\varepsilon = n \tag{7}$$

Using these results, we may define coherence and incoherence as follows:

If the phase of a wave is random and (directly or in effect) uniformly distributed over an interval of length 2π, it will be called incoherent; it follows from (7) that the mean power densities of incoherent waves may be added algebraically. If the phase of a wave is constant, we shall call it a coherent wave; the total power density of coherent waves is obtained by summing the individual fields vectorially and determining the total power from the resulting total field, in particular, if n coherent waves of equal power are all in phase, the total power is proportional to n^2.

Note that this definition does not exhaust all possibilities: if the phase of the wave is random, but not uniformly distributed over the basic phase cycle, the wave is neither coherent nor incoherent. As we shall presently see, there is a continuous transition from pure coherence to pure incoherence.

We may also note that since the phase of incoherent waves is random and uniformly distributed over an interval 2π, the amplitude of the sum of many incoherent waves is Rayleigh-distributed.

Now consider the elementary waves scattered by a rough surface. The phase difference between two waves 1 and 2 scattered at the points $\zeta(x_1)$

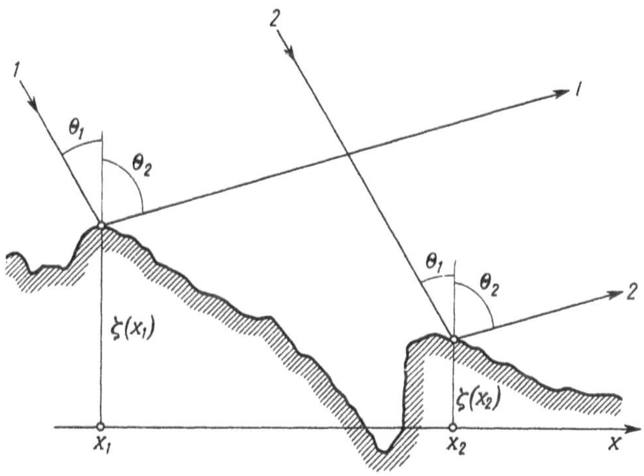

Fig. 7.9.

and $\zeta(x_2)$ of the surface (Fig. 7.9) is, from geometric considerations, or by using the relation $\Delta\varphi = \mathbf{v}\cdot(\mathbf{r}_1-\mathbf{r}_2)$,

$$\Delta\varphi_{12} = \frac{2\pi}{\lambda}\left[(x_1-x_2)(\sin\theta_1-\sin\theta_2)+(\zeta_1-\zeta_2)(\cos\theta_1+\cos\theta_2)\right] \tag{8}$$

where $\zeta_1 = \zeta(x_1)$, $\zeta_2 = \zeta(x_2)$. Therefore the phases of the elementary waves scattered in the direction θ_2 will vary in the interval

$$\Delta\varphi = \frac{4\pi L}{\lambda}(\sin\theta_1-\sin\theta_2)+F(\zeta) \tag{9}$$

where F is a random function of ζ. Since $\zeta_{max} \ll 2L$, the term $F(\zeta)$ will always be small compared to the first term if $\theta_1 \neq \theta_2$. Hence

$$\Delta\varphi \gg 2\pi \quad \text{for} \quad \frac{2L}{\lambda}(\sin\theta_1 - \sin\theta_2) \gg 1 \tag{10}$$

Thus for $\theta_1 \neq \theta_2$, $\Delta\varphi$ will vary over many intervals 2π, so that whatever the distribution of ζ, the phases of the individual scattered waves are in effect uniformly distributed over the basic phase cycle. Thus the field scattered by a rough surface in the non-specular directions is always incoherent and we now see the physical reason why it is Rayleigh-distributed.

In the specular direction $\theta_1 = \theta_2$ the phase of each elementary wave has by (8) only a random component:

$$\varphi = \frac{4\pi}{\lambda}\zeta(x)\cos\theta_1 \tag{11}$$

If $D\{\zeta\} \gg \lambda^2/4\cos^2\theta_1$ (very rough surface with grazing incidence excluded), then $D\{\varphi\} \gg 4\pi^2$ and for the same reasons as above the waves scattered in the specular direction will again be incoherent (and therefore the amplitude of the field will be Rayleigh-distributed).

Now let $\zeta(x) \equiv 0$ (smooth plane). Then from (11) it follows that all φ are equal (to zero) and the field in this one case is coherent.

There remains the gap between the coherent field reflected by a smooth surface and the incoherent field scattered by a very rough surface in the specular direction. Writing the mean square value of the scattering coefficient

$$\langle\rho\rho^*\rangle = D\{\rho\} + \langle\rho\rangle\langle\rho\rangle^* \tag{12}$$

we see that $\langle\rho\rho^*\rangle$ consists of two parts. When the surface is smooth, reflecting only coherent waves, $D\{\rho\} = 0$; when the surface is very rough, scattering only incoherent waves, $|\rho|$ is Rayleigh-distributed, making $\langle\rho\rho^*\rangle = D\{\rho\}$. Between these two extremes there is a continuous transition from coherence to incoherence, corresponding to a decrease of the parameter (Section 7.1)

$$B^2 = \frac{\alpha^2}{s_1 + s_2} = \frac{\langle\rho\rangle\langle\rho\rangle^*}{D\{\rho\}} \tag{13}$$

from ∞ to 0.

It is interesting to note the decrease of the mean square value of the vector sum of n unit vectors from n^2 to n, i.e. from coherence to incoherence, as the variance of the phase increases from 0 to ∞; the variance of the phase is in the present application (for $\theta_1 = \theta_2$) proportional to the variance of ζ, as may be seen from (11).

From (31), Section 7.1, we have

$$\langle r^2 \rangle = \alpha^2 + s_1 + s_2 \tag{14}$$

Referring to Table 7.1, p. 123, we find the following instructive examples:

If ζ (and hence φ in the specular direction) is distributed normally with variance σ^2, then

$$\langle r^2 \rangle = n^2 e^{-\sigma^2} + n(1 - e^{-\sigma^2}) \tag{15}$$

yielding n^2 for $\sigma^2 = D\{\zeta\} = 0$, and n for $\sigma^2 \to \infty$ ($\sigma^2 \gg \pi$).

If ζ is distributed uniformly from $-a$ to a (here $D\{\zeta\} = a/3$), then

$$\langle r^2 \rangle = n^2 \operatorname{sinc}^2 a + n(1 - \operatorname{sinc}^2 a) \tag{16}$$

yielding n^2 for $a = 0$, and n for $a = m\pi$ ($m \neq 0$) or for $a \to \infty$ ($a \gg \pi$).

If ζ is Simpson-distributed [cf. (3), Appendix C], then

$$\langle r^2 \rangle = n^2 \operatorname{sinc}^4 a + n(1 - \operatorname{sinc}^4 a) \tag{17}$$

yielding n^2 or n under the same conditions as in (16).

From (11) we have for $\theta_1 = \theta_2$

$$D\{\varphi\} = \frac{4\pi^2}{\lambda^2} \cos^2\theta_1 D\{\zeta\} \tag{18}$$

Comparing this with (15) to (17), we note how rapidly the field becomes incoherent and hence Rayleigh-distributed as the roughness (proportional to $D\{\zeta\}$) increases or the wavelength decreases, unless of course $\cos\theta_1 \to 0$ (grazing incidence).

Summarizing this note on the coherence and incoherence of scattered waves, we may say that for non-grazing incidence and $2L \gg \lambda$, the amplitude of the field scattered by a rough surface in the specular direction has a constant, coherent component and a random, Hoyt-distributed component. When the surface is very rough, the latter component becomes incoherent (Rayleigh-distributed) and the former vanishes. In the non-specular directions the field (if any) is always incoherent. Grazing incidence is by (18) equivalent to a slightly rough ($\theta_1 \to \pi/2$) or smooth ($\theta_1 = \pi/2$) surface.

CHAPTER 8

THE DEPOLARIZATION OF ELECTROMAGNETIC WAVES SCATTERED FROM A ROUGH SURFACE

WE have so far almost invariably considered the scalar value of the scattered field, assuming the incident field to be either vertically ($\mathbf{E}_1 \perp \mathbf{y}_0$) or horizontally ($\mathbf{E}_1 \parallel \mathbf{y}_0$) polarized, and have hitherto ignored the question of whether the polarization of the incident wave is changed in the process of scattering, i.e. whether the polarization of a scattered wave is the same as that of the incident wave or whether, in the contrary case, "depolarization" sets in. If the incident wave is either purely vertically or purely horizontally polarized, and the surface is of the "locally flat" type (Section 3.3), it is reasonable to assume that for reasons of symmetry, the field scattered in the plane of incidence will not be depolarized, and we shall find this assumption confirmed. But in the general case depolarization must be expected, as even a smooth, perfectly conducting surfaces depolarizes, for it reverses the sense of rotational polarization and in the side lobes it changes circular into elliptic polarization.†

The problem is important for several applications such as polarization discrimination for television channels, circularly polarized radar or microwave relays, etc., but far from being solved, it has as yet apparently never been seriously attacked.

In the present chapter we attempt at least a first step towards the solution of the problem. It would seem that since our general method of Chapter 3 is a Kirchhoff method, we are not well equipped for polarization problems. This, unfortunately, is partly true. On the other hand, we have not used the original Kirchhoff method which gives no clue to the polarization of the diffracted field, but a Kirchhoff method with boundary conditions expressed in terms of the plane-surface "local" reflection coefficient $R^{\pm}$, which differs for the two polarizations "+" and "−" in amplitude and phase for finite conductivity and only in phase (by 180°) for infinite conductivity; this difference may be utilized to obtain an approximate solution.

† From a mathematical point of view, linear and circular polarization are only special cases of elliptical polarization, but we find it more instructive to classify polarization into linear and rotational, the latter including circular and "truly" elliptic polarization.

As in polarization problems we deal with quantities that are space vectors and at the same time complex time vectors, we start by introducing a workable symbolism and defining a polarization and depolarization coefficient. We then find the depolarization caused by a plane surface; this provides us with the relations necessary to solve polarization problems of rough surfaces with large radii of curvature. Their solution turns out to be interesting only in the case of finite conductivity. Finally, we shall investigate the depolarization of laterally scattered waves (out of the plane of incidence), where the main difficulty is the complicated geometry.

8.1. The Depolarization Factor

In problems of polarization we deal with quantities that are space vectors, but the "length" of these space vectors (at the moment we purposely avoid the terms "absolute value" or "amplitude") is a harmonic function of time, represented by the real (or imaginary) part of a complex number or time vector $A \exp(i\omega t + \varphi)$, where A is a real scalar. Contrary to circuit theory which works with complex scalars (time vectors) or electrostatics which work with real space vectors, we must work with complex space vectors, and the first thing is to introduce a useful system of symbols and terminology. (If $\mathbf{E}$ is a complex vector, then the term "absolute value", denoted by $|\mathbf{E}|$, is ambiguous, since it is not clear whether this applies to the space vector, converting it to a scalar, or to its complex value, converting it to a real value.)

We shall therefore temporarily indicate a complex number by a circumflex accent (^); when omitted, the symbol denotes the "absolute value" or "amplitude" of the complex number, i.e.

$$\hat{z} = z e^{i \arg \hat{z}}; \qquad z = |\hat{z}| \tag{1}$$

where

$$\arg \hat{z} = \arctan \frac{\operatorname{Im} \hat{z}}{\operatorname{Re} \hat{z}}$$

and the terms "absolute value" or "amplitude" are thus applied to a complex number, not to a space vector.

Space vectors will, as usual, be indicated by bold print; the same symbol printed in ordinary print denotes the "scalar value" of the vector, i.e.

$$\mathbf{E} = \mathbf{e}E; \quad E = \mathbf{E} \cdot \mathbf{e} \tag{2}$$

where $\mathbf{e}$ is a unit vector in the direction of $\mathbf{E}$.

Suppressing the time factor $\exp(i\omega t)$, a complex vector is thus expressed as

$$\hat{\mathbf{E}} = \hat{E}\mathbf{e} = E e^{i\varphi}\mathbf{e} = |\hat{E}| e^{i \arg \hat{E}}\mathbf{e} \tag{3}$$

so that

$$\hat{\mathbf{E}} = \hat{E}\mathbf{e} \text{ is a complex vector}$$
$$\mathbf{E} = E\mathbf{e} = |\hat{E}|\,\boldsymbol{e} \text{ is a real vector}$$
$$\hat{E} = \hat{\mathbf{E}}\cdot\mathbf{e} = Ee^{i\varphi} \text{ is a complex scalar}$$
$$E = |\hat{\mathbf{E}}| = |\hat{\mathbf{E}}\cdot\mathbf{e}| \text{ is a real scalar}$$

Now let $\mathbf{e}^+$ and $\mathbf{e}^-$ be unit vectors in the direction of vertical (+) and horizontal (−) polarization. To define these polarizations, we first define the plane $\mathbf{k}_1$, $\mathbf{z}_0$ (using the notation of Section 3.1) as the plane of incidence and the plane $\mathbf{k}_2$, $\mathbf{z}_0$ as the plane of scattering (Fig. 8.1); we again use the index 1 for quantities associated with the incident wave and the index 2 for quantities associated with a scattered wave. Then $\mathbf{e}_1^+$ lies in, and $\mathbf{e}_1^-$ is

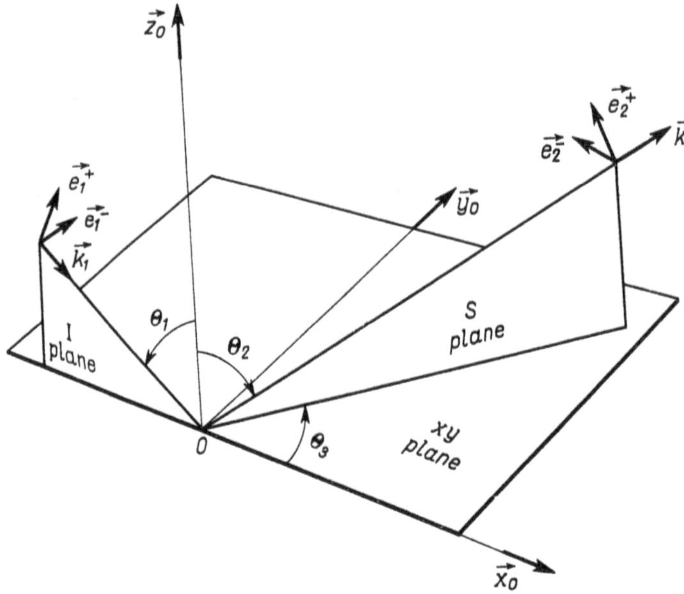

Fig. 8.1. The plane of incidence (*I*), the scattering plane (*S*), horizontal (−) and vertical (+) polarization.

perpendicular to, the plane of inicidence, $\mathbf{e}_1^+$, $\mathbf{e}_1^-$, $\mathbf{k}_1$ being mutually perpendicular. Similarly, $\mathbf{e}_2^+$ lies in, and $\mathbf{e}_2^-$ is perpendicular to, the plane of scattering, $\mathbf{e}_2^+$, $\mathbf{e}_2^-$, $\mathbf{k}_2$ being mutually perpendicular. ($\mathbf{e}_1^+ \perp \mathbf{y}_0$ and $\mathbf{e}_1^- \parallel \mathbf{y}_0$ always; $\mathbf{e}_2^+ \perp \mathbf{y}_0$ and $\mathbf{e}_2^- \parallel \mathbf{y}_0$ only for $\theta_3 = 0$.)

Now let $\hat{\mathbf{E}}_1$ be arbitrarily polarized. Resolving the field into its vertical and horizontal components, we have

$$\hat{\mathbf{E}}_1 = \hat{E}_1^+\mathbf{e}_1^+ + \hat{E}_1^-\mathbf{e}_1^- \tag{4}$$

where $\hat{E}_1^+$ and $\hat{E}_1^-$ are the scalar values of the electric field with which we have hitherto been working (previously there was no need for the circumflex accents).

We now define the "polarization factor" $\hat{p}$ of an electromagnetic wave as

$$\hat{p} = \frac{\hat{E}^+}{\hat{E}^-} \tag{5}$$

so that (4) may be written as

$$\mathbf{E}_1 = \hat{E}_1^- (\hat{p}_1 \mathbf{e}_1^+ + \mathbf{e}_1^-) \tag{6}$$

The complex polarization factor given by (5) uniquely defines any polarization of an electromagnetic wave: the absolute value $|\hat{p}|$ gives the ratio of the amplitudes of the two components, and arg $\hat{p}$ gives the phase shift between them. Hence

$\text{Im}\,\hat{p} = 0$ implies linear polarization,
$\text{Im}\,\hat{p} > 0$ implies right-handed rotational polarization,
$\text{Im}\,\hat{p} < 0$ implies left-handed rotational polarization; in particular
$\hat{p} = 0$ implies horizontal polarization,
$\hat{p} = \infty$ implies vertical polarization,
$\hat{p} = i$ implies right-handed circular polarization,
$\hat{p} = -i$ implies left-handed circular polarization.

To each possible polarization there corresponds one value of $\hat{p}$ and vice versa.

As may be seen from (6), the complex scalar $\hat{p}$ not only uniquely determines the polarization, but also rids us of the need for space vectors and therefore also of the need to distinguish between space vectors and complex quantities; we therefore drop the circumflex accent again, as now confusion is no longer possible.

From p other quantities relating to polarization, such as the parameters of the polarization ellipse, may be obtained. The latter is inscribed in a rectangle (which we shall call the "polarization rectangle") with sides $2E^-$ and $2pE^-$ (Fig. 8.2), with the axes of the ellipse shifted with respect to the axes of the rectangle by an angle ψ.† This angle is related to p through

$$\tan 2\psi = \frac{1-|p|^2}{2|p|^2} \cos(\arg p) \tag{7}$$

The diagonals of the polarization rectangle are determined by

$$\tan \alpha = \pm |p| \tag{8}$$

If E^+ and E^- are in phase ($\arg p = 0$), α is simply the angle between $\mathbf{e}^-$ and $\mathbf{E}$.

† Stratton, J. A.: *Electromagnetic Theory*. McGraw-Hill, New York, 1941, p. 280.

The relation between p and the points on the Poincaré sphere† is simple: the complex p plane is the stereographic projection of the Poincaré sphere in the same way as any complex plane is the stereographic projection of its corresponding Riemann sphere (Fig. 8.3).

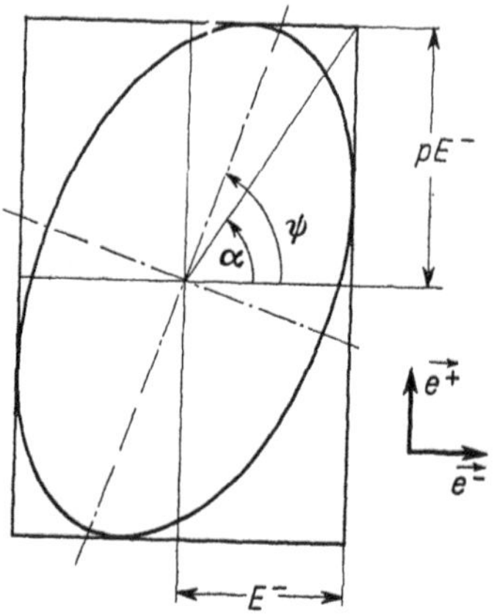

Fig. 8.2. The polarization rectangle circumscribed about the polarization ellipse.

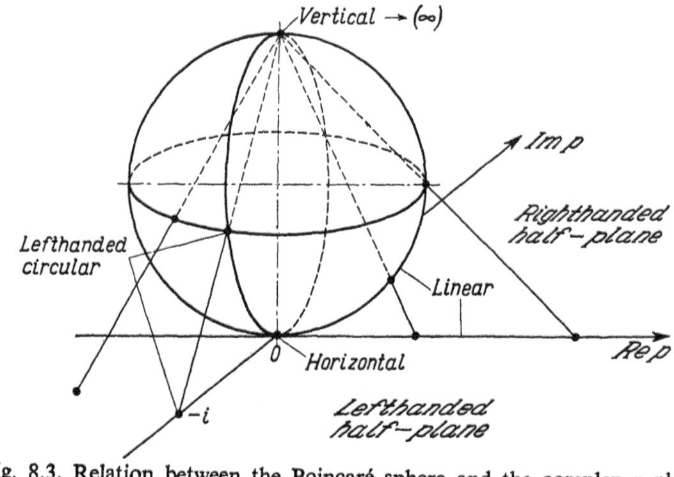

Fig. 8.3. Relation between the Poincaré sphere and the complex p plane.

Now let the polarization of the incident wave be described by p_1 and that of a scattered wave by p_2. Then we define the "depolarization factor" q by the relation

$$p_2 = qp_1 \tag{9}$$

† Born, M. and Wolf, E.: *Principles of Optics*. Pergamon Press, London–New York, 1959, p. 30.

From (9) and (5) we have

$$q = \frac{\mathcal{R}^+}{\mathcal{R}^-} \tag{10}$$

where

$$\mathcal{R}^+ = \frac{E_2^+}{E_1^+}; \qquad \mathcal{R}^- = \frac{E_2^-}{E_1^-} \tag{11}$$

are generalized reflection coefficients reducing to the Fresnel coefficients R^+ and R^- when the scatterer is a properly oriented smooth plane.

We are of course working under the assumption that all media of propagation are linear, so that the coefficient $\mathcal{R}^+$ is independent of E_1^+, and $\mathcal{R}^-$ is independent of E_1^-. But now we must be cautious. It may happen that the scatterer is such that a purely horizontally polarized incident wave E_1^- is depolarized and gives rise to a scattered wave with a vertical component E_2^+. (It should not be imagined that such a scatterer must be complicated: any plane element of a rough surface that is not perpendicular to the plane of incidence will accomplish this kind of depolarization.) Then varying E_1^- will affect $\mathcal{R}^+$ (though not $\mathcal{R}^-$). Similarly, varying E_1^+ may affect $\mathcal{R}^-$.

To respect this effect, we write E_{2c}^+ for the vertically polarized scattered field caused by a horizontally polarized incident field (the index c stands for "cross-polarized"); similarly E_{2c}^- is due to E_1^+. For the scattered field due to the incident field of the same polarization we write $E_{2\parallel}^+$ and $E_{2\parallel}^-$. Thus the total scattered field E_2^+, which is vertically polarized and due to an incident wave with both a vertical and a horizontal component, is

$$E_2^+ = E_{2\parallel}^+ + E_{2c}^+ \tag{12}$$

Similarly,

$$E_2^- = E_{2\parallel}^- + E_{2c}^- \tag{13}$$

Substituting this in (11), we have

$$\left.\begin{aligned} \mathcal{R}^+ &= \mathcal{R}_\parallel^+ + \frac{1}{p_1}\mathcal{R}_c^+ \\ \mathcal{R}^- &= \mathcal{R}_\parallel^- + p_1\mathcal{R}_c^- \end{aligned}\right\} \tag{14}$$

where

$$\left.\begin{aligned} \mathcal{R}_c^+ &= \frac{E_{2c}^+}{E_1^-} = \frac{p_1 E_{2c}^+}{E_1^+} \\ \mathcal{R}_c^- &= \frac{E_{2c}^-}{E_1^+} = \frac{E_{2c}^-}{p_1 E_1^-} \end{aligned}\right\} \tag{15}$$

and

$$\mathcal{R}_\parallel^\pm = \frac{E_{2\parallel}^\pm}{E_1^\pm} \tag{16}$$

Hence from (10) and (14),

$$q = \frac{p_1 \mathcal{R}_{\|}^{+} + \mathcal{R}_c^{+}}{p_1 \mathcal{R}_{\|}^{-} + p_1^2 \mathcal{R}^{-}} \tag{17}$$

From now until Section 8.4 we assume

$$\mathcal{R}_c^{+} = \mathcal{R}_c^{-} = 0 \tag{18}$$

i.e. that the surface or the scattering geometry is such that purely horizontal or purely vertical polarization is not depolarized. This will obviously be true for scattering in the plane of incidence (longitudinal scattering), but will not hold, as we shall see, for lateral scattering. Under these conditions, i.e. if (18) holds, it follows from (17) that q will be independent of p_1, and we find

$$q = \frac{\mathcal{R}_{\|}^{+}}{\mathcal{R}_{\|}^{-}} = \frac{E_{2\|}^{+}}{E_{2\|}^{-}} = \frac{\rho^{+}}{\rho^{-}} \tag{19}$$

This is the formula we shall use to calculate the depolarization factor q when (18) holds. The first two equalities in (19) may be used to determine q when the scattered field is not normalized through ρ as in (29), Section 3.1, e.g. in the case of Twersky's "Scattering Amplitude" of Section 6.1.

If (18) holds, it follows from (19) that the depolarization factor q depends only on the properties of the scatterer (which, incidentally, need not be a rough surface) and the scattering geometry θ_1, θ_2, θ_3, but not on the polarization of the incident wave.

If, however, (18) does not hold, then q will by (17) be a function of p_1 and though it still uniquely determines the depolarization, it thus loses its main advantage and it is then more convenient to work directly with the polarization of the scattered wave p_2. Multiplying (17) by p_1 and using (9) we find the basic depolarization formula

$$p_2 = \frac{p_1 \mathcal{R}_{\|}^{+} + \mathcal{R}_c^{+}}{\mathcal{R}_{\|}^{-} + p_1 \mathcal{R}_{c|}^{-}} \tag{20}$$

valid for any rough surface or any other scatterer.

Other quantities associated with depolarization such as the depolarization angle and the cross-polarization may be determined from q. The depolarization angle ε is defined by

$$\varepsilon = \alpha_2 - \alpha_1 \tag{21}$$

and thus in the special case of $\arg p_2 - \arg p_1 = 0$ it determines the angle through which the polarization plane (**k**, **E**) has turned in the process of scattering; more generally it determines the angle through which the diagonal of the polarization rectangle (Fig. 8.2) has turned. From (8), (9) and (21) we find

$$\varepsilon = \arctan(q \tan \alpha_1) - \alpha_1 \tag{22}$$

relating q to the depolarization angle ε.

The "cross-polarization" is defined as

$$D = \left| \frac{E_2^{\perp}}{E_2^{\parallel}} \right| \tag{23}$$

where $E_2^{\parallel}$ and $E_2^{\perp}$ are the scalar values of the components parallel and perpendicular respectively to the incident field $\mathbf{E}_1$. The cross-polarization is useful only if $\mathbf{E}_1$ is linearly polarized and is related to q through

$$D = \left| \frac{q-1}{\cot \alpha_1 + q \tan \alpha_1} \right| \tag{24}$$

The depolarization factor q uniquely defines the depolarization to which an electromagnetic wave is subjected. Thus $q = 1$ implies that no depolarization has set in; $q = -1$ implies that the polarization rectangle is inverted (Fig. 8.4), e.g. the sense of rotational polarization is reversed and for linear polarization we find from (22) that the polarization plane is turned through an angle $-2\alpha_1$. If $q = 0$ (or ∞) for any p_1, this implies that the surface

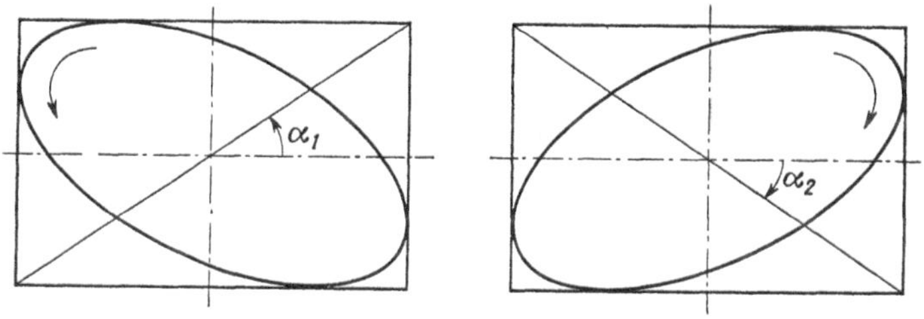

Fig. 8.4. Depolarization for $q = -1$.

will act as a "filter", scattering only horizontal (or vertical) polarization regardless of how the incident wave is polarized. Thus for reflection by a dielectric plane under the Brewster angle, we have $q = 0$ for any p_1; in the case of a "surface" consisting of long thin wires with their ends in the plane $z = 0$ and not spaced too close together [Ament, 1956], $q = \infty$ for any p_1. (In both cases the other component is totally transmitted into the space $z < 0$.)

In conclusion we may say that any polarization of an electromagnetic wave is uniquely determined by the complex polarization factor p defined by (5), and any depolarization of this wave is uniquely determined by the complex depolarization factor q as given by (9) or (17). We may therefore consider any problem of depolarization solved if we determine the corresponding value of q; if the depolarization is random, we shall require the statistical distribution of q (two-dimensional since q is complex).

8.2. Depolarization by a Smooth Plane

The problem of depolarization by a smooth plane is easily solved, but still not trivial, as we shall see within a few lines.

From (34), Section 3.1, the scattering coefficient of a finite smooth plane is

$$\rho^{\pm} = -\frac{(1+R^{\pm})\cos\theta_2-(1-R^{\pm})\cos\theta_1}{2\cos\theta_1}\operatorname{sinc} v_x L \tag{1}$$

Hence

$$q = \frac{\rho^+}{\rho^-} = \frac{(1+R^+)\cos\theta_2-(1-R^+)\cos\theta_1}{(1+R^-)\cos\theta_2-(1-R^-)\cos\theta_1} \tag{2}$$

This general expression applies to the side lobes of the scattering diagram, determined by the maxima of the function $\operatorname{sinc} v_x L$; in the direction of specular reflection, (2) reduces to

$$q = \frac{R^+}{R^-} \tag{3}$$

If the plane is perfectly conducting, then $R^+ = 1$, $R^- = -1$, and (2) simplifies to

$$q = -\frac{\cos\theta_2}{\cos\theta_1} \tag{4}$$

and for the specular direction this becomes

$$q = -1 \tag{5}$$

Thus even in the simplest of all cases, reflection by a perfectly conducting plane, q does not equal unity. Since

$$p_2 = q p_1 \tag{6}$$

it is thus seen that a smooth plane will in general depolarize all incident polarizations except horizontal ($p = 0$) and vertical ($p = \infty$). There will also be no depolarization if $R^+ = R^-$, as may be seen from (3); the only important case of this is near grazing incidence onto a plane of finite conductivity, when

$$R^+ \approx R^- \approx -1 \tag{7}$$

In practically all other cases a smooth plane will depolarize. For example, right-handed circular polarization ($p_1 = i$) will, by (6) and (3), be depolarized to

$$p_2 = i\frac{R^+}{R^-} \tag{8}$$

i.e. to elliptic polarization for $R^+ \neq R^-$; for

$$\cos(\arg R^+ - \arg R^-) < 0 \tag{9}$$

the sense of rotation will moreover be reversed (since $\operatorname{Im} p_2$ will differ in sign from $\operatorname{Im} p_1$, cf. Section 8.1).

A large, perfectly conducting plane, whose depolarization factor is given by (5), will always turn the polarization plane through an angle $\varepsilon = -2\alpha_1$, which will leave only vertical ($\alpha_1 = \pi/2$) and horizontal ($\alpha_1 = 0$) polarization unaffected. It will always reverse the sense of rotational polarization and change the orientation of the polarization ellipse (Fig. 8.4, p. 157), but will leave the shape of the ellipse unaltered.

The above results will enable us to determine the depolarization for lateral scattering.

8.3. Depolarization by a Rough Surface in the Plane of Incidence

As the general solution for a two-dimensionally rough surface (10), Section 3.2, reduces for $\theta_3 = 0$ to the solution of a one-dimensionally rough surface, we may use (40), Section 3.1, obtaining

$$q = \frac{\rho^+}{\rho^-} = -\frac{F_2 I + e^+(L)}{F_2 I - e^-(L)} \tag{1}$$

where F_2 and $e(L)$ are given by (45) and (41), Section 3.1, respectively, and

$$I = \int_{-L}^{L} e^{i\mathbf{v}\cdot\mathbf{r}}\,dx \tag{2}$$

From (1) it follows that the "edge effect" (Section 3.3) is responsible for the depolarization of longitudinally scattered waves in the side lobes if the surface is not very long. If $L \gg \lambda$, then as we know from Section 3.3, $e(L) \ll F_2 I$ and by (1) the depolarization factor again reduces to

$$q = -1 \tag{3}$$

Since (1) is based on the general solution (40), Section 3.1, it follows that (3) will hold for any perfectly conducting surface, no matter whether it is periodic, random or (as in the preceding section) smooth. However, this result applies only to surfaces with radii of curvature much larger than the wavelength of the incident radiation, and our result (3) in essence only expresses the fact that for a perfectly conducting surface $R^+ = -R^-$. For radii of curvature comparable to, or smaller than the wavelength, we must restrict ourselves to a more particular model of a rough surface for which the known solution is sufficiently exact to be applied to polarization problems. This practically leaves only the Deryugin model of Section 4.3D, and the Twersky model of Section 6.1, and of these only the latter leads to a solution in a closed form. If we again keep only to scattering in the plane of incidence, condition (18) of Section 8.1 will be satisfied and we may again determine q as E_2^+/E_2^-.

Using Twersky's results [1957a] for the reflection coefficient and differential scattering cross-section in the case of perfectly conducting semicylindrical or hemispherical bosses on a plane, we find for semicylindrical bosses of radius $a \ll \lambda$, with the aid of (19), Section 8.1,

$$|q|^2 = \frac{1+\cos 2\theta_1}{2-\cos 2\theta_1} \tag{4}$$

valid for $\theta_1 = \theta_2$ and θ_1 not near $\pi/2$.

As Twersky's results are calculated for power, not fieldstrength, they do not include (explicitly) the phase of the scattered field. We therefore obtain only $|q|$, not q, i.e. the polarization rectangle, but not the ellipse inscribed in it. In any case, (4) shows that our result (3) is not valid when $a \ll \lambda$, as might be expected from the limits of validity of the Kirchhoff approximation. For $a \gg \lambda$, we obtain from Twersky's results [1957a]

$$|q| = 1 \tag{5}$$

in agreement with (3).

For a surface consisting of hemispherical bosses of radius a, we find from Twersky's results [1957a] for $\theta_1 = \theta_2$ when $a \ll \lambda$ and θ_1 is not near $\pi/2$,

$$|q|^2 = \frac{3+4\pi Na^2(ka)^4(1+4\sin^2\theta_1)/\cos\theta_1}{3+4\pi Na^2(ka)^4\cos\theta_1} \tag{6}$$

where N is the number of bosses per unit area. For $a \gg \lambda$, Twersky's results [1957a] again make $|q|$ tend to unity in agreement with (3).

For a random surface, ρ^+ and ρ^- are both random, but if the surface is a perfect conductor (and complies to the conditions of Section 3.3), these two random quantities are fully correlated by the relation

$$\rho^+ = -\rho^- \tag{7}$$

which is a consequence of $R^+ = -R^- = 1$ for any local angle of incidence and leads to the constant value $q = -1$. If the surface is not a perfect conductor, so that R^+ and R^- will vary over the surface in accordance with the local angle of incidence (and even with the electrical properties of the scattering medium if it is not homogeneous), q will be a random quantity, equal as before to the ratio of the two random quantities ρ^+ and ρ^-. However, for finite conductivity these quantities will not be related by the "deterministic" relation (7), they may be statistically correlated. In the general case one would then have to determine the correlation between ρ^+ and ρ^- in terms of R^+, R^- and the statistical properties of the surface and then calculate q as the ratio of two correlated random variables. In principle, one could perform this calculation starting from (33), Section 3.1, but this would be a very difficult task unless some short-cut procedure is found. Also, the integrations necessary to find the distribution of two correlated random variables are usually rather involved.

In view of these difficulties we shall assume that the field scattered into a direction θ_2, θ_3 is due only to local specular reflections by favourably oriented tangent plane sheets; we shall neglect the diffraction effects (side lobes) produced by tangent sheets reflecting specularly into other directions, but contributing to the considered direction θ_2, θ_3 by their side lobes. For a surface having irregularities of dimensions and radii of curvature very much larger than the wavelength of the incident radiation, this will evidently be a good approximation, since in this case the surface may be considered to consist of large, randomly oriented plane mirrors. Since the field reflected by a finite sheet is proportional to its area, the depolarization caused by a number of sheets of equal electrical properties and orientation, but different areas and positions is

$$p_2 = \frac{E_2^+}{E_2^-} = \frac{\sum_j A_j E_{20}^+}{\sum_j A_j E_{20}^-} = \left|\frac{E_{20}^+}{E_{20}^-}\right| \frac{e^{i\varphi_0^+} \sum_j e^{i(\varphi_j^+ - \varphi_0^+)}}{e^{i\varphi_0^-} \sum_j e^{i(\varphi_j^- - \varphi_0^-)}} \tag{8}$$

where $E_{20} = |E_{20}| \exp i\varphi_0$ is the field reflected by any one arbitrarily selected sheet of area A_0', and $A_j = A_j'/A_0'$. Since the phase shift $\varphi_j - \varphi_0$ depends only on the position of the sheet S_j with respect to S_0, we have

$$\varphi_j^+ - \varphi_0^+ = \varphi_j^- - \varphi_0^- \tag{9}$$

so that the sums in (8) cancel and we have

$$p_2 = |p_{20}|\ e^{i \arg p_{20}} = p_{20} \tag{10}$$

Hence the field scattered by the rough surface into the direction (θ_2, θ_3) *is depolarized in the same way as the field reflected by a plane of the same electrical properties inclined in such a way that it will reflect the incident field into the same direction* (θ_2, θ_3).

For irregularities with large radii of curvature our problem therefore reduces to determining the depolarization caused by inclined planes, i.e. to finding p_2 as a function of p_1 for any given inclination of the plane (determined by the scattering geometry θ_1, θ_2, θ_3) and arbitrary electrical properties.

The general solution of this remaining problem will be derived in Section 8.4; in the present section, where we are interested only in scattering in the plane of incidence ($\theta_3 = 0$), the above result immediately yields

$$q = \frac{R^+(\vartheta)}{R^-(\vartheta)} \tag{11}$$

where

$$\vartheta = \frac{\theta_1+\theta_2}{2} \tag{12}$$

This result shows that unless $R^+ = R^-$, all polarizations except purely vertical ($p_1 = \infty$) or purely horizontal ($p_1 = 0$) will be depolarized, as we have already seen in Section 8.2. The only case of practical importance when $R^+ = R^-$, is the case of finite conductivity at grazing incidence; we then have $R^+ = R^- = -1$, hence $q = 1$ and no depolarization will set in for waves scattered in the plane of incidence.

Note that (8) holds regardless of whether the surface is deterministic or random; thus (11) and (12) are the general solution for a surface with radii of curvature very much larger than the wavelength in the plane of incidence. The reason why a random surface will depolarize in the same way as a deterministic (e.g. periodic) surface is that for a random surface E_2^+ and E_2^- are random, but fully correlated, so that their ratio remains constant.

If the radii of curvature of the irregularities are small compared with the wavelength, the situation is much more complicated, since local diffraction effects will set in and there will in general be no equivalent of (8). We must therefore calculate q directly from its definition $q = E_2^+/E_2^-$, which restricts us to particular models of rough surfaces such as the Twersky model, for which we have already obtained the depolarization factors (4) and (6).

In the absence of a more general theory for irregularities with small radii of curvature we must assume that the mean scattered powers $\langle W^+\rangle$ and $\langle W^-\rangle$ are known either from theory or from measurement. If the surface is deterministic, the problem is of course immediately solved from the definition of q:

$$|q| = \frac{1}{|p_1|}\sqrt{\frac{W^+}{W^-}} \tag{13}$$

though $|q|$ gives only the polarization rectangle circumscribed about the polarization ellipse; to find the latter, the phase difference $\arg E_2^+ - \arg E_2^-$ must be known.

If the surface, and hence E_2, is random, E_2 will be Rayleigh-distributed in all non-specular directions; if moreover, the variance $D\{\xi\} \gg \lambda$, i.e. the surface is very rough, then as we know from Section 7.2, the field will be Rayleigh-distributed everywhere. Examples of rough surfaces with large irregularities but small radii of curvature are the following: the surface formed by the tree-tops of a forest at VHF; grass at SHF; roof tops including chimneys, television antennae and other sharp irregularities; inhabited areas, etc.

For the sake of briefness we write

$$|E_2| = A \tag{14}$$

and denote its mean square value by

$$s = \langle A^2 \rangle = \langle E_2 E_2^* \rangle = 120\,\pi \langle W \rangle \tag{15}$$

The distribution of A will then be

$$w(A) = \frac{2A}{s}\,e^{-A^2/s} \tag{16}$$

where w, A and s are to be superscripted by $+$ or $-$.

Now if the irregularities are not all of the same shape (as in the Twersky model) so that each of them depolarizes in a different way and if we moreover assume that the irregularities are more or less symmetrical about a plane parallel to the plane of incidence (so that for reasons of symmetry there will be no cross-polarized component), than A^+ and A^- may be assumed independent. Let

$$|q| = \frac{A^+}{A^-} = Q \tag{17}$$

The distribution of Q will then be that of the ratio of two independent Rayleigh-distributed quantities. From (57), Appendix C, we have

$$W(Q) = \frac{Q}{s^+s^-}\int_0^\infty x^3 \exp\left[-\frac{x^2(s^+ + s^-Q)}{2s^+s^-}\right]dx =$$

$$= \frac{2s^+s^-Q}{(s^+ + s^-Q^2)^2} = \frac{2G^2Q}{(G^2+Q^2)^2} \tag{18}$$

where

$$G^2 = \frac{s^+}{s^-} = \frac{\langle W^+ \rangle}{\langle W^- \rangle} \tag{19}$$

is the ratio of the mean powers scattered with the two polarizations.

The distribution function of Q is

$$P\{Q<Q_0\} = \int_0^{Q_0} \frac{2G^2 Q \,\mathrm{d}Q}{(G^2+Q^2)^2} = \frac{1}{2}\left[1-\cos\left(2\arctan\frac{Q_0}{G}\right)\right] \tag{20}$$

Fig. 8.5 shows the distribution function of the normalized quantity Q/G:

$$P\left\{\frac{Q}{G}<Q_0\right\} = \tfrac{1}{2}[1-\cos(2\arctan Q_0)] \tag{21}$$

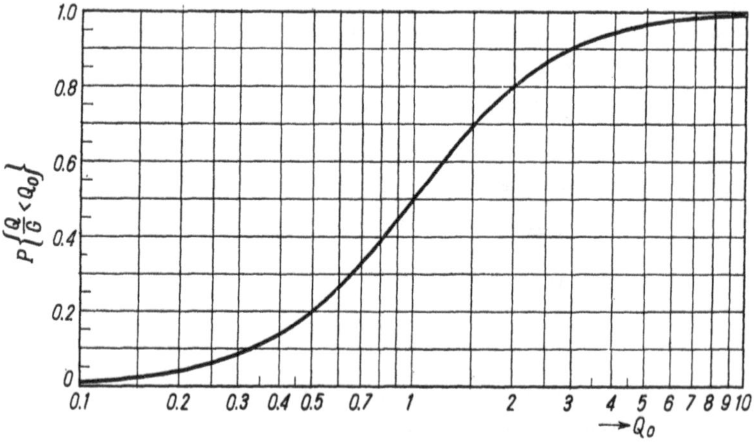

Fig. 8.5. The distribution function of Q/G for scattering by a surface with finite conductivity.

The mean value of Q is from (8), Appendix C,

$$\langle Q\rangle = \int_0^{\infty} \frac{2G^2 Q^2 \,\mathrm{d}Q}{(G^2+Q^2)^2} = \frac{\pi G}{2} \tag{22}$$

Finally, since

$$\arg q = \arg \rho^+ - \arg \rho^- \tag{23}$$

and the arg ρ^+ are distributed uniformly from $-\pi$ to $+\pi$, $\arg q$ will be Simpson-distributed from -2π to 2π, which, as explained in Section 7.3, is exactly equivalent to a uniform distribution from $-\pi$ to $+\pi$ (or any other chosen interval of length 2π).

Thus for a surface of large irregularities with small radii of curvature the distributions of $|q| = Q$ and $\arg q$ are given by (21) and a uniform distribution respectively, $|q|$ and $\arg q$ being independent.

For a surface with large radii of curvature, q is constant and given by (11) and (12).

Regardless of the conductivity or the kind of surface, purely horizontal ($p_1 = 0$) or purely vertical ($p_1 = \infty$) polarization is not depolarized in the plane of incidence, for when $p_1 = 0$ or ∞, then $p_2 = p_1 q = p_1$ for any (finite) value of q.

8.4. The Depolarization of Laterally Scattered Waves

We have already defined, in Chapter 3, the plane of incidence as the plane determined by $\mathbf{k}_1$ and $\mathbf{z}_0$, and the "local" plane of incidence as the plane determined by $\mathbf{k}_1$ and $\mathbf{n}$, where $\mathbf{k}_1$ lies in the direction of propagation of the incident wave, $\mathbf{z}_0$ is a unit vector normal to the xz plane and $\mathbf{n}$ is the normal to a local surface element, i.e. the normal to the tangent plane at the considered point (Fig. 8.6). Similarly, the scattering plane is defined as the plane $\mathbf{k}_2$, $\mathbf{z}_0$ with respect to the xy plane, whilst the "local" scattering plane, given by $\mathbf{k}_2$ and $\mathbf{n}$, is identical with the "local" plane of incidence, since Snell's law of reflection (from the tangent plane) requires $\mathbf{k}_1, \mathbf{k}_2$ and $\mathbf{n}$ to lie in the same plane.

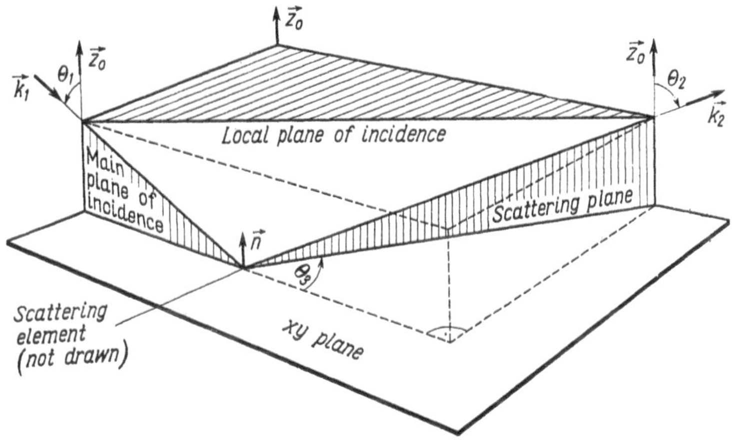

Fig. 8.6. The local plane of incidence for lateral scattering out of the main plane of incidence. The normal to the scattering element lies in the local, not the main plane of incidence.

As long as $\theta_3 = 0$, the normal to the reflecting elements lies in the plane of incidence, which is thus identical with the "local" plane of incidence. For the reasons given in preceding sections, purely horizontal or purely vertical polarization is never depolarized in the plane of incidence. This enabled us to calculate the depolarization factor q from (19), Section 8.1.

If, however, the normal to an element of the surface does not lie in the plane of incidence, so that the elementary wave will be scattered laterally out of the plane of incidence, $\mathbf{E}_1^+$ will give rise to a component $\mathbf{E}_1^-$, i.e. purely horizontal or purely vertical polarization will be depolarized. This may be seen from a simple example, where a vertically polarized incident wave is scattered with horizontal polarization, i.e. the polarization plane is in the process of scattering turned through an angle of 90°. Consider an element of surface or its tangent plane tilted about the x axis by 45° as shown in Fig. 8.7, where we are looking into the oncoming vertically polarized incident wave. We consider near grazing incidence so that we need not complicate matters by the forward-tilt of $\mathbf{E}_1$ (out of the plane of the drawing in Fig. 8.7). Let the surface be perfectly conducting. Resolving $\mathbf{E}_1$ into its normal and tangent components, the boundary conditions require $E_{2n} = E_{1n}$, $E_{2t} = -E_{1t}$, and on recomposing the components we find that $\mathbf{E}_2$ is horizontally polarized. A horizontally polarized wave will similarly be depolarized into a vertically polarized wave.†

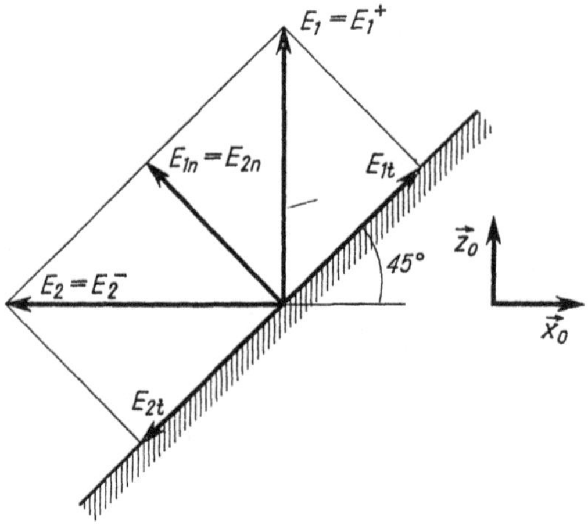

Fig. 8.7. Depolarization by a tilted plane element.

Under these conditions (18) of Section 8.1 will not hold and we shall first find the depolarization caused by a perfectly conducting plane inclined at arbitrary and independent angles with respect to the direction of propagation and with respect to the polarization plane (Fig. 8.8.). For a plane incident wave, the geometry is uniquely defined by two independent angles, the local grazing angle γ and the "polarization angle" β, which we define as the

† This example shows that oblique conducting planes such as sloping tin-plated roofs are responsible for much of the depolarization of television waves in cities.

angle concluded by $\mathbf{E}_1$ and the intersection of the wave front with the reflecting plane (measured from the plane to $\mathbf{E}_1$) as shown in Fig. 8.8.†

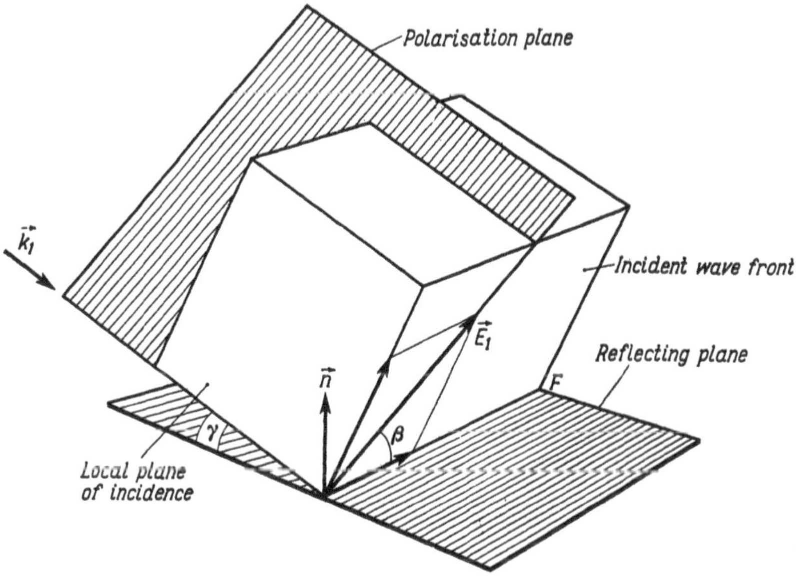

Fig. 8.8. General orientation of the direction of propagation and the polarization plane with respect to the reflecting plane.

We temporarily assume $\mathbf{E}_1$ linearly polarized, but we state nothing about the orientation of the reflecting plane with respect to the coordinate axes x, y, z ($\mathbf{z}_0$ vertical, $\mathbf{x}_0$, $\mathbf{y}_0$ horizontal), i.e. about the orientation of Fig. 8.8 in space. This enables us to proclaim the polarization plane as vertical (identical with the xz plane) or horizontal (identical with the xy plane) at our convenience and the inclination of the reflecting plane in Fig. 8.8 will change accordingly.

Let then $\mathbf{E}_1$ be horizontal, i.e. $\mathbf{E}_1 = \mathbf{E}_1^-$, and consider Fig. 8.9, where the wave fronts of the incident and reflected waves are shown simultaneously

---

† In general, the angle β will be a function of time. If the reflecting plane in Fig. 8.8 is horizontal, then

$$\tan\beta(t) = \operatorname{Re} p_1 - \operatorname{Im} p_1 \tan\omega t$$

where $t = 0$ corresponds to a moment when $\operatorname{Re} E^-$ is maximum. Thus for linear polarization ($\operatorname{Im} p_1 = 0$) we have from (8), Section 8.1 and the above equation $\beta(t) = \alpha = \text{const}$, and for circular polarization ($p_1 = \pm i$), we find $\beta(t) = \pm\omega t$. Fortunately, this complication need not worry us, as we shall temporarily assume $\mathbf{E}_1$ to be linearly polarized and then construct any rotational field from two mutually perpendicular, linearly polarized components with the necessary phase shift between them.

by turning the two planes about OF in Fig. 8.8 into the common plane of the drawing. If the plane is tilted simultaneously about both the x and y axes, one has to consider an additional angle β_2, concluded by $\mathbf{E}_2^-$ and the intersection of the wave front with the reflecting plane OF. We shall return

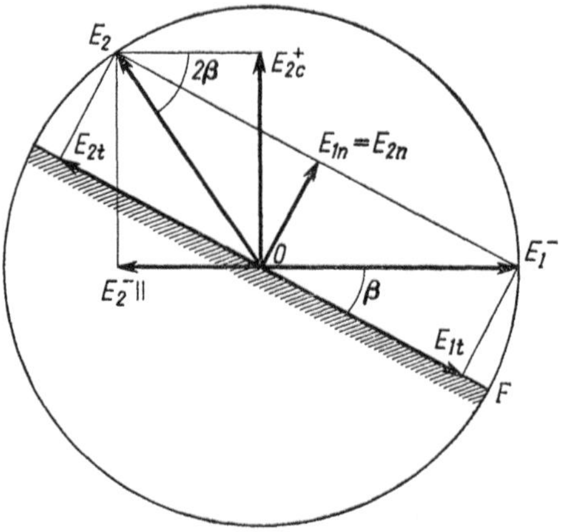

Fig. 8.9. Decomposition of E_1 and E_2. OF in this figure corresponds to OF in Fig. 8.8.

to this complication later, but unless the inclination of the plane with respect to both the x and y axes is large, $\beta_2 \approx \beta$, so that we may use Fig. 8.9 even in the general case.

Let the reflecting plane be perfectly conducting, so that $E_{1n} = E_{2n}$, $E_{1t} = -E_{2t}$; then resolving $\mathbf{E}_2$ into a component $\mathbf{E}_{2\|}$ parallel to $\mathbf{E}_1^-$ and a cross-polarized component $\mathbf{E}_{2c}$ perpendicular to $\mathbf{E}_1^-$, we find

$$E_{2\|} = -E_1 \cos 2\beta; \quad E_{2c} = E_1 \sin 2\beta \tag{1}$$

or substituting this in (15) and (16), Section 8.1,

$$\mathcal{R}_{\|}^- = -\cos 2\beta; \quad \mathcal{R}_c^- = \sin 2\beta \tag{2}$$

The polarization angle of $\mathbf{E}^+$ with respect to the same reflecting plane is

$$\beta^+ = \pi/2 - \beta \tag{3}$$

Hence

$$\mathcal{R}_{\|}^+ = \cos 2\beta; \quad \mathcal{R}_c^+ = \sin 2\beta \tag{4}$$

Comparing (2) and (4) we have

$$\mathcal{R}_c^+ = \mathcal{R}_c^-; \quad \mathcal{R}_{\|}^+ = -\mathcal{R}_{\|}^- \tag{5}$$

The first of these equalities (which states a kind of reciprocity theorem for polarization) enables us to drop the superscript $\pm$ at $\mathcal{R}_c$ as superfluous.

It follows from (2) and (4) that for a perfectly conducting plane the coefficients $\mathcal{R}$ are real.

Note also that we simply "proclaimed" the polarization plane in Fig. 8.8 as horizontal and the definition of β does not depend on any polarization; therefore (2) and (4) will apply to the reflection by any perfectly conducting plane with the polarizations "+" and "−" mutually perpendicular, but otherwise arbitrarily defined, and with β the angle concluded by the incident $\mathbf{E}_1^-$ and the intersection of the (plane) wave front with the reflecting plane, i.e. by $\mathbf{E}_1^-$ and its "locally horizontal" component.

For example, in the case of TV waves reflected by a conducting roof (footnote, p. 168) and the example of Fig. 8.7, vertical and horizontal polarization is defined with respect to the surface of the earth and $\beta = 45°$ so that (2) and (4) yield $\mathcal{R}_c = 1$, $\mathcal{R}_{||}^{\pm} = 0$ in agreement with our previous result. For $\beta = 0$, we find $\mathcal{R}_c = 0$, whilst $\mathcal{R}_{||}^{+}$ and $\mathcal{R}_{||}^{-}$ reduce to the Fresnel coefficients $+1$ and -1 respectively; for $\beta = \pi/2$ (vertical plane), the roles of $\mathcal{R}^+$ and $\mathcal{R}^-$ are interchanged. For any β, (2) and (4) satisfy the principle of the conservation of energy

$$(\mathcal{R}_{||}^{\pm})^2 + (\mathcal{R}_c^{\pm})^2 = 1 \tag{6}$$

Substituting (2) and (4) in (14) of Section 8.1, we obtain

$$\begin{aligned} \mathcal{R}^+ &= \cos 2\beta + \frac{1}{p_1} \sin 2\beta \\ \mathcal{R}^- &= -\cos 2\beta + p_1 \sin 2\beta \end{aligned} \tag{7}$$

and substituting (2) and (4) in (20), Section 8.1, we obtain the general depolarization equation of a perfectly conducting plane inclined with respect to the "+" and "−" polarizations:

$$p_2 = \frac{\tan 2\beta + p_1}{p_1 \tan 2\beta - 1} \tag{8}$$

If the approximation $\beta_2 \approx \beta$ is not admitted, a similar calculation (of the kind that presents no major problems except keeping one's patienee through the tedious geometry) leads to the more exact formula

$$p_2 = \frac{p_1(1 - \tan\beta \tan\beta_2) + \tan\beta + \tan\beta_2}{p_1(\tan\beta + \tan\beta_2) + \tan\beta \tan\beta_2} \tag{9}$$

whilst for a plane of arbitrary conductivity one obtains

$$p_2 = \frac{p_1(R^- \tan\beta \tan\beta_2 + R^+) - R^- \tan\beta + R^+ \tan\beta}{R^- + R^+ \tan\beta \tan\beta_2 - p_1(R^- \tan\beta_2 - R^+ \tan\beta)} \tag{10}$$

It is easily verified that for $\beta_2 = \beta$ and $R^+ = 1$, $R^- = -1$, (10) reduces to (8).

For $\beta = 0$ or $\pi/2$, we find from (8)

$$p_2 = -p_1 \qquad (\beta = 0 \text{ or } \pi/2)$$

making $q = -1$ for any p_1; for all other β $(-\pi \leqslant \beta \leqslant \pi)$, the ratio $q = p_2/p_1$ depends on p_1 and it is therefore more convenient to work directly with p_2 instead of q.

For $\beta = \pi/4$

$$p_2 = \frac{1}{p_1}$$

so that horizontal polarization $(p_1 = 0)$ will be depolarized into vertical $(p_1 = \infty)$ and vice versa (cf. Fig. 8.7).

Varying p_1, we find from (8) that for vertical polarization

$$p_2 = \cot 2\beta \quad (p_1 = \infty) \tag{11}$$

whilst for horizontal polarization

$$p_2 = -\tan 2\beta \quad (p_1 = 0) \tag{12}$$

and for circular polarization

$$p_2 = \mp i \quad (p_1 = \pm i) \tag{13}$$

i.e. the sense of rotation is reversed, but the polarization remains circular. In general, if

$$p_1 = s + it \tag{14}$$

then from (8) we find

$$\operatorname{Re} p_2 = \frac{(s^2+t^2-1)\tan 2\beta + s(\tan^2 2\beta - 1)}{(1+s\tan 2\beta)^2 + t^2 \tan^2 + 2\beta} \tag{15}$$

$$\operatorname{Im} p_2 = -\frac{t(1+\tan^2 2\beta)}{(1+s\tan 2\beta)^2 + t^2\tan^2 2\beta} \tag{16}$$

It follows from (16) that $\operatorname{Im} p_2$ and $t = \operatorname{Im} p_1$ always have opposite signs; hence the sense of rotational polarization is reversed for any β.

We have thus found the generalized reflection coefficient (7) and the depolarization (8) for a perfectly conducting, arbitrarily inclined plane and for an incident wave of arbitrary polarization p_1 in terms of the polarization angle β and β_2. The next step is to express β in terms of the scattering geometry in the case of a rough surface. This could be done by analytical geometry or vector algebra; there is, however, a short-cut by which β, and hence the depolarization, may be determined directly in terms of the angles $\theta_1, \theta_2, \theta_3$ without considering the surface $\zeta(x, y)$. Consider Fig. 8.10, in which a wave is incident along AO at an angle of incidence θ_1. The wave front of the incident wave is given by the plane EVW. For horizontal polarization, E_1 lies along OE, for vertical polarization it lies along OV, for circular polarization, EVW is the polarization circle. At O, the wave strikes a conducting plane (not shown in the drawing) inclined in such a way that the wave will be specularly reflected to B, given by the scattering angles θ_2 and θ_3. Then the ("local") plane of incidence with respect to the inclined reflecting plane must be $OACB$; the

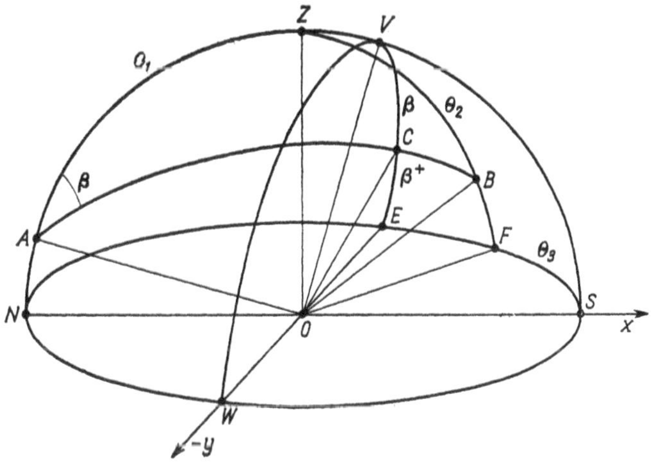

Fig. 8.10. Calculation of β by spherical trigonometry. $NWSE$ is the xy plane, AO the incident ray with $ECVW$ the wave front. At O there is a plane element (not shown) oriented so that the ray AO is reflected to B.

vector $\mathbf{E}_1^-$, lying along OE, concludes an angle $\beta^+ = \widehat{CE}$ with this plane and hence $\beta = \pi/2-\beta^+ = \widehat{VC}$. This is easily calculated by spherical trigonometry and yields†

$$\sin\beta = \frac{\sin\theta_3 \sin\theta_2}{\sqrt{[1-(\cos\theta_1\cos\theta_2-\sin\theta_1\sin\theta_2\cos\theta_3)^2]}} \tag{17}$$

If one wishes to use also the angle β_2, this may be determined as follows. It will be seen from Fig. 8.1 (p. 154) that the vectors $\mathbf{e}_1$ and $\mathbf{e}_2$ make an angle θ_3, and since $\mathbf{e}_1^-$ is perpendicular to $\mathbf{x}_0$,

$$\mathbf{e}_2^- = \mathbf{e}_1^-\cos\theta_3 - \mathbf{x}_0\sin\theta_3 \tag{18}$$

If we direct a unit vector $\mathbf{t}$ along the intersection of the wave front with the reflecting plane (OF in Fig. 8.8, p. 169), then

$$\cos\beta_2 = \mathbf{e}_2^-\cdot\mathbf{t} = \mathbf{e}_1^-\cdot\mathbf{t}\cos\theta_3 - \mathbf{x}_0\cdot\mathbf{t}\sin\theta_3 \tag{19}$$

Similarly,

$$\cos\beta = \mathbf{e}_1^-\cdot\mathbf{t} \tag{20}$$

and $\mathbf{x}_0\cdot\mathbf{t}$ is found from Fig. 8.10 as the cosine of the angle $\widehat{FS}$:

$$\mathbf{x}_0\cdot\mathbf{t} = \sin\beta\cos\theta_1 \tag{21}$$

† In astronomic terminology, we are calculating the right ascension of the "star" B of given zenith distance (θ_2) and azimuth ($-\theta_3$); the geographic latitude is $90°-\theta_1$.

Substituting (21) and (20) in (19), we obtain the required relation for β_2:

$$\cos\beta_2 = \cos\beta\cos\theta_3 - \sin\beta\cos\theta_1\sin\theta_3 \tag{22}$$

Thus for a smooth conducting plane inclined in such a way that it will reflect the radiation incident from $A(\theta_1)$ to $B(\theta_2, \theta_3)$, the depolarization of p_1 to p_2 is given by (8) [or more exactly by (9)] with β given by (17) [and β_2 by (22)].†

In general, a smooth plane of arbitrary conductivity, inclined in such a way as to reflect the wave incident from $A(\theta_1)$ to $B(\theta_2)$ will depolarize the wave from p_1 to p_2 according to (10), p. 171, with β and β_2 given by (17) and (22) respectively.

A rough surface will by (10), p. 163 depolarize in the same way, provided it is "locally flat", i.e. that its radii of curvature are sufficiently large for the basic Kirchhoff approximation (11) and (12) p. 20, or (1) and (2), pp. 28–29, to be valid. For surfaces of this kind the general depolarization formula (10), p. 171, or some of its simpler special cases may be used in practical applications. Thus for example, it should be possible to eliminate FM distortion and TV "ghost pictures" caused by certain types of reflection by using circular polarization.‡

For the special case of a perfectly conducting rough surface the general expressions simplify and the depolarization caused by such a surface may be summarized as follows:

Let a wave with polarization factor p_1 strike a rough surface (or any other scatterer) with radii of curvature large compared to the wavelength at an angle of incidence θ_1; then the polarization factor of a wave scattered into the direction θ_2, θ_3 will be given by (9), *or approximately by*

$$p_2 = \frac{\tan 2\beta + p_1}{p_1\tan 2\beta - 1} \tag{18}$$

where β is given by (17) *and β_2 by* (22).

For a scatterer other than a rough surface, we are free to choose what we define as "horizontal" or "—" polarization; our choice does not affect (18), as the scattering geometry θ_1, θ_2, θ_3 will change accordingly (we must of course choose $\mathbf{E}_1$ perpendicular to the direction of propagation $\mathbf{k}_1$).

---

† For an element of surface with inclinations determined by ζ'_x and ζ'_y, one finds

$$\tan\beta = \frac{\zeta'_y}{\sin\theta_1 - \zeta'_x\cos\theta_1}$$

‡ Beckmann, P.: On a possibility of eliminating FM distortion due to multipath propagation. *Acta Technica ČSAV* **6**, 271–285 (1961).

Equation (18) is general and includes our previous result for longitudinally scattered waves: when $\theta_3 = 0$, we find from (17) $\beta = 0$ (or π), and hence from (18) we obtain $q = p_2/p_1 = -1$.

We have thus found the depolarization caused by a "locally flat" rough surface of arbitrary conductivity and have gained a certain insight into special cases of small radii of curvature including finite conductivity. (Section 8.3.) This leaves a long way to go to the general solution of the problem of depolarization, where finite conductivity and small radii of curvature are the cases that are evidently the most interesting. From this point of view, our results are rather limited, leaving only the consolation of the Czech saying, "Even a small fish is a fish."

CHAPTER 9

CONCLUSION

We have found it possible to derive explicit expressions for the field scattered into any direction by a perfectly conducting rough surface with large radii of curvature under an arbitrary angle of incidence. For a random surface given by its two-dimensional statistical distribution (or equivalent), we have derived the statistical distribution of the field, including the mean field and mean power scattered into any direction for a given angle of incidence. We have also derived the depolarization caused by a rough surface of arbitrary conductivity.

All these results should hold when the radii of curvature of the irregularities of the rough surface are large compared to the wavelength and when mutual electromagnetic interaction among the irregularities can be neglected.

When the conductivity of the surface is finite, we have found the resulting integrals too difficult to be evaluated exactly in the general case; we have then obtained an approximate solution by averaging over the local reflection coefficients.

When the radii of curvature of the irregularities forming the rough surface are not large compared with the wavelength of the incident radiation, the above method becomes approximate and may lead to large errors when edges or sharp points are abundantly present in the surface. If the roughness is slight, one may then use a Rayleigh method, but for large roughness the error incurred in the numerical approximation of the solution consisting of infinitely many terms may be even greater than that inherent in the Kirchhoff method. However, for certain special profiles, a good approximation or a practically exact solution is possible [Deryugin, Twersky].

This leaves a gap for large irregularities with small radii of curvature and for finitely conducting surfaces. There are reasons to believe that finite conductivity will not radically alter the qualitative scattering picture except for depolarization effects. For large irregularities with small radii of curvature (such as grass for microwaves or forests for meter waves) no general and exact theory appears to be worked out at present. In all cases the statistical distribution of the amplitude and phase of the scattered field must be that of Chapter 7, but its constants ($\langle \rho \rangle$, $D\{\rho\}$ and K) must be measured where there is no theory to predict them.

In summary one may say that the theory of scattering from rough surfaces has not yet reached the point where an explicit and mathematically manageable solution can be obtained for any type of surface; such solutions are, however, obtainable in the majority of physically important cases, in particular for random surfaces with large correlation distances and periodic surfaces of almost any type.

APPENDIX A

A NOTE ON THE HELMHOLTZ INTEGRAL

LET V be a volume bounded by a surface S and let $\mathbf{A}$ be any continuous vector function. According to the Divergence Theorem,

$$\iiint_V \nabla\cdot\mathbf{A}\,\mathrm{d}V = \oiint_S \mathbf{A}\cdot\mathbf{n}\,\mathrm{d}S \tag{1}$$

where $\mathbf{n}$ is the normal to the surface.

Substituting

$$\mathbf{A} = E\,\nabla\psi \tag{2}$$

where E and ψ are continuous scalar functions, in (1) and using the vector identities

$$\nabla\cdot(E\,\nabla\psi) = \nabla E\cdot\nabla\psi + E\,\nabla^2\psi \tag{3}$$

$$\nabla\psi\cdot\mathbf{n} = \frac{\partial\psi}{\partial n} \tag{4}$$

we obtain Green's First Theorem

$$\iiint_V \nabla E\cdot\nabla\psi\,\mathrm{d}V = \oiint_S E\frac{\partial\psi}{\partial n}\,\mathrm{d}S - \iiint_V E\,\nabla^2\psi\,\mathrm{d}V \tag{5}$$

Interchanging E and ψ in this identity and subtracting the result from (5), we obtain Green's Second Theorem

$$\iiint_V [E\,\nabla^2\psi - \psi\nabla^2 E]\,\mathrm{d}V = \oiint_S \left[E\frac{\partial\psi}{\partial n} - \psi\frac{\partial E}{\partial n}\right]\mathrm{d}S \tag{6}$$

Let now E be some quantity that satisfies the wave equation

$$\nabla^2 E + k^2 E = 0 \tag{7}$$

.g. a re ctangular component of the electric field strength $\mathbf{E}$. Further let

$$\psi = \frac{\mathrm{e}^{ikr}}{r} \tag{8}$$

where r is the distance from a fixed point P to a variable point P', both inside the volume V. Then ψ will also satisfy the wave equation (7) everywhere inside V except at the point P, where it becomes infinite since $r = 0$. We therefore surround the point P by a small sphere of radius R and surface S_1. Then in the space between S_1 and S (Fig. A1) both E and ψ satisfy the wave equation (7) and therefore the left side of (6) will vanish, whilst the right side should now be integrated over both S_1 and S. (One could keep the

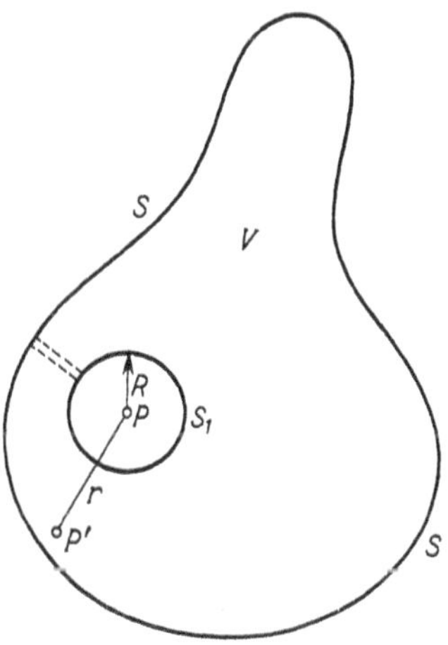

Fig. A1. Derivation of the Helmholtz integral.

entire surface simply connected by boring a "corridor" from S_1 to S as shown by the dashed lines in Fig. A1, but there is no need for this as the integrals on opposite walls of the corridor will cancel.) Thus

$$\oiint_S \left[E \frac{\partial}{\partial n} \frac{e^{ikr}}{r} - \frac{e^{ikr}}{r} \frac{\partial E}{\partial n} \right] dS + \oiint_{S_1} \left[E \frac{\partial}{\partial n} \frac{e^{ikr}}{r} - \frac{e^{ikr}}{r} \frac{\partial E}{\partial n} \right] dS_1 = 0 \quad (9)$$

Calculating $\partial/\partial n$ at $r = R$ and using the mean value theorem, the second integral becomes

$$\oiint_{S_1} [\ldots]\, dS_1 = 4\pi R^2 \left(\frac{ik}{R} - \frac{1}{R^2} \right) e^{ikR} \langle E \rangle - 4\pi R^2 \frac{e^{ikR}}{R} \left\langle \frac{\partial E}{\partial n} \right\rangle \quad (10)$$

where the angular brackets denote the average over the sphere S_1.

Now let $R \to 0$; then since

$$\lim_{R \to 0} \langle E \rangle = E(P) \quad (11)$$

we find from (10)

$$\lim_{R \to 0} \oiint_{S_1} [\ldots]\, dS_1 = -4\pi E(P) \quad (12)$$

Substituting this in (9) we obtain the Helmholtz integral

$$E(P) = \frac{1}{4\pi} \oiint_S \left(E \frac{\partial}{\partial n} \frac{e^{ikr}}{r} - \frac{e^{ikr}}{r} \frac{\partial E}{\partial n} \right) dS \tag{13}$$

giving the solution of the wave equation (7) at an interior point of a region in terms of the values of the function and its normal derivative on the boundary of the region.

The Helmholtz integral is easily applied to surfaces that are not closed; e.g. in our application to a rough surface in Chapter 3, one may regard the rough surface S as part of any closed surface S', of which only S has a non-vanishing reflection coefficient; the remaining integral over $S'-S$ is then easily shown to vanish so that there is no need to introduce the surface S'.

We shall not dwell on the point that the Helmholtz integral "over-determines" the solution, which is already uniquely determined by *either* the function *or* its normal derivative on S alone. For this point, and also a more rigorous derivation of (13), the reader is referred to any textbook on mathematical physics or the special literature on the subject;† here we shall only note that to find $E(P)$ from $(E)_S$ or $(\partial E/\partial n)_S$ alone, one has to know the Green's function of the surface.

The point we wish to pursue here is an incorrect application of the Helmholtz integral to rough-surface scatter. The integral (13) may be further developed as follows: Let E be a component of the field strength due to a point source at some point Q, so that

$$E = \frac{e^{ikr_0}}{r_0} \tag{14}$$

where r_0 is the distance of Q from a variable point on the surface S (Fig. A2). Then

$$\frac{\partial E}{\partial n} = (\nabla E) \cdot \mathbf{n} = \frac{\mathbf{r}_0 \cdot \mathbf{n}}{r_0} \frac{\partial E}{\partial r_0} \tag{15}$$

with a similar expression for $\partial\psi/\partial n$. When the distances of the source Q and the point of observation P are large compared with the wavelength, we may neglect $1/r$ and $1/r_0$ in comparison with $k = 2\pi/\lambda$ and substituting (14) and (15) in (13) we then find after elementary manipulations

$$E(P) = \frac{i}{2\lambda} \iint_S \left[\frac{\mathbf{r} \cdot \mathbf{n}}{r} + \frac{\mathbf{r}_0 \cdot \mathbf{n}}{r_0} \right] \frac{e^{i(r+r_0)}}{r r_0} dS \tag{16}$$

† BAKER, B. B. and COPSON E. T.: *The Mathematical Theory of Huyghen's Principle.* Clarendon Press, Oxford 1950.

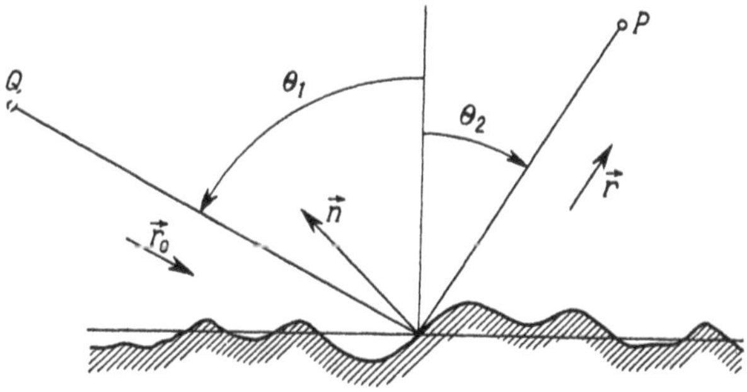

Fig. A2. Incorrect application of the Helmholtz integral to rough surfaces.

Now if S is a *plane* surface, e.g. a plane screen with slits or other apertures, then since $\mathbf{r}/r$ and $\mathbf{r}_0/r_0$ are unit vectors, we have

$$\frac{\mathbf{r}_0 \cdot \mathbf{n}}{r_0} = \cos\theta_1; \qquad \frac{\mathbf{r} \cdot \mathbf{n}}{r} = \cos\theta_2 \tag{17}$$

so that

$$E(P) = \frac{i}{2\lambda} \iint_S (\cos\theta_1 + \cos\theta_2)\, e^{i(r+r_0)} \cdot \frac{dS}{r r_0}, \tag{18}$$

a well-known formula from physical optics.†

But for a *rough* surface, where $\mathbf{n}$ is tilted with respect to the normal to the mean plane (Fig. A2), (17) will obviously not hold and the application of (18) to a rough surface will be incorrect unless the surface has very gentle slopes, a point overlooked in several papers on rough-surface scatter.

† The unfamiliar plus sign between the cosines s due to our definition of θ_2.

APPENDIX B

FORMULAE FROM THE THEORY OF BESSEL FUNCTIONS

In the course of our derivations we have used the following formulae, the proof of which will be found in any textbook of Special Functions or in books specially devoted to the theory of Bessel functions.†

The Bessel Function of the first kind (the second and third kinds do not occur in this monograph) of order m is defined as

$$J_m(x) = \sum_{s=0}^{\infty} \frac{(-1)^s}{s!\,(m+s)!} \left(\frac{x}{2}\right)^{m+2s} \tag{1}$$

where

$$(m+s)! = \Gamma(m+s+1) = \int_0^{\infty} e^{-t} t^{m+s} dt \tag{2}$$

so that (1) holds also for fractional m.

For integral m,

$$J_m(x) = (-1)^m J_{-m}(x) \tag{3}$$

From (1) we find

$$J_m(0) = \begin{cases} 1 & \text{for } m = 0 \\ 0 & \text{for integral } m \neq 0 \end{cases} \tag{4}$$

and

$$J_0'(x) = -J_1(x) \tag{5}$$

Note that when m is odd (even), $J_m(x)$ is an odd (even) function of x.

On several occasions we have used the expansion

$$e^{ix\sin\theta} = \sum_{m=-\infty}^{\infty} J_m(x)\, e^{im\theta} \tag{6}$$

from which we find

$$e^{ix\cos\theta} = \sum_{m=-\infty}^{\infty} i^m J_m(x)\, e^{im\theta} \tag{7}$$

† Watson, G. N.: *A Treatise on the Theory of Bessel Functions.* Cambridge University Press, Cambridge, 1944; Gray, A. and Mathews, G. B.: *A Treatise on Bessel Functions.* Macmillan, London, 1931; McLachlan, N. W.: *Bessel Functions for Engineers.* Clarendon Press, Oxford, 1934.

For integral m,

$$J_m(x) = \frac{1}{2\pi} \int_{\theta}^{\theta+2\pi} e^{ix\sin t - imt}\, dt \tag{8}$$

where θ is an arbitrary constant.

From (7) and (3) we find

$$\frac{1}{2\pi} \int_{\theta}^{\theta+2\pi} e^{ix\sin t + imt}\, dt = (-1)^m J_m(x) \tag{9}$$

$$\frac{1}{2\pi} \int_{\theta}^{\theta+2\pi} e^{ix\cos t + imt}\, dt = i^m J_m(x) \tag{10}$$

Formulae (8) to (10) hold for integral m only; for any, including fractional m,

$$J_m(x) = \frac{1}{2\pi} \int_{\theta}^{\theta+2\pi} e^{ix\sin t - imt}\, dt - \frac{\sin m\pi}{\pi} \int_0^{\infty} e^{-mt - x\sinh t}\, dt \tag{11}$$

Hence

$$\frac{1}{2\pi} \int_{\theta}^{\theta+2\pi} e^{ix\cos t + imt}\, dt = i^m \left[J_m(x) + \frac{\sin m\pi}{\pi} \int_0^{\infty} e^{-mt - x\sinh t}\, dt \right] \tag{12}$$

In Chapter 5 we have used the formulae

$$\frac{1}{2\pi} \int_{\theta}^{\theta+2\pi} e^{\pm ix\cos\varphi \pm iy\sin\varphi}\, d\varphi = J_0(\sqrt{[x^2+y^2]}) \tag{13}$$

and

$$\int_0^{\infty} J_0(v\tau)\, e^{-m\tau^2}\, \tau\, d\tau = \frac{1}{2m} e^{-v^2/4m} \qquad (\mathrm{Re}\, m > 0) \tag{14}$$

The modified Bessel function of the first kind of order m is defined as

$$I_m(x) = i^{-m} J_m(ix) \tag{15}$$

or

$$I_m(x) = \sum_{s=0}^{\infty} \frac{1}{s!\,(m+s)!} \left(\frac{x}{2}\right)^{m+2s} \tag{16}$$

From this it follows that $I_m(x)$ is real when x is real, is an increasing function of x, and is even or odd with respect to x as m is even or odd.

In analogy to $J_m(x)$ we have for integral m

$$I_m(0) = \begin{cases} 1 & \text{for } m = 0 \\ 0 & \text{for } m \neq 0 \end{cases} \tag{17}$$

but unlike (3) we now have

$$I_{-m}(x) = I_m(x) \tag{18}$$

regardless of whether m is even or odd.

In Chapter 7 we apply the formulae

$$I_m(x) = \frac{(-1)^m}{2\pi} \int_{\theta}^{\theta+2\pi} e^{-x\cos t + imt}\, dt \tag{19}$$

$$e^{-x\cos\theta} = \sum_{m=-\infty}^{\infty} (-1)^m I_m(x)\, e^{im\theta} \tag{20}$$

Modified Bessel functions of fractional order do not occur in the main text.

APPENDIX C

FORMULAE FROM PROBABILITY THEORY

LET ζ be a variable assuming random values z; let the probability

$$P\{z < \zeta < z+\mathrm{d}z\} \text{ (per element d}z)$$

be given by a function $p(z)$; this function is called the "probability density" or simply the "distribution" of ζ.

The distributions most often met in this monograph are the normal (Gaussian) distribution (Fig. C1)

$$p(z) = \frac{1}{\sigma\sqrt{(2\pi)}}\, \mathrm{e}^{-(z-a)^2/2\sigma^2} \tag{1}$$

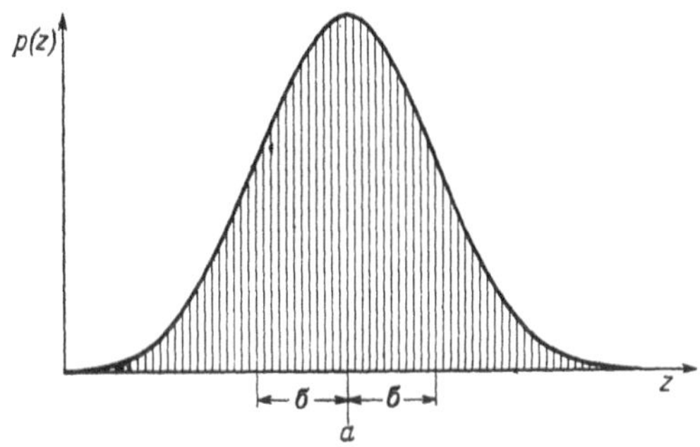

Fig. C1. The normal distribution.

the uniform distribution (Fig. C2)

$$p(z) = \begin{cases} \dfrac{1}{b-a} & \text{for} \quad a < z < b \\ 0 & \text{for} \quad z < a \text{ and } z > b \end{cases} \tag{2}$$

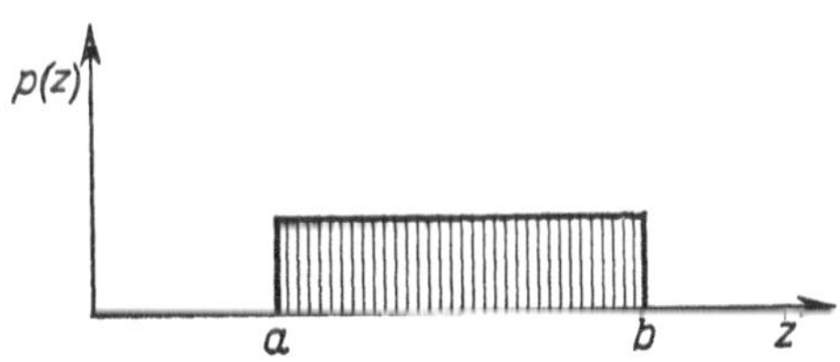

Fig. C2. The uniform distribution.

the Simpson distribution (Fig. C3)

$$p(z) = \begin{cases} \dfrac{z-2a}{(b-a)^2} & \text{for} \quad 2a \leqslant z \leqslant a+b \\ \dfrac{2b-z}{(b-a)^2} & \text{for} \quad a+b \leqslant z \leqslant 2a \\ 0 & \text{for} \quad z \leqslant 2a \text{ and } z \geqslant 2b \end{cases} \tag{3}$$

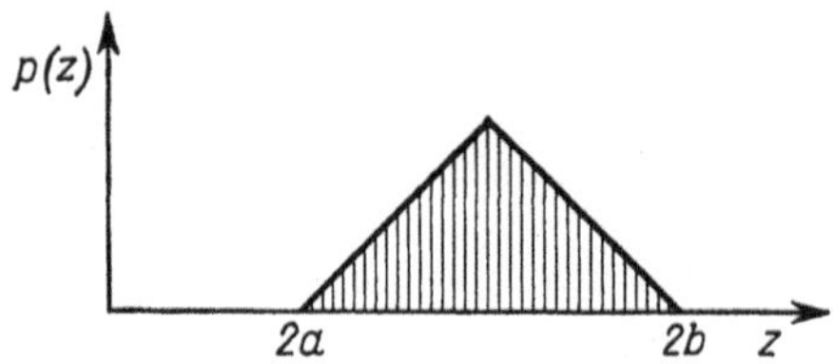

Fig. C3. The Simpson distribution.

and the Rayleigh distribution (Fig. C4)

$$p(z) = \begin{cases} \dfrac{z}{\sigma^2} e^{-z^2/2\sigma^2} & \text{for} \quad z > 0 \\ 0 & \text{for} \quad z < 0 \end{cases} \tag{4}$$

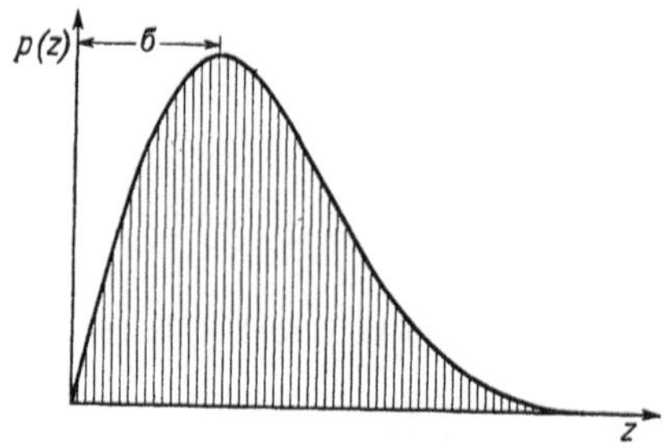

Fig. C4. The Rayleigh distribution.

The probability density $p(z)$ is always non-negative and bounds unit area.

The distribution function $F(z_0)$ is defined as

$$F(z_0) = P\{\zeta < z_0\} = \int_{-\infty}^{z_0} p(z)\,dz \tag{5}$$

Hence

$$F(-\infty) = 0, \quad F(\infty) = \int_{-\infty}^{\infty} p(z)\,dz = 1 \tag{6}$$

The distribution function is always non-negative and a non-decreasing function of z_0.

Note that in communication engineering it is customary to give the probability with which a random variable (e.g. signal strength) will *exceed* a given value. This probability is related to the distribution function by

$$P\{\zeta > z_0\} = 1-P\{\zeta < z_0\} = 1-F(z_0) \tag{7}$$

The mean value of ζ is

$$\langle\zeta\rangle = \int_{-\infty}^{\infty} z\,p(z)\,\mathrm{d}z \tag{8}$$

and the mean value of a function of ζ is

$$\langle f(\zeta)\rangle = \int_{-\infty}^{\infty} f(z)\,p(z)\,\mathrm{d}z \tag{9}$$

If A is a constant,

$$\langle A\zeta\rangle = A\langle\zeta\rangle; \quad \langle A+\zeta\rangle = A+\langle\zeta\rangle \tag{10}$$

and

$$\langle\zeta_1+\zeta_2+\ldots\rangle = \langle\zeta_1\rangle+\langle\zeta_2\rangle+\ldots \tag{11}$$

regardless of whether the ζ_i are mutually independent or not. Also,

$$\left\langle\int g(x,\zeta)\,\mathrm{d}x\right\rangle = \int\langle g(x,\zeta)\rangle\,\mathrm{d}x \tag{12}$$

where the mean value is taken with respect to ζ.

The mean values of (1), (2), (3), and (4) are a; $(a+b)/2$; $a+b$; $\sigma\sqrt{(\pi/2)}$; respectively.

The variance of a random variable is defined as

$$D\{\zeta\} = \langle(\zeta-\langle\zeta\rangle)^2\rangle \tag{13}$$

i.e. it is the average square of the deviation from the mean value. The variance is characteristic of the range of the values that the random variable assumes: a large variance implies widely scattered values; zero variance implies a single value assumed by the variable, which is thus no longer random.

The square root of the variance

$$\sigma = \sqrt{(D\{\zeta\})} \tag{14}$$

is called the "standard deviation"; if the mean value of ζ is zero, then the standard deviation equals the root mean square value, as may be seen from (13) and (14).

Removing the brackets in (13) by rule (11) we find the important formula

$$D\{\zeta\} = \langle\zeta^2\rangle-\langle\zeta\rangle^2 \tag{15}$$

or

$$D\{\zeta\} = \int_{-\infty}^{\infty} z^2 p(z)\,\mathrm{d}z - \langle\zeta\rangle^2 \tag{16}$$

The variance of the distribution (1) is σ^2.

If A is a constant, then

$$D\{A\zeta\} = A^2 D\{\zeta\} \tag{17}$$

If $\zeta_1, \zeta_2, \ldots$ are independent random variables (cf. below), then

$$D\{\zeta_1+\zeta_2+\ldots\} = D\{\zeta_1\}+D\{\zeta_2\}+\ldots \tag{18}$$

Let ζ and η be two random variables; then

$$P\{z<\zeta<z+\mathrm{d}z;\ y<\eta<y+\mathrm{d}y\} = p_2(z, y)\,\mathrm{d}z\,\mathrm{d}y \tag{19}$$

where $p_2(z, y)$ is the two-dimensional probability density of the random variables ζ and η. Two variables ζ and η are called independent if

$$p_2(z, y) = p(z)\,w(y) \tag{20}$$

where

$$p(z) = \int_{-\infty}^{\infty} p_2(z, y)\,\mathrm{d}y; \quad w(y) = \int_{-\infty}^{\infty} p_2(z, y)\,\mathrm{d}z \tag{21}$$

are the (one-dimensional) probability densities or ζ and η respectively.

If ζ and η are dependent, i.e. their two-dimensional probability density $p_2(z, y)$ cannot be expressed in the form (20), they may be correlated. Their covariance is defined as

$$K\{\zeta, \eta\} = \int_{-\infty}^{\infty}\int_{-\infty}^{\infty} zy\,p_2(z, y)\,\mathrm{d}z\mathrm{d}y - \langle\zeta\rangle\langle\eta\rangle \tag{22}$$

and their correlation coefficient is

$$C_{\zeta\eta} = \frac{K(\zeta, \eta)}{\sqrt{(D\{\zeta\}\,D\{\eta\})}} = \frac{\langle\zeta\eta\rangle-\langle\zeta\rangle\langle\eta\rangle}{\sigma_\zeta\sigma_\eta} \tag{23}$$

$|C| = 1$ implies full correlation (for $C = +1$, ζ and η will vary in conformity and for $C = -1$ in opposition); $C = 0$ implies no correlation between ζ and η.

Note that independence is stronger than non-correlation: it is easily verified that independent variables must always be uncorrelated, but uncorrelated variables may still be interdependent, e.g. if $\zeta = \sin t$, $\eta = \cos t$, where t is distributed uniformly from 0 to 2π, then from (23) we find $C_{\zeta\eta} = 0$, although ζ and η are obviously not independent, since $\zeta = \sqrt{(1-\eta^2)}$.

For any two random variables ζ and η

$$\langle\zeta\eta\rangle = \langle\zeta\rangle\langle\eta\rangle + K\{\zeta, \eta\} \tag{24}$$

$$D\{\zeta\pm\eta\} = D\{\zeta\}+D\{\eta\}\pm 2K\{\zeta, \eta\} \tag{25}$$

For a complex random variable

$$\rho = \mathrm{Re}\,\rho + i\,\mathrm{Im}\,\rho \tag{26}$$

we define

$$\langle\rho\rangle = \langle\mathrm{Re}\,\rho\rangle + i\langle\mathrm{Im}\,\rho\rangle \tag{27}$$

and

$$D\{\rho\} = \langle|\rho - \langle\rho\rangle|^2\rangle = D\{\mathrm{Re}\,\rho\} + D\{\mathrm{Im}\,\rho\} \tag{28}$$

On expressing the absolute value by means of complex conjugate quantities and using (11) we obtain the important formula

$$D\{\rho\} = \langle\rho\rho^*\rangle - \langle\rho\rangle\langle\rho^*\rangle \tag{29}$$

If follows from (27) that

$$\langle\rho^*\rangle = \langle\rho\rangle^* \tag{30}$$

The characteristic function $\chi(v)$ of a distribution $p(z)$ is

$$\chi(v) = \int_{-\infty}^{\infty} p(z)\, e^{ivz}\, dz \tag{31}$$

Hence

$$|\chi(v)| \leqslant 1; \quad \chi(0) = 1 \tag{32}$$

$$\chi(-v) = \chi^*(v) \tag{33}$$

$$\langle\zeta\rangle = \frac{1}{i}\chi'(0) \tag{34}$$

The characteristic function of the normal distribution (1) is

$$\chi(v) = e^{iav - \sigma^2 v^2/2} \tag{35}$$

If the characteristic functions associated with the random variables $\zeta_1, \zeta_2, \ldots$ are $\chi_1, \chi_2, \ldots$, and the characteristic function of

$$\zeta = \zeta_1 + \zeta_2 + \ldots$$

is χ, then

$$\chi = \chi_1\ \chi_2 \cdots \tag{36}$$

The characteristic function $\chi_2(v_1, v_2)$ of a two-dimensional distribution $p_2(z, y)$ is defined as

$$\chi_2(v_1, v_2) = \int_{-\infty}^{\infty}\int_{-\infty}^{\infty} p_2(z, y)\, e^{iv_1 z + iv_2 y}\, dz\, dy \tag{37}$$

If ζ and η are independent, then

$$\chi_2(v_1, v_2) = \chi_\zeta(v_1)\,\chi_\eta(v_2) \tag{38}$$

Let ζ and η be normally distributed with mean value zero and variance σ^2; let their correlation coefficient be C; then their two-dimensional normal distribution is

$$p_2(z, y) = \frac{1}{2\pi\sigma^2\sqrt{(1-C^2)}} \exp\left[-\frac{z^2-2Czy+y^2}{2\sigma^2(1-C^2)}\right] \tag{39}$$

and its characteristic function is from (37)

$$\chi_2(v_1, v_2) = \exp\left[-\tfrac{1}{2}\sigma^2(v_1^2+2Cv_1v_2+v_2^2)\right] \tag{40}$$

and if $v_1 = -v_2 = v$,

$$\chi_2(v, -v) = \exp\left[-\sigma^2 v^2(1-C)\right] \tag{41}$$

Let $\zeta(t)$ be a function of time and let it assume random values z with a two-dimensional probability density $p_2(z_1, z_2;\ t_1, t_2)$, where z_1 and z_2 are the values assumed at the moments t_1 and t_2 respectively. Then the correlation function of the process is defined as

$$B(t_1, t_2) = \langle\zeta(t_1)\zeta(t_2)\rangle = \int_{-\infty}^{\infty}\int_{-\infty}^{\infty} z_1 z_2 p_2(z_1, z_2;\ t_1, t_2)\,\mathrm{d}z_1\mathrm{d}z_2 \tag{42}$$

A random process is called "stationary" in a wider sense if $\langle\zeta\rangle$ and $D\{\zeta\}$ do not depend on t, and B depends only on

$$\tau = t_2 - t_1 \tag{43}$$

Then

$$B(\tau) = \langle\zeta(t)\zeta(t+\tau)\rangle \tag{44}$$

Stationary processes are under certain very general conditions "ergodic", which, loosely formulated, means that averages over the distribution at a certain moment may be replaced by time averages over a single realization of the process; in particular,

$$\langle\zeta(t)\rangle = \int_{-\infty}^{\infty} zp(z)\,\mathrm{d}z = \lim_{T\to\infty}\frac{1}{T}\int_0^T \zeta(t)\,\mathrm{d}t \tag{45}$$

$$D\{\zeta(t)\} = \int_{-\infty}^{\infty} z^2p(z)\,\mathrm{d}z - \langle\zeta\rangle^2 = \lim_{T\to\infty}\frac{1}{T}\int_0^T [\zeta(t)]^2\,\mathrm{d}t - \langle\zeta\rangle^2 \tag{46}$$

and (44) is replaced by

$$B(\tau) = \lim_{T\to\infty}\frac{1}{T}\int_0^T \zeta(t)\zeta(t+\tau)\,\mathrm{d}t \tag{47}$$

From (46) and (47) we find

$$B(0) = D\{\zeta\} + \langle\zeta\rangle^2 \tag{48}$$

and from (44) and (48) we obtain

$$\frac{B(\tau)}{B(0)} = \frac{\langle \zeta_1 \zeta_2 \rangle}{\langle \zeta_1^2 \rangle} \tag{49}$$

where $\zeta_1 = \zeta(t)$, $\zeta_2 = \zeta(t+\tau)$.

Now consider the correlation coefficient of ζ_1 and ζ_2, or the "auto-correlation coefficient" $C(\tau)$ of the process $\zeta(t)$; from (23) we find, since ζ_1 and ζ_2 are distributed with the same distribution,

$$C(\tau) = \frac{\langle \zeta_1 \zeta_2 \rangle - \langle \zeta_1 \rangle^2}{\langle \zeta_1 \rangle - \langle \zeta_1 \rangle^2} \tag{50}$$

where we have used (15).

Finally let $\zeta(t)$ be normalized so that $\langle \zeta \rangle = 0$; then

$$C(\tau) = \frac{\langle \zeta_1 \zeta_2 \rangle}{\langle \zeta_1^2 \rangle} = \frac{B(\tau)}{B(0)} \tag{51}$$

The important properties of $B(\tau)$ and $C(\tau)$ are

$$|C(\tau)| \leqslant C(0) = 1; \quad |B(\tau)| \leqslant B(0) \tag{52}$$

$$\lim_{\tau \to \infty} C(\tau) = 0; \quad \lim_{\tau \to \infty} B(\tau) = \langle \zeta \rangle^2 \tag{53}$$

$$C(-\tau) = C(\tau); \quad B(-\tau) = B(\tau) \tag{54}$$

Finally we shall mention the Central Limit Theorem, which states under what conditions the sum

$$\zeta = \zeta_1 + \zeta_2 + \dots + \zeta_n \tag{55}$$

will be normally distributed when $n \to \infty$. The least stringent conditions under which this will be the case have only been found relatively recently† and need more explanations than are in place in this appendix. We shall, therefore only (loosely) state a special case of the Central Limit Theorem namely Lyapunov's Theorem: Let $\zeta_1, \zeta_2, \dots$ in (55) be independent and distributed with the same distribution $p(z)$ having a finite mean value a and a finite variance σ^2; then as $n \to \infty$, the distribution of ζ will tend to the normal distribution with mean value na and variance $n\sigma^2$, regardless of the distribution $p(z)$ of the individual terms.

In practice the distribution of ζ approaches the normal distribution very rapidly, especially when $p(z)$ is symmetrical. For example, when $p(z)$ is uniform, the distribution of ζ differs nowhere from that of the normal distri-

† Cf. Gnedenko, B. V. and Kolmogorov, A. N.; *Limit Distributions for Sums of Independent Random Variables*, Gostekhizdat, Moscow, 1949 (English translation: Addison-Wesley Publ. Co., Cambridge, Mass., 1954).

bution by more that 0.5 % when $n = 7$. Lyapunov's theorem is needlessly stringent, for in most practical applications the ζ_t do not have to be distributed with the same distribution.

In Chapter 8 we need the distribution of the ratio of two independent random variables, i.e. of

$$\zeta = \frac{\eta}{\xi} \tag{56}$$

where η and ξ are independent and distributed with probability densities $p_\eta(z)$ and $p_\xi(z)$ respectively. Then the distribution of ζ is

$$p_\zeta(z) = \int_0^\infty t p_\xi(tz)\, p_\eta(t)\,\mathrm{d}t - \int_{-\infty}^0 t p_\xi(tz)\, p_\eta(t)\,\mathrm{d}t \tag{57}$$

In the special case that we need in (12), Section 8.3, $p_\xi(t) = p_\eta(t) = 0$ for $t < 0$ (Rayleigh distribution), leaving only the first term of (57).

APPENDIX D

THE DISTRIBUTION OF SLOPES OF A NORMALLY DISTRIBUTED SURFACE

In Part II of this book, it is often convenient to work with the slopes of a rough surface rather than with its heights above a mean level. This is particularly useful in the case of reflections from the sea or from the surface of the Moon. In Part I, however, we define a random rough surface by the distribution of its heights $\zeta(x)$ above its mean level $\langle\zeta\rangle = 0$ and by its correlation function $C(\tau)$. In particular, we investigate the scattering by a normally distributed surface (ζ distributed normally with mean value zero and standard deviation σ) with the normalized correlation function

$$C(\tau) = \mathrm{e}^{-\tau^2/T^2} \tag{1}$$

i.e. with the correlation function [cf. (18b), Section 5.2]

$$B(\tau) = \sigma^2 \mathrm{e}^{-\tau^2/T^2} \tag{2}$$

To derive the distribution of the slopes ψ, defined by

$$\tan\psi = \zeta'(x) \tag{3}$$

of a normally distributed surface, we use a theorem from the theory of stochastic processes, according to which the distribution of the derivative of a normal process is again normal with mean value zero and variance

$$D\{\zeta'\} = -B''(0) \tag{4}$$

From (2) we have

$$-B''(0) = \frac{2\sigma^2}{T^2} \tag{5}$$

Hence the distribution of ζ' is

$$w(\zeta') = \frac{T}{2\sigma\sqrt{\pi}} \exp\left(-\frac{\zeta'^2 T^2}{4\sigma^2}\right) \tag{6}$$

On transforming to the random variable ψ by (3), we obtain the required distribution of the slopes ψ:

$$p(\psi) = \frac{T}{2\sigma\sqrt{\pi}\cdot\cos^2\psi} \exp\left(-\frac{T^2\tan^2\psi}{4\sigma^2}\right) \tag{7}$$

In Part II, the author introduces the quantity β_0 defined by $\tan \beta_0 = 2\sigma/T$, so that (7) may be written as

$$p(\psi) = \frac{1}{\sqrt{\pi} \cdot \tan \beta_0 \cos^2 \psi} \exp\left(-\frac{\tan^2 \psi}{\tan^2 \beta_0}\right) \tag{8}$$

Many of the rough surfaces occurring in nature, e.g. the surface of the sea, and evidently also that of the moon, are gently undulated ($2\sigma \ll T$), i.e. their slopes ψ are mostly small; for these surfaces we may well approximate the tangents by their arguments and the cosine by unity, obtaining

$$p(\psi) \approx \frac{1}{\beta_0 \sqrt{\pi}} e^{-\psi^2/\beta_0^2} \tag{9}$$

which is the normal distribution with standard deviation $\beta_0/\sqrt{2}$.

Hence the slopes of a gently undulated normally distributed surface ($\sigma \ll T$) are themselves normally distributed. Normally distributed slopes have been measured in the case of the surface of the sea (cf. Part II, Chapter 18) and seem very probable in the case of the surface of the moon (cf. Chapter 20).

The general distribution of the slopes of a normally distributed surface is given by (7) or (8).

APPENDIX E

EVALUATION OF AN INTEGRAL

IN Chapter 7 we encountered the integral

$$I = \int_0^{2\pi} e^{-P\cos^2\varphi + Q\cos\varphi}\, d\varphi \tag{1}$$

We first write this as

$$I = e^{-P/2} \int_0^{2\pi} \exp\left[-\frac{P}{2}\cos 2\varphi + Q\cos\varphi\right] d\varphi \tag{2}$$

Using (20), Appendix B, we find on integrating term by term

$$I = e^{-P/2} \sum_{m=-\infty}^{\infty} (-1)^m I_m\left(\frac{P}{2}\right) \int_0^{2\pi} e^{2im\varphi + Q\cos\varphi} d\varphi \tag{3}$$

Evaluating the integral by (19), Appendix B, we have

$$I = 2\pi e^{-P/2} \sum_{m=-\infty}^{\infty} (-1)^m I_m\left(\frac{P}{2}\right) I_{2m}(-Q) \tag{4}$$

We may omit the negative sign of Q since Bessel functions of even order are even functions of their argument; combining the terms with $-m$ and m by (18) of Appendix B we finally obtain

$$I = 2\pi e^{-P/2} \sum_{m=0}^{\infty} (-1)^m \varepsilon_m I_m\left(\frac{P}{2}\right) I_{2m}(Q) \tag{5}$$

where $\varepsilon_m = 2$ for $m \neq 0$ and $\varepsilon_0 = 1$.

A discussion of this solution is given in [Beckmann and Schmelovsky, 1958].

An alternative evaluation of (1) is the following:

Replacing $\cos^2\varphi$ by $(1-\sin^2\varphi)$ and expanding the first factor in an exponential series, we obtain

$$I = e^{-P} \sum_{m=0}^{\infty} \frac{P^m}{m!} \int_0^{2\pi} e^{Q\cos\varphi} \sin^{2m}\varphi\, d\varphi \tag{6}$$

where the term-by-term integration is permissible owing to the uniform convergence of the series.

Using the integral representation of $I_m(Q)$†

$$I_m(Q) = \frac{2\left(\frac{Q}{2}\right)^m}{\sqrt{\pi}\cdot\Gamma(m+\frac{1}{2})} \int_0^{2\pi} e^{\pm\cos\varphi} \sin^{2m}\varphi \, d\varphi \tag{7}$$

we obtain from (6)

$$I = 2\pi e^{-P} \sum_{m=0}^{\infty} \frac{2^m \Gamma(m+\frac{1}{2})}{m!} P^m Q^{-m} I_m(Q) \tag{8}$$

This result was derived by B. Chytil‡ and has the advantage of expressing the modified Bessel function directly in the form

$$Q^{-m} I_m(Q) = i_m(Q) \tag{9}$$

used in the most extensive tables at present available.§

Since

$$\Gamma(m+\tfrac{1}{2}) = \frac{1\cdot 3\cdot 5 \ldots (2m-1)}{2^m} \sqrt{\pi} \tag{10}$$

we find from (8)

$$I = 2\pi e^{-P} \sum_{m=0}^{\infty} 2^m A_m P^m i_m(Q) \tag{11}$$

where the coefficients

$$A_m = \frac{1\cdot 3\cdot 5 \ldots (2m-1)}{2\cdot 4\cdot 6 \ldots 2m} \tag{12}$$

are tabulated in Table E1.

For large Q we may use the asymptotic formula

$$I_m(Q) \approx \frac{e^Q}{\sqrt{(2\pi Q)}} \tag{13}$$

or

$$i_m(Q) \approx \frac{e^Q}{Q^m \sqrt{(2\pi Q)}} \tag{14}$$

Substituting this in (11), we find for $Q > 2P \gg 1$

$$I \approx \sqrt{\left(\frac{2\pi}{Q}\right)} e^{Q-P} \sum_{m=0}^{\infty} A_m \left(\frac{2P}{Q}\right)^m \tag{15}$$

† Watson, G. N.: *Theory of Bessel Functions*. University Press, Cambridge, 1944.

‡ Personal communication.

§ *Tables of Bessel Functions*, vol. II, Committee of Mathem. Tables, Cambridge 1952 (contains i_m up to order and argument 20).

TABLE E1—Values of the Coefficients A_m

A_0	1.0000000	A_{16}	0.1399500
A_1	1.0000000	A_{17}	0.1358338
A_2	0.3750000	A_{18}	0.1320606
A_3	0.3125000	A_{19}	0.1285853
A_4	0.2734375	A_{20}	0.1253707
A_5	0.2460938	A_{21}	0.1223857
A_6	0.2255859	A_{22}	0.1196042
A_7	0.2094726	A_{23}	0.1170041
A_8	0.1963806	A_{24}	0.1145665
A_9	0.1854706	A_{25}	0.1122752
A_{10}	0.1761971	A_{26}	0.1101161
A_{11}	0.1681881	A_{27}	0.1080769
A_{12}	0.1611803	A_{28}	0.1061470
A_{13}	0.1549810	A_{29}	0.1043168
A_{14}	0.1494460	A_{30}	0.1025782
A_{15}	0.1444645		

a result obtainable also directly from (1) by saddle-point integration [Beckmann and Schmelovsky, 1958].

From (12) we have

$$\frac{A_m}{A_{m-1}} = 1 - \frac{1}{m} \tag{16}$$

For large m the ratio (16) will tend to unity, so that if the series (15) is broken off after the mth term, the remainder R_{m+1} may be estimated as the sum of a geometric series:

$$R_{m+1} \approx A_{m+1}\left(\frac{2P}{Q}\right)^{m+1} \frac{Q}{Q-2P} \tag{17}$$

Finally we consider the evaluation of the integral (83), Section 7.1, for $\beta \neq 0$. Performing the squares in the exponent and replacing $\sin^2\psi$ by $(1 - \cos^2\psi)$, one obtains a constant factor multiplied by the integral

$$I = \int_0^{2\pi} e^{-P\cos^2\psi + Q\cos\psi + R\sin\psi} d\varphi \tag{18}$$

where R is a further constant with respect to the integration (a function only of r, α, β, s_1, s_2).

As shown by B. Chytil in an as yet unpublished paper,† the integral (18) may be expressed by the series

† Personal communication.

$$I = 2\pi e^{-P/2} \sum_{m=0}^{\infty} (-1)^m \varepsilon_m I_m(\tfrac{1}{2}P) I_{2m}(\sqrt{[Q^2+R^2]}) \cos\left(2m \arctan \frac{R}{Q}\right) \quad (19)$$

reducing to (5) for $R = 0$. For numerical computations it is obviously quicker to perform the numerical integration of (18) directly rather than to sum the series (19).†

† Note added in proof: Some of the results of Section 7.1 and this Appendix have also been derived independently in a series of Japanese papers by M. Nagakami and colleagues; they are summarized in English in NAGAKAMI, M.: The *m*-distribution — a general formula of intensity distribution of rapid fading. *Statistical Methods in Radio Wave Propagation* (Ed. W.C. Hofman), Pergamon Press 1960. The author is indebted to Prof. Nagakami for this information and the original Japanese papers.

APPENDIX F

CURVES OF THE SCATTERED-FIELD DISTRIBUTION

CURVES of $p(u)$ as given by (35), Section 7.1, are shown in Figs. F1 to F5 for various values of B and K. The complement of the distribution function [cf. Appendix, equation (7)] is

$$P\{u > z; B, K\} = 1 - \int_0^z p(u; B, K)\,\mathrm{d}u \tag{1}$$

Curves of the probability (1) are plotted on probability paper in Figs. F6 to F10 for various values of B and K.

For comparison of theoretical with measured curves it is convenient to normalize the resulting amplitude with respect to its RMS value. From (31), Section 7.1, we have

$$r_{\mathrm{RMS}} = \sqrt{(s_1 + s_2 + \alpha^2)} \tag{2}$$

or by (32) to (34), Section 7.1,

$$u_{\mathrm{RMS}} = \sqrt{(1 + B^2)} \tag{3}$$

Hence

$$\frac{r}{r_{\mathrm{RMS}}} = \frac{u}{u_{\mathrm{RMS}}} = \frac{u}{\sqrt{(1+B^2)}} \tag{4}$$

and from (1),

$$P\left\{\frac{r}{r_{\mathrm{RMS}}} > z\right\} = 1 - \int_0^{z\sqrt{(1+B^2)}} p(u)\,\mathrm{d}u = P\{r > z\sqrt{(1+B^2)}\} \tag{5}$$

Thus if the probability (1) is known as a function of z (Figs. F6 to F10), the probability (5) may be calculated from it. This has been done in Figs. F11 to F15, where the functions $P\{r/r_{\mathrm{RMS}} > z\}$ are plotted on Rayleigh paper for various values of B and K.†

† Figures F1 to F15 are taken from BECKMANN, P., The statistical distribution of the amplitude and phase of a multiply scattered electromagnetic field, *Institute of Radio Engineering and Electronics*, Czechoslovak Academy of Sciences, Report No. 18, 1961. The curves in Figs. F1 to F10 were calculated by direct integration of the equivalent of (21), Section 7.1 and (1) of this section. This was carried out on a punch-card installation; the numerical integration was directed by Mr. Robert Vích, C.Sc.

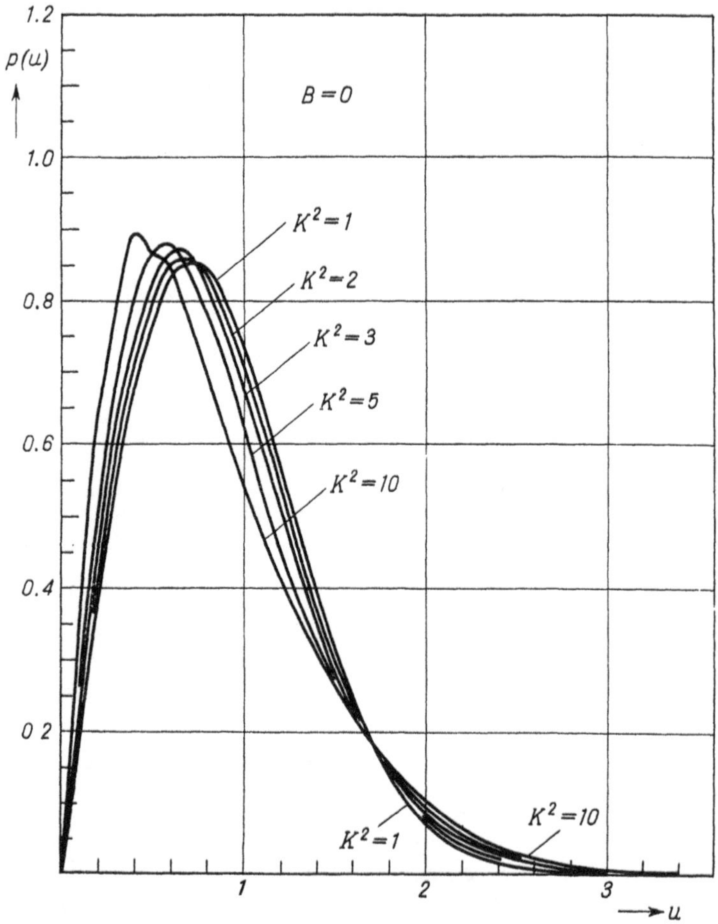

Fig. F1. Probability densities $p(u)$ for $B = 0$.

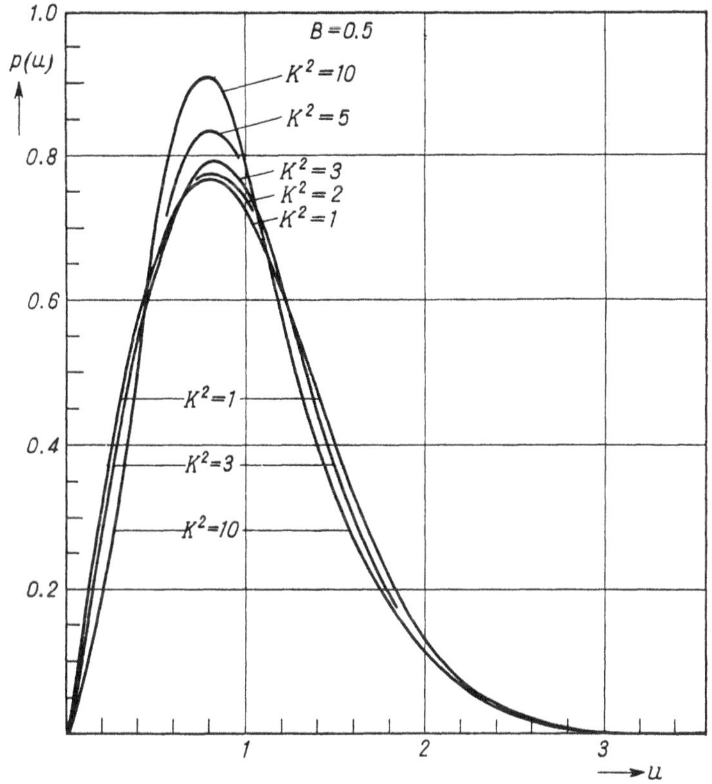

Fig. F2. Probability densities $p(u)$ for $B = 0.5$.

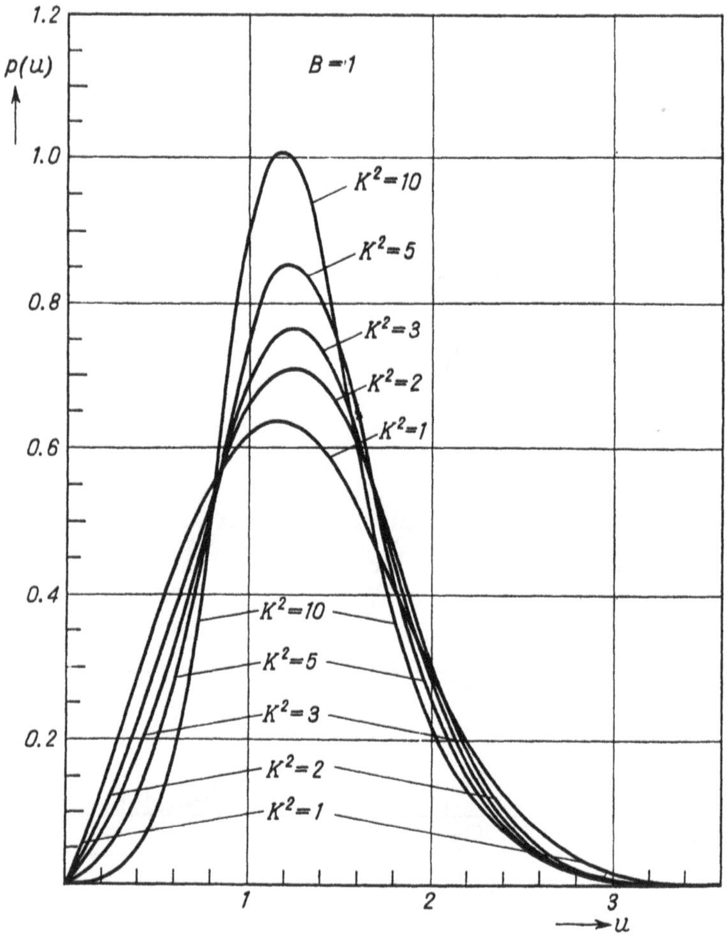

Fig. F3. Probability densities $p(u)$ for $B = 1$.

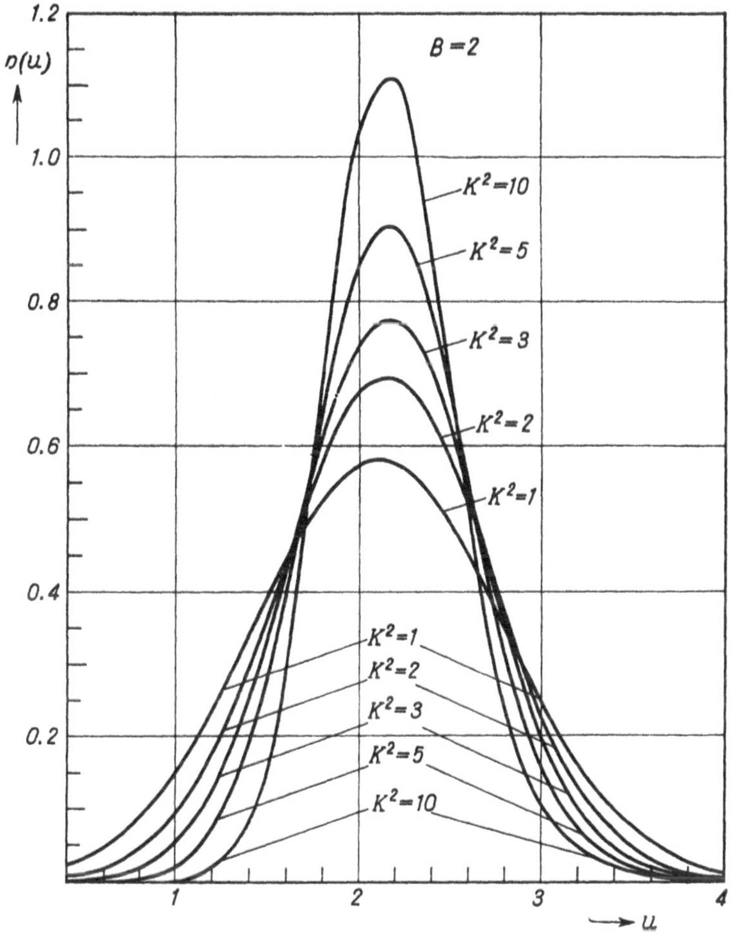

Fig. F4. Probability densities $p(u)$ for $B = 2$.

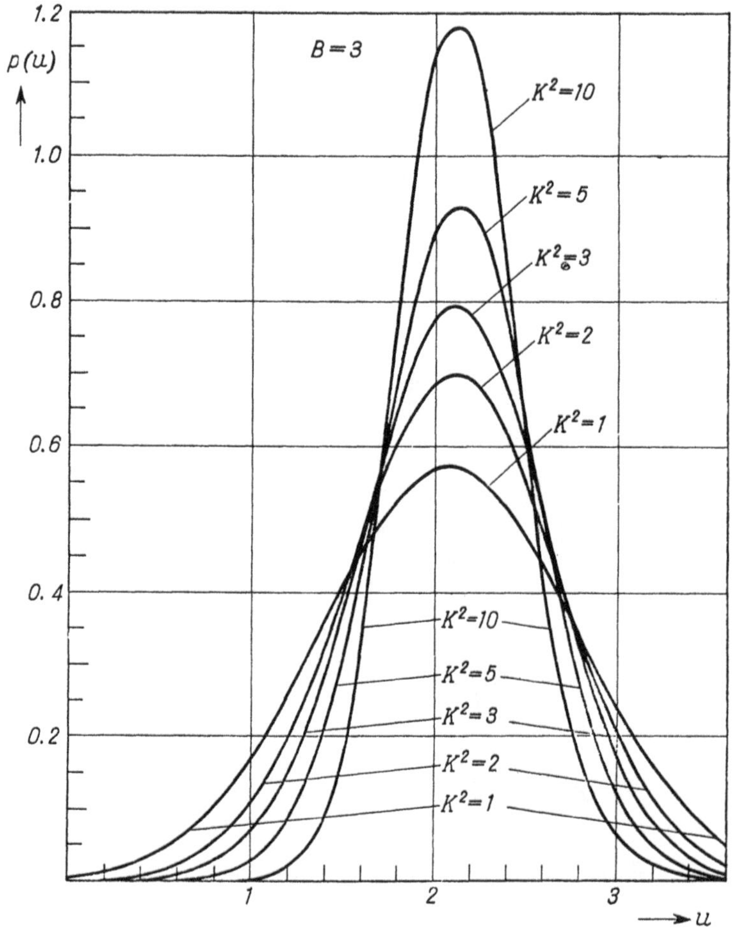

Fig. F5. Probability densities $p(u)$ for $B = 3$.

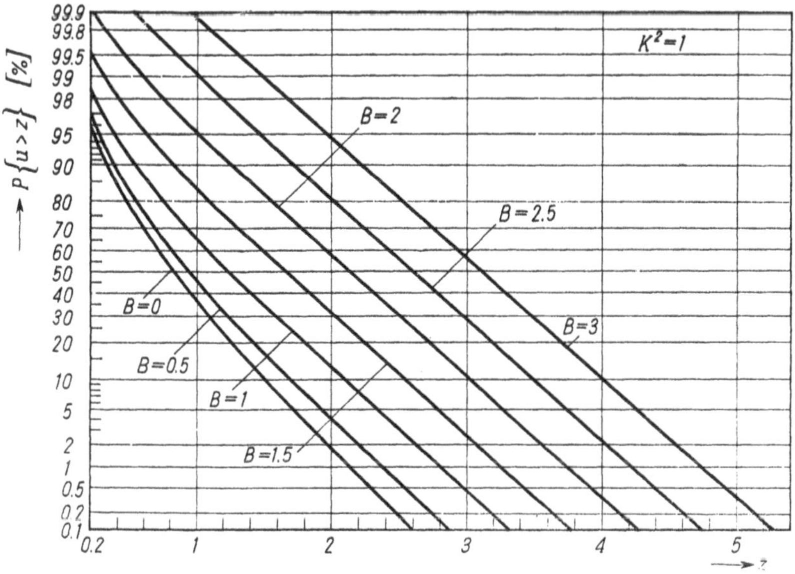

Fig. F6. Probabilities $P\{u > z\}$ for $K^2 = 1$. The curve with $B = 0$ is the Rayleigh distribution.

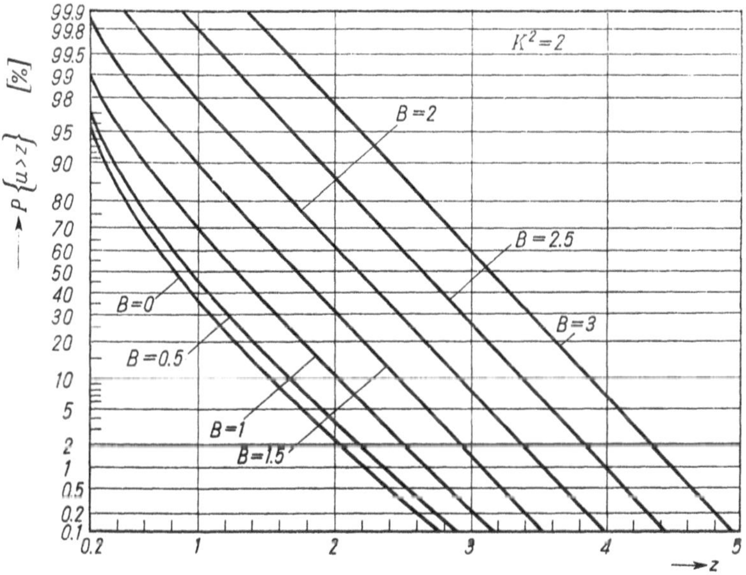

Fig. F7. Probabilities $P\{u > z\}$ for $K^2 = 2$.

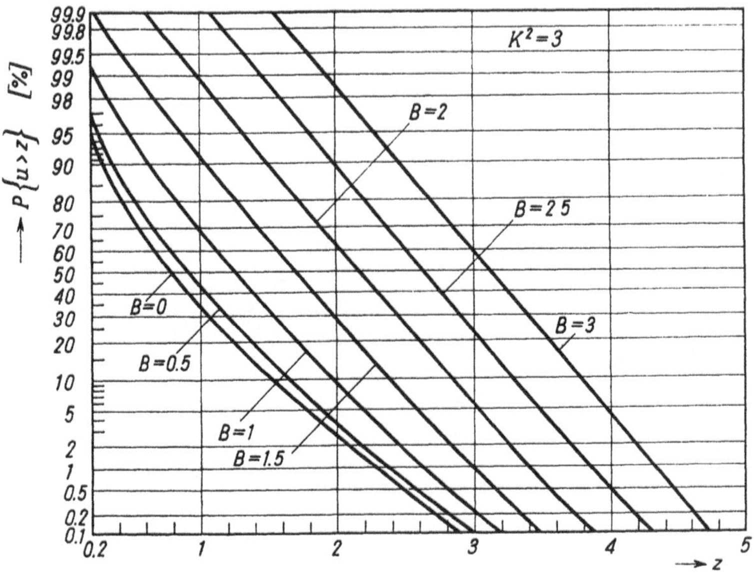

Fig. F8. Probabilities $P\{u>z\}$ for $K^2 = 3$.

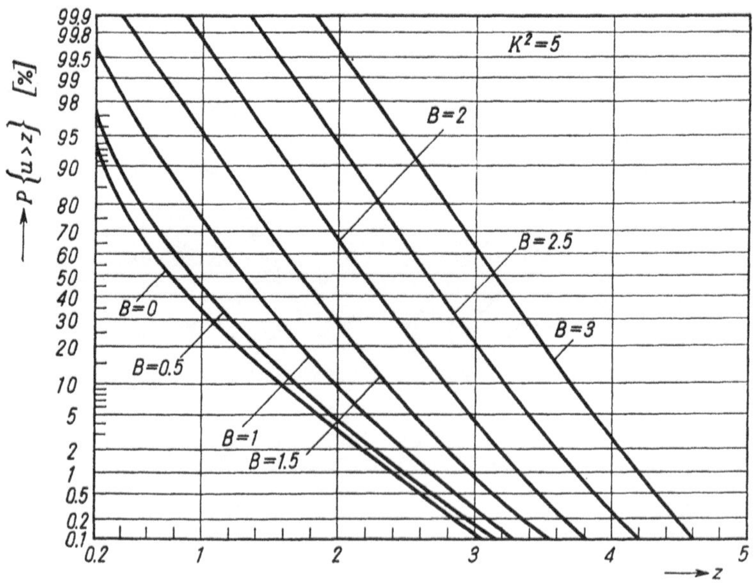

Fig. F9. Probabilities $P\{u>z\}$ for $K^2 = 5$.

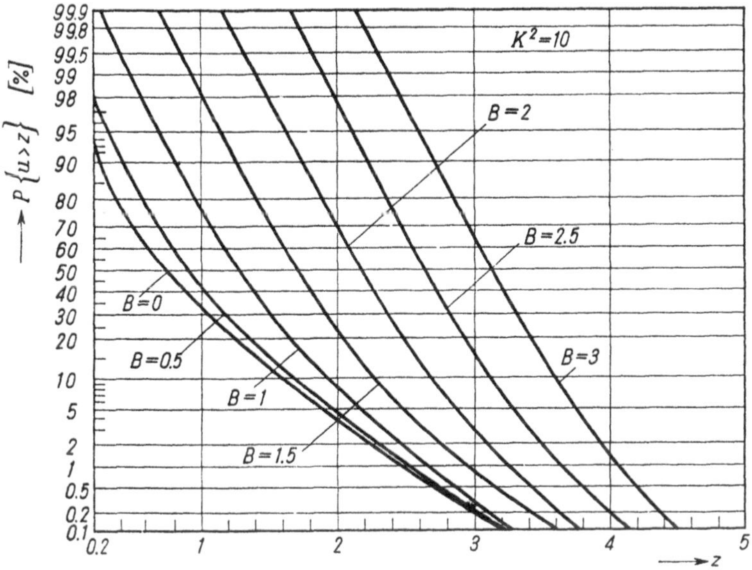

Fig. F10. Probabilities $P\{u > z\}$ for $K^2 = 10$.

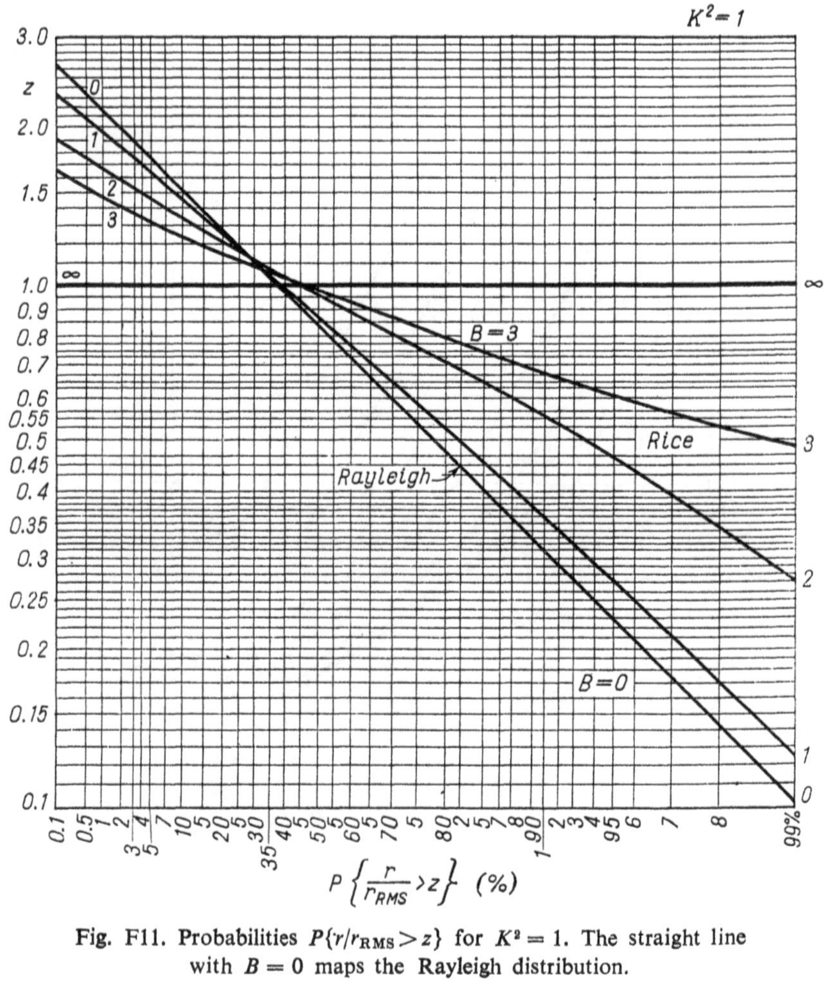

Fig. F11. Probabilities $P\{r/r_{RMS} > z\}$ for $K^2 = 1$. The straight line with $B = 0$ maps the Rayleigh distribution.

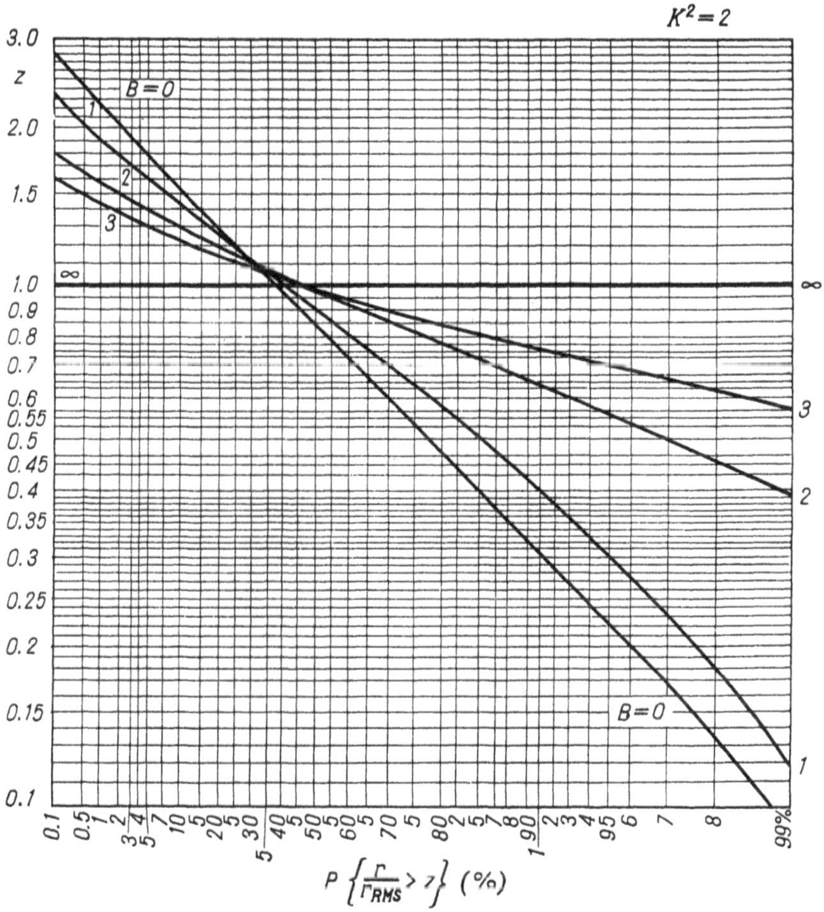

Fig. F12. Probabilities $P\{r/r_{RMS} > z\}$ for $K^2 = 2$.

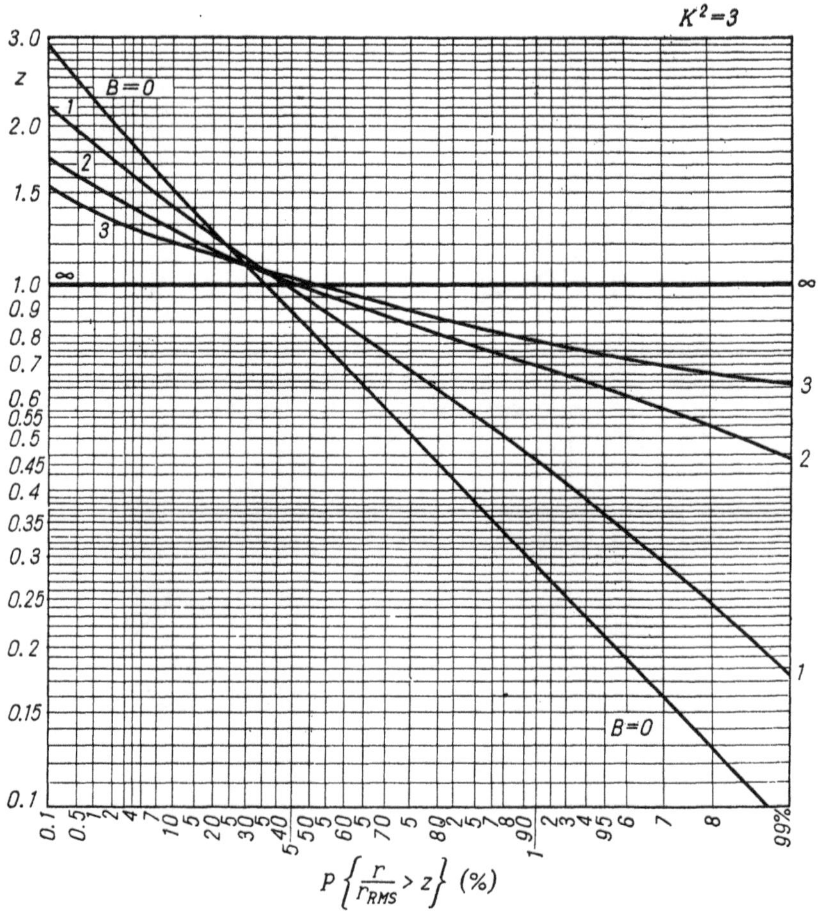

Fig. F13. Probabilities $P\{r/r_{\mathrm{RMS}} > z\}$ for $K^2 = 3$.

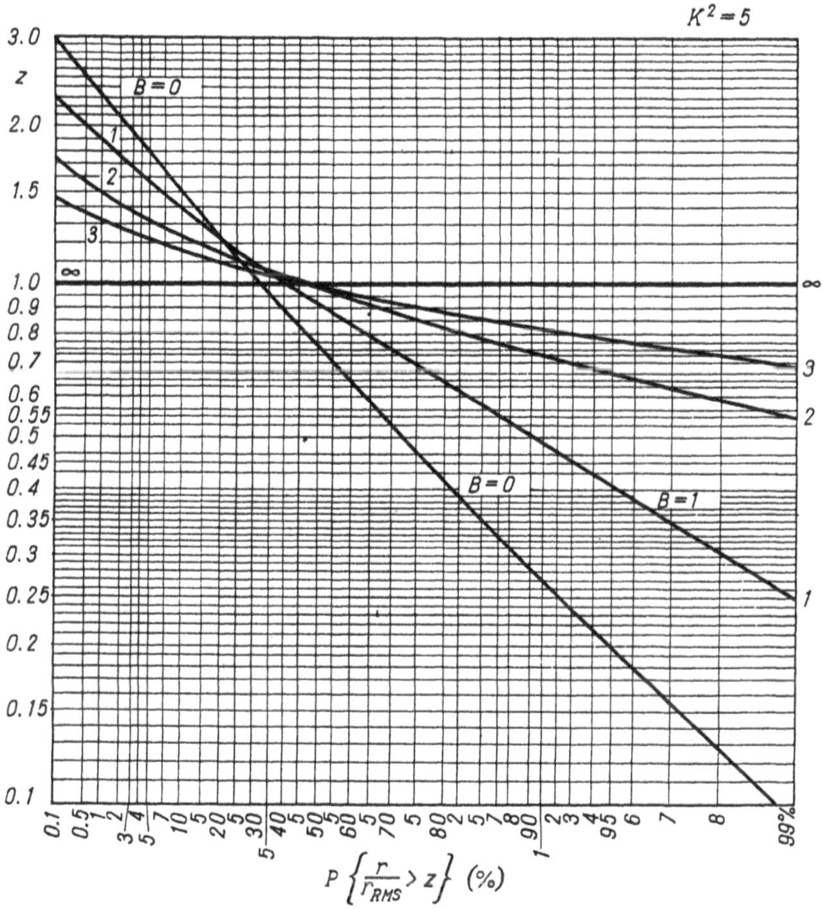

Fig. F14. Probabilities $P\{r/r_{RMS} > z\}$ for $K^2 = 5$.

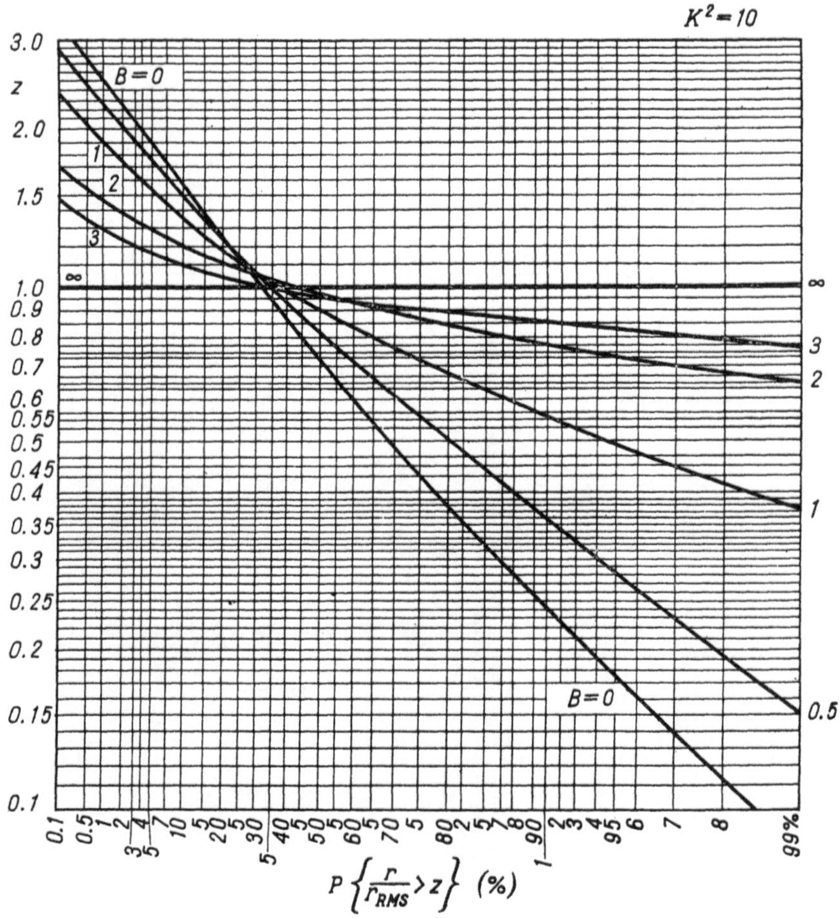

Fig. F15. Probabilities $P\{r/r_{\text{RMS}} > z\}$ for $K^2 = 10$.

Part II

APPLICATION

by

ANDRÉ SPIZZICHINO

Translated by

PETR BECKMANN

CHAPTER 10

INTRODUCTION

In Part I of this monograph, the reader has become acquainted with the general theory of scattering by rough surfaces. We shall now apply the results of this theory to certain concrete problems.

Because many types of waves exist and all surfaces are in a certain sense rough, almost innumerable applications of the above theory are possible and it has been out of the question to treat them all here Part II of this book is therefore restricted to applications in the field of radio waves, though waves of the infrared and visible part of the electromagnetic spectrum will occasionally also be considered. But if the second part of this book had been entrusted to an optician or to an acoustician, possibly he would have written an equally long treatise on quite different applications.

We shall succesively consider the reflection of radio waves by the surface of the earth, by layers in the troposphere and ionosphere, and by the Moon, the Sun and the planets.

The term "applications" should not be taken to mean exclusively practical applications for the immediate solution of concrete problems. Thus the application of the general theory to the case of reflections by the earth's surface consists of three parts:

(1) application of the general equations to the case of a ground-to-ground radio path (Chapter 11 and 12);

(2) experimental data on the reflection coefficient of the earth and confrontation of experimental with theoretical results (Chapters 13, 14, 15);

(3) practical applications (Chapters 16, 17, 18).

The same structure will be found in the study of atmospheric reflections (Chapter 19) and of reflection by the Moon and the planets (Chapter 20), though these subjects are not subdivided into several chapters.

In view of the diversity of these subjects it was found advisable to enable each chapter of Part II to be read independently. This is made possible by frequently recalling previous results (at the risk of being somewhat repetitious). In particular, the reader who is primarily interested in experimental results and does not wish to follow the theoretical developments in great detail may conveniently start with Chapter 13.

Insofar as possible the notation of Part I has been followed. But in some cases this would have been detrimental to the clarity of our account. For example, the standard deviation of the heights of the points of a rough surface, denoted by σ in Part I of this book (as in most theoretical treatises), is denoted by Δh in Part II, as this notation is used in almost all papers dealing with practical applications.

In Part II, the letter R stands for the reflection coefficient of a rough surface, and R_0 for the reflection coefficient of a smooth surface of the same material. We have assumed that R is related to R_0 by

$$R = \rho R_0 \tag{1}$$

where ρ depends only on the shape of the irregularities of the surface. Thus ρ is the reflection coefficient of a rough surface identical in shape but infinite in conductivity. To avoid any confusion, distances are denoted by the letter r (or d).

The definition of the scattering coefficient (1) is not identical with that given in Part I Section 3.1, equation (29), pp. 22–23. However, for most practical purposes (1) is equivalent to the definition given in Part I.

To demonstrate this, we temporarily denote the scattering coefficient as defined in Part I, formula (29), p. 22, by ρ^{I}, and ρ as defined in (1) by ρ^{II}. Then for a smooth, plane sheet with reflection coefficient R_0 we find from (34), p. 23, for the direction of specular reflection

$$\theta_1 = \theta_2, \qquad v_x = 0 \tag{2}$$

the relation

$$\rho^{\mathrm{I}}_{sp} = -R_0 \tag{3}$$

where the index sp stands for "smooth plane". (The negative sign is due to the horizontal polarization of the reference field, cf. footnote on p. 23.)

We now define the reflection coefficient of a rough surface as

$$R = \frac{E_2}{E_1} \tag{4}$$

where E_1 is the incident, and E_2 the scattered field.

The quantity R will be constant if E_2 is a glane wave that does not decrease with distance (in Part I, E_2 was a wave with a plane wave front, but decreasing with distance). Then from (29), p. 21,

$$\rho^{\mathrm{I}} = \frac{E_2}{E_{20}} = \frac{RE_1}{E_{20}} \tag{5}$$

If E_{sp} is the field E_2 reflected by a smooth, plane sheet in the specular direction, we have from (5), (3) and (1)

$$\rho^{I} = -\frac{RE_1}{E_{sp}R_0} = -\rho^{II}\frac{E_1}{E_{sp}} \quad (6)$$

or since $E_{sp}/E_1 = R_0$,

$$\rho^{II} = -R_0\rho^{I} \quad (7)$$

Thus ρ^{I} and ρ^{II} differ only by the factor $-R_0$. Moreover, for $R_0 = -1$, ρ^{I} and ρ^{II} will be exactly equal; this will be the case not only for perfect conductivity at horizontal polarization, but also for the most commonly met application, namely grazing incidence over irregular ground (land or sea).

Alternatively, one may define both ρ^{I} and ρ^{II} by (29), p. 22, but instead of defining the reference field E_{20} as on p. 23, it may be defined as the specularly reflected field reflected by a smooth plane *of the same electrical constants* and of the same dimensions as the rough surface illuminated under the same angle of incidence and measured at the same distance as E_2. When the scattered waves may be considered plane (not weakening with distance), we may set $E_{20} = R_0E_1$, from which (1) follows.

We therefore omit the superscript II, writing ρ for ρ^{II}.

CHAPTER 11

REFLECTION OF ELECTROMAGNETIC WAVES BY A PERFECTLY SMOOTH EARTH

BEFORE beginning the study of the reflection of electromagnetic waves by irregular ground, we shall recall in this chapter some results concerning the reflection of waves by a smooth earth. First we shall see (Section 1) how the reflection coefficient of a perfectly plane earth is determined by the use of electromagnetic theory and by what are now classical experimental results. We shall then consider the effect of the curvature of the earth on its reflection coefficient (Section 2). Finally in Section 3, we shall examine the interference of the ray reflected by a smooth earth and the direct ray, and the effects of this interference on propagation.

11.1. THE REFLECTION COEFFICIENT OF A PERFECTLY PLANE EARTH

A theoretical expression for the reflection coefficient of a plane earth may be obtained from Maxwell's equations. The earth is defined by its relative complex dielectric constant

$$\varepsilon_{rc} = \frac{\varepsilon}{\varepsilon_0} - 60\, i\lambda\sigma$$

where ε is the dielectric constant and σ the conductivity in mho/m, ε_0 the dielectric constant of free space and λ the wavelength in metres.†

Theoretically one should also introduce the relative magnetic permeability

$$\mu_{rc} = \frac{\mu}{\mu_0}$$

where μ is the magnetic permeability of the earth and μ_0 that of free space. In the following we shall always assume $\mu_{rc} = 1$.

The reflection coefficient of the earth is then

$$R_0^+ = \frac{Y^2 \sin\gamma - \sqrt{(Y^2 - \cos^2\gamma)}}{Y^2 \sin\gamma + \sqrt{(Y^2 - \cos^2\gamma)}} \tag{1}$$

† In Part II the time factor is $\exp(+i\omega t)$ to conform with engineering practice.

$$R_0^- = \frac{\sin\gamma - \sqrt{(Y^2-\cos^2\gamma)}}{\sin\gamma + \sqrt{(Y^2-\cos^2\gamma)}} \tag{2}$$

where R_0^+ and R_0^- are the reflection coefficients for vertical and horizontal polarizations respectively; γ is the grazing angle of the incident wave (i.e. the complement of the angle of incidence Θ) and Y is the normalized admittance of the medium:

$$Y = \sqrt{\frac{\varepsilon_{rc}}{\mu_{rc}}}$$

We shall not repeat the derivation of the relations (1) and (2) which may be found in the standard treatises of electromagnetic wave theory.†

Numerous investigators [Ford and Oliver, 1946; Straiton and Tolbert, 1947; and others] have measured the reflection coefficient R_0 of different types of plane ground for various values of the grazing angle γ. They have confirmed that R_0 obeys the laws of the type (1) and (2) when γ is varied. From these measurements, and from measurements of the absorption coefficient of the ground, they have calculated the characteristic ground constants.

The results obtained for the ground vary widely: $\varepsilon/\varepsilon_0$ varies from about 2 in the case of very dry ground to 24 in the case of very wet ground, whereas σ varies from 10^{-4} to several units of 10^{-3} mho/m.‡ We shall use the mean values

$$\frac{\varepsilon}{\varepsilon_0} = 10 \quad \text{and} \quad \sigma = 10^{-3}\,\text{mho/m} \tag{3}$$

In the case of the sea, we shall assume

$$\frac{\varepsilon}{\varepsilon_0} = 80 \quad \text{and} \quad \sigma = 4\ \text{mho/m} \tag{4}$$

Using the values (3) and (4) for $\varepsilon/\varepsilon_0$ and σ, we obtain curves of the absolute value and phase of the reflection coefficient R_0 shown in Figs. 11.1 to 11.4. These curves give only approximate values of the reflection coefficient corresponding to average conditions.

Certain important properties of the expressions (1) and (2) for the reflection coefficient become evident from these figures.

† The reader will find a derivation of (1) and (2) in STRATTON, J. A.: *Electromagnetic Theory*. McGraw-Hill, New York (1941).; graphs of R_0^+ and R_0^- as functions of the angle of incidence θ, the wavelength λ and the ground constants ε and σ may be found, e.g. in MacPetrie [1938].

‡ See, for example, Ford and Oliver [1946] and Kerr [1951].

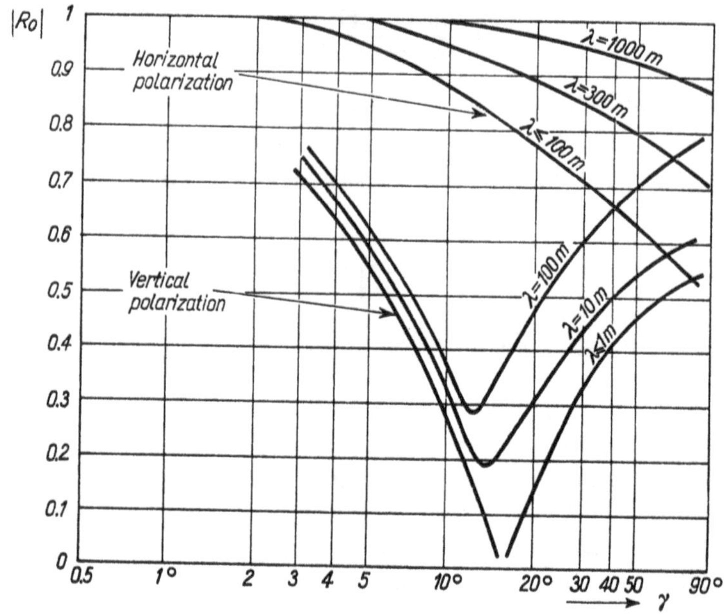

Fig. 11.1. Reflection coefficient of a perfectly plane earth with $\varepsilon/\varepsilon_0 = 10$, $\sigma = 10^{-3}$ mho/m.

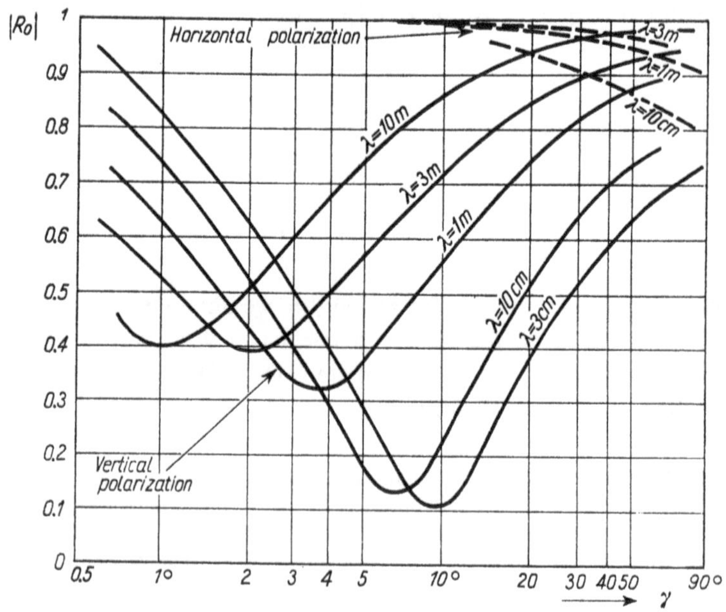

Fig. 11.2. Reflection coefficient of a very smooth sea. $\varepsilon/\varepsilon_0 = 80$, $\sigma = 4$ mho/m.

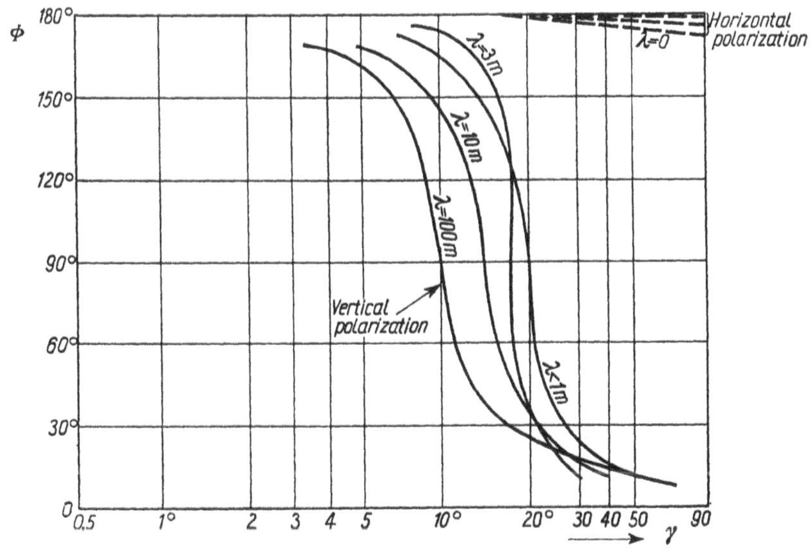

Fig. 11.3. Phase Φ of the reflection coefficient of a perfectly plane earth. $\varepsilon/\varepsilon_0 = 10$, $\sigma = 10^{-3}$ mho/m.

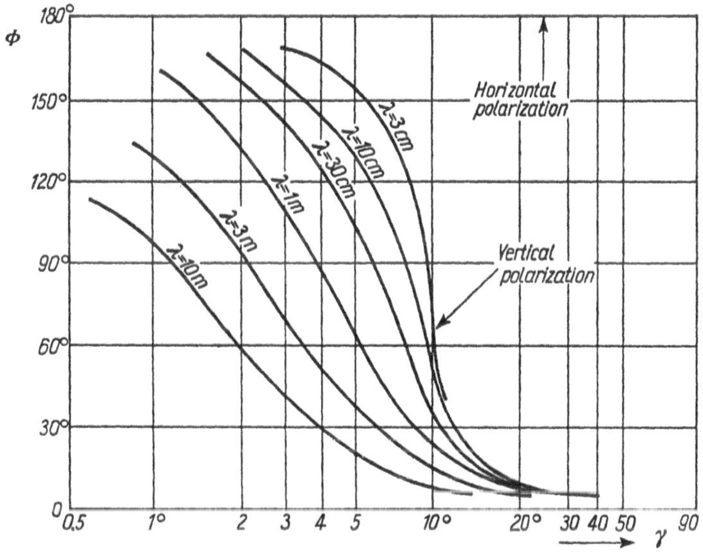

Fig. 11.4. Phase Φ of the reflection coefficient of a very smooth sea. $\varepsilon/\varepsilon_0 = 80$, $\sigma = 4$ mho/m.

When γ tends to zero, the expressions (1) and (2) for R_0 tend to -1. In the majority of radio links used in practice, the angle γ is very small, usually smaller than 1°.

In many practical applications $|R_0| = 1$, as may be verified from Figs. 11.1 to 11.4. However, this does not always apply to a vertically polarized wave reflected from the surface of the sea.

Since the relative complex dielectric constant is by (3) and (4) large compared with unity, Y is also large compared with unity. Hence R_0^- as given by (2) is near to -1. In practice we are therefore only slightly in error if we assume that *the reflection coefficient of a plane earth or a plane surface of the sea equals* -1.

At the wavelengths in which we shall be interested in the present work (metric or shorter), the real part of ε_{rc}, i.e. $\varepsilon/\varepsilon_0$, is much larger than its imaginary part $60\lambda\sigma$, but only in the case of the earth: the earth behaves as a dielectric, its conductivity is negligible. In the case of the sea, a similar statement holds only in the case of the shorter wavelengths. If the earth were a perfect dielectric (Y real and much larger than unity), R_0^+ would vanish for

$$\sin\gamma = \frac{1}{Y}$$

(Brewster angle). Since, however, Y has a finite imaginary part, $|R_0^+|$ passes through a non-vanishing minimum in the neighbourhood of the Brewster angle. This minimum is more pronounced in the case of the earth (Fig. 11.1) than in the case of the sea (Fig. 11.2).

11.2. Effect of the Earth's Curvature

The expressions (1) and (2) for the reflection coefficient will be strictly valid only if the surface of the earth is plane. In practice they may be used for radio paths the length of which is small compared with the distance to the horizon.

The effect of the curvature of the earth on its reflection coefficient may be derived by an argument founded on the laws of geometrical optics. Let T be the transmitter, R the receiver, I the point of reflection on the earth's surface and dS and element of this surface in the neighbourhood of I (Fig. 11.5). The energy received by the element dS from the transmitter T will be

$$dW = \frac{4\pi W_t}{r_1^2} dS \sin\gamma \tag{5}$$

where W_t is the product of the transmitted power and the transmitting antenna gain, r_1 is the distance TI, and γ the grazing angle of the ray incidence onto the surface of the earth. A fraction dW' of this energy is reflected:

$$dW' = |R_0|^2 dW \tag{6}$$

where R_0 is the reflection coefficient of a plane earth defined by (1) or (2). This reflected energy $\mathrm{d}W'$ is contained in a conical volume (Fig. 11.5), which is intercepted by a sphere with centre I and radius r_2 over an area

$$\mathrm{d}\Sigma = r_2^2\left[\left(\frac{1}{r_1}+\frac{1}{r_2}\right)\sin\gamma+\frac{2}{a}\right]\left(\frac{1}{r_1}+\frac{1}{r_2}+\frac{2}{b}\right)\mathrm{d}S \tag{7}$$

(as is easily shown by giving $\mathrm{d}S$ the form of a small rectangle as in Fig. 11.5). In this expression, a and b are the radii of curvature of the intersections

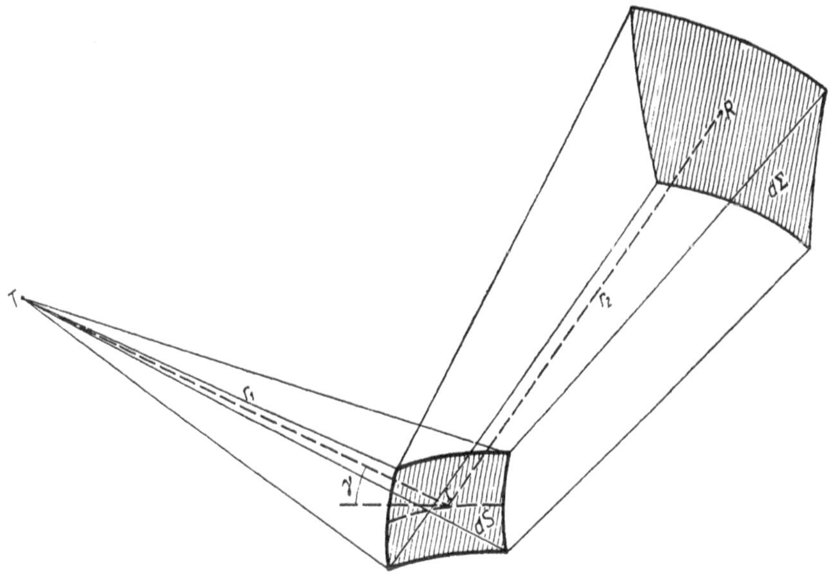

Fig. 11.5. Geometry of reflection from a curved surface.

of the surface of the earth and two vertical planes perpendicular to the direction of propagation. The relation (7) is quite general: it may be applied to any form of the earth's surface (mountainous profiles). In the case when the earth is assumed perfectly spherical, a and b equal the radius of the earth.

The reflecton coefficient R of the earth may be obtained by comparing the energy flux in the neighbourhood of the receiver w_r to the energy flux that would be received in free space w_{fs} at a distance r_1+r_2:

$$|R|^2 = \frac{w_r}{w_{fs}}$$

where $w_r = dW'/d\Sigma$ is given by (5), (6) and (7), and where

$$w_{fs} = \frac{4\pi W_t}{(r_1+r_2)^2}$$

We thus obtain an expression for R, which may be written in the form

$$R = DR_0 \tag{8}$$

where D, the divergence coefficient, is given by

$$D = \left[1+\frac{2r_1r_2}{a(r_1+r_2)\sin\gamma}\right]^{-1/2}\left[1+\frac{2r_1r_2}{b(r_1+r_2)}\right]^{-1/2}$$

In practical applications, when $a = b$ and γ is a small angle, this may be written as

$$D \approx \left[1+\frac{2r_1r_2}{a(r_1+r_2)\sin\gamma}\right]^{-1/2} \tag{9}$$

11.3. Interference of the Direct and Reflected Rays

In the majority of radio links the receiver receives simultaneously the direct wave transmitted by the transmitter and the wave reflected from the ground. The amplitude of the resulting field is

$$E = E_0\sqrt{[(1+|R|)^2+4|R|\cos^2\tfrac{1}{2}\varphi]} \tag{10}$$

where E_0 is the field received in free space, R is the reflection coefficient of the ground and φ is the phase shift between the direct wave and the wave reflected from the ground, both measured at the receiver. The phase shift φ is given by

$$\varphi = 2\pi\frac{\delta}{\lambda}+\Phi \tag{11}$$

where δ is the path difference between the direct and reflected waves, and Φ is the phase of the reflection coefficient, so that

$$R = |R|e^{i\Phi}$$

The received field may be determined through (10) and (11) after finding δ. In general, if the receiver (or transmitter) is displaced along a straight line (e.g. horizontal or vertical), δ, and hence φ, will vary monotonously, $\cos^2(\varphi/2)$ will oscillate between 0 and 1, and the received field E will pass through a series of alternate maxima and minima with respective values

$$E_M = E_0(1+|R|) \quad \text{and} \quad E_m = E_0(1-|R|)$$

To determine the exact positions of these interference fringes, δ must be expressed in terms of the geometrical characteristics of the path. This is particularly simple when the path is sufficiently short for the surface of the earth to be approximated by a plane. To simplify the problem still further, we assume the atmosphere to be homogeneous. The radio path is then defined

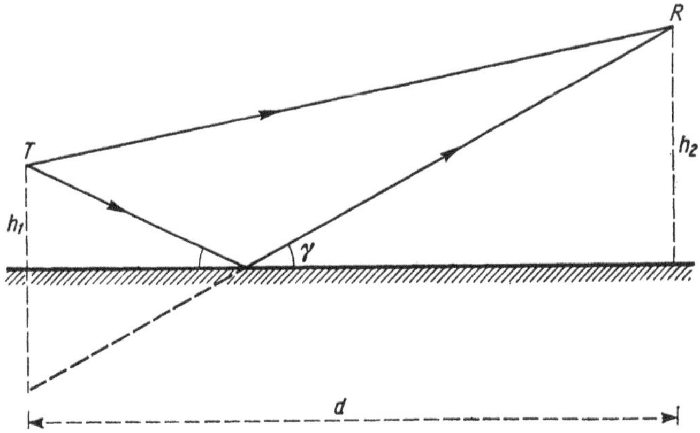

Fig. 11.6. Geometry of a radio link.

by its length d and the antenna heights h_1 and h_2 of the transmitter and receiver (Fig. 11.6). In this case it is easily shown that

$$\delta = 2\frac{h_1 h_2}{d} \tag{12}$$

If in addition $\Phi = \pi$, it follows from (11) and (12) that

$$\varphi = \frac{4\pi h_1 h_2}{\lambda d} + \pi \tag{13}$$

The maxima of the received field will appear at the points for which

$$\frac{4h_1 h_2}{\lambda d} = 2K+1$$

where K is an integer, whilst the corresponding expression for the minima is

$$\frac{2h_1 h_2}{\lambda d} = K$$

Figures 11.7 to 11.9, computed for certain numerical values, show the kind of field received in space in this particularly simple case. To simplify this case still further, it has been assumed that the absolute value of the reflection coefficient remains near unity. In Fig. 11.7 the transmitter is assumed fixed, whereas the receiver is displaced along a horizontal straight line ($h_1 = h_2 = 10$ m, $\lambda = 10$ cm, d variable). The diagram has been limited to distances within which the approximation of a plane earth and that of geometrical optics are valid. In Fig. 11.8 the transmitter is assumed fixed and the receiver is displaced along a vertical straight line ($h_1 = 10$ m,

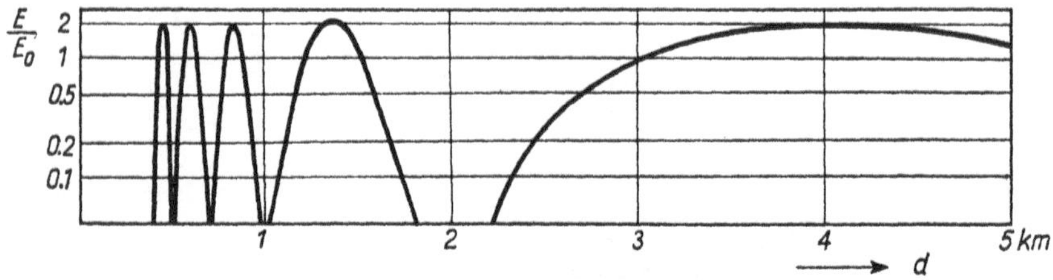

Fig. 11.7. Ratio of the received field E to the free-space field E_0 as a function of distance for $h_1 = h_2 = 10$ m, $\lambda = 10$ cm. The earth is assumed to be perfectly plane and the reflection coefficient R_0 remains near -1.

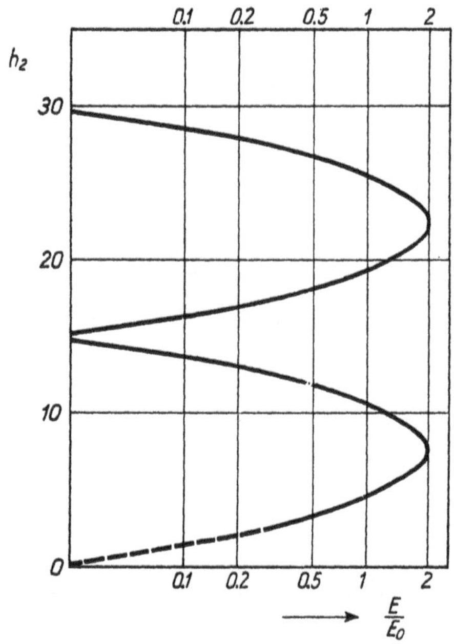

Fig. 11.8. Ratio of the received field E to the free-space field E_0 as a function of the height of the receiving antenna h_2. $h_1 = 10$ m, $d = 3$ km, $\lambda = 10$ cm. The reflection coefficient is assumed to remain near -1.

$d = 3$ km, $\lambda = 10$ cm, h_2 variable). The method of representation adopted in Fig. 11.9 is often convenient in practical applications.† The curves along which the received field has a constant value are plotted on a net with the

† We shall use it later (Chapter 16) to investigate the problem of radar coverage.

distance d as abscissa and the receiving-antenna height h_2 as ordinate. It will be noted that the received field is a function of the site angle α under which the receiver is seen from the foot of the transmitting antenna:

$$\tan \alpha = \frac{h_2}{d}$$

In Fig. 11.9 it was assumed that $h_1 = 10$ m and $\lambda = 10$ cm.

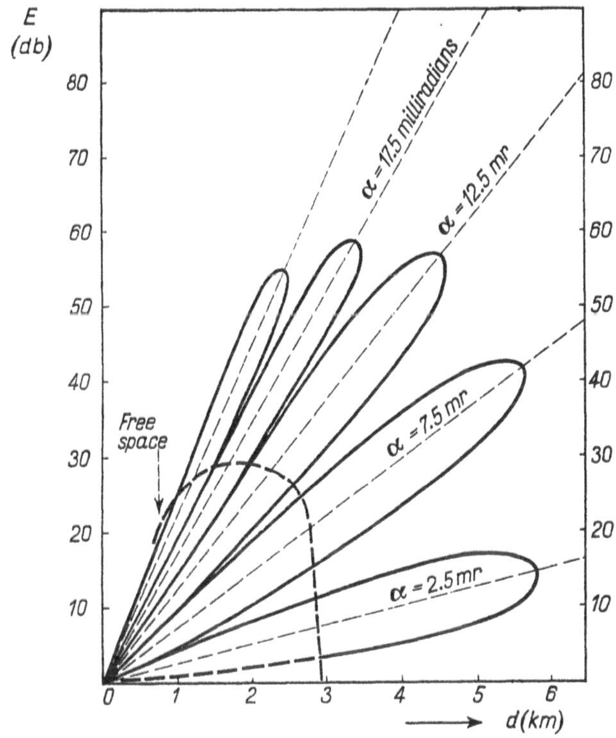

Fig. 11.9. Curves of equal values of the received field. $h_1 = 10$ m, $\lambda = 10$ cm. Plane earth and reflection coefficient equal to -1 is assumed. Broken curve: The same curve in free space.

Although the case of reflection by a plane earth can be treated by very simple equations, the general case of a spherical earth is much more complicated. We cannot describe the mathematical treatments of this problem in detail in this work and shall restrict ourselves to the use of a set of nomograms, shown in the annex to this chapter, through which one may easily determine the position of the point of reflection from a spherical earth and the angle of incidence when the transmitter and receiver are fixed, the position of the interference lobes (maxima and minima of the field) when the transmitter is fixed, and finally the divergence coefficient and the received field at any point. These nomograms have been constructed by L. Boithias.†

Figure 11.10 gives an example of the variations of fieldstrength as a function of distance over a spherical earth, setting $h_1 = h_2 = 100$ m, $\lambda = 1$ m.

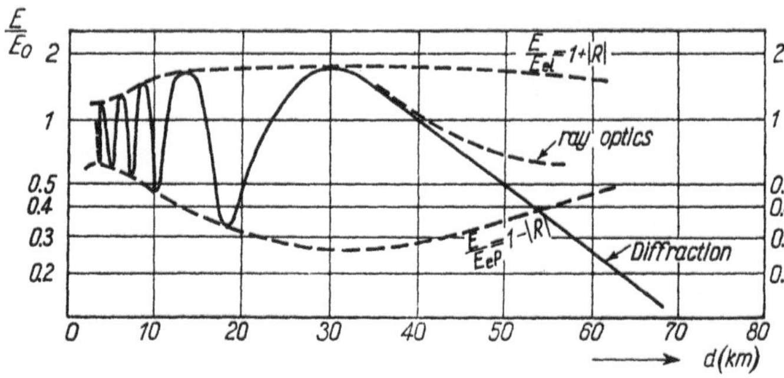

Fig. 11.10. Ratio of received field E to free-space field E_0 as a function of the distance. $h_1 = h_2 = 100$ m, $\lambda = 1$ m. Vertical polarization above the surface of the sea. The spherical shape of the earth has been taken into account.

This figure may be compared with Fig. 11.7 which applies to h_1, h_2 and λ 10 times smaller: if the approximation of a flat earth were still valid, Fig. 11.10 would be similar to Fig. 11.7 after multiplying the distance scale by 10. The comparison of these two figures gives some idea of the effect of the earth's curvature. It will be seen that the positions of the maxima and minima are roughly the same in both cases when the distance d is small, whereas they differ for large distances. A second difference concerns the amplitude of the fluctuations of the field. The latter equals $2|R|$, since E/E_0 varies between $1-|R|$ and $1+|R|$. In Fig. 11.10 the variation of $|R|$ with the distance d was taken into account as it was assumed that the reflection is produced by the surface of the sea and that the wave is vertically polarized. According to (8), $|R|$ is the product of two factors, $|R| = D|R_0|$.

For small distances, D remains equal to 1. $|R_0|$, which is a decreasing function of the grazing angle γ (Fig. 11.2), increases with increasing d and the same applies to the reflection coefficient $|R|$. For large distances d, on

† Engineer at the C.N.E.T. (National Centre of Telecommunication Research, Paris). Cf. Boithias, L.: Nomogrammes relatifs à la propagation des ondes radioélectriques de longueur inferiéure à 10 m. *Centre National d'études des Télécommunications, Note technique d'information No. PR 3104.* 3rd edition, 1962. Other solutions of this problem have been proposed (see, for example, David, P. and Voge, I.: *Cours de radioélectricité générale.* Vol. 4: Propagation des ondes, Paris (1961)).

the other hand, $|R_0|$ remains equal to 1, whereas D decreases with increasing d. The reflection coefficient $|R|$ is then a decreasing function of the distance. These variations of $|R|$ may be seen from Fig. 11.10, where the broken curves

$$\frac{E}{E_0} = 1-|R|$$

and

$$\frac{E}{E_0} = 1+|R|$$

envelop the maxima and minima of E/E_0. It may be seen that $|R|$ attains a maximum at a distance near 30 km. If the curve of Fig. 11.10 had been computed for reflection by dry ground or for horizontal polarization, the decrease of $|R_0|$ as a function of γ would have been much less pronounced: the field would have much deeper minima and the maximum of $|R|$ would be attained at a much shorter distance, much shorter, in fact, than that of current radio links.

A third difference between Figs. 11.7 and 11.10 is associated with the presence of a zone in the latter figure where the laws of geometrical optics are no longer valid. This is the "diffraction zone" which appears near the horizon. We shall return to this question in Chapter 19.

Annex to Chapter 11

Nomograms of Quantities Determining the Reflection of Electromagnetic Waves by a Smooth, Spherical Earth†

Nomogram A—Calculation of m, Δ_0 and γ_0

These three quantities are intermediate parameters to be used in the following. They are defined by

$$m = \frac{d^2}{4Ka(h_1+h_2)}$$

$$\gamma_0 \approx \tan\gamma_0 = \frac{h_1+h_2}{d}$$

$$\Delta_0 = \frac{(h_1+h_2)^2}{2d}$$

† The nomograms contained in this annex are taken from Boithias, L.: Nomogrammes relatifs à la propagation des ondes radioélectriques de longueur inférieure à 10 mètres. *Note Technique d'Information No. PR 3104*, C.N.E.T., Issy-les-Moulineaux (France) 3rd edition 1962. They will be included in a publication by the same author on problems of radio wave propagation, to be published in the near future.

Nomogram A gives these three quantities as a function of

the heights h_1 and h_2 of the transmitting and receiving antennae,

the pathlength d of the link,

the equivalent radius of the earth Ka which takes account of atmospheric refraction ($K = 4/3$ in the nomogram, but since m is inversely proportional to K, its values for $K \neq 4/3$ may easily be derived).

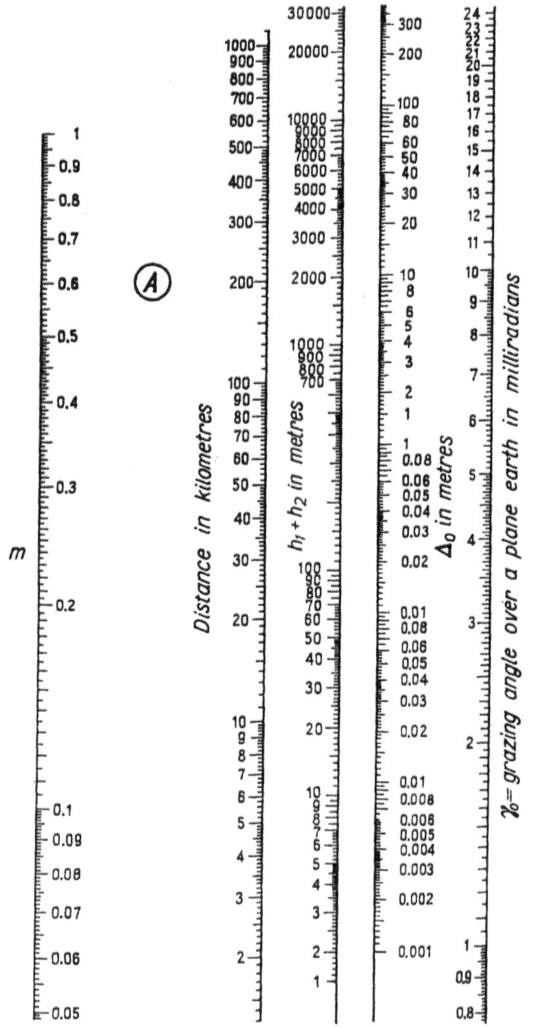

Fig. 11.11. Nomogram A. The quantity m for $K = 4/3$.

Nomogram B—Determination of the Point of Reflection

This nomogram enables the point of reflection on the spherical surface of the earth to be determined if the pathlength d and the heights h_1 and h_2 above the ground are known. The point of reflection is defined by distance d_1 from the transmitter or d_2 from the receiver. The distances d_1 or d_2 are obtained from Nomogram **B** by using the ratio h_1/h_2 and the coefficient m as given by Nomogram A.

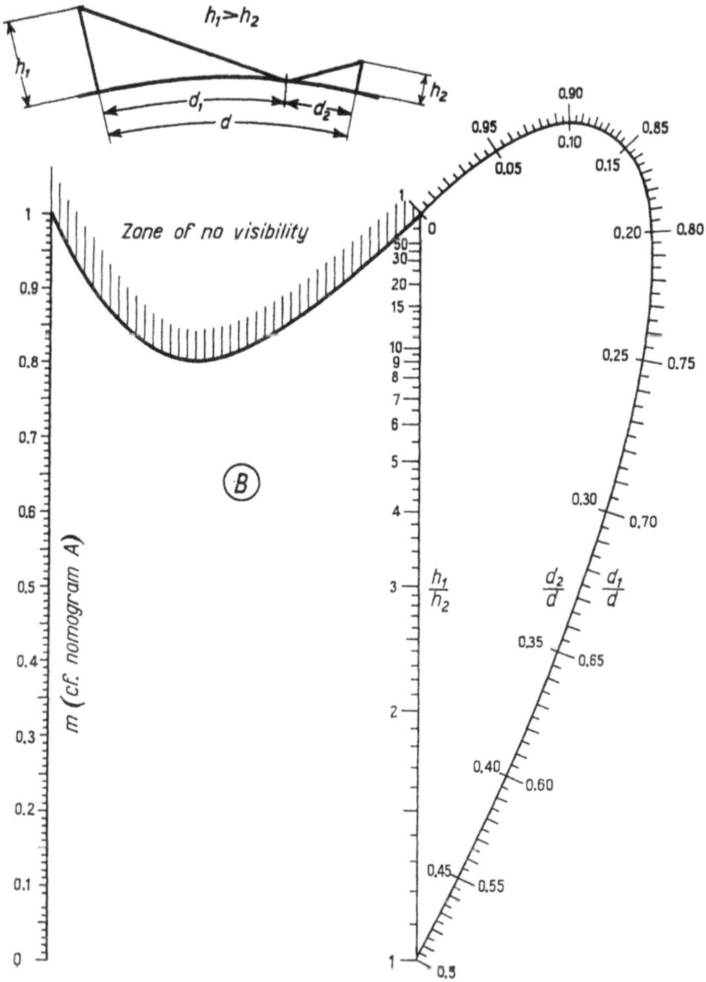

Fig. 11.12. Nomogram B. The point of reflection.

Since the point of reflection does not change when the transmitter and receiver are interchanged, it is sufficient to consider the values of h_1/h_2 greater than one.

If the straight line of alignment cuts the shaded part of the curve marked "zone of no visibility", the value thus obtained is not correct: in this case the transmitter and receiver are not within sight of each other.

Nomogram C—The Grazing Angle γ over a Spherical Earth

Nomogram A yields the grazing angle γ_0 with respect to a plane earth. The grazing angle γ for a spherical earth is given by

$$\gamma \approx Z\gamma_0$$

where Z is given by Nomogram B.

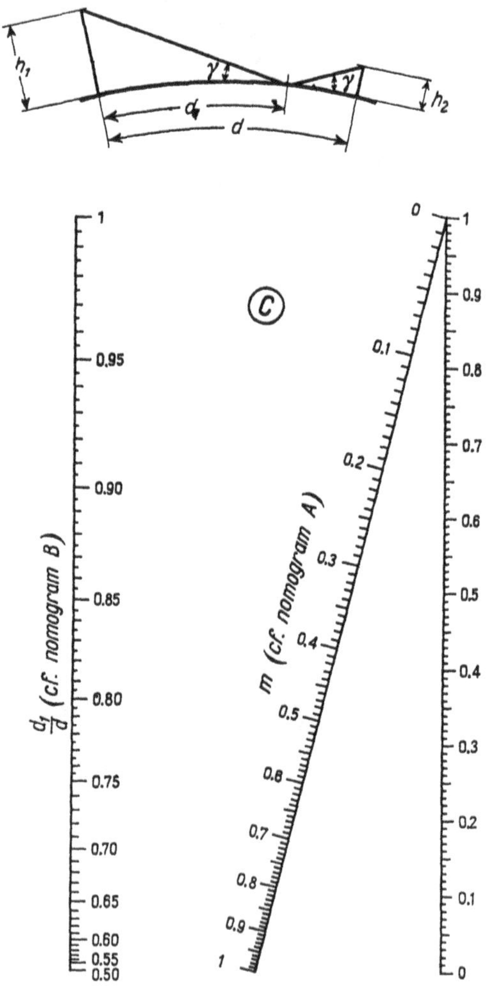

Fig. 11.13. Nomogram C. The factor Z, by which the grazing angle must be reduced in the case of a spherical earth.

Nomogram D—*Path Difference* δ *Between the Direct and Reflected Rays*

The quantity Δ_0 is first determined from Nomogram A. (This quantity has no immediate physical significance.) The path difference δ between the direct and reflected rays is the product of Δ_0 and a coefficient μ, which is given by Nomogram D as a function of m (Nomogram A) and of d_1/d (Nomogram B).

According to (10) and (11) of Chapter 1, the resultant field E at any point may be obtained from δ (cf. Nomograms G and H).

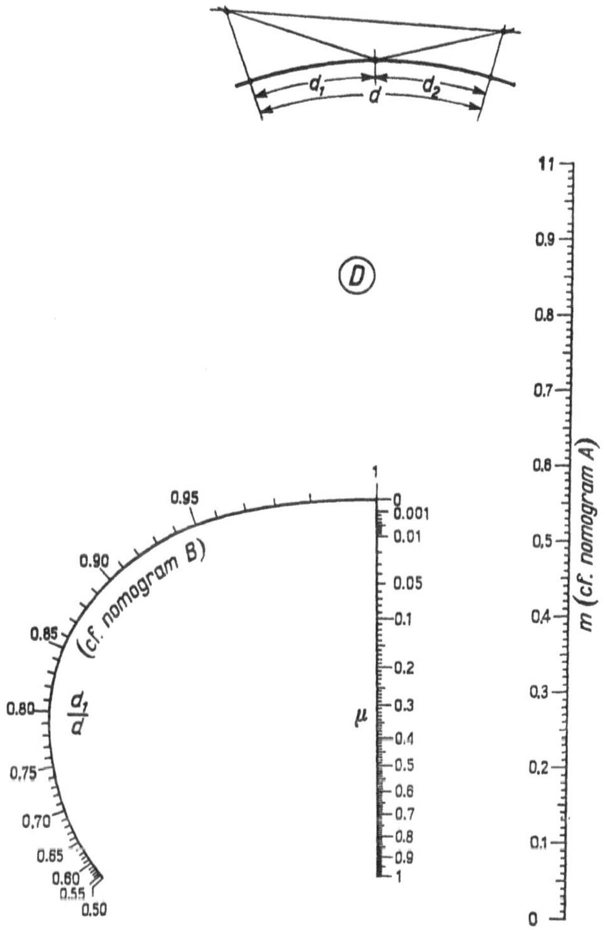

Fig. 11.14. Nomogram D. The path difference $\delta = \mu\Delta_0$ between the direct and reflected rays.

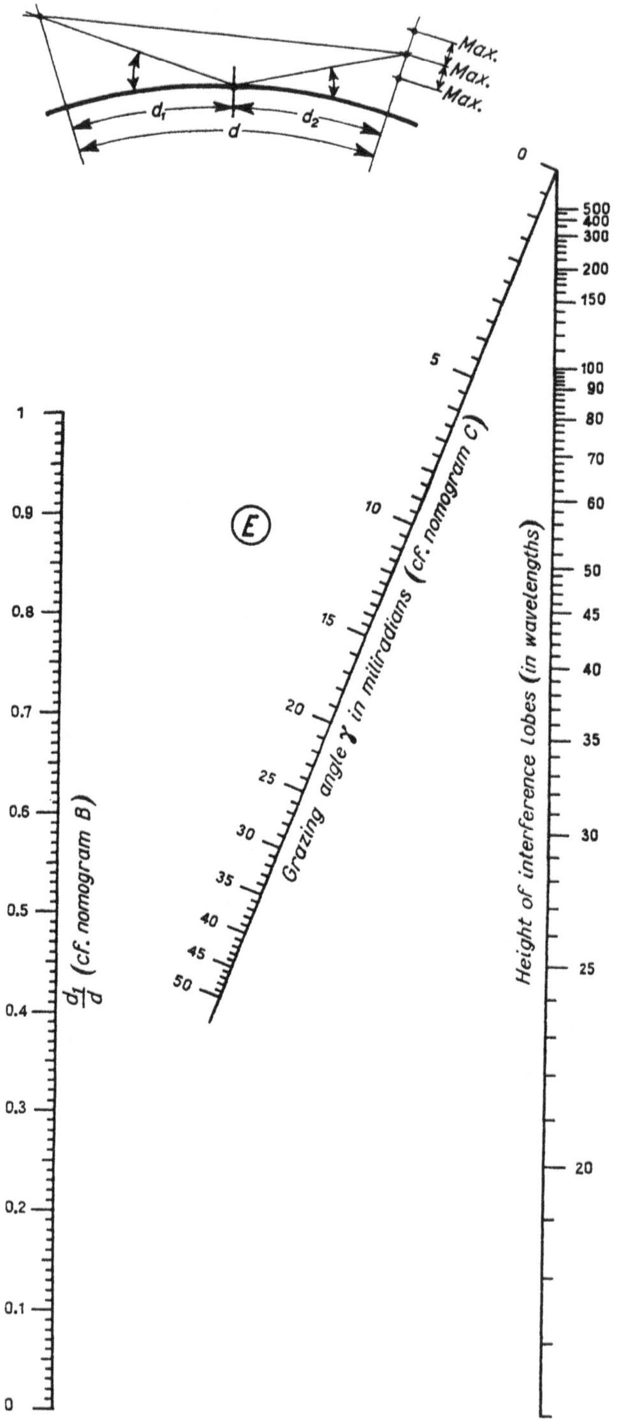

Fig. 11.15. Nomogram E. Position of interference lobes.

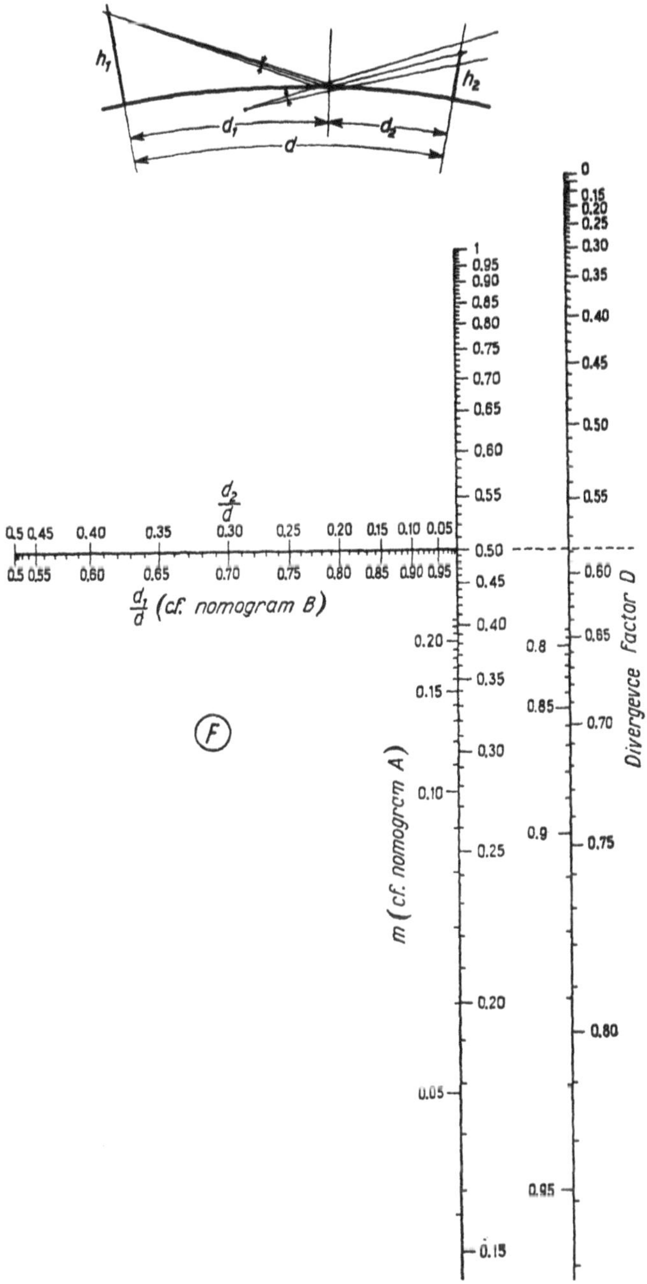

Fig. 11.16. Nomogram F. The divergence factor D. For small values of m use the values indicated on the left of the m and D scales.

Nomogram E—*Position of the Interference Lobes*

The position of the interference lobes is defined by the distance between two neighbouring maxima (or minima) along a vertical straight line. It is given by Nomogram E as a function of d_1/d and γ, which are themselves given by Nomograms B and C.

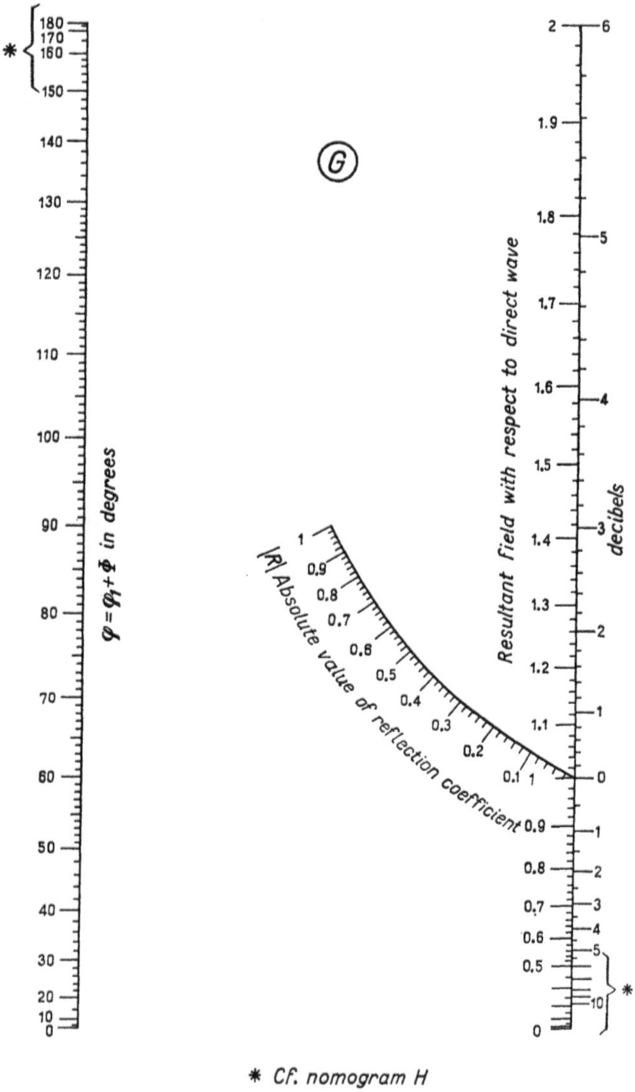

Fig. 11.17a. Nomogram G. φ_1 is the phase shift due to a path difference δ, i.e. $\varphi_1 = 360\,\delta/\lambda$ degrees. Φ is the phase of the reflection coefficient $R = |R|\exp(i\Phi)$. For the ranges marked by an asterisk use Nomogram H (Fig. 11.17b).

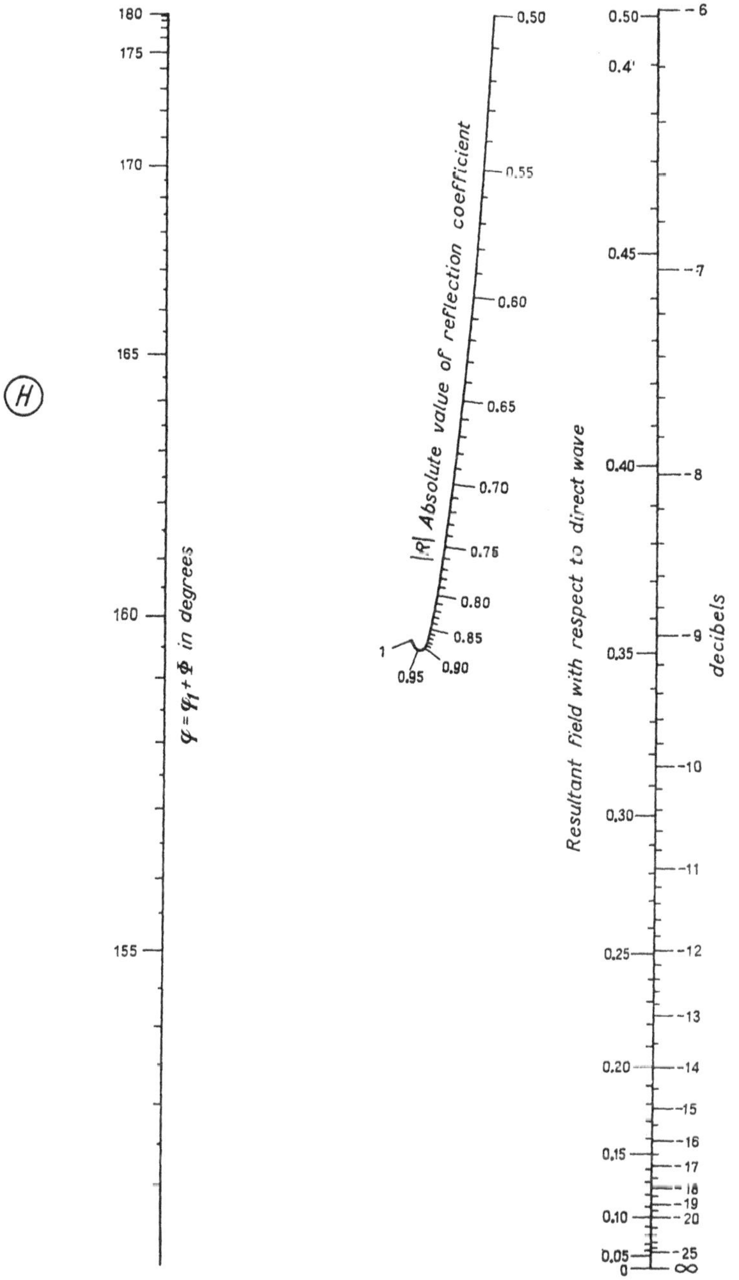

Fig. 11.17b. Nomogram H. Enlarged scales of the ranges marked by an asterisk in Nomogram G (Fig. 11.17b).

Nomogram F—*Divergence Factor*

According to (9) and (14), the divergence factor D is a function of m and d_1/d (or d_2/d), which are given by Nomograms A and B.

The reflection coefficient $|R|$ is obtained from

$$|R| = D|R_0|$$

where $|R_0|$, the reflection coefficient of a plane earth, is given by Fig. 11.1 or 11.2.

Nomograms G *and* H—*The Resultant Field*

As shown above, the absolute value of the reflection coefficient $|R|$ may be obtained from Nomogram F, its phase from Figs. 11.3 and 11.4, and the path difference between the direct and reflected rays from Nomogram D. The value of the resultant field may then be obtained by successively using equations (11) and (10). We shall in fact use only (11):

$$\varphi = 2\pi \frac{\delta}{\lambda} + \Phi$$

where φ is the phase shift between the direct and reflected rays; the value of (10) may be derived from Nomograms G and H.

With the aid of this set of nomograms, the principal quantities occurring in line-of-sight paths may thus be rapidly determined: the geometry of the link, the direct and reflected fields, the resultant field and its variations in space.

CHAPTER 12

THE REFLECTION OF WAVES FROM IRREGULAR GROUND

Application of Theoretical Results to the Surface of the Earth

The laws of reflection by a perfectly smooth earth as we have recalled them in the preceding chapter cannot, in general, be directly applied to the real earth: in many cases they would contradict experimental results.

The irregularities of the earth's surface must thus be taken into account. In order to do this, we shall make use of the theoretical results derived in the first part of this monograph. These results are very general and may be applied to almost any rough surface. In the present chapter we shall see how they may be applied to the special case of ground reflections in a radio link.

12.1. Qualitative Description of Scattering Phenomena: Criterions of Roughness

The first problem that arises is to decide whether the surface of the earth may be considered smooth or whether, on the contrary, the role of the irregularities cannot be neglected. This question has been treated in detail in Part I (Chapter 2). We shall therefore restrict ourselves to recalling a few results.

Let us first suppose that the surface of the earth is perfectly smooth. To a first approximation it may be assumed that the reflected energy originates in a zone of the earth's surface, the "first Fresnel zone" defined as the set of points P of this surface such that

$$\delta(P) - \delta(I) = \frac{\lambda}{2}$$

where $\delta(P)$ is the optical path of a light ray travelling from transmitter to receiver by way of reflection at a point P; I is the point of reflection given by the laws of geometrical optics (angle of incidence equal to the angle of reflection), and λ is the wavelength. It may be shown that the first Fresnel zone is bounded by an ellipse (first Fresnel ellipse) the centre of which lies approximately at I.†

Let us now introduce the irregularities of the reflecting surface. These irregularities should be defined by their vertical dimension Δh: if there is a height difference Δh between two points of the surface, then the waves reflected at those two points will be shifted in phase with respect to each other by

$$\Delta\Phi = \frac{4\pi\Delta h \sin\gamma}{\lambda}$$

where λ is the wavelength and γ the grazing angle (complement of the angle of incidence θ_1). If the surface is to behave like a smooth surface, then the waves reflected at all points that participate in the reflection, i.e. the points of the first Fresnel zone, must be only very slightly shifted in phase with respect to each other. Rayleigh proposed to consider a surface smooth if

$$\Delta\Phi < \frac{\pi}{2}$$

or according to (1),

$$\Delta h < \frac{1}{8}\frac{\lambda}{\sin\gamma}$$

where Δh is the mean height of the irregularities *within the First Fresnel ellipse*. The quantity $\lambda/(\sin\gamma)$ plays an important part in many problems of wave propagation and we shall meet it frequently in the course of our investigations. It is called the *space wavelength.*

The Rayleigh criterion (2) supplies only a qualitative indication and uses only orders of magnitude: the vertical dimension of the irregularities Δh is not very exactly defined‡ and the coefficient 1/8 is according to some authors§ better replaced by other values: 1/16, 1/32, etc. In reality there is no well-defined dividing line between a smooth and a rough surface; as a quantitative investigation will show,‡‡ the transition from one to the other

† We shall return later (Section 12.3) to the question of how to determine this ellipse in practice.

‡ According to some authors, Δh is the maximum height difference between two points of the surface, according to others, it is the standard deviation of this height, etc. At this point we shall regard Δh as an order of magnitude without attempting to give a more exact definition.

§ See, for example, Kerr [1951], Burrows and Attwood [1959], etc. The extreme values ascribed to this coefficient seem to be 1/40 and 1/8.

‡‡ Cf. below, Section 12.3.

is continuous. Therefore there is little sense in trying to find a more exact definition of the parameters and the value of the coefficient in the Rayleigh criterion.

Other criterions of roughness such as that of Bullington differ little from the Rayleigh criterion.†

12.2. Specular Reflection and Diffuse Scattering

The Rayleigh criterion establishes an approximate dividing line between truly rough surfaces and those that may be regarded as smooth, but this does not suffice for a quantitative description of the reflection by a rough surface.

It has been shown in Part I of this book that the field scattered by a rough surface should be considered as the sum of two components: the specular component and the diffuse component. This was apparent in the analysis of the field scattered by a normally distributed surface‡ and also in the investigation of other models of rough surfaces.§ The existence of these two components is thus not simply a matter of the choice of a model of a particular rough surface, but a more general phenomenon which appears in any type of rough surface.

Specular reflection is a reflection of the same type as caused by smooth surface: it is directional and obeys the laws of classical optics. Its phase is coherent and it is the result of the radiation of the points on the Fresnel ellipse, which transmit waves of approximately equal phases towards the receiver. Its fluctuations‡‡ have a relatively small amplitude.

Diffuse scattering is a phenomenon that has little directivity† and which consequently takes place over a much larger area of the surface than the first Fresnel zone. Its phase is incoherent, and its fluctuations, which have a large amplitude, are Rayleigh-distributed.‡

---

† Cf. Bullington [1954]. In the case of a link between two points of equal height h above flat ground, the expression (1) for $\Delta\Phi$ becomes

$$\Delta\Phi = \frac{2\pi h}{H_0^2}\Delta h$$

where H_0 is the width of the first Fresnel zone. The Bullington criterion is obtained by setting $\Delta\Phi < \pi/2$.

‡ Cf. Part I, Section 5.3.

§ Cf. Part I, Sections 6.4 and 6.5.

‡‡ Cf. Part I, Chapter 7.

† Cf. Part I, Section 5.3.

‡ Cf. Part I, Chapter 7.

The existence of these two components and their properties is not only shown in various theoretical investigations of scattering by rough surfaces, but they have also been demonstrated experimentally by very varied measurements.

Let us first note that in the region of visible light waves this phenomenon can be currently observed. Everyone knows that a well polished surface (mirror) reflects light in a directive manner according to the laws of classical optics, whereas a plane, but more rough surface (unpolished glass, white wall, ceiling, etc.) will scatter the incident light into all directions. The same surface (e.g. a piece of glossy paper) may appear white when viewed under a small angle of incidence (diffuse scattering), whereas at grazing incidence it may show the image of an object (specular reflection, which sets in when γ is small, in agreement with the Rayleigh criterion). This image is always a little pale as though it were covered by a white veil (superposition of the diffuse onto the specular component).

In radio communication, these two types of reflection can hardly appear in the propagation of long and medium waves: it is sufficient to examine the Rayleigh criterion (2) to see that for decametric and longer waves, the ground nearly always behaves like a perfectly smooth surface.† As against this, in the ultra-short wave band the effect of the irregularities of the ground and the two components of the reflected field have been clearly evident ever since the first measurements performed in this wave band.

By different methods and using slightly different notations and vocabularies, several investigators have arrived at the same result: the existence of specular reflection and diffuse scattering, the two phenomena sometimes being superposed.

Thus Ford and Oliver [1946] studied ground reflections using a very wide beam‡ and found that when the irregularities of the ground are sufficiently large, the reflection ceases to be directional. Other authors§ distinguished the two components by their difference in phase coherence rather than their difference in directivity: if the surface of the earth is sufficiently smooth for the reflection to be specular, the reflected wave interferes with

---

† In addition, since the antenna heights are usually smaller than the wavelength, the laws of geometrical optics may rarely be applied and the concept of reflection often loses its meaning.

‡ Their measurements are described in greater detail in the annex to Chapter 13.

§ Sherwood and Ginzton [1955], Straiton and Tolbert [1956], MacGavin and Maloney [1959], Tomlinson and Straiton [1959] and others. Their measurements are described in the annex to Chapter 13.

the direct wave, giving rise to regularly spaced interference fringes.† On the other hand, if the surface is rough (by the Rayleigh criterion), diffuse scattering sets in: instead of regular interference fringes, random variations of the received field are observed.‡

Specular reflection and diffuse scattering often occur simultaneously. An analysis of the variations of fieldstrength in space should therefore permit the separation and study of the two components of the reflected field. This has been done in two independent investigations, one by Beard, Katz and Spetner [1956], and the other by MacGavin and Maloney [1959].§ Studying simultaneously the variations of fieldstrength in space and its fluctuations in time, the Soviet authors Braude, Komarov and Ostrovskiy [1958]‡‡ have arrived at the same conclusion.

The simultaneous existence of specular reflection and diffuse scattering has been demonstrated by Boithias and Spizzichino† in a completely different way: the fact that diffuse scattering takes place over a region that substantially differs from the Fresnel ellipse (where specular reflection is produced) was utilized and the two components were then separated by goniometry.†

The existence of these two components of the reflected field is thus nct simply an hypothesis: it is a soundly established experimental fact that becomes evident in any method used to approach the problem of the reflection of electromagnetic waves from the surface of the earth. Moreover, these two components having very different properties, they also have very different practicel effects. For example:‡

(a) *Interference with the direct ray fadings.* The specularly reflected field may, if its amplitude is comparable to that of the directly propagated field, produce deep fades of the resultant field through interference. The diffusely scattered field, the amplitude of which is usually smaller, produces more rapid and less deep fading which, though not negligible, is less irksome when it occurs by itself.

(b) *The effect of ground reflections on the antenna-gain pattern.* If the earth reflects specularly in the neigbourhood of an antenna, the total field will be equivalent to the directly propagated field with the field of the image of the antenna superposed. The original (free-space) antenna pattern is then

† Cf. Section 11.3.

‡ Cf. below, Section 12.6.

§ These experiments are described in greater detail in the annex to Chapter 13.

‡‡ Cf. Spizzichino [1959].

† We shall return to these experiments and treat them more fully in Chapter 13.

‡ The problems listed here in summary will be treated in more detail in Chapters 16 and 17.

modified in a way that can easily be predicted. Diffuse scattering affects the antenna pattern to a lesser extent, but this effect is random and cannot be exactly predicted.

(c) *Mutual interference of radio links.* In a radio link with beamed antennae the specularly reflected ray is always directed towards the receiving antenna. If, however, the reflection is diffuse, the reflected energy is scattered into all directions; the region of the earth's surface in the neighbourhood of the transmitting antenna becomes a source of interference for other links. Back-scattering by the ground or the sea, which is detrimental to radar operation, is a special case of this interference caused only by the diffuse component of reflection.†

Regardless of whether one is interested in the theoretical or experimental aspect or in the practical applications of the problem, the study of ground reflections always brings to light these two fundamentally different phenomena: specular reflection and diffuse scattering. We shall always investigate them separately.

12.3. Quantitative Investigation of Specular Reflection in Radio Links

So far our aim has been to separate the principal *qualitative* concepts concerning reflections by the rough surface of the earth. Having done this, we may now proceed with a *quantitative* investigation of the problem.

We begin with specular reflection. Let us recall that certain properties of specular reflection do not differ from those of the reflection by a perfectly smooth surface:

the region of the surface where specular reflection is produced is the first Fresnel zone;

the phase of the reflected ray, and hence the form of the interference lobes are the same as if the reflecting surface were perfectly smooth.

Everything that has been said in the chapter devoted to the reflection of waves by a smooth earth (Chapter 11) may thus be applied to the study of specular reflection with the following two restrictions:

the reflection coefficient is smaller than that of a smooth earth;

the reflection coefficient fluctuates.

Let us set

$$R_s = \rho_s D R_0$$

† In view of the importance of this problem, a special chapter (Chapter 18) has been devoted to it.

where R_s is the reflection coefficient for specular reflection by a rough earth, R_0 is the reflection coefficient of a smooth and plane earth and D is the divergence coefficient caused by the curvature of the earth as introduced in Section 11.2: thus DR_0 represents the reflection coefficient of smooth but spherical earth.

ρ_s is the coefficient by which the reflection coefficient R_0D must be multiplied to respect the effect of the irregularities of the earth's surface.

In Part I of this monograph, exact results have been given for the case when the absolute value of the reflection coefficient DR_0 of a smooth earth equals unity. The coefficient ρ_s, called "scattering coefficient" is then equal to the reflection coefficient R_s. We shall assume that the expressions for ρ_s derived in Part I remain valid for the case when $DR_0 \neq 1$. This is equivalent to assuming that the coefficient ρ_s depends only on the geometrical form of the irregularities of the reflecting surface, but is independent of its chemical nature. This assumption, which has been justified theoretically in Part I,† is in good agreement with experimental results‡ and has been used, though implicitly, by several authors.§

With this hypothesis, the study of specular reflection reduces to that of the "scattering coefficient" ρ_s which has been developed in Part I: its mean value and variance were derived in Chapters 5 and 6, and its statistical distribution in Chapter 7. A study of ρ_s should also include some considerations of the depolarization of the reflected wave (Chapter 8).

Expression for the Mean Value of the Reflection Coefficient

According to (3), the root mean square value of the coefficient of specular reflection is given by

$$(R_s)_{\text{RMS}} = (\rho_s)_{\text{RMS}}\, DR_0 \tag{4}$$

where $(\varrho_s)_{\text{RMS}}$ is the root mean square value of ρ_s. In Part I of this book its author has preferred using the root mean square value $(\rho_s)_{\text{RMS}}$ of ρ_s

† Sections 4.4 and 5.4. If we restrict ourselves to a plane earth ($D = 1$) and assume the undulations of the earth's surface to be much larger than the space wavelength, we may interpret the mathematical treatment given there as follows: the radiation reflected by the entire surface may be considered as the sum of reflections at each of its points. Owing to the undulations of the surface, the grazing angle of a plane wave varies in an interval (γ_1, γ_2). If R_0 may be considered independent of γ in this interval, the field reflected at any point of the surface is reduced by the same factor $|R_0|$, whilst its phase is shifted by the same amount $\arg R_0$: hence the resulting field is also reduced by a factor $|R_0|$ and turned through a phase angle $\arg R_0$.

‡ Cf. Section 14.2.

§ Gough [1958] and others.

rather than its mean amplitude $\langle|\rho_s|\rangle$ both for reasons of mathematical expedience and for physical reasons: the square of $(\rho_s)_{RMS}$ gives the mean reflected power. In practice $\langle|\rho_s|\rangle$ and $(\rho_s)_{RMS}$ are always very near in value so that within the accuracy of measurements they may be used equivalently.†

Various theoretical expressions for $(\rho_s)_{RMS}$ have been derived in Part I: they correspond to different theoretical models of rough surfaces. Thus for a Gaussian model (Part I, Section 5.3)

$$\langle|\rho_s|^2\rangle = e^{-(\Delta\Phi)^2} \tag{5}$$

with

$$\Delta\Phi = \frac{4\pi\Delta h \sin\gamma}{\lambda} \tag{6}$$

where Δh is the standard deviation of the normal distribution of heights,‡ γ the grazing angle and λ the wavelength. It will be noted that (6) is an analogy of the expression (1) for the phase shift appearing in the Rayleigh criterion: the only difference is that Δh, which represents only an order of magnitude in the Rayleigh criterion, is now precisely defined.

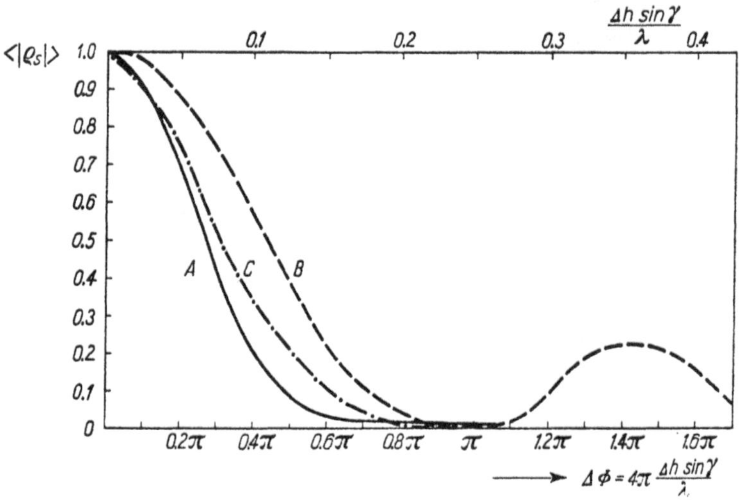

Fig. 12.1. Dependence of the mean square of the scattering coefficient on the phase variance (cf. Part I, Chapter 5). A—normally distributed surface: $\langle|\rho_s|^2\rangle = \exp(-\Delta\Phi)^2$; B—uniformly distributed surface: $\langle|\rho_s|^2\rangle = \mathrm{sinc}^2\Delta\Phi$; C—First Fresnel zone of a uniformly distributed surface: $\langle|\rho_s|^2\rangle = (1-\Delta\Phi/\pi)\mathrm{sinc}^2\Delta\Phi$.

† Cf. Beckmann [1957].

‡ The distribution of heights of the points of the earth's surface is usually the Gauss or normal distribution. We have preferred to denote the standard deviation of this distribution by the symbol Δh as is usual in papers on propagation problems, rather than by the symbol σ used in theoretical investigations and used in Part I.

Other models of rough surfaces treated in Part I† lead to similar results: $\langle|\rho_s|^2\rangle$ is always a function of $\Delta\Phi$ given by (6) in terms of a parameter Δh which defines the height of the irregularities. Some of these results are compared in Fig. 12.1: for values of $\Delta\Phi$ smaller than 2π (the only values for which specular reflection is important) they differ very little; considering the errors of measurement and the statistical scatter of experimental results, the difference between the curves corresponding to different models is not appreciable. We shall adopt the normally distributed model: the normal (Gauss) distribution seems to us less artificial than the other distributions, which often include discontinuities, and we shall see‡ that it accounts well for the experimental results.

Practical Determination of the Mean Reflection Coefficient

The mean reflection coefficient is now given by (4), (5) and (6). Among the quantities occurring in these equations, D, R_0 and γ can easily be obtained by using Figs. 11.1 and 11.2 and the nomograms in the annex to Chapter 11. The only difficulty that remains is the determination of Δh, the standard deviation of the heights of the points of the reflecting surface inside the first Fresnel ellipse.

The determination of Δh must therefore begin with that of the first Fresnel ellipse. If the surface of the earth may be regarded as plane, the reader may use the relations (3), (4) and (5) of Part I, Section 2.2, which give the position and dimensions of the Fresnel ellipse. However, in the case of a radio link over the surface of the earth, these general relations may often be replaced by simple approximate expressions. Two special cases, treated in the annex to this chapter, cover the majority of practical problems. They are the following:

the case where the antenna heights and their distances from the point of reflection are much larger than the wavelength (this applies to most shortwave radio links);

the case where the transmitter and receiver are at very different heights (this applies to various problems connected with the antenna pattern: radar coverage, antenna diagrams for ionospheric communication, etc.)

---

† The reader will find for example in Part I, Section 6.4, the result

$$(\rho_s)_{\mathrm{RMS}} = \frac{\sin(\Delta\Phi/2)}{\Delta\Phi/2}$$

derived in the case of a surface, the points of which are uniformly distributed over all heights contained between h_0 and $h_0+\Delta h$.

‡ Cf. Section 14.2.

In the annex, these two problems are first treated assuming the earth to be plane. Simple formulae and graphs then give the position and dimensions of the Fresnel ellipse. The effect of the earth's curvature is then analysed and practical methods of calculating the dimensions of the Fresnel ellipse are given for reflection from a surface of arbitrary profile.

Once the first Fresnel ellipse has been determined, Δh, the root mean square value of the height of the points of the earth, can easily be obtained: by means of a profile section in case of terrain with hills and valleys; on the basis of information on buildings, vegetation, etc., if the terrain is less perturbed;

in the case of reflections from the sea, the height of the waves Δh may be appreciated by direct observation, or approximately deduced from the "state of the sea". Tables correlating the "sea state" as used by seamen and the wave height Δh have been proposed† and we shall use the values given in Table 12.1.

TABLE 12.1—Sea State and Wave Height

No.	Sea state	Wave height Δh
0	Calm	0
1	Smooth	0–30 cm
2	Slight	30 cm–1 m
3	Moderate	1 m–1.50 m
4	Rough	1.50 m–2.50 m
5	Very rough	2.50 m–4 m
6	High	4 m–6 m
7	Very high	> 6 m

Fluctuations of the Specularly Reflected Field

From what we have considered so far, the mean amplitude of the specular-reflection coefficient may be calculated. The reflection coefficient fluctuates strongly and we shall see that it is impossible to interpret the characteristics of the resultant field correctly without taking account of these fluctuations. It is a weakness of many analyses that have hitherto been published that these fluctuations have been neglected and the specular-reflection coefficient considered constant, all fluctuations of the received wave being interpreted as the result of diffuse scattering.

The analysis performed in Part I of this monograph (Chapter 7) shows that the fluctuations of the specular-reflection coefficient R_s are far from being negligible and provide the means for calculating the distribution of

† Such tables in a simple form may be found in Kerr [1951], Court [1955] and others.

its instantaneous values. The results are very general and may be applied to different models of rough surfaces.

We shall here restrict ourselves to recalling certain particular results that will be useful in the following chapters.

1. Whatever the model of a rough surface,† the distribution of R_s depends on the same parameter Δh, which occurs in the expression of its mean amplitude.‡ The variance of R_s is zero if $\Delta h = 0$ and increases with growing Δh; if Δh tends to infinity, the (normalized) variance tends to its maximum value which is that of the Rayleigh distribution.

2. Whatever the model of a rough surface, the reflected field fluctuates as though it resulted from the superposition of a constant field and a field with mean value zero, whose real and imaginary parts are normally distributed with variances s_1, and s_2 respectively.

3. The case of any normally distributed rough surface, which we have found to be best suited for investigations of reflections from the earth's surface, is treated; the values of s_1 and s_2 are given in Table 7.1 (Part I, p. 123) as functions of Δh; the distribution $|R_s|$ may hence be derived.§

Depolarization of the specularly reflected wave. The reader is referred to Section 8.3, Part I. For the applications in which we are interested, we recall that a wave reflected in the plane of incidence — which is the case for the specular component of ground reflections — is not depolarized if the wave is polarized purely vertically or horizontally.

12.4. Quantitative Investigation of Diffuse Scattering: Mean Value of the Scattered Field

This problem has been studied theoretically in Part I of this monograph. In Chapters 5 and 6 the author has derived the expressions giving the mean value of the amplitude of the scattering coefficient for the component of diffuse scattering. However, to be applied to the concrete problem of scattering by the surface of the earth, these general expressions must be further mathematically treated. Throughout Part I of this book, only the scattering of a *plane wave* by a rough surface has in effect been studied; in other words, it was assumed that the dimensions of the reflecting surface are much smaller than its distance from the transmitter, so that the grazing angle γ (or its complement, the angle

† Cf. Part I, Table 7.1 (p. 123). Δh intervenes through the intermediate parameters σ and a.

‡ Cf. equation (5) and (6) of this chapter.

§ Using (10), (11) and (16) of Section 7.2 and (35) or (86) of Section 7.1 (Part I), one obtains the distribution of ρ_s, which is related to R_s by (3). According to (3), the distribution of R_s is, apart from a constant factor, identical with that of ρ_s.

of incidence θ_1) remain constant over the surface. It turns out that in the majority of applications that will be treated here, this condition is fulfilled. Thus in the case of the reflection of electromagnetic waves by tropospheric sheets, or by the Moon, we shall see that the reflecting region of the surface is always small compared with its distance from the transmitter. This is also the case for specular reflection by the surface of the earth, where the region producing the reflection is the Fresnel ellipse, and the dimensions of the latter are small compared with its distance from the transmitter. On the other hand, diffuse scattering by the surface of the earth is a phenomenon with little directivity and the region of the earth's surface producing it extends right up to the foot of the transmitting and receiving antenna. Thus, for a given position of the transmitting and receiving antennae, a very extensive region of the surface of the earth may, by diffuse scattering, re-radiate energy from the transmitter towards the receiver: this region may extend from the neighbourhood of the transmitter to the neighbourhood of the receiver. The grazing angle γ cannot be considered constant over this "shining" or "glistening surface".†

This does not mean that the results derived in Chapters 5 and 6 of Part I are useless here. If the angle γ cannot be considered constant over the entire "glistening surface", it is nearly always possible, in the practical cases we are going to study, to divide this surface into smaller elements such that

(a) each element of surface δS is large compared with the horizontal dimensions of the irregularities: it is then possible to treat these irregularities by statistical methods inside each of the surface elements;

(b) each element of surface δS is sufficiently small for the angle γ and the distance to the transmitter not to vary appreciably from one point of the element to another.‡

Thus the results of Chapters 5 and 6 (Part I) may be applied to each element δS. Subsequent integration then enables the diffuse-scattering coefficient to be determined for the entire surface. It is this integration which will be the subject of the rest of this section; but first we must go back to the basic equations that we intend to use.

---

† French original: *surface éclairante* (Transl. note).

‡ By this it should be understood that inside a surface element δS the variations of γ are so small that they entail only a small variation of the reflection coefficient R_0 of a smooth earth and of the "scattering coefficient" ρ_d for diffuse scattering, which we shall investigate below. The distance to the transmitter r_1 should vary so little that the energy flux at each point, proportional to $1/r_1$, varies little.

12.4.1. *Theoretical Foundations*

In Chapters 5 and 6 (Part I), the laws of diffuse scattering were derived for different models of rough surfaces. To bring the calculation of the scattering coefficient of the earth's surface to a conclusion and to obtain concrete and useful results, we must choose among these models. Moreover, we shall want to compare the results of integration performed on a particular model with those that would have been obtained by using another model. We therefore begin by comparing the general expressions of the scattering coefficient in the case of the earth's surface corresponding to several models of a rough surface.

Choice of a model. Let us first examine the model treated in Chapter 5: the surface of the earth is rough, the heights of its points are normally distributed with standard deviation σ.† Its horizontal autocorrelation distance is identical for all directions and denoted by T; its autocorrelation function at a distance τ is of the form‡

$$C(\tau) = e^{-\tau^2/T^2}$$

To these two assumptions, made in Chapter 5, we add another, namely that the standard deviation of the height of the irregularities σ is large compared with the space wavelength $\lambda/\sin\gamma$; in other words we shall investigate diffuse scattering only in the case when specular reflection is negligible. Finally we assume, as was done in Sections 5.1 to 5.4 (Part I), that the reflection coefficient of a perfectly smooth, plane, and infinite earth would be equal to -1.

Under these assumptions, the mean square of the reflection coefficient δR_1 of a surface element δS is given by§

$$\langle |R_1|^2 \rangle = \frac{\delta S}{\pi r_2^2} \frac{1}{\cos^4 \beta} \cot^2 \beta_0 \exp\left(-\frac{\tan^2 \beta}{\tan^2 \beta_0}\right) \tag{7}$$

where β is the angle made by the bisector of the incident and scattered rays with the z axis; β_0 is given by

$$\tan \beta_0 = \frac{2\sigma}{T} \tag{8}$$

$\tan \beta_0$ may be interpreted as a mean value of the ratio of the vertical and the horizontal dimensions of the irregularities, and represents a mean-square value of the slope of the irregularities;‡‡ r_2 is the distance from the surface

† In Part II this quantity is denoted by Δh.

‡ Part I, Section 5.3, equation (4).

§ Cf. Part I, Section 5.3, equation (63), p. 89.

‡‡ Cf. Part I, Appendix D, p. 193.

element δS to the receiver. In (7), the reflection coefficient δR_1 is defined as the ratio of the reflected field to the incident field on the reflecting plane. If we call δR the ratio of the reflected field to the incident field received by the same receiver (which is more often used in practical problems), δR is given by

$$\delta R = \frac{r}{r_1}\delta R_1 \tag{9}$$

where r is the distance from the transmitter to the receiver, and r_1 is the distance from the transmitter to the reflecting plane. Using (7) and (9), we find:

$$\langle|\delta R|^2\rangle = \delta S \frac{r^2}{\pi r_1^2 r_2^2} \frac{1}{\cos^4\beta} \cot^2\beta_0 \exp\left(-\frac{\tan^2\beta}{\tan^2\beta_0}\right) \tag{10}$$

Since the vertical dimension σ of the irregularities is in general much smaller than the horizontal dimension T, the angle β_0 is small. In practical links, β is also a small angle, so that (10) becomes

$$\langle|\delta R|^2\rangle \approx \delta S \frac{r^2}{\pi r_1^2 r_2^2} \cot^2\beta_0 \exp\left(-\frac{\tan^2\beta}{\tan^2\beta_0}\right) \tag{11}$$

Let us now consider another method of a rough surface briefly described in Part I (Sections 6.4 and 6.5): the surface, which we shall assume to be very rough, is composed of small elements, each of which acts as a small mirror in reflecting the incident wave according to the laws of geometrical optics. This model is completely defined by the distribution of slopes α of these elementary mirrors: it is assumed that α does not exceed a certain value β_{max}, its probability distribution being uniform for $\alpha < \beta_{max}$. Under these assumptions, the field at the point of reception is caused by those of the elementary mirrors where the normal to the surface bisects the angle between the incident ray and the ray reflected towards the receiver; at these points we have

$$\alpha = \beta$$

The energy reflected by the surface thus originates at a region defined by

$$\beta < \beta_{max}$$

Using this model of a rough surface, one finds the following expression for the mean field reflected by unit area of the surface:

$$\langle|\delta R|^2\rangle \approx \frac{1}{4\pi}\left(\frac{r}{r_1 r_2}\right)^2 G_0(\beta, \beta_{max})\ \delta S \tag{17}$$

where $G_0(\beta, \beta_{max})$ is given by

$$G_0(\beta, \beta_{max}) = \begin{cases} \cot^2\beta_{max} & \text{for} \quad \beta \leqslant \beta_{max} \\ 0 & \text{for} \quad \beta > \beta_{max} \end{cases} \tag{18}$$

This expression for $\langle|\delta R|^2\rangle$ is similar to (11) obtained for a normally distributed surface.

The function $G_0(\beta, \beta_0)$ plays the same part as the function

$$\cot^2\beta_0 \cdot \exp\left(-\frac{\tan^2\beta}{\tan^2\beta_0}\right)$$

contained in (11). These two functions, in which β_0 and β_{max} play the same part, are equal to $\cot^2\beta_0$ (or $\cot^2\beta_{max}$) for $\beta = 0$ and decrease rapidly when β approaches the value of β_0 (or β_{max}). They are compared in Fig. 12.2.

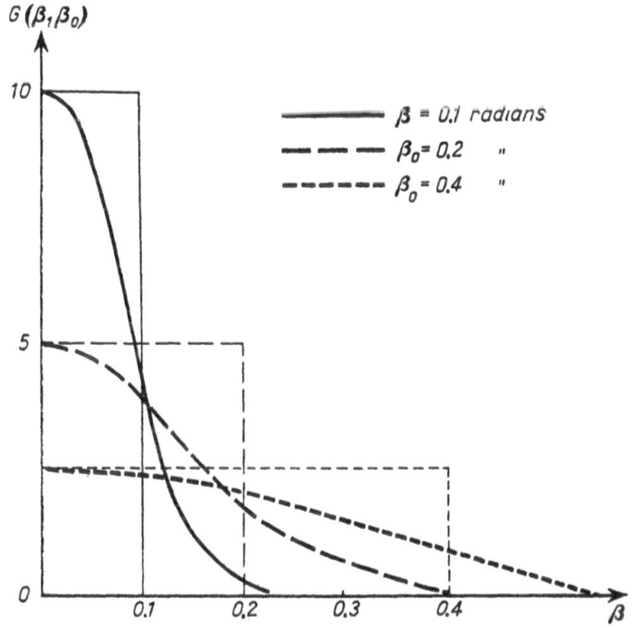

Fig. 12.2. Qualitative indication of the dependence of $\langle|\delta R|^2\rangle$ on β for a particular value of β_0. Full curve: dependence given by (15). Broken curve: dependence to be expected by a rigorous derivation.

This result may be generalized as follows:

1. The above model may be generalized by using a rough surface consisting of elementary mirrors the slopes of which are described by almost

any† statistical distribution with standard deviation β_0: one then finds a general expression of the form

$$\langle|\delta R|^2\rangle = \frac{1}{4\pi}\left(\frac{N}{r_1 r_2}\right)^2 G(\beta, \beta_0)\,\delta S$$

where $G(\beta, \beta_0)$ is a decreasing function of β of the form

$$G(\beta, \beta_0) = \frac{1}{\tan^2\beta_0}\, g\left(\frac{\tan\beta}{\tan\beta_0}\right) \approx \frac{1}{\beta_0^2}\, g\left(\frac{\beta}{\beta_0}\right) \quad \text{for} \quad \beta_0 \ll 1 \tag{20}$$

2. An analysis of the mean reflection coefficient in the general case‡ shows that although (19) and (20) are not valid for all possible cases, they apply to all models that will scatter diffusely regardless of the wavelength (e.g. Bachynski's model). We shall use these expressions in the ensuing analysis.

Integration of the elementary reflection coefficients. The mean square $\langle|E_d|^2\rangle$ of the field scattered by the surface of the earth is obtained by adding the mean squares of the fields scattered by the elements of the surface:

$$\langle|E_d|^2\rangle = \sum \langle|\delta E_d|^2\rangle \tag{21}$$

In doing so we are assuming that the field is completely incoherent, i.e. that the phases of the elementary fields δE_d are independent and uniformly distributed over a basic interval of length 2π (cf. Part I, Section 7.3).

If the transmitting and receiving antennae were isotropic, then we should have

$$\langle|\delta E_d|^2\rangle = \langle|\delta R|^2\rangle E_0^2$$

† It need only be assumed that the probability density is maximum for $\beta = 0$ and decreases monotonously with increasing β.

‡ The general expression for the mean scattering coefficient is given in Part I, Section 5.2, equation (14). Assuming β to be small, so that $F \approx 1$ and $v_{xy} = k\beta$, and that the surface is given by its two-dimensional distribution (or equivalent, e.g. for a normally distributed surface, the one-dimensional distribution and auto-correlation function) then it is easily shown that $\langle\rho\rho^*\rangle$ is of the form

$$\langle\rho\rho^*\rangle = \frac{1}{\beta^2} H(v_{xy} T,\ v_z \sigma)$$

Since this is by assumption to be independent of λ, we may multiply v_{xy} and v_z by the same factor, obtaining

$$\langle\rho\rho^*\rangle = \frac{1}{\beta^2}\, h\left(\frac{v_{xy}}{v_z} \cdot \frac{T}{\sigma}\right) = \frac{1}{\beta^2}\, h\left(\frac{\tan\beta}{\tan\beta_0}\right)$$

with

$$\tan\beta_0 = \frac{2\sigma}{T}$$

E_0 being the field of the direct wave. If, however, the transmitting antenna has different gains g_{10} and g_1 in the directions of the receiver and that of the element δS respectively, and if the gains of the receiving antenna in these two directions are g_{20} and g_2, then we find

$$\langle|\delta E_d|^2\rangle = \langle|\delta R|^2\rangle \left(\frac{g_1 g_2}{g_{10} g_{20}}\right)^2 E_0^2 \tag{22}$$

So far we have assumed that the reflection coefficient of a smooth earth equals unity; the mean square of the scattering coefficient which we shall denote by ρ_d, will then by (21) and (22) be given by

$$\langle|\rho_d|^2\rangle = \sum \left(\frac{g_1 g_2}{g_{10} g_{20}}\right)^2 \langle|\delta R|^2\rangle$$

or, on substituting for $\langle|\delta R|^2\rangle$ from (19),

$$\langle|\rho_d|^2\rangle = \frac{1}{4\pi} \int \left(\frac{g_1 g_2}{g_{10} g_{20}}\right)^2 \left(\frac{r}{r_1 r_2}\right)^2 G(\beta, \beta_0)\, \mathrm{d}\rho \tag{23}$$

where the integral is to be taken over the entire reflecting surface.

Finally we assume, as we have done in the case of specular reflection (Section 12.3), that if the reflection coefficient of a perfectly smooth earth differs in absolute value from unity, the coefficient of diffuse scattering R_d is given by

$$\langle|R_d|^2\rangle = \langle|\rho_d|^2\rangle R_0^2 \tag{24}$$

which is quite analogous to (4) except for the fact that this expression does not contain the divergence coefficient D: in diffuse scattering, the curvature of the earth has a negligible effect on the reflection coefficient.

Conclusions. The relations (23) and (24) for the mean diffuse-scattering coefficient are general and are an approximation that may be applied to the majority of rough surfaces that have so far been considered. In these expressions, $G(\beta, \beta_0)$ is a decreasing function of β which becomes small when $\beta \gg \beta_0$. It is of the form (20).

We shall now apply these expressions to the case of a radio link in the vicinity of the earth: We shall give a general method and then apply it in detail for the special case when $G = G_0$ given by (18); the different possible functions G have similar properties, so that the results obtained for $G = G_0$ are, in fact, more general.

12.4.2. *Practical Application*

The glistening surface. Equation (23) for the mean absolute square of the scattering coefficient contains a function $G(\beta, \beta_0)$ which becomes very small when $\beta \gg \beta_0$. For given positions of the transmitter and receiver above the surface of the earth, one may say for a first approximation that the reflected energy comes from a region of the earth's surface where $\beta < \beta_0$.

This result becomes more exact if we adopt a particular model:†

$$G(\beta, \beta_0) = G_0(\beta, \beta_0)$$

where G_0 is defined by (19). The above statement assumes a physical significance if one uses a model where the reflecting surface consists of small mirrors: β is the slope one of these mirrors must have to reflect a transmitted ray towards the receiver; the reflected energy originates, on the average, at that part of the earth's surface where this slope is sufficiently small.

We shall use the term "glistening surface" for that part of the earth's surface which can participate in the reflection of waves for a given position of the transmitter and receiver. If the gains of the transmitting and receiving antennae are so small that they illuminate the entire region of the surface where

$$\beta < \beta_0 \tag{25}$$

then it is this part of the surface that constitutes the "glistening surface". If, on the contrary, the antennae are very directional, the "glistening surface" may be limited by the antenna beams.

The three practical cases. Three practical cases will be treated in the next three sections, i.e. Sections 12.4.3, 12.4.4, and 12.4.5. They more or less cover all the situations that may arise in radio links.

Sections 12.4.3 and 12.4.5 correspond to the case of a link between two points in the vicinity of the earth (on buildings, towers, etc). The heights of the transmitting and receiving antennae, h_1 and h_2, are assumed small with respect to the distance d between the terminal points of the link.

In Section 12.4.3, the "glistening surface" is assumed limited by condition (25), i.e. by the irregularities of the earth's surface: this will be the case when the antennae are not strongly beamed, e.g. nearly always in the metre wave band.

In Section 12.4.4, the "glistening surface" is assumed limited solely by the antenna pattern: this will be the case if the surface is very rough (β_0 relatively large) or if the antennae are strongly beamed (radio relay links, radars, millimetre-wave links, etc).

In Section 12.4.5, one of the terminals of the link is assumed near the ground, and the other is far removed from it and assumed at infinity; the "glistening surface" is limited by condition (25). This will be the case of radio links with aircraft, or radars observing aircraft or meteor trails (if the antenna is not too directional).

† We shall use the notations β_{max} and β_0 equivalently.

It may also be applied to a tropospheric or ionospheric scatter link in order to consider the effect of ground reflections on the "line-of-sight" part of the path, which extends from one terminal of the link to the scattering medium. It may also be applied to calculations of the solar energy reflected by the earth (albedo).†

12.4.3. *Case I: Link between Two Points in the Vicinity of the Earth, "Glistening Surface" Limited by the Irregularities of the Ground*

Assumptions. We assume

(1) the heights of the transmitting and receiving antennae small with respect to the distance between them:

$$h_1, h_2 \ll d \tag{26}$$

(2) the antenna gains sufficiently small to be practically constant over the entire "useful surface", which is limited by the condition $\beta < \beta_0$. Equation (23) then reduces to

$$\langle|\rho_d|^2\rangle = \frac{1}{4\pi}\int\left(\frac{r}{r_1 r_2}\right)^2 G(\beta, \beta_0)\,\mathrm{d}S \tag{27}$$

We recall that ρ_d *is the coefficient by which the reflection coefficient* R_0 *of a smooth earth must be multiplied with to respect the roughness*:

$$\langle|R_d|^2\rangle = \langle|\rho_d|^2\rangle R_0^2$$

Condition for Case I to arise. Let the transmitting antenna T be beamed towards the receiving antenna and vice versa. These antennae illuminate cones with apertures α_1 and α_2 respectively about the direction TR along which they are beamed: for Case I to arise, the condition

$$\measuredangle RTP \leqslant \alpha_1, \quad \measuredangle TRP \leqslant \alpha_2 \tag{28}$$

must hold for all points P of the "glistening surface" defined by $\beta < \beta_0$ (Fig. 12.3).

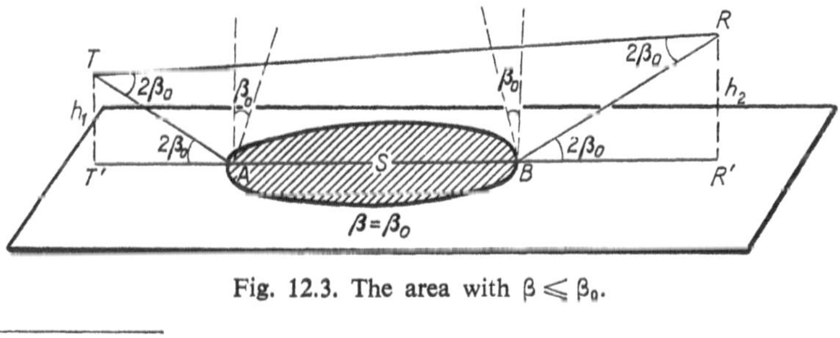

Fig. 12.3. The area with $\beta \leqslant \beta_0$.

† Misme, Villevieille, Robert, Du Castel and Spizzichino [1961].

Using (26) it may be shown that the angles $\sphericalangle RTP$ and $\sphericalangle TRP$ are maximum at the points A and B where the curve $\beta = \beta_0$ bounding the "glistening surface" S is intersected by the vertical plane passing through the transmitter T and the receiver R. The bisectors of the angles $\sphericalangle TAR$ and $\sphericalangle TBR$ conclude an angle β_0 with the vertical. Since by (26) the directions AR, TB and TR may be regarded as horizontal, we have for a first approximation

$$\sphericalangle RTA = \sphericalangle TRB = 2\beta_0$$

Considering (28), we may write the condition for Case I as

$$\alpha_1, \alpha_2 \geqslant 2\beta_0 \tag{29}$$

Determination of the "glistening surface" S. In Case I, the "glistening surface" is bounded by the curve C corresponding to the equation $\beta = \beta_0$. We shall attempt to determine the equation of the curve C in the Cartesian coordinates shown in Fig. 12.4: the axis Oz is vertical and passes through the transmitter T, the plane x_1Oy is the (mean) plane of the earth's surface.

For reasons of symmetry, we may use equivalently either x_1 as the abscissa, or else

$$x_2 = d - x_1$$

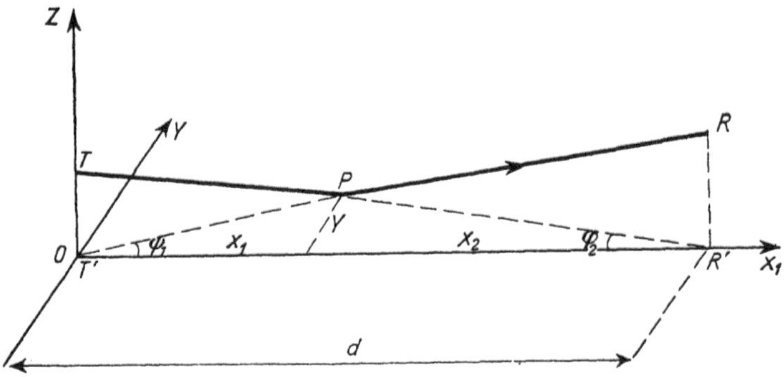

Fig. 12.4. Coordinate system used in the derivation of the "glistening surface".

Since the angle β_0 is assumed small and (26) applies, we have for all points of the "glistening surface" S and its boundary C

$$J \ll x_1, x_2$$
$$h_1 \ll x_1; \quad h_2 \ll x_2$$

Respecting the irregularities and setting β, expressed as a function of x_1, x_2 and y, equal to β_0, we find the equation of C:

$$J = \pm \frac{x_1 x_2}{d}\left(\frac{h_1}{x_1} + \frac{h_2}{x_2}\right)\sqrt{\left[\tan^2 \beta_0 - \frac{1}{4}\left(\frac{h_1}{x_1} - \frac{h_2}{x_2}\right)^2\right]} \tag{30}$$

The shape of this curve is shown in Fig. 12.5 in the plane x_1Oy of the earth's surface. At the terminals of the link, (30) reduces to the equation of the ellipse

$$y = \pm h_i \sqrt{\left[\tan^2 \beta_0 - \left(\frac{h_i}{2x_i}\right)^2\right]}; \quad i = 1, 2$$

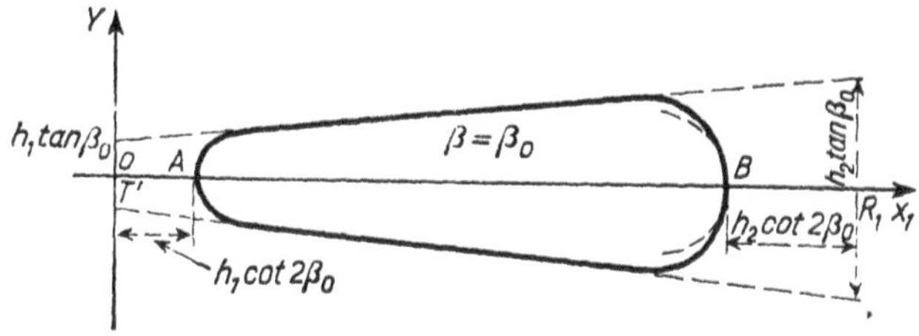

Fig. 12.5. The "glistening surface".

For the greater part of the path we have

$$J = \pm \frac{h_1 x_2 - h_2 x_1}{p} \tan \beta_0$$

If $h_1 = h_2 = h$, the equation of the boundary reduces to the simple form

$$y = \pm h \sqrt{\left[\tan^2 \beta_0 - \frac{h^2}{4}\left(\frac{1}{x_1} - \frac{1}{x_2}\right)^2\right]}$$

It may be seen that—as a result of our assumption (26)—the "glistening surface" has in general the form of a narrow strip extending along the axis of the link.

Expression for the mean scattering coefficient. The mean absolute square of the scattering coefficient for diffuse scattering is given by (27). Whatever the expression for $G(\beta, \beta_0)$, the simplest method of integration consists in first expressing β in terms of the Cartesian coordinates x_1 (or x_2) and y:

$$\tan^2 \beta = \frac{1}{4}\left(\frac{h_1}{x_1} - \frac{h_2}{x_2}\right)^2 + \frac{d^2 y^2}{x_1^2 x_2^2} \frac{1}{\left(\frac{h_1}{x_1} + \frac{h_2}{x_2}\right)^2}$$

which may for simplicity be written as

$$\tan^2 \beta = A(x_1) + B(x_1) y^2$$

Substituting this expression for β in (27), the integration may be performed with respect to the variable y: according to (20), G is a function of β through $\tan^2 \beta$. For simple expressions of G, the integrand is thus a simple function of y, and the integral may be easily evaluated. We introduce the following transformation of variables:

$$\left.\begin{aligned} x_1 &= d\sin^2\frac{\xi}{2} \\ x_2 &= d\cos^2\frac{\xi}{2} \end{aligned}\right\} \tag{31}$$

ξ is an angle that varies between 0 and π when the abscissa point x passes from one terminal of the link to the other. Similarly, we set

$$\left.\begin{aligned} h_1 &= 2h\sin^2\frac{\mu}{2} \\ h_2 &= 2h\cos^2\frac{\mu}{2} \end{aligned}\right\} \tag{32}$$

h and μ are defined by

$$\left.\begin{aligned} h &= \frac{h_1+h_2}{2} \\ \tan^2\frac{\mu}{2} &= \frac{h_1}{h_2} \end{aligned}\right\} \tag{33}$$

Let us apply this general procedure to the case when $G = G_0$ defined by (19): G_0 equals $\cot^2\beta_0$ in the interior of the "glistening surface" S: the integral (27) is thus taken over S, bounded by the contour C given by (30). We then obtain

$$\langle|\rho_d|\rangle^2 = \frac{2}{\pi}K_\beta\int\limits_{\xi_A}^{\xi_B}\frac{1-\cos\mu\cos\xi}{\sin^3\xi}\sqrt{\left[1-\left(K_\beta\frac{\cos\xi-\cos\mu}{\sin^2\xi}\right)^2 d\xi\right]} \tag{34}$$

where K_β is defined by

$$K_\beta = \frac{2h}{d\tan\beta} = \frac{h_1+h_2}{d\tan\beta_0} \tag{35}$$

and ξ_A and ξ_B are the values of ξ at the points A and B (Fig. 12.3). They are also the values of ξ for which the expression under the root of (34) will vanish. For the frequent case when $K_\beta \ll 1$,

$$\xi_A = \sqrt{2K_\beta}\cdot\sin\frac{\mu}{2};\quad \xi_B = \pi-\xi_A$$

Using other models, one arrives at expressions for $\langle|\rho_d|^2\rangle$ with properties very similar properties to those of (34).†

† Cf. footnote, p. 254.

Physical significance of the parameters K_β and μ. Whether one uses (34) or whether one adopts another model, the mean scattering coefficient depends only on the two parameters K_β and μ instead of the four parameters d, h_1, h_2 and β_0 initially introduced to define the radio path.

The parameter K_β is "symmetrical", i.e. it does not change if the two terminals of the link are interchanged. There are several simple physical interpretations which follow from its definition (35).

According to (35) we may write

$$K_\beta = \frac{\tan \gamma_s}{\tan \beta_0} \approx \frac{\gamma_s}{\beta_0} \tag{35a}$$

Thus K_β is the ratio of the grazing angle γ_s of a ray that is specularly reflected to the maximum slope β_0 of the irregularities of the terrain.

It is easily shown that for small values of K_β,

$$1 - \frac{K_\beta}{2} = \frac{\overline{AB}}{d}$$

(Fig. 12.3). Thus the expression $1-K_\beta/2$ represents the section of the radio path (projected onto the ground plane) intercepted by to the "glistening surface": we see that the parameter K_β plays a dominating role in the study of diffuse scattering.

The parameter μ, defined by (33) depends only on the ratio h_1/h_2 of the heights of the transmitting and receiving antennae. When h_1/h_2 varies from 0 to ∞, the angle μ varies from 0 to π, assuming the value $\mu = \pi/2$ for $h_1 = h_2$. Unlike the parameter K_β, which is "symmetrical", the parameter μ represents the asymmetry of the link.

To separate the effects of the parameters K_β and μ, we shall successively consider the case of a symmetrical link ($h_1 = h_2$, or $\mu = \pi/2$) and then the effect of the "asymmetry parameter" μ.

Symmetrical link ($h_1 = h_2 = h$). For simplicity, we shall use the model with $G = G_0$; the scattering coefficient is then given by (34), which, for $\mu = \pi/2$, is equal to

$$\langle|\rho_d|^2\rangle = \frac{2}{\pi} K_\beta \int_{\xi_A}^{\pi-\xi_A} \frac{1}{\sin^3 \xi} \sqrt{\left[1 - \left(K_\beta \frac{\cos \xi}{\sin^2 \xi}\right)^2\right]}\, d\xi \tag{36}$$

with

$$\frac{\sin^2 \xi_A}{\cos \xi_A} = K_\beta \tag{37}$$

Consider first the integrand in (36):

$$f(\xi) = \frac{2}{\pi} K_\beta \frac{1}{\sin^3 \xi} \sqrt{\left[1-\left(K_\beta \frac{\cos \xi}{\sin^2 \xi}\right)^2\right]} \tag{38}$$

$f(\xi)$ represents the participation of the energy scattered by the points with abscissa ξ, i.e. the distribution of the energy scattered by the points of the earth's surface along the link.† It is the product of two factors:

The first factor,

$$f_1(\xi) = K_\beta \sqrt{\left[1-\left(K_\beta \frac{\cos \xi}{\sin^2 \xi}\right)^2\right]}$$

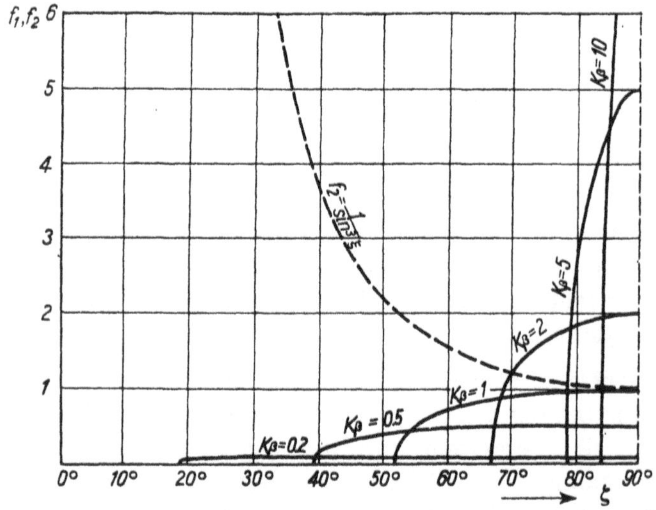

Fig. 12.6. The functions $f_1(\xi)$ (full curve) and $f_2(\xi)$ (broken curve). For $\xi > 90°$ the symmetry of the curves with respect to $\xi = 90°$ may be used.

the variations of which are shown in Fig. 12.6, is due to the boundaries of the "glistening surface", i.e. the result of the condition $\beta < \beta_0$. It has a meaning only if ξ is included in the interval of integration $[\xi_A, \pi-\xi_A]$. If the slope of the irregularities β_0 is small (K_β large), this interval is very small and the scattered energy comes from a small region of the surface about the point of specular reflection. In the limit as $\beta_0 \to 0$ ($K_\beta \to \infty$) we again find the results corresponding to specular reflection.

The second factor, $f_2(\xi) = \sin^{-3}\xi$, originates from the factor $(r/r_1 r_2)^2$

† We recall that ξ is a function of the abscissa x of a point on the earth's surface; thus $f(\xi)$ represents also the contribution to the reflection by points with abscissa x along the radio link.

in the general equation (27). It tends to infinity near the two terminals of the link ($\xi = 0$ and $\xi = \pi$), where r_1 and r_2 become very small; it assumes its minimum value at midpoint on the path. Its variations are shown in Fig. 12.6 (broken curve).

This factor becomes dominating if β_0 is large (K_β small): the "glistening surface" is then extensive, the factor $f_1(\xi)$ varies slowly and $f(\xi)$ is maximum near the terminals of the link.

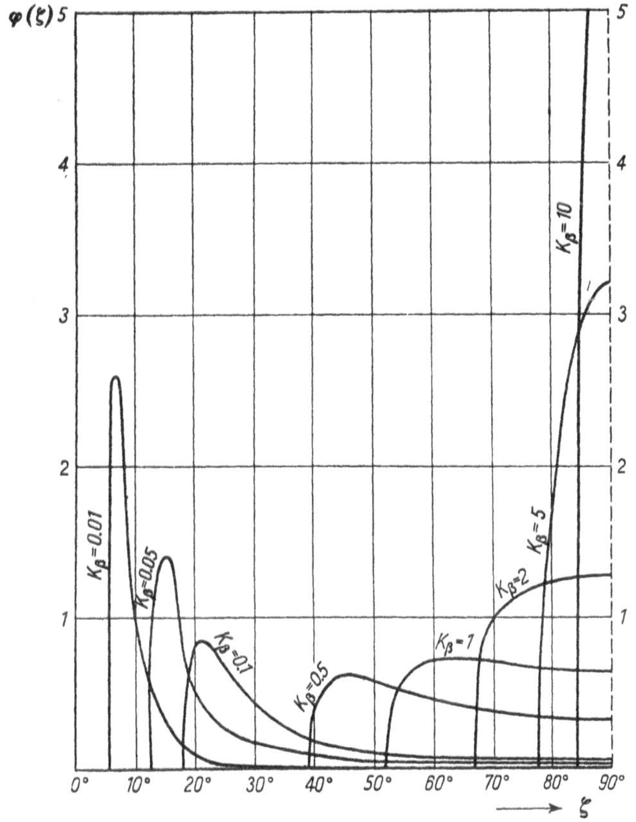

Fig. 12.7. The function $f(\xi)$. The curves are symmetrical with respect to the line $\xi = 90°$.

What we have just said about the variations of these two factors explains the variations of the function $f(\xi)$ plotted in Fig. 12.7: *Depending on the value of* K_β, *diffuse scattering may occur in two forms*:

(a) *if* $K_\beta > 1, f(\xi)$ *is maximum for* $\xi = \pi/2$: *the scattered energy originates in a zone surrounding the point of reflection as given by geometrical optics*;

(b) *if* $K_\beta \ll 1$, $f(\xi)$ *has two maxima located symmetrically with respect* *to $\xi = \pi/2$; the scattered energy comes principally from the two regions in the neighbourhood of the terminals of the link.*

The mean absolute square of the scattering coefficient for diffuse scattering, $\langle|\rho_d|^2\rangle$, is obtained by integrating the functions $f(\xi)$ in Fig. 12.7 from 0 to π. The result is shown in Fig. 12.8.

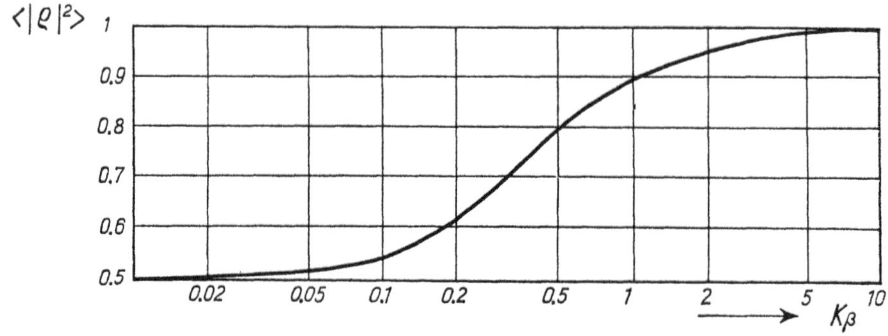

Fig. 12.8. The reflection coefficient in the case of a symmetrical link.

For small values of K_β the scattered energy originates at the two regions in the vicinity of the terminals of the link, so that the integration of (36) may be confined only to those two regions. In practice one may write (36) in the form

$$\langle|\rho_d|^2\rangle \approx \frac{2}{\pi} K_\beta \left[\int_{\xi_A}^{\infty} \frac{1}{\xi^3} \sqrt{\left(1-\frac{K_\beta^2}{\xi^4}\right)}\, d\xi + \int_{-\infty}^{\pi-\xi_A} \frac{1}{(\pi-\xi)^3} \sqrt{\left(1-\frac{K_\beta^2}{(\pi-\xi)^4}\right)}\, d\xi\right]$$

$$= \frac{4}{\pi} K_\beta \int_{\xi_A}^{\infty} \frac{1}{\xi^3} \sqrt{\left(1-\frac{K_\beta^2}{\xi^4}\right)}\, d\xi \qquad (39)$$

where $\xi_A \approx \sqrt{K_\beta}$. It is easily shown that this integration yields

$$\langle|\rho_d|^2\rangle = \tfrac{1}{2}$$

For large values of K_β, (36) yields

$$\langle|\rho_d|^2\rangle = 1$$

For the above cases (a) and (b) we thus find

(a) if $K_\beta > 1$, $\langle|\rho_d|^2\rangle \approx 1$

(b) if $K_\beta \ll 1$, $\langle|\rho_d|^2\rangle \approx 1/2$ (the scattering coefficient of each of the two zones near the terminals of the link equals 1/4).

The above results derived for the particular model with $G = G_0$ do not change qualitatively if another model of a rough surface is adopted (provided a decreasing function $G(\beta)$ is chosen). Only the numerical value of $\langle|\rho_d|^2\rangle \approx 1/2$ for $K_\beta \ll 1$ may change: for small values of K_β, the scattered energy comes from the two zones near the terminals of the link; the expression for $\langle|\rho_d|^2\rangle$ of the same form as (35) may be decomposed into two equal integrals similar to (39) which are always independent of K_β: thus the numerical value of $\langle|\rho_d|^2\rangle$ is constant if $K_\beta \ll 1$, but this value may change with the adopted model.

Finally we note that in the majority of practical radio links, K_β as given by (35) or (35a) is small with respect to unity; thus the most frequent case is that of the scattered energy coming from the regions near the terminals of the link with $\langle|\rho_d|^2\rangle$ constant (equal to 1/2 for the model with $G = G_0$).

Asymmetry effects. Consider now an asymmetrical link with $\mu \neq \pi/2$.

(a) if $K_\beta > 1$, it is easily shown that

$$\langle|\rho_d|^2\rangle \approx 1$$

for any μ.

(b) if $K_\beta \ll 1$ (which is the most interesting case for practical applications), (36) may be developed similarly as in (39):

$$\langle|\rho_d|^2\rangle = \frac{2}{\pi} K_\beta (1-\cos\mu) \int_{\xi_A}^{\infty} \frac{1}{\xi^3} \sqrt{\left(1 - \frac{K_\beta^2 (1-\cos\mu)^2}{\xi^4}\right)} d\xi +$$

$$+ \frac{2}{\pi} K_\beta (1+\cos\mu) \int_{-\infty}^{\xi_B} \frac{1}{(\pi-\xi)^3} \sqrt{\left(1 - \frac{K(1+\cos\mu)^2}{(\pi-\xi)^4}\right)} d\xi \qquad (40)$$

with

$$\xi_A = \sqrt{[K_\beta(1-\cos\mu)]}$$

$$\xi_B = \sqrt{[K_\beta(1+\cos\mu)]}$$

To convert (40) to an expression similar to (39), it is sufficient to introduce the following variables:

$$\xi = \xi_1 \sqrt{(1-\cos\mu)}$$

and

$$\pi - \xi = \xi_2 \sqrt{(1+\cos\mu)}$$

It may then be seen that whatever the value of μ, $\langle|\rho_d|^2\rangle = 1/2$, so that each of the reflecting zones in the vicinity of the terminals of the link has the same scattering coefficient, 1/4.

The parameter μ has no effect on the scattering coefficient either for very large or very small values of K_β. A more detailed analysis would show that it also has little effect on the intermediate values of K_β (the scattering coefficient increases slightly when μ deviates from $\pi/2$). Moreover, since in the majority of practical applications $K_\beta \ll 1$, the effect of the parameter μ may be neglected.

Conclusions. In a link with small antenna gains diffuse scattering depends practically on a single parameter K_β defined by (35) or (35a):

(a) If $K_\beta > 1$, the scattered energy comes from a zone on the earth's surface in the neighbourhood of the point of specular reflection and the value of $\langle|R_d|^2\rangle$ is near to that of $|R_0|^2$ for a smooth surface (i.e. $\langle|\rho_d|^2\rangle \approx 1$).

(b) If $K_\beta \ll 1$, the scattered energy comes from two zones on the earth's surface in the vicinities of the transmitter and receiver. The mean absolute square of the reflection coefficient for the diffusely scattered component, $\langle|R_d|^2\rangle$ is given by

$$\langle|R_d|^2\rangle = \langle|\rho_d|^2\rangle R_0^2$$

where $\langle|\rho_d|^2\rangle$ is constant and depends on the chosen model of the rough surface: a particular model ($G = G_0$) gives

$$\langle|\rho_d|^2\rangle = \tfrac{1}{2}$$

If the link is symmetrical ($h_1 = h_2$), a more exact description is contained in Figs. 12.7 and 12.8. If $h_1 \neq h_2$, this description is only approximate.

12.4.4. *Case II: Terminals of the Link near the Earth, "Glistening Surface" Bounded by the Antenna Patterns*

We assume that the transmitting antenna T and the receiving antenna R are beamed towards each other in the direction TR which is nearly horizontal. If the antennae are very directional and do not comply with condition (29), they will only partly illuminate the zone on the earth's surface where scattering would take place if the antennae were isotropic; the scattering coefficient will thus be smaller than that obtained with weakly beamed antennae.

Approximate solution. The approximate solution follows from the preceding section (12.4.3): When $K_\beta \ll 1$ (i.e. in the majority of practical applications), the scattered energy originates in the two narrow regions near the ends A and B of the "glistening surface" S (Fig. 12.3). These two regions, which consist in practical applications of two points, have the same scattering coefficient if they are illuminated by antennae of small directivity: for the particular model with $G = G_0$ we found that the mean absolute square of the scattering coefficient of each region equals 1/4. If the antenna gains are $g_1(\alpha)$ and $g_2(\alpha)$ along the direction making an angle α with the line TR, then the gains in the direction of point A are $g_1(2\beta_0)$ and $g_2(0)$ and in the direction of point B they are $g_1(0)$ and $g_2(2\beta_0)$.

From this the mean absolute square of the scattering coefficient is found to be

$$\langle|\rho_d|^2\rangle = \frac{1}{4}\frac{g_1(2\beta_0)}{g_1(0)} + \frac{1}{4}\frac{g_2(2\beta_0)}{g_2(0)} \tag{41}$$

The mean reflection coefficient for diffuse scattering is then given by (24).

Exact solution: *Assumptions*. Although the approximate solution (41) suffices for most practical applications, we shall give the outline of a more general solution.

According to (24), the reflection coefficient for the diffusely scattered component is given by

$$\langle|R_d|^2\rangle = \langle|\rho_d|^2\rangle R_0^2$$

with $\langle|\rho_d|^2\rangle$ given by (23). In view of the complexity of the problem† we shall not consider (23) in the general case, but only for the simple model where the function G equals G_0 given by (18). We define the antennae by their gains $g_1(\alpha)$ and $g_2(\alpha)$ in the direction concluding an angle α with their axis, and we assume the "glistening surface" entirely bounded by the antenna patterns, i.e. the integration of (23) is extended to that region of the earth's surface where

$$g_1(\alpha)\cdot g_2(\alpha) \neq 0$$

which is assumed to lie entirely inside the contour $\beta = \beta_0$, which bounds the area of integration in the case of isotropic antennae.

We further take $g_1(\alpha)$ and $g_2(\alpha)$ of the particular kinds

$$\begin{aligned} g_1(\alpha) &= g_{10}\left(1-\frac{\alpha^2}{2\alpha_1^2}\right) \\ g_2(\alpha) &= g_{20}\left(1-\frac{\alpha^2}{2\alpha_2^2}\right) \end{aligned} \tag{42}$$

which is equivalent to limiting these expressions to second-order terms in α, α_1 and α_2 being the 3db half-widths of the transmitting and receiving antenna beams. The above assumption of the "glistening surface" being bounded by the antenna patterns implies

$$\alpha_1, \alpha_2 \ll 1$$

Finally, as in the preceding section, we assume

$$h_1, h_2 \ll d$$

a condition that is practically always fulfilled in actual radio links.

---

† To the five parameters d, h_1, h_2, β_0 and λ, on which $\langle|\rho_d|^2\rangle$ depends for non-directional antennae, we must add two more, α_1 and α_2 (antenna-directivities) for beamed antennae.

The above assumptions imply that inside the surface illuminated by the antennae,

$$y \ll x_1, x_2$$
$$h_1 \ll x_1; \quad h_2 \ll x_2 \tag{43}$$

where we have used the same coordinate system as in the preceding section (Fig. 12.4).

Expression for the reflected power. It is convenient to express (23) by using the angles

$$\psi_1 = \sphericalangle R'T'P; \qquad \psi_2 = \sphericalangle T'R'P$$

(Fig. 12.4) as coordinates of a point P of the surface of integration; these angles are connected with Cartesian coordinates by the relations

$$x_1 = d\frac{\psi_2}{\psi_1+\psi_2}; \qquad x_2 = d\frac{\psi_1}{\psi_1+\psi_2}$$
$$y = d\frac{\psi_1\psi_2}{\psi_1+\psi_2} \tag{44}$$

Then (23) becomes

$$\langle|\rho_d|^2\rangle = \frac{d^2}{4\pi\beta_0^2}\iint_S \left\{1-\frac{\psi_1^2}{2\alpha_1^2}\left[1-\left(\frac{h_1\psi_1+h_2\psi_2}{d\psi_1\psi_2}\right)^2\right]\right\}^2 \times$$
$$\times\left\{1-\frac{\psi_2^2}{2\alpha_2^2}\left[1-\left(\frac{h_1\psi_1+h_2\psi_2}{d\psi_1\psi_2}\right)^2\right]\right\}^2 \frac{\psi_1+\psi_2}{\psi_1\psi_2}\,d\varphi_1\,d\psi_2$$

Neglecting 4th-order terms in ψ_1/α_1 and ψ_2/α_2 [as we have already done in (42)] we find

$$\langle|\rho_d|^2\rangle = \frac{d^2}{4\pi\beta_0^2}\iint_S \left\{1-\left(\frac{\psi_1^2}{\alpha_1^2}+\frac{\psi_2^2}{\alpha_2^2}\right)\left[1-\left(\frac{h_1\psi_1+h_2\psi_2}{d\psi_1\psi_2}\right)^2\right]\right\}\frac{\psi_1+\psi_2}{\psi_1\psi_2}\,d\varphi_1\,d\psi_2$$

Introducing

$$\left.\begin{aligned}\psi_1 &= \alpha_1\psi\cos\varepsilon\\ \psi_2 &= \alpha_2\psi\sin\varepsilon\\ K_1 &= \frac{h_1}{d\alpha_2}; \quad K_2 = \frac{h_2}{d\alpha_1}\end{aligned}\right\} \tag{45}$$

we obtain

$$\langle|\rho_d|^2\rangle = \frac{1}{4\pi\beta_0^2}\iint_S \left[1-\psi^2+\left(\frac{K_1}{\sin\varepsilon}+\frac{K_2}{\cos\varepsilon}\right)^2\right]\frac{\alpha_1}{\sin\varepsilon}+\frac{\alpha_2}{\cos\varepsilon}\Big)\,d\psi\,d\varepsilon$$

where the contour of the surface of integration S is given by the equation

$$\beta^2 = 1-\left(\frac{K_1}{\sin\varepsilon}+\frac{K_2}{\cos\varepsilon}\right)^2$$

Finally, it is convenient to set

$$K_1 = K_\alpha \sin\nu$$
$$K_2 = K_\alpha \cos\nu$$

and

$$\alpha_1 = 2\alpha \sin\chi$$
$$\alpha_2 = 2\alpha \cos\chi$$

The expression for $\langle|\rho_d|^2\rangle$ may then be written as

$$\langle|\rho_d|^2\rangle = \frac{\alpha}{2\pi\beta_0^2}\int_{\varepsilon_1}^{\varepsilon_2}\int_0^{\psi(\varepsilon)} (\psi_0^2-\psi^2)F(\chi, \varepsilon)\, d\varepsilon d\psi$$

or

$$\langle|\rho_d|^2\rangle = \frac{\alpha}{3\pi\beta_0^2}\int_{\varepsilon_1}^{\varepsilon_2} \psi_0(\varepsilon)^3 F(\chi, \varepsilon)\, d\varepsilon \tag{46}$$

where

$$\psi_0(\varepsilon) = \sqrt{[1-K_\alpha^2 F(\nu, \varepsilon)]} \tag{47}$$

and where $F(x, y)$ stands for

$$F(x, y) = \frac{\sin x}{\sin y}+\frac{\cos x}{\cos y} = 2\frac{\sin(x+y)}{\sin 2y} \tag{48}$$

ε_1 and ε_2 are the roots of the equation

$$\psi_0(\varepsilon) = 0$$

Significance of the parameters determining the reflected power. Equation (46) shows that the scattering coefficient $\langle|\rho_d|^2\rangle$ depends in fact on four independent parameters: the ratio α/β_0^2, K_α, χ and ν instead of the six parameters initially introduced: d, h_1, h_2, α_1, α_2 and β_0.

(1) *The parameter* α defined by

$$4\alpha^2 = \alpha_1^2+\alpha_2^2$$

represents an average value of the directivities of the transmitting and receiving antennae.

(2) *The parameter* K_α defined by

$$K_\alpha^2 = K_1^2+K_2^2 \tag{49}$$

with

$$K_1 = \frac{h_1}{d\alpha_2}; \qquad K_2 = \frac{h_2}{d\alpha_1}$$

has a simple physical meaning in the case of a symmetrical link ($h_1 = h_2 = h$, $\alpha_1 = \alpha_2 = \alpha$). In this case we have

$$K = \sqrt{2}\cdot\frac{h}{d\alpha}$$

and it is seen that apart from a factor $1/\sqrt{2}$, the expression for K_α is similar to (35) for K_β, which we met when S was bounded by the irregularities of the earth's surface: α replaces β_0 (Section 12.4.3). The properties of K_α are similar to those of K_β mentioned earlier. As in (35a), we have

$$K = \frac{1}{\sqrt{2}} \frac{\gamma_s}{\alpha} \tag{49a}$$

with γ_s the grazing angle for specular reflection. Also,

$$1 - \sqrt{2} \cdot K_\alpha = \frac{d_u}{d}$$

where d_u is the section of the radio path which, when projected onto the ground, lies inside the "glistening surface".

Finally it should be noted that the condition for the "glistening surface" to be bounded by the antenna patterns may be written as

$$K_\alpha > \frac{1}{2\sqrt{2}} K_\beta$$

(3) *The parameters* ν *and* χ defined by

$$\tan\nu = \frac{K_1}{K_2} = \frac{h_1}{h_2} \frac{\alpha_2}{\alpha_1} \tag{50}$$

and by

$$\tan\chi = \frac{\alpha_1}{\alpha_2} \tag{51}$$

equal $\pi/4$ when the link is symmetrical ($h_1 = h_2$, $\alpha_1 = \alpha_2$). Now since the parameters α and K_α are symmetrical, i.e. they are unaffected by an interchange of transmitter and receiver, ν *and* β *define the asymmetry of the link* (ν plays a similar part to that of μ in Section 12.4.3).

As we have done in Section 12.4.3, we shall first consider the effect of the symmetrical parameters α/β_0^2 and K_α by taking the case of a symmetrical link ($h_1 = h_2$, $\alpha_1 = \alpha_2$) and then the effect of the asymmetry (parameters ν and χ).

Symmetrical link ($h_1 = h_2 = h$, $\alpha_1 = \alpha_2 = \alpha$). $\langle|\rho_d|^2\rangle$ is given by (46), (47) and (48). Since the link is symmetrical, we have $\chi = \nu = \pi/4$ and (48) may be written in the form

$$F(\chi, \varepsilon) = F(\nu, \varepsilon) = \frac{1}{\sqrt{2}} \left(\frac{1}{\sin\varepsilon} + \frac{1}{\cos\varepsilon} \right)$$

Let the integrand of (46) be denoted by $g(\varepsilon)$:

$$g(\varepsilon) = \frac{\alpha}{3\pi\beta_0^2} \psi_0^3(\varepsilon) F(\chi, \varepsilon) = \frac{\alpha}{3\pi\sqrt{2}\cdot\beta_0^2} \psi_0^3(\varepsilon) \left(\frac{1}{\sin\varepsilon} + \frac{1}{\cos\varepsilon} \right)$$

$g(\varepsilon)$ represents the contribution of the points with abscissa ε on the earth's surface to the total scattered energy.† It plays the same part as the function $f(\xi)$ of Section 12.4.3; like the latter, it is the product of two factors:

1. The factor $\psi_0^3(\varepsilon)$, where ψ_0 is given by (48), originates from the limits of the "glistening surface", i.e. the illumination by the antennae. It has a meaning only in the interval of integration $[\psi_1, \psi_2]$; outside this interval it may be considered equal to zero. If α is small (K_α large), this interval is small and may even vanish. In the case of a symmetrical link, this interval is located in the centre of the path.

2. The factor

$$g_2(\varepsilon) = \frac{\alpha}{3\pi\sqrt{2}\cdot\beta_0^2}$$

has its origin in the coefficient $(r/r_1 r_2)^2$ of the general equation (23). It assumes its minimum value at midpoint on the path and tends to infinity near the link terminals when r_1 or r_2 become small.

The variations of these two factors are plotted in Fig. 12.9. Figure 12.10 shows the variation of $g(\varepsilon)$ and Fig. 12.11 gives the reflection coefficient as a function of K_α.

If $K_\alpha > 0.5$, there is no reflection (the surface is not illuminated by the antennae).

If $0.1 < K_0 < 0.5$, the reflection coefficient is small and the scattered energy comes from an extensive part of the earth's surface.

If K_α is smaller than about 0.1, most of the scattered energy comes from the extremities of the link.

Asymmetrical links. As it is not possible to give numerical values of $\langle|\rho_d|^2\rangle$ for all values of the four parameters on which it depends in the case of an asymmetrical link (α/β_0^2, K_α, ν and χ), we shall confine the following to some qualitative remarks.

Using (46), (47) and (48), we obtain $\langle|\rho_d|^2\rangle$. In applying these expressions to a symmetrical link, we have introduced no additional assumptions nor resorted to any important simplifications; what we have said about the effect of the parameters K_α and α/β_0^2 remains roughly valid when the link is asymmetrical. In particular, the ratio α/β_0^2 always plays the role of a constant

† We recall that the curvilinear abscissa ε, defined by (45) and (44), depends only on the Cartesian abscissa x_1, since

$$\tan \varepsilon = \frac{\alpha_1}{\alpha_2}\frac{x_1}{d-x_1}$$

As x_1 increases from 0 to d, ε increases from 0 to $\pi/2$.

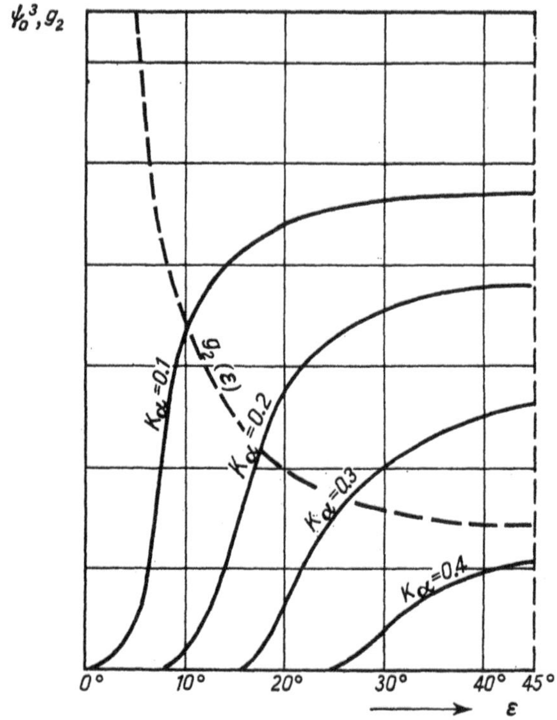

Fig. 12.9. The two factors contained in the expression for $g(\varepsilon)$: ψ_0^3 (full curve) and $g_2(\varepsilon)$ (broken curve). The coordinates give relative values (unspecified constant factor). The curves may be completed by using their symmetry with respect to the straight line $\varepsilon = 45°$.

of proportionality. The condition for all reflections to vanish may be written in the form†

$$K_\alpha > (\cos^{2/3}\nu + \sin^{2/3}\nu)^{2/3}$$

where the right side varies only between the values 0.5 (for $\nu = \pi/4$, symmetrical link) and 1 (for $\nu = 0$ or $\pi/2$).

Some qualitative results concerning the effect of the "asymmetry parameters" ν and χ may be established. It may be shown that for $\nu = \chi = \pi/4$,

$$\frac{\partial}{\partial \nu}\langle|\rho_d|^2\rangle = \frac{\partial}{\partial \chi}\langle|\rho_d|^2\rangle = 0$$

It may also be shown that

$$\frac{\partial^2}{\partial \chi}\langle|\rho_d|^2\rangle = -\langle|\rho_d|^2\rangle$$

† All reflections will vanish if $\psi_0(\varepsilon)$, given by (47), is imaginary for all values of ε.

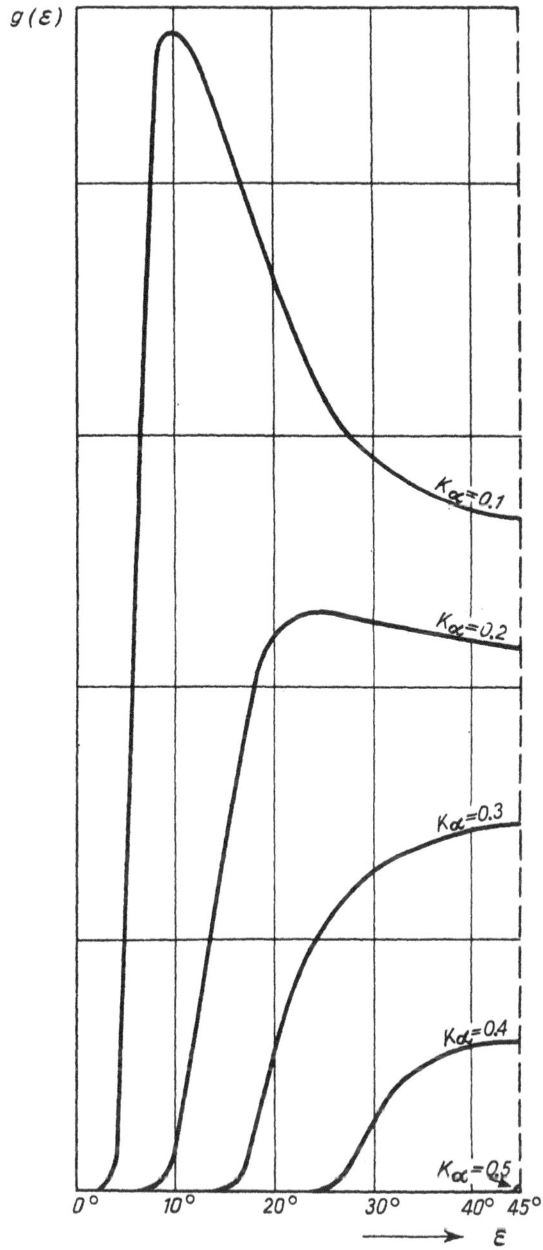

Fig. 12.10. The functions $g(\varepsilon)$. The ordinate contains an unspecified constant factor. The curves may be completed by using their symmetry with respect to the line $\varepsilon = 45°$.

Thus $\langle|\rho_d|^2\rangle$ is of the form

$$\langle|\rho_d(\nu, \chi)|^2\rangle = \langle|\rho_d(\nu, \pi/4)|^2\rangle \cos\left(\chi - \frac{\pi}{4}\right)$$

$\langle|\rho_d|^2\rangle$ *decreases when* χ *deviates from* $\pi/4$; on the other hand, it may be shown that $\langle|\rho_d|^2\rangle$ *increases when* ν *deviates from* $\pi/4$.

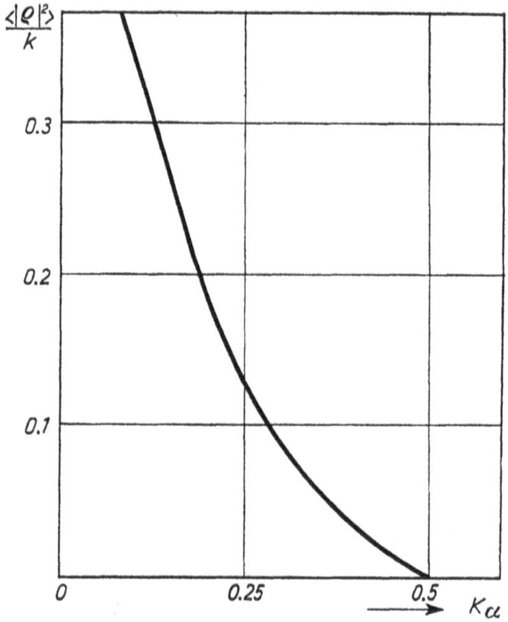

Fig. 12.11. The dependence of the mean square of the scattering coefficient $\langle|\rho_d|^2\rangle$ on K_α in the case of a symmetrical link. The ordinates contain an unspecified constant factor. To obtain $\langle|\rho_d|^2\rangle$, the ordinate should be multiplied by $k = \alpha/\beta_0^2$.

Conclusions. If the antennae are sufficiently strongly beamed to bound the "glistening surface", diffuse scattering is principally determined by the parameter

$$K_\alpha = \frac{1}{d}\sqrt{\left(\frac{h_1^2}{\alpha_2^2} + \frac{h_2^2}{\alpha_1^2}\right)}$$

If K_α exceeds a certain limit, equal to 0.5 for a symmetrical link and lying between 0.5 and 1 for an asymmetrical link, then the reflection coefficient is negligibly small.

If K_α decreases from this limit to about 0.1, $\langle|\rho_d|^2\rangle$ will increase rapidly (Fig. 12.11); the scattered energy comes from an increasingly extensive area of the earth's surface. As K_α decreases, the regions naar the link terminals participate increasingly in the reflection (Fig. 12.10).

If $K_\alpha < 0.1$, the reflected energy originates practically only near the extremities of the link.

If K_α is of the order of K_β, but $K_\alpha > K_\beta$, the approximate solution at the beginning of this section may be applied.

In all of these cases, the reflection coefficient increases rapidly with decreasing K_α.

If, on the contrary, $K_\alpha < K_\beta$, the directivity of the antennae is no longer important for diffuse scattering and the reflection coefficient remains constant.

12.4.5. *Case III: One Terminal near the Ground, the Other at Infinity*

We restrict the following to be used in deriving the expression for the reflection coefficient when the antenna near the ground has little directivity and the function G of the general expression (23) for $\langle|\rho_d|^2\rangle$ is equal to G_0 as defined by (18). In this simple case the mean absolute square of the reflection coefficient for the diffusely scattered component $\langle|R_d|^2\rangle$ is given by

$$\langle|R_d|^2\rangle = \langle|\rho_d|^2\rangle R_0^2$$

where R_0 is the reflection coefficient of a smooth earth and

$$\langle|\rho_d|^2\rangle = \frac{1}{4\pi\beta_0^2}\int\frac{dS}{r^2} \tag{52}$$

where r is the distance from a point P on the surface of the earth to the terminal R near the ground.

The surface of integration is the "glistening surface" (Section 12.4.2).

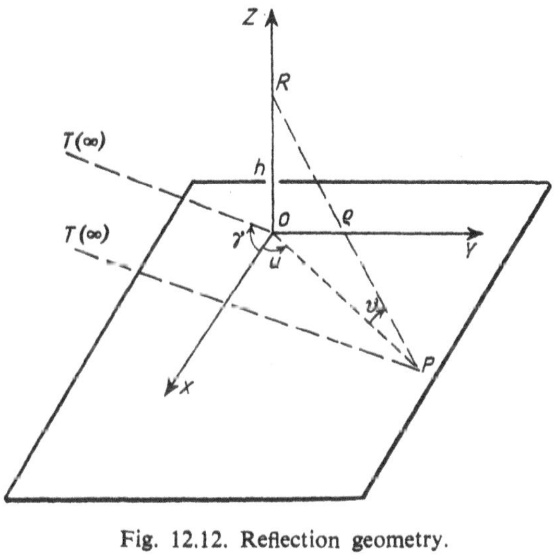

Fig. 12.12. Reflection geometry.

Equation of the contour Γ *of the surface* S. We first assume the earth, plane (though not smooth), and let its plane coincide with the xy plane of Cartesian coordinates. Let one terminal be at infinity in a direction lying in the xz plane and making an angle γ with the x-axis. Let this terminal be the transmitter (Fig. 12.12). We choose the origin O so that the receiver R lies on the z axis. Let h be its height above the ground.

It is convenient to define a point P on the surface of the earth by curvilinear coordinates:

$$u = \sphericalangle(Ox, OP) \qquad v = \sphericalangle OPR$$

The equation of the boundary Γ of S is obtained by using the angle β concluded by the bisector of the angle $\sphericalangle TPR$ with the vertical. In curvilinear coordinates u, v this yields

$$2\cos u = \frac{\cos v}{\cos\gamma} + \frac{\cos\gamma}{\cos v} - \tan^2\beta_0 \frac{\sin\gamma + \sin v}{\cos\gamma \quad \cos v} \tag{53}$$

This is the equation of the boundary; to obtain Γ, we confine u and v to the following intervals:

$$\left.\begin{array}{c} \gamma - 2\beta_0 < v < \gamma + 2\beta_0 \\ 0 < v < \dfrac{\pi}{2} \\ -\pi < u < \pi \end{array}\right\} \tag{54}$$

Figure 12.13 shows some of the possible shapes of the curve Γ in u, v coordinates and in Cartesian coordinates x, y. In case A, v is limited by the first of the inequalities (54). In case B, the limit $v = 0$ intervenes: in the xy

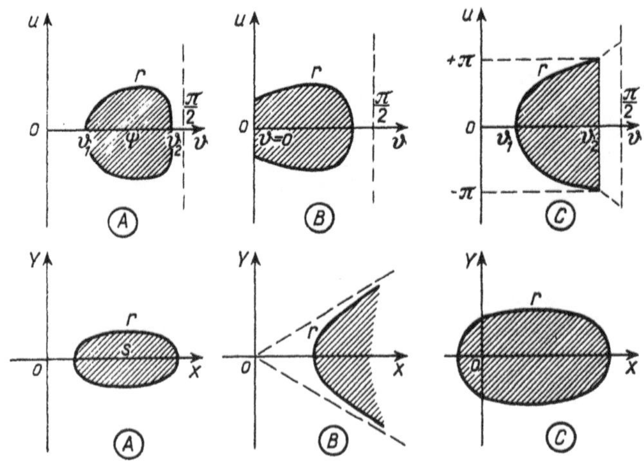

Fig. 12.13. Possible shapes of the curve Γ.

plane, the surface S extends to infinity (in this case the curvature of the earth should obviously be respected by limiting S by the horizon). In case C, S is also limited by the third of the inequalities (54). This inequality defines an area surrounding the origin in the xy plane; it is necessary to prevent certain regions of the surface S to be counted several times in the integration over the uv plane.

The three examples given above are intended to illustrate the meaning of the inequalities (54) and do not exhaust all possibilities.

In the following we shall denote by

$$u = \pm u(v, \gamma)$$

the expression (53) made explicit in u in the interval $V_1 < V < V_2$, V_1 and V_2 being the limits of variation imposed by the inequalities (54).

Calculation of the reflection coefficient. Using the above notation, (52) may be written as

$$\langle|\rho_d|^2\rangle = \frac{1}{2\pi h^2 \beta_0^2} \int_{v_1}^{v_2} u(v, \gamma) \sin^2 v \left| \frac{\partial x}{\partial u}\frac{\partial y}{\partial v} - \frac{\partial x}{\partial v}\frac{\partial y}{\partial u} \right| dv$$

which yields

$$\langle|\rho_d|^2\rangle = \frac{1}{2\pi \beta_0^2} \int_{v_1}^{v_2} u(v, \gamma) \cot v \, dv \tag{55}$$

where $u(v, \gamma)$ is given by (53) and the interval (v_1, v_2) by (54).†

12.5. The Statistical Distribution of the Scattered Field

It has been shown in Part I of this monograph that the diffusely scattered component of the amplitude of the resultant fieldstrength fluctuates in accordance with the Rayleigh distribution

$$dP\left\{\frac{R}{R_{\text{RMS}}} > u\right\} = 2e^{-u^2} u \, du$$

This distribution (probability density) is shown in Fig. C4 (p. 186); The complement of the distribution function

$$P\left\{\frac{R}{R_{\text{RMS}}} > u\right\} = e^{-u^2}$$

is shown in Fig. F11 (p. 208) plotted on Rayleigh paper, and in Fig. 12.14 plotted on probability paper.

† A more detailed analysis of this expression is given in Du Castel, Misme, Voge and Spizzichino [1961].

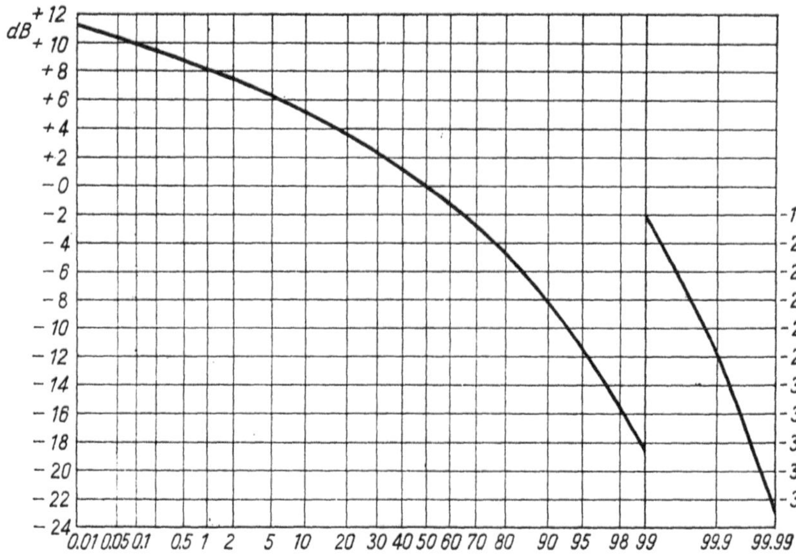

Fig. 12.14. The Rayleigh distribution (cf. also Fig. F11, p. 208).

12.6. Interference of the Direct Wave with the Diffusely Scattered Component

We have seen that when the specularly reflected wave interferes with the direct wave, a regular pattern of interference lobes is produced, i.e. when the receiver is moved along a straight line (e.g. horizontal or vertical), the mean field will pass through a regular series of maxima and minima. The variations of this field may be accurately predicted even if the statistical characteristics of the reflecting surface are not known. This is due to the fact that the phase of the reflected wave is coherent, i.e. at every point in space the mean phase of the reflected wave is known.

If, however, diffuse scattering sets in, the phase of the scattered wave is incoherent, i.e. at any point in space it may assume any value with equal probability. If a receiver is now moved along a straight line, the received field will vary in a random manner and may be described by (a) its statistical distribution, and (b) by its autocorrelation radius (correlation distance) when the receiver changes its position along this line.

12.6.1. *The Statistical Distribution of the Received Field*

The received fieldstrength is the result of the superposition of a field E_0 with constant amplitude and phase (the direct field) and a field with random phase and amplitude (scattered field), which is Rayleigh-distributed.

The distribution law of a field of this kind may be derived from the work of Rice [1954]. It has been treated by Blake [1950], Norton, Vogler, Mansfield and Short [1955] and Zuhrt [1957], tables of numerical values being contained in the latter two papers.

The amplitude $|E|$ of the resultant field E is then distributed with the probability density

$$dP\left\{\frac{|E|}{E_0} > \varepsilon\right\} = \frac{2\varepsilon}{\langle R_d^2\rangle} \exp\left[-\frac{1+\varepsilon^2}{\langle R_d^2\rangle}\right] I_0\left(\frac{2\varepsilon}{\langle R_d^2\rangle}\right) d\varepsilon \tag{56}$$

where I_0 is the modified Bessel function of order zero.†

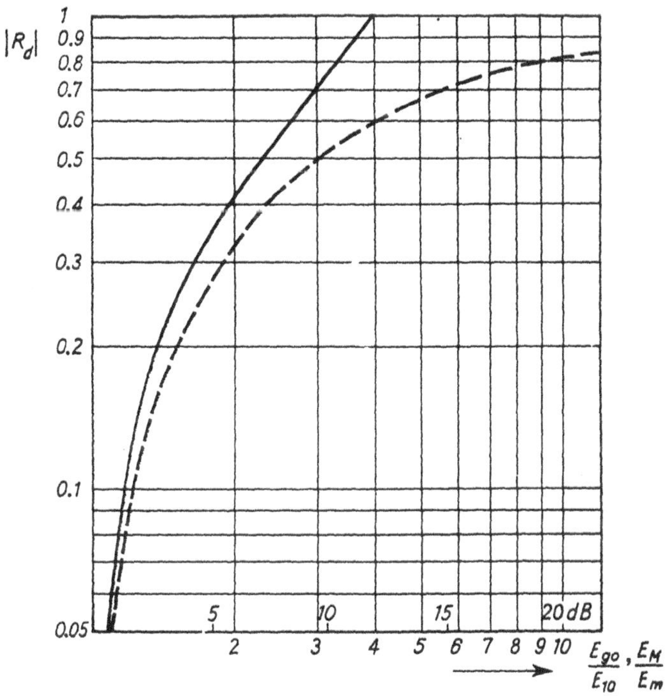

Fig. 12.15. Dependence of the reflection coefficient of the diffusely scattered component $|R_d|$ on E_{g0}/E_{10} (full curve) and of the reflection coefficient of the specularly reflected component $|R_s|$ on E_M/E_m (broken curve).

Figure 12.15 shows the difference in db between the 10 and 90% levels of the distribution (56) as a function of $(R_d)_{\mathrm{RMS}}$.

† This result follows from Part I, Section 7.1, (37), by setting $R = 1/(R_d)_{\mathrm{RMS}}$, $u = \varepsilon/(R_d)_{\mathrm{RMS}}$, remembering that, by definition, $(R_d)^2_{\mathrm{RMS}} = \langle R_d^2\rangle$. The function I_0 is defined in Part I, Appendix B, (15) to (19) (p. 183).

12.6.2. *The Radius of Autocorrelation*

We introduce Cartesian coordinates x, y, z with x along the radio link and z along the vertical. For an analysis of the fluctuations of the field in directions parallel to each of these axes, three radii of autocorrelation (correlation distances) may be defined:

Δx, the longitudinal correlation radius,

Δy, the transversal correlation radius,

Δz, the vertical correlation radius.

We shall calculate the values of Δx, Δy and Δz in the neighbourhood of a point R_0 with coordinates x_0, y_0, z_0.

The received field E is the sum of the direct wave $E_0 \exp(i\omega t)$ and a scattered wave engendered by a region S of the earth's surface:

$$E_0 e^{i\omega t} \int_S \frac{dR}{dS} e^{2\pi i\delta/\lambda} dS$$

where dR is the reflection coefficient of an element dS of the earth's surface and δ is the path difference between the direct and reglected rays.

The absolute value of the field at a point with coordinates x, y, z is

$$|E(x, y, z)| = E_0 \left|1 + \int_S \frac{dR}{dS} e^{2\pi i\delta/\lambda} dS\right|$$

This may be written as

$$|E(x, y, z)|$$
$$= E_0 \left[1 + 2\int_S \frac{dR}{dS} \cos\left(2\pi \frac{\delta}{\lambda}\right) dS + \int_{S_1}\int_{S_2} \frac{dR}{dS_1} \frac{dR}{dS_2} \cos\left(2\pi \frac{\delta_1 - \delta_2}{\lambda}\right) dS_1 dS_2\right] \quad (57)$$

where the last integral is the autocorrelation function of the random variable

$$\frac{dR}{dS} e^{2\pi i\delta/\lambda}$$

and represents the fluctuations of $|E|$ caused by the mutual interference of the elementary scattered waves.

The term

$$2E_0 \int_S \frac{dR}{dS} \cos\left(2\pi \frac{\delta}{\lambda}\right) dS \quad (58)$$

represents the fluctuations of $|E|$ due to the interference of the direct ray with each of the waves scattered by the elements dS of the earth's surface.

In the neighbourhood of the point x_0, y_0, z_0,

$$\delta(x, y, z) \approx \delta(x_0, y_0, z_0) + \frac{\partial\delta}{\partial x}(x - x_0) + \frac{\partial\delta}{\partial y}(y - y_0) + \frac{\partial\delta}{\partial z}(z - z_0) \quad (59)$$

Substituting this in (57), we obtain an expression for $|E(x, y, z)|$, in which the different Fourier components with respect to the variables x, y, z become apparent:

To each element of the reflecting surface dS there corresponds, in accordance with the term (58), a component of frequency

$$\frac{1}{\lambda}\frac{\partial \delta}{\partial x_i}; \qquad i = 1, 2, 3 \tag{60}$$

(x_1, x_2, x_3 stand for the coordinates x, y, z) and amplitude dR. These components are due to the interference between the direct ray and the ray reflected by dS.

To each combination of surface elements dS_1, dS_2, there corresponds, according to (57), a component of frequency

$$\frac{1}{\lambda}\frac{\partial}{\partial x_i}(\delta_1 - \delta_2) \tag{61}$$

and amplitude $dR_1 dR_2$. These components are due to the mutual interference of the waves reflected by dS_1 and dS_2.

Let us now apply the results of Sections 12.4.3 and 12.4.4. In most cases the scattered energy originates in two small zones of the earth's surface near the link terminals in the neighbourhood of points A and B, defined by the angle $2\beta'$ concluded by TA and BR with the horizontal plane (Fig. 12.16). In order to make this definition applicable to the results of both Sections 2.4.3 and 2.4.4, we define β' as the smaller of the two angles β_0 and $\alpha/2$ (we recall that β_0 is the maximum inclination of the irregularities and α is the angle of the transmitting and receiving antenna beams, which we assume equal).

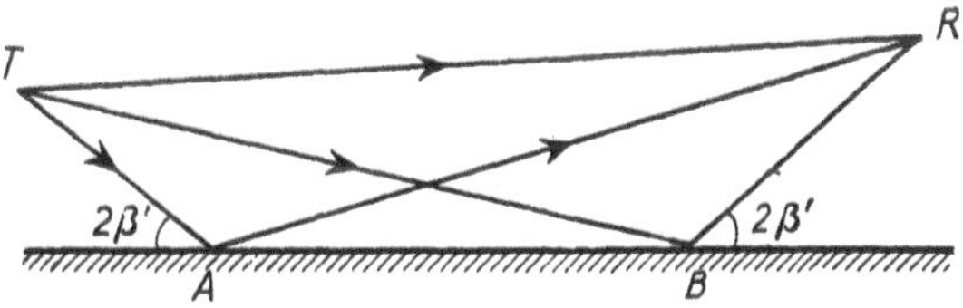

Fig. 12.16. Geometry of diffuse scattering.

Since the scattered radiation includes rays originating at the set of elements dS_A of the earth's surface in the vicinity of point A, and rays originating at the surface elements dS_B near the point B, the Fourier spectrum of $|E(x, y, z)|$ contains the following components:

1. The components of the type (60) are distributed over two narrow bands near the frequencies

$$\frac{1}{\lambda}\frac{\partial \delta_A}{\partial x_i} \quad \text{and} \quad \frac{1}{\lambda}\frac{\partial \delta_B}{\partial x_i} \tag{62}$$

where δ_A and δ_B are the values of δ for the rays reflected at A and B respectively.

2. The components of the type (61), obtained by combining an element dS_A with an element dS_B, are distributed over a narrow band near the frequency

$$\frac{1}{\lambda}\frac{\partial}{\partial x_i}(\delta_A-\delta_B)$$

3. The components of the type (61) obtained by combining the elements dS_A, or those of dS_B, with each other, result in much lower frequencies.

If we assume that the square of the reflection coefficient of each of the zones near A and B equals at most 1/4 (cf. Section 2.4.3 and 2.4.4), then the total amplitude of the components of the first group (expressed as *energy*) equals at most 1/2 of the energy of the direct wave; that of the second group equals at most 1/8; and that of the third group also at most 1/8. In fact, the reflection coefficient of each zone remains smaller than 1/4, the disproportion between the total amplitudes of the components of the first group and those of the other two always being very great.† Moreover, since the components of the first group are of much higher frequencies than those of the other two, it may be assumed that the autocorrelation radius of the field depends practically only on this first group of components, the frequencies of which are given by (62) with

$$\left.\begin{aligned}\delta_A(x,z) &= h_1\tan\beta'+\frac{z^2}{2(x-h_1\cot 2\beta')}\\ \delta_B(x,z) &= z\tan\beta'+\frac{h_1^2}{2(z-x\cot 2\beta')}\end{aligned}\right\}\qquad(63)$$

where, as before, at the point of reception $x = d$ and $z = h$ (we have set $y = 0$). From the two components derived in (62), the autocorrelation radii Δx and Δz may always be obtained.

As in the following we shall need only simple expressions giving the *order of magnitude* of the autocorrelation radii Δx and Δz, we shall replace them by the period of the most rapid of the components (62), which we shall denote by $\Delta' x$ and $\Delta' z$. It is then easily shown from (62) and (63) that

$$\Delta' x = 2\frac{d^2}{h^2}\lambda \qquad (64)$$

† As we shall see in Chapter 15, the experimentally measured values are $\langle|R_d|\rangle \approx 0.3$ or $\langle|R_d|\rangle \approx 0.09$ for the combined two zones, each of which has a coefficient of the order of 0.05. Accepting this value, the total amplitude of the components of the first group equals 0.1 times the energy of the direct wave, whereas the corresponding factor for the other two groups equals 0.005.

$$\Delta' z = \frac{\lambda}{\tan\beta'} \tag{65}$$

where h is the greater of the heights h_1 and h_2.

Physical interpretation of the results. It is interesting to compare $\Delta' x$ and $\Delta' z$ with the correlation radii obtained in specular reflection; in this case (Section 11.3) the variations of the resultant field along the horizontal direction may in the neighbourhood of each point be replaced by sinusoidal variations with periods

$$\Delta x_s = \frac{1}{2}\frac{d^2\lambda}{h_1 h_2} \tag{64a}$$

$$\Delta z_s = \frac{1}{2}\frac{d\lambda}{h_1} \tag{65a}$$

In comparing (64) with (64a), two cases arise. Consider first the case when h_1 and h_2 are very different: $\Delta x_s \gg \Delta' x$. This case will arise for example if the transmitter is located on the ground and the receiver is airborne, the aircraft moving in the direction of the link. The inequality $\Delta x_s \gg \Delta' x$ means that the field will vary along the route of the aircraft much more rapidly if the surface of the earth scatters diffusely than if it reflects specularly. (Further on, this will enable us to explain certain experiments.) If, on the other hand, the heights h_1 and h_2 do not differ considerably, we find Δx_s and $\Delta' x$ of the same order of magnitude. We shall come back to this point later.

Let us now compare (65) with (65a): we always have $\Delta z_s \gg \Delta' z$, the two quantities being, under the assumptions we have made, of quite different orders of magnitude.† For example, if $\lambda = 10$ cm, $d = 20$ km, $h_1 = h_2 = 50$ m and $\beta' = \alpha/2 = 2°$ (directional antennae), we obtain $\Delta z_s = 20$ m and $\Delta' z = 3$ m. This means that if the receiving antenna is moved in a vertical direction (e.g. up or down a mast), the received field will fluctuate much more quickly if the reflection is diffuse than if it is specular.

We thus see from these two comparisons that if the receiver is moved along a horizontal or vertical straight line, the received field will fluctuate much more if the earth's surface scatters diffusely than if it reflects specularly. The only exception to this rule seems to be the case of a symmetrical link ($h_1 - h_2$) with the receiver moving along the horizontal.

Let us take an example of a symmetrical link: $\lambda = 3$ cm, $d = 10$ km, $h_1 = h_2 = 50$ m, $\beta' = \alpha/2 = 2°$ (directional antennae). This gives $\Delta' z$

† In deriving the expression (65) for $\Delta' z$ we have assumed that the scattered energy originates near the points A and B of Fig. 12.16. This implies that these points are both near the link terminals, i.e. $\beta' \gg h_1/d$ and h_2/d.

$= 0.5$ m, $\Delta' x = 1200$ m. Thus it would seem that the field, which fluctuates very rapidly along the vertical, is practically constant along the horizontal, and that the energy of the electromagnetic field is concentrated in narrow, horizontal beams of about 1 m width and several kilometres' length (Fig. 12.17a). One might also think that if the reflecting surface is fixed (unlike

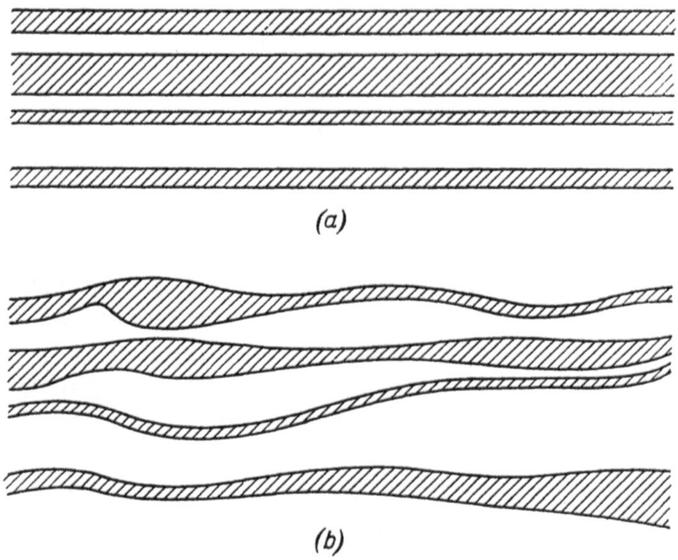

Fig. 12.17. Theoretical structure of the field in the presence of diffuse scattering (a) and its deformation by the atmosphere (b).

the surface of the sea), these narrow beams are immobile and the field is constant in time at every point. In reality, these results are physically absurd: we have obtained them only because in the model we have hitherto adopted, the atmosphere was assumed homogeneous, or of the "standard" type. But in the actual atmosphere there are small inhomogenities in constant motion. From experiments with strongly beamed antennae it is known that a beam of 1 m width is rapidly deformed after having passed through several hundreds of metres of the atmosphere: it is displaced, concentrated (focused) or widened (defocused) in a manner that varies constantly in time (Fig. 12.17b). Also, the existence of rapid fluctuations of the field along the vertical necessarily entail rapid fluctuations along the horizontal and also fluctuations in time. This does not apply to the scattering of electromagnetic waves by the earth's surface, but is a rule that is often met in the study of the propagation of very short waves through the atmosphere: the field that results from any propagation mechanism always depends on the space coordinates x, y, z, on the time t, and also on the frequency f; if a propagation mechanism gives rise to a field that fluctuates rapidly as a function of one of these five variables, it will almost invariably fluctuate rapidly also as a function of the four others.†

We may thus say in summary that *in interfering with the direct wave, diffuse scattering gives rise to a more rapidly fluctuating field than specular reflection*; this means that it varies more rapidly in space, in time and as a function of frequency.

12.7. The Depolarization of the Diffusely Scattered Component

It has been shown in Part I, Section 8.4, that a horizontally or vertically polarized wave is strongly depolarized if it is scattered out of the plane of incidence. This concerns the diffusely scattered component of the field, which originates partly at points on the earth's surface located off the axis of the link.

The role of the irregularities may be illustrated by a simple case. Let the transmitted wave be polarized horizontally. From (4), Section 8.4 (Part I), we have $\mathcal{R}_c = \sin 2\beta$ where $\mathcal{R}_c$ is the ratio of the cross-polarized (vertical) reflected field to the originally (horizontally polarized) incident field. The points of the surface which contribute most to diffuse scattering are those for which β is at most of the order of β_0.

It may thus be expected that $\mathcal{R}_c$ will be at most of the order of $\sin 2\beta_0$.

Annex to Chapter 12

Practical Determination of the First Fresnel Ellipse in Some Special Cases

A. Plane Earth

In Part I, Section 2.2, general expressions for the centre of the Fresnel ellipse and its axes are given; simplified expressions for the case when the antenna heights and the path differences are much smaller than the path length will also be found in Section 2.2. However, these general expressions cannot be applied immediately to numerical calculations, as the path difference δ must first be determined,‡ and the properties of the solution are

† We shall meet this rule again in connection with line-of-sight (Chapter 16) and tropospheric-scatter propagation (Chapter 20).

‡ It may be determined from the nomograms in the Annex to Chapter 11.

not immediately apparent. In practical applications, one of the following cases often arises:

Case I: Antenna Heights Much Larger than the Space Wavelength

Let us *a priori* assume that the dimensions of the Fresnel ellipse are small compared with the distances of the link terminals from the point of reflection I as given by geometrical optics.

We take I as the origin of a coordinate system. Let T be the transmitter, R the receiver and the xy plane the surface of a plane, smooth earth (Fig. 12.18), the x axis being parallel to the projection of TR onto the horizontal

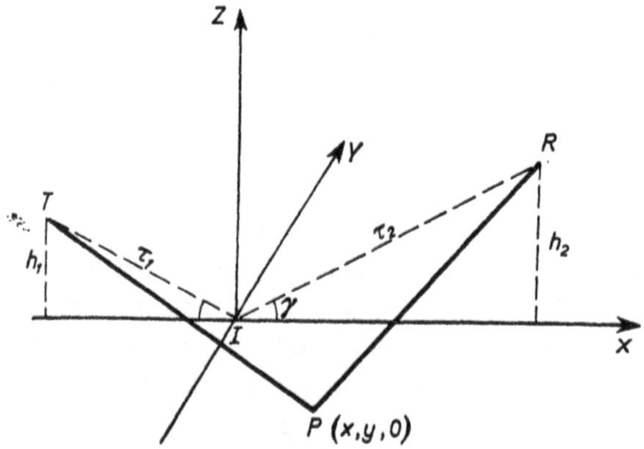

Fig. 12.18. Geometry of the derivation of the Fresnel ellipse.

plane. Let P be a point on the earth's surface with coordinates x, y and let δ_1 be the optical path difference

$$\delta_1(x, y) = (TP+PR)-(TI+IR)$$

One then finds

$$\delta_1(x, y) = \frac{r_1+r_2}{r_1 r_2}(x^2 \sin^2\gamma+y^2) \tag{66}$$

with $r_1 = TI$, $r_2 = IR$ and γ the grazing angle at I. The first Fresnel ellipse is the locus of all points for which $\delta_1 = \lambda/2$, hence the lengths of the semi-axes X (parallel to the path) and Y (perpendicular to the path) are

$$\left.\begin{aligned} X &= \frac{1}{\sin\gamma}\sqrt{(\lambda r)} \\ Y &= \sqrt{(\lambda r)} \end{aligned}\right\} \tag{67}$$

with

$$r = \frac{r_1 r_2}{r_1+r_2}$$

We have *a priori* assumed $X \ll r_1$, $Y \ll r_2$; this condition is according to (67) fulfilled if the transmitting and receiving antenna heights are much larger than the space wavelength $\lambda/\sin\gamma$. This is frequently the case in ultra-short wave links.

Case II: Very Different Antenna Heights

Let us assume $h_1 \ll h_2$. It is then easily shown that the entire Fresnel ellipse lies in the vicinity of the transmitter. In order to determine it, the receiver may be assumed at infinity.

The choice of coordinate axes is that of Fig. 12.19, where T stands for the transmitter; the receiver R lies in the xz plane in a direction concluding an angle γ with the x axis.

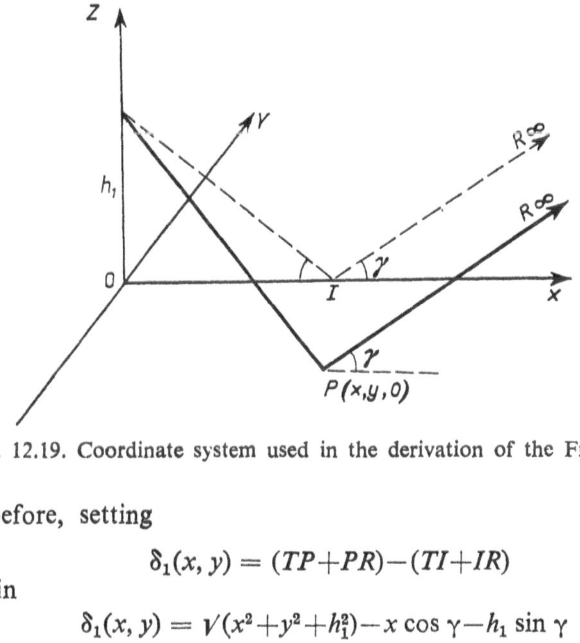

Fig. 12.19. Coordinate system used in the derivation of the Fresnel zones.

As before, setting

$$\delta_1(x, y) = (TP+PR)-(TI+IR)$$

we obtain

$$\delta_1(x, y) = \sqrt{(x^2+y^2+h_1^2)}-x\cos\gamma-h_1\sin\gamma \qquad (68)$$

The first Fresnel ellipse is the locus of all points for which

$$\delta_1(x, y) = \frac{\lambda}{2} \qquad (69)$$

Intercept of the major axis. The extremities of the major axis of the Fresnel ellipse on the x axis are found from (69) with $y = 0$. We may set

$$\frac{x}{\lambda} = \rho\cos\Omega; \quad \frac{y}{\lambda} = \rho\sin\Omega$$

It is then easily shown that $\delta_1 = \lambda/2$ for

$$\rho = \frac{1}{4\sin^2\frac{\gamma-\Omega}{2}} \tag{70}$$

In the plane of coordinates x/λ, h_1/λ, equation (70) is that of a set of confocal parabolae derived from each other by rotation about the origin; they are shown in Figs. 12.20a and 12.20b. These figures give the abscissae x of the two extremities of the major axis of the Fresnel ellipse as a function of the transmitting (or receiving) antenna height h_1 for various values of the grazing angle γ.

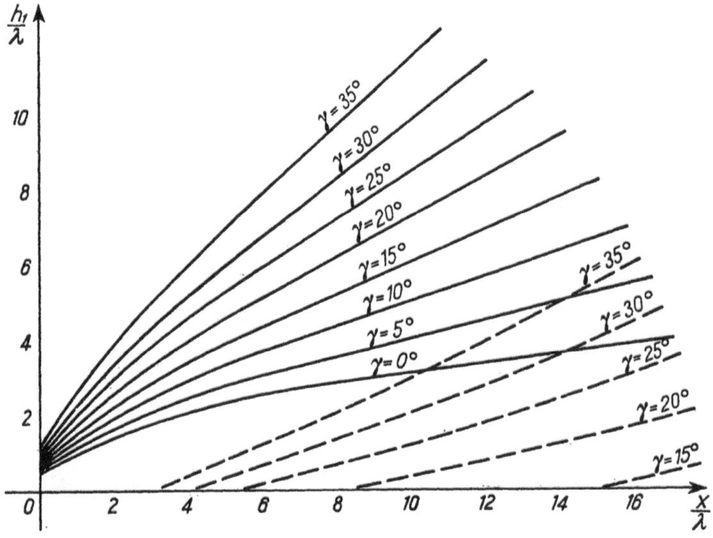

Fig. 12.20a. The abscissae x of the intercepts of the major axis of the Fresnel ellipse with the x axis as a function of the transmitting antenna height h_1. Receiver (at infinity) in a direction making an angle γ with the x axis. The full curves give the intercept nearer to the transmitter, the broken curves the further intercept.

Centre of the Fresnel ellipse. The abscissa x_0 of the centre of the ellipse is given as a function of h_1 by the curves in Fig. 12.21. They are a set of straight lines, diameters of the parabolae in Fig. 12.20 conjugate with the direction of the x axis. Their equation is

$$\frac{h_1}{\lambda} = \frac{x_0}{\lambda}\tan\gamma - \frac{1}{2}\frac{1}{\sin\gamma} \tag{71}$$

Minor axis of the Fresnel ellipse. The abscissa of the minor axis of the

Fresnel ellipse being given by Fig. 12.21, it remains to determine the ordinates of its extremities $\pm y$. From (68), (69) and (71) we obtain

$$\left(\frac{y}{\lambda}\right)^2 = \frac{1}{\sin \lambda}\,\frac{h_1}{\lambda} + \frac{1}{4}\,\frac{1}{\sin^2 \gamma} \tag{72}$$

which in the plane of coordinates y/λ, h_1/λ represents a set of confocal parabolae whose axis is identical with the ordinate axis (Fig. 12.22).

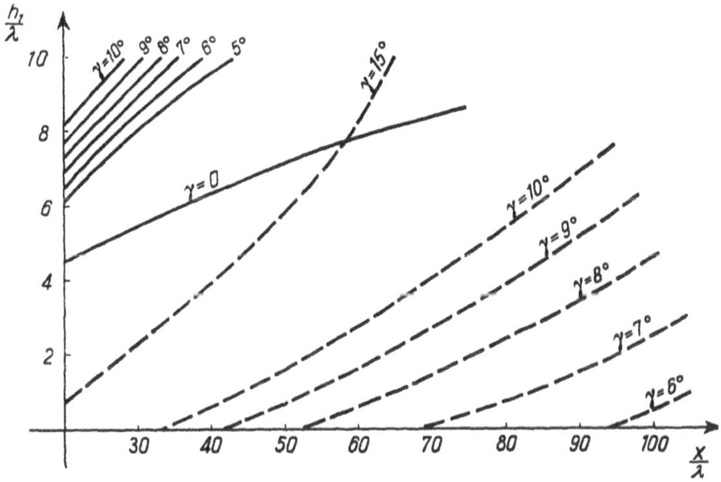

Fig. 12.20b. Curves as Fig. 12.20a: higher values of x.

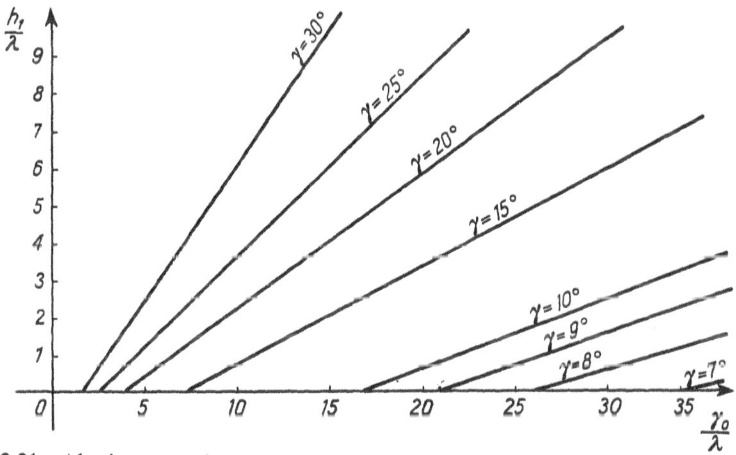

Fig. 12.21. Abscissa x_0 of the centre of the Fresnel ellipse as a function of the transmitting antenna height h_1. The receiver is at infinity, in a direction making an angle γ with the x axis.

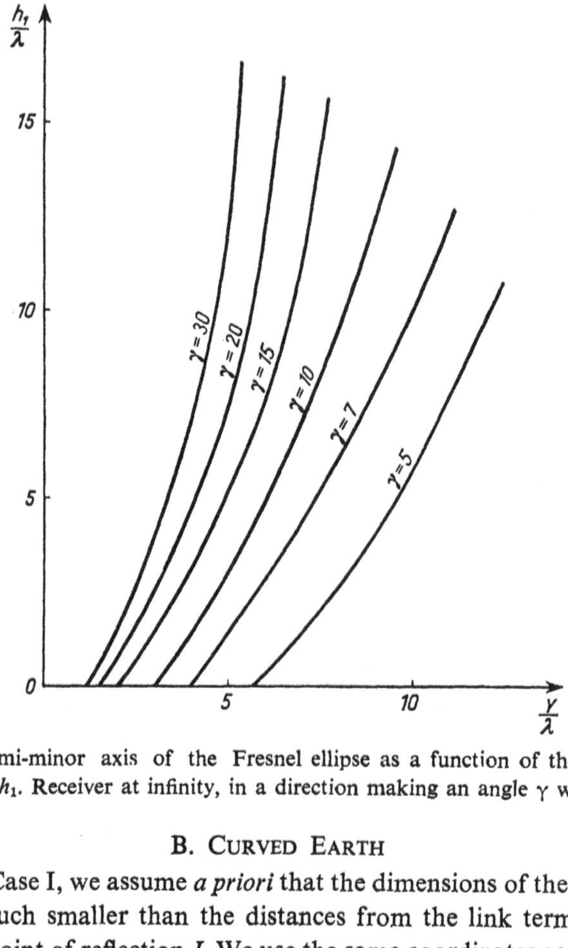

Fig. 12.22. Semi-minor axis of the Fresnel ellipse as a function of the transmitting antenna height h_1. Receiver at infinity, in a direction making an angle γ with the x axis.

B. Curved Earth

As in A, Case I, we assume *a priori* that the dimensions of the first Fresnel ellipse are much smaller than the distances from the link terminals to the geometrical point of reflection I. We use the same coordinates as in A, Case I (Fig. 12.18); but the xy plane, instead of being the plane of the earth's surface, is now the plane tangent to the earth's surface at the point of reflection I. Let a and b be the radii of curvature of the earth's surface at I in the x and y directions respectively; for a spherical earth, a and b equal the radius of the earth.

The optical path difference δ_1, which was given by (66) in the case of a plane earth, is now given by

$$\delta_1 = \left(\frac{r_1+r_2}{2r_1r_2}+\frac{1}{a\sin\gamma}\right)x^2\sin^2\gamma+\left(\frac{r_1+r_2}{r_1r_2}+\frac{\sin\gamma}{b}\right)y^2$$

The first Fresnel ellipse is the locus of all points for which $\delta_1 = \lambda/2$. It will be seen that in most cases the earth's curvature does not appreciably modify the minor axis $2Y_c$ of the Fresnel ellipse, whereas its major axis $2X_c$ is given by

$$X_c = \sqrt{\left(\lambda \frac{r_1 r_2}{r_1+r_2} \frac{1}{1+\dfrac{2 r_1 r_2}{(r_1+r_2)\, a \sin\gamma}}\right)} \tag{73}$$

It should be noted that one may write

$$X_c = DX \tag{74}$$

where X is the major axis of the first Fresnel ellipse in the case of a smooth earth, given by (67), and D is the divergence coefficient given by (9), Section 11.2. The latter may be obtained from the nomograms in the annex to Chapter 11.

The relations (73) and (74) may be applied equally well to the reflection by a mountainous relief of any form, provided its radius of curvature a has been (graphically) determined.

CHAPTER 13

METHODS OF MEASUREMENT OF THE REFLECTION COEFFICIENTS OF THE EARTH; SEPARATION OF THE SPECULAR AND DIFFUSE COMPONENTS

WE shall always speak of the reflection coefficients of the earth (in the plural), because there is always a coefficient of specular reflection R_s and a coefficient of diffuse scattering R_d.

We shall first deal with measurements of the reflection coefficients in two simple cases: when the reflection is solely specular ($R_d = 0$), and when it is solely diffuse ($R_s = 0$). We shall then consider the case when both components are present; in this case methods of measuring R_s and R_d separately are necessary.

13.1. MEASUREMENT OF THE SPECULAR REFLECTION COEFFICIENT OF THE EARTH (IN THE ABSENCE OF DIFFUSE SCATTERING)

13.1.1. *Separation of the Reflected Ray*

If the antenna beams are sufficiently narrow to separate the direct ray from the reflected one, the reflection coefficient of the earth may be measured directly: the antenna beams are first pointed directly towards each other and the direct field is measured; the beams are then directed to the point of reflection and the reflected field is obtained. Comparison of the two fields then yields the reflection coefficient.

This method has obvious advantages: it is easy to operate, requires only modest equipment and may be very accurate. Compared with other, indirect methods of measuring the reflection coefficient, it has the advantage of being independent of all physical assumptions. Finally, it yields not only the mean value of the reflected field, but also its instantaneous values, permitting its fluctuations to be studied.

Unfortunately the method may be applied only in a small number of cases: it requires the beamwidth of one of the two antennae to be smaller than the angle concluded by the direct with the reflected ray. This angle is of the same order as the grazing angle,† which rarely exceeds 1° in practical radio links.‡ Even in the centimetre wave band the realization of beamwidths smaller than 1° is not easy, and if the beams are narrowed down to some tenths of a degree, they may be deflected by atmospheric refraction.

This method has therefore been applied only with very short pathlengths amounting to some tens or hundreds of metres [Ford and Oliver, 1946; Megaw, 1946; Archer-Thomson, Brooke, Gold and Hoyle, 1943§], or in the case of links in practical operation, but having special profiles.‡‡

In the last two of these papers the authors obtained interesting results concerning the fluctuations of the coefficient of specular reflection, which could not have been obtained by indirect methods of measurement.

N.B. In order not to overburden this outline of experimental work, a more detailed description of the experiments quoted in Chapters 13, 14 and 15 is given in the annex to this chapter. There the reader will find a list (which does not claim to be complete) of experiments performed to measure the reflection coefficient of the surface of the earth.

13.1.2. *Interference of the Direct with the Reflected Wave*

It has been shown in Section 11.3 that the field that results from the interference between the direct and reflected waves varies sinusoidally along a straight line: if the receiver is moved along the vertical, the received field is a sinusoidal function of the height, with period

$$Z = \frac{\lambda d}{2h_1} \tag{1}$$

as follows from (13), Chapter 11. Similarly, along a straight line parallel to the axis of the link, the field may, in the neighbourhood of any point, be replaced by a sinusoidal function of the distance, with period

---

† We denote the grazing angle, which is the complement of the angle of incidence, by γ and the largest angle concluded by the direct and reflected rays by ψ. It is easily shown that $\gamma \leqslant \psi \leqslant 2\gamma$.

‡ The grazing angle is given by

$$\tan \gamma = \frac{h_1 + h_2}{d}$$

where h_1 and h_2 are the transmitting and receiving antenna heights above the ground and d is the pathlength. If the transmitter and receiver are on the ground (on a building, a tower, etc.), h_1 and h_2 are at most of the order of 100 m. If $d > 10$ km, γ is alway smaller than 1°; it may exceed 1° in links with aircraft, but then the use of very directional antennae is delicate and the method of measurement by separation of the reflected ray is difficult to apply.

§ Described in Kerr [1951], where also other experiments of this kind will be found.

‡‡ Boithias and Spizzichino, cf. Spizzichino [1959].

$$X = \frac{\lambda d^2}{2h_1 h_2} \tag{2}$$

In these two cases the reflection coefficient $|R_s|$ may be derived from the variations of the amplitude by means of the relation

$$|R_s| = \frac{E_M/E_m - 1}{E_M/E_m + 1} \tag{3}$$

where E_m and E_M are the minimum and maximum values of the received field. Figure 13.1 shows the dependence of $|R_s|$ on the ratio E_M/E_m. Thus

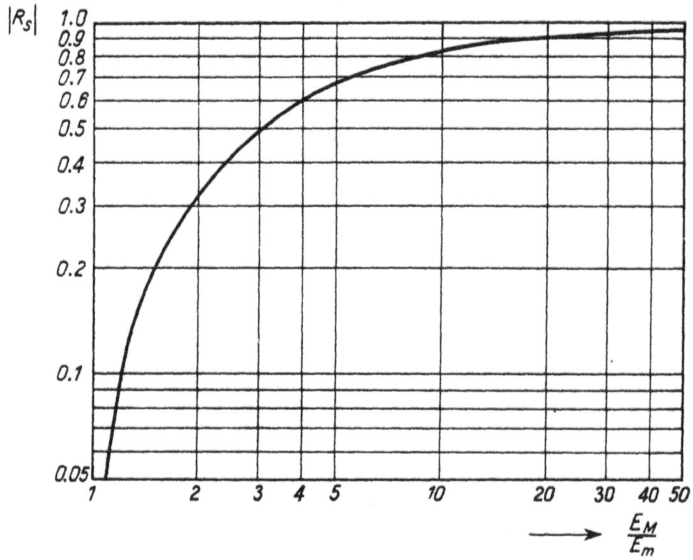

Fig. 13.1. Dependence of the reflection coefficient of the specularly reflected component $|R_s|$ on the ratio E_M/E_m.

one may measure $|R_s|$ by determining E_M and E_m and using (3). There are two possibilities to be considered:

(a) *Analysis of vertical field variations.* Several antennae may be placed (or one antenna may be moved) along the height of a tower, a mast, etc., and the variations of the field with the height may be analysed, but it is necessary that the period Z of the variations as given by (1) should not exceed some tens of metres.

For links of some 10 km length, this condition restricts the above method to frequencies exceeding about 1000 Mc/s.†

† However, the method might be applied to lower frequencies if a helicopter were employed. This solution does not seem to have been attempted.

Such investigations have been made by many authors, of whom we quote Straiton, Gerhardt, La Grone and Tolbert [1952], who worked at 4000, 1000 and 3000 Mc/s; Day and Trolese [1950] at 24,000, 9375 and 3300 Mc/s; Hamlin and Gordon [1948] at 10,000 Mc/s; Tomlinson and Straiton [1959] at 10,000 Mc/s; Sherwood and Ginzton [1955] and Sherwood [1943] at 3000 Mc/s; and Epstein [1953] at 850 Mc/s.† In the experiments described in the last paper, a tower of 230 m height was used.

Figure 13.2 gives an example of a curve showing the dependence of the field on the height as given by Stration and Tolbert [1960]. As may be seen from this figure, the sinusoidal variation of the field is superposed onto other irregular components caused by diffuse scattering or other phenomena.

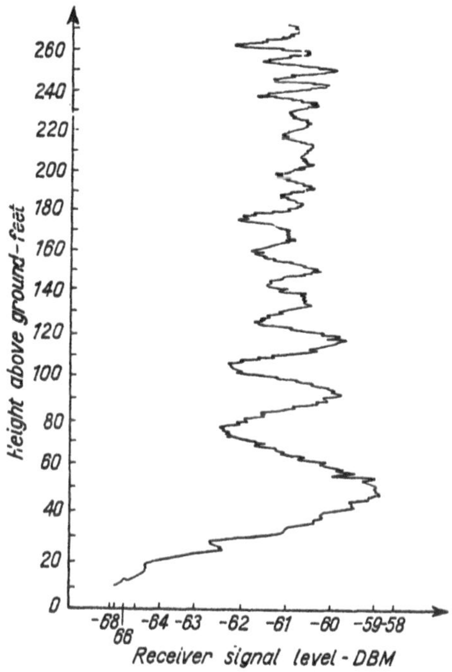

Fig. 13.2. Superposition of random and sinusoidal components in the height-gain pattern [Straiton, 1959]. (*By permission of the I.R.E.*)

The latter inevitably limit the accuracy of measurement, particularly when E_m is very small, i.e. when $|R_s|$ is near 1, and when E_m is very near to E_M (an error is introduced in the difference $E_M - E_m$), i.e. when $|R_s|$ is small.

If several receivers spaced vertically are available, a discontinuous series of measurements is obtained, which may be difficult to interpret. If a single

† Cf. Annex to this chapter for data of these experiments.

receiver is moved up and down, the measurements are more reliable and accurate, but then the rapid fluctuations in time may not be studied, and only its slow variations can be investigated.

(b) *Analysis of horizontal field variations.* The receiver may be installed in a vehicle: aircraft, ship, truck, etc., and the variations of the field along a horizontal line, parallel to the axis of the link, may be measured. But this method may be used only if the period X of the field variations, given by (2), is sufficiently small in comparison with the pathlength d of the link. This condition limits the method to the centimetre (SHF) wave band if the antennae are on the ground (on towers, masts, etc.);† airborne receivers permit the method to be applied in the decimetre (UHF) band.

Thus one finds among the experiments listed in the annex a large number of experiments performed with the aid of aircraft: Rubinstein and Fishback [1943] at 3000 Mc/s above dry ground; Fishback and Rubinstein [1944] at 3000 and 10,000 Mc/s above the sea; Day and Trolese [1950], MacGavin and Maloney [1949], etc. Experiments with receivers installed on ships have been performed in the millimetre wave band [Lamont and Watson, 1946] or in the centimetre wave band at short distances in the last interference lobes [Megaw, 1946].

The use of a mobile receiver excludes the possibility of a simultaneous investigation along a horizontal straight line:‡ this method does therefore not permit the study of the fluctuations of the reflection coefficient, but only gives a mean value.

As in the case of the vertical variations, irregular components are superimposed onto the sinusoidal variations of the field and these may give rise to important errors when $|R_s|$ is near 0 or 1. To these errors one must add technological errors, particularly when an aircraft is used: since the latter moves very fast, the variations that are to be recorded are very rapid and difficulties may be encountered in finding a recording instrument with a sufficiently small time constant for the variations of the field to be correctly reproduced; in addition, there is always the risk that the flight is not perfectly horizontal; finally only slightly directional antennae may be employed, thus allowing unwanted phenomena to interfere (diffuse scattering, scattering by distant obstacles, etc.)

† Choosing extreme values, $d = 10$ km, $h_1 = 100$ m, $h_2 = 30$ m (a mobile antenna can hardly be higher), we obtain $X = 2$ km for $\lambda = 6$ cm.

‡ The simultaneous operation of several receivers spaced along a horizontal straight line is difficult: there would have to be a rather large number of them and they would have to be well aligned at intervals of several kilometers.

Thus regardless of whether the analysis concerns horizontal or vertical variations, the same difficulties are encountered: inaccuracy of the measurements, especially near the minima of the field; impossibility of exploring the field simultaneously in an extensive region; limited frequency band where the method may be applied. One or the other restriction may sometimes be overcome by certain modifications of the method. Thus for instance a radar has certain advantages over a point-to-point link: one may simultaneously gain information coming from an extensive region in space. Unfortunately only qualitative, or at least inaccurate information may be gained in this way: to make the measurement quantitative, specially designed targets with known scattering characteristics must be located in places where their echo can clearly be distinguished from that of parasitic echoes. Moreover, for the echo to be sufficiently strong, one is often forced to use targets whose dimensions are not small with respect to the interference fringes. The measurement of the field variations may thus be inaccurate, especially near a minimum.†

On the other hand the accuracy of the measurement may be improved by using not only the maxima and minima of the field (which, as has already been said, are always very ill-defined), but the entire sinusoidal variation. Anderson, Smyth, Abbott and Revelle [1942]‡ use the relation (10) of Chapter 11 in the form

$$\left|\frac{E}{E_0}\right|^2 = 1+|R_s|^2+2|R_s|\cos kz \tag{4}$$

where E is the received field, E_0 the free-space field, R_s the reflection coefficient and

$$k=\frac{4\pi h_1}{\lambda d} \tag{5}$$

In order to measure R_s, the authors use several values of the received field E at different heights z and plot the curve of $|E/E_0|^2$ as a function of $\cos kz$ point by point: this should be a straight line with slope equal to $2|R_s|$.

Another interesting method consists in studying the variations of the field as a function of frequency instead of investigating them in space. According to (4) and (5), these variations are sinusoidal and the reflection coefficient R_s is connected with the maximum E_M and minimum E_m of the field by the same relation (3) as in the case of variations in space.

The period of these variations is

$$F=\frac{cd}{2h_1 h_2}$$

† Such experiments are reported by Megaw [1946], Kerr [1951], etc.

‡ The method is also given by Kerr [1951].

where c is the velocity of light. The method can obviously be applied only to frequencies $f \gg F$. For the usual values of h_1 and h_2 and d about 10 km, f must be greater than about 1000 Mc/s: the limits of application of the method are thus similar to those of the analysis in space. However, it is more convenient for investigations of the instantaneous values of the field, a frequency sweep being much more rapid than mechanical displacement of the antenna.†

13.2. Measurement of the Diffuse-Reflection Coefficient of the Earth (in the Absence of Specular Reflection)

As in measuring the specular reflection coefficient, there are two basic methods of measurement: the reflected wave may be studied either by separating it from the direct wave or by letting it interfere with the direct wave.

13.2.1. *Separation of the Scattered Wave*

If sufficiently directional antennae are available, they may be pointed first towards each other to measure the direct wave, and then towards the ground to measure the scattered field: the ratio of these two fields will equal the reflection coefficient.

For diffuse scattering, one may repeat what has been said in the case of specular reflection: the method is easy to operate, accurate and provides information on the fluctuations of the reflection coefficient. But it may be applied only in the case of short links where the grazing angle γ is large; in links of the usual length it is very difficult to apply and requires highly directional antennae which can only be realized in the band of centimetre or shorter waves.‡ However, the latter inconvenience is less irksome than in the case of specular reflection: diffuse scattering, which occurs more often at the shorter wavelengths,§ is of particular practical interest in the centimetre and millimetre wave bands.

But this method presents a difficulty in the case of diffuse scattering which was not met in the case of specular reflection. We have seen (Section 12.4) that the reflection coefficient for the diffuse component depends on the

† This method has already been used for the study of other propagation phenomena: multipath propagation [Kaylor, 1953], propagation beyond the horizon [Biggi, Du Castel Simon, Spizzichino and Voge, 1960]. Various methods of analysing the frequency dependence of the field using a synchronized frequency sweep in transmission and reception have been applied by the Compagnie Générale de Télégraphie sans fil (Corbeville, France); they permit a quasi-instantaneous investigation of the field.

‡ Cf. Section 13.1.1.

§ This is evident from the Rayleigh criterion.

beamwidths of the antennae, particularly if the latter are very directional (Section 12.4.4). In using very directional antennae to suppress the direct ray, we inevitably restrict ourselves to a particular case: since the antennae will illuminate only a limited region of the earth's surface, only a small part of the phenomenon called "diffuse scattering" is observed, although an extensive region on the surface of the earth participates in it.

Nevertheless, this difficulty becomes an advantage if it is possible, by successively changing the orientation of the antennae, to illuminate different parts of the earth's surface and to measure the energy scattered by each of them: in this way one may determine the contribution of each part of the surface to the total scattered energy. This has been done for example by Ford and Oliver [1946], who thus experimentally verified that when the height of the irregularities of the ground is greater than one half of the space-wavelength (Rayleigh criterion), the scattered field may come from a very extensive region of the surface of the earth.†

13.2.2. *Interference of the Direct and Scattered Waves*

As we have shown in Section 12.6, the interference between the direct and diffusely scattered wave leads to rapid fluctuations of the resultant field in time, space and as a function of frequency. The rapidity and irregular form of these fluctuations immediately show that diffuse scattering is present. The reflection coefficient of the diffusely scattered component may be derived from the amplitude of these fluctuations by using Fig. 12.15 or relation (57) of Chapter 12.

Note. The full curve in Fig. 12.15 shows the variation of $|R_d|$ as a function of E_{00}/E_{10}; the broken curve shows the dependence of the specular reflection coefficient on the ratio E_M/E_m, where E_M is the maximum, and E_m the minimum amplitude of the resultant field as in (3). For small values of the reflection coefficient, the two curves are very near each other: we may say that if the reflection coefficients are equal, specular reflection and diffuse scattering will, on interfering with the direct wave, give rise to fluctuations of about the same amplitudes. This justifies (*a posteriori*) the approximation made by many authors who, intending to measure the reflection coefficient of the diffusely scattered component, try to evaluate a maximum and minimum of the resultant field and apply the relation (3) as though the reflection were specular.

By the above method the reflection coefficient of the diffusely scattered component R_d may be derived from the fluctuations in time, space or as a function of frequency. In practice the investigation of the field fluctuations in time is the most easily accomplished because it requires only a fixed receiver. But it will not show up other propagation phenomena that may be superim-

† In Section 13.3.1 we shall consider other results obtained by an analogous method used by the C.N.E.T. (Paris).

posed onto diffuse scattering and that may cause the results to be in error (specular reflection, for example). Most measurements reported in the literature are performed by analysing the variations of the field in space.

Sherwood and Ginzton [1943, 1955] have analysed the field variations along a vertical using a very short experimental link and have obtained results that concern both diffuse scattering and specular reflection. Similar experiments have been carried out by Tomlinson and Straiton [1959] who observed regular interference lobes caused by specular reflection and random fluctuations caused by diffuse scattering in accordance with the conditions of the experiment. Rubinstein and Fishback [1943] and especially MacGavin and Maloney [1959] report interesting results on diffuse scattering obtained by a horizontal analysis of the field using an airborne receiver.

13.3. Separation of the Specular and Diffuse Components

When the resultant field contains both a specular and a diffuse component, the separation of the two components and the measurement of the reflection coefficients R_s and R_d becomes more delicate, but the methods that have hitherto been used fall into one of the two schemes that we have met when only one of the two components was present: separation of the reflected or scattered wave and interference of the reflected or scattered wave with the direct wave.

13.3.1. *Separation of the Reflected Wave; Goniometry*

By the use of a very directional antenna, the reflected wave may be separated from the direct wave. But in general it is not possible to separate the two components of the reflected wave in this way.

This separation could be obtained, for example, by performing a spectral analysis of the fluctuations of the reflected field, either in time or in space, along a straight line (horizontal or vertical): the rapid components would always be caused by diffuse scattering and would permit R_d to be measured, whereas R_s could be derived from the mean value of the reflected field.

We shall describe a series of experiments where the two components were separated by using a goniometric method. These measurements were carried out by the C.N.E.T in the south of France (La Turbie, Alpes-Maritiones).†

A radio link was established above the sea between two points on the coast located on hills. The profile of the path is shown in Fig. 13.3.

† These experiments are described in outline by Spizzichino [1959], but will be the subject of a more detailed publication ([Boithias and Spizzichino], to be published).

The path runs over a rather calm sea in a sheltered bay. Both polarizations at 3 cm wavelength were used. Under these conditions the reflection should, according to the Rayleigh criterion, be specular.

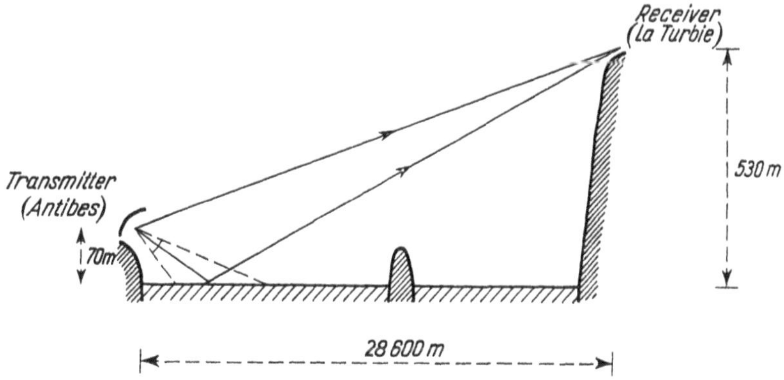

Fig. 13.3. Profile of the path Antibes–La Turbie.

A very directional transmitting antenna enabled the wave reflected by the surface of the sea to be isolated and only this wave was observed.

At the receiving terminal, two horn antennae with small gains were located at a vertical distance of 2 m from each other. The fields received by these two antennae are fed into two rotating dephasing devices which rotate in opposite senses (Fig. 13.4). They are then added in a "magic T". The field is fed from the output of the magic T to a receiver and after detection the signal controls the radial motion of the spot on a cathode ray tube.

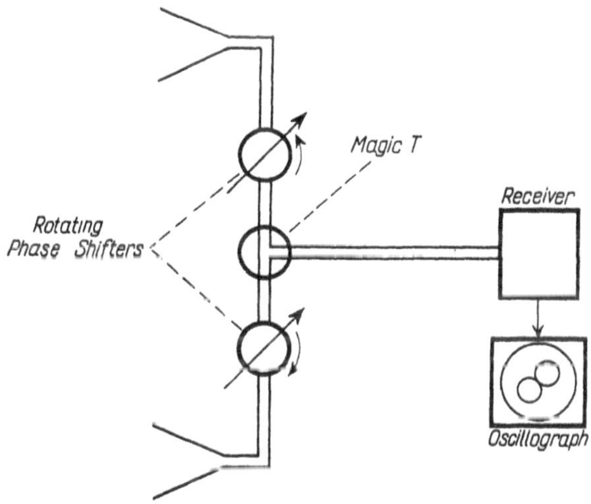

Fig. 13.4. Principle of the goniometer.

The rotational motion of this spot and that of the rotating dephasing devices is controlled by the same motor.

Let us suppose that this equipment receives a plane wave, the equiphase plane of the wave making an angle α with the straight line joining the centres of the horn antennae. The fields incident onto the latter are of equal amplitude, but shifted in phase by

$$\varphi = \frac{2\pi l\alpha}{\lambda} \tag{6}$$

where l is the distance between the horns. The fields entering the magic T are equal to

$$E_1 = E\,e^{i(\omega t+\Omega t+\omega)} \quad \text{and} \quad E_2 = E\,e^{i(\omega t-\Omega t)}$$

where Ω and $-\Omega$ are the angular velocities of the rotating dephasing devices. The radial distance of the spot is thus proportional to

$$|E_1+E_2| = E|1+\cos(2\Omega t+\varphi)+i\sin(2\Omega t+\varphi)|$$

whilst its polar angle equals Ωt. It is easily shown that the spot will describe a figure 8 consisting of two circles (Fig. 13.5) the radius of which is proportional to the amplitude $|E|$ of the field, whilst the polar angle of their centres equals $\varphi/2$.

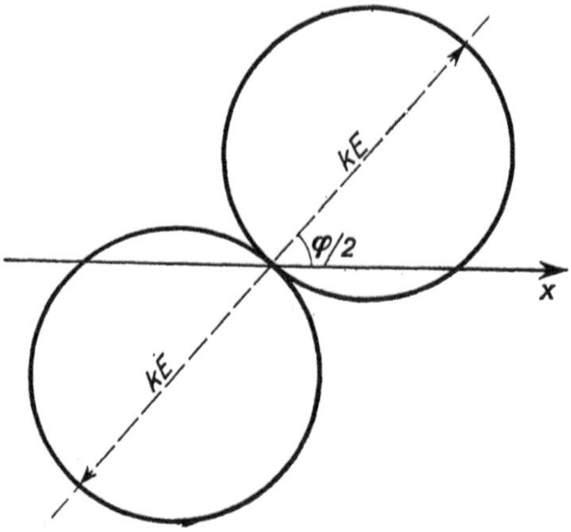

Fig. 13.5. Figure "8" obtained on the CRT screen.

For $\lambda = 3$ cm and $l = 1.8$ m one obtains

$$\frac{\varphi}{2} = 180\alpha$$

The figure 8 will thus turn through an angle of 180° if the angle of arrival of the reflected ray changes by 1°: the apparatus forms a very sensitive goniometer.

If the incident wave is not plane, the curve obtained on the CRT screen will no longer have the form of a figure 8; it will be a curve of the fourth degree without a double point.

In using this goniometer to analyse the wave reflected by the surface of the sea, it might be expected that the received wave will not be plane when

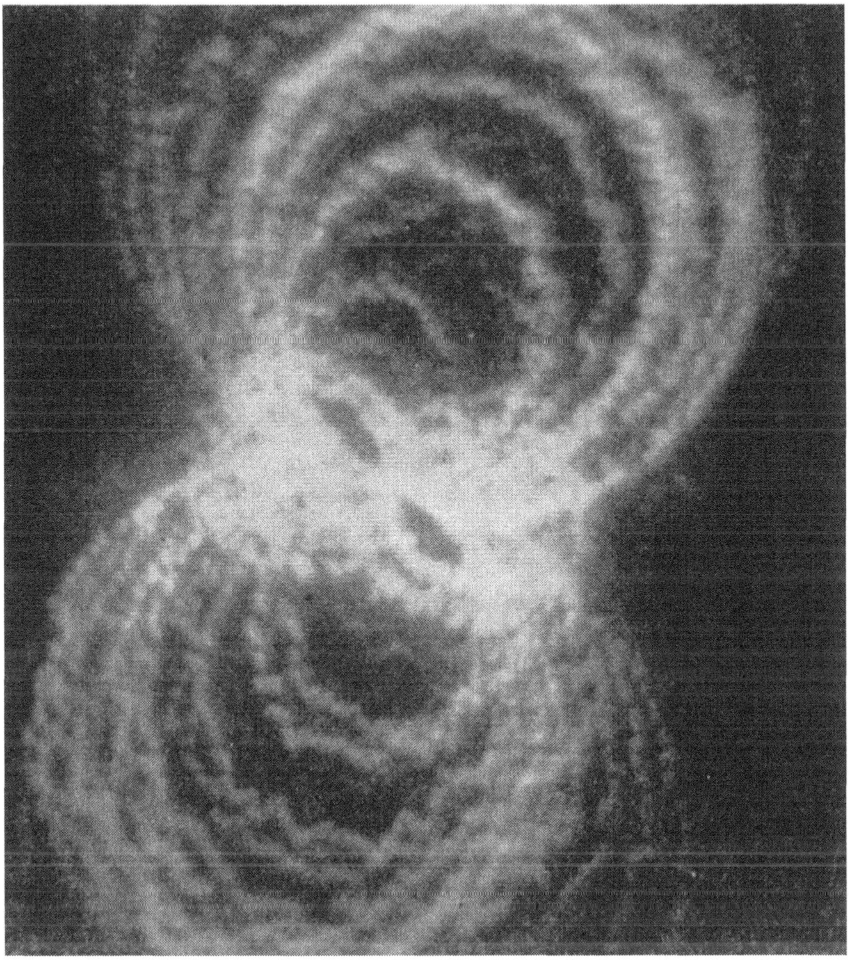

Fig. 13.6. Typical recordings of figures 8 by the goniometer.

it is the result of several components. In reality, this is not the case: each component fluctuates and at any instant only one component dominates. Only curves that have the form of a figure 8 may in fact be observed on the

CRT screen; however, at times these eights have a large amplitude and a fixed direction, which *coincides exactly with the direction of arrival of a specularly reflected wave.*† At other times, on the contrary, they have a smaller amplitude and come from different directions (diffuse scattering).

Figure 13.6 shows an example of the obtained results: the photo shows several successive positions of the figure-8 curves. In Fig. 13.7 a large number of observations of this kind are shown in the form of a diagram. Each point represents an observation: the abscissa of each point corresponds to the abscissa of the point of the surface where the scattered wave seemed to originate in accordance with the orientation of the figure 8 observed on the

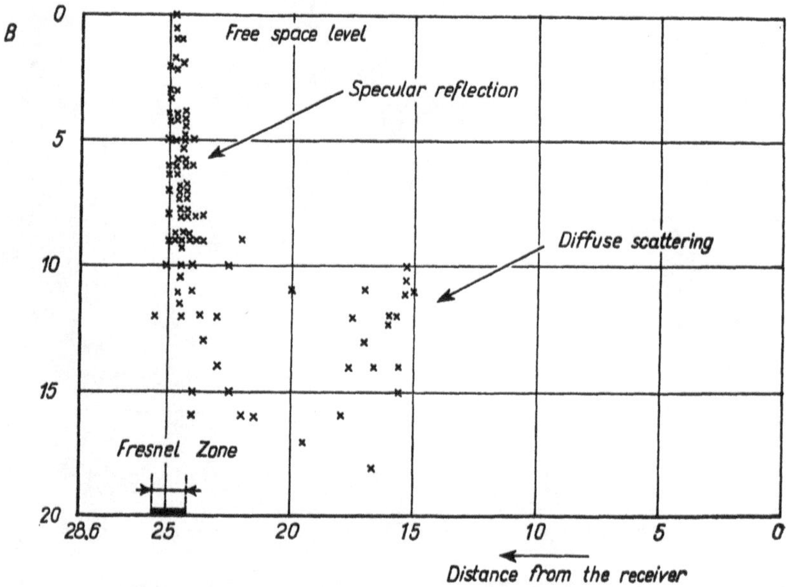

Fig. 13.7. Reflections by different parts of the path.

CRT screen; the ordinate of each point corresponds to the size of this figure 8. The diagram clearly shows the existence of a specular component coming from the first Fresnel zone, and a diffusely scattered component. Owing to natural screens, the latter cannot come from the extremities of the link as predicted by theory. It originates mainly at the points on the surface of the

† To measure the absolute value of the angle of arrival of the reflected wave (i.e. to eliminate the unknown constant of proportionality), one measures the angles $\varphi/2$ (in accordance with the orientation of the figure 8 observed on the CRT screen) for the reflected ray and for the direct ray; this is accomplished by changing the orientation of the transmitting antenna.

sea that are not screened and are most distant from the geometrical point of reflection.

This method, which makes the two components of the reflected field appear distinctly, can be used to measure the corresponding reflection coefficients and to study their fluctuations.

In addition, the contribution of the distant regions of the sea surface to the reflected energy have been studied by deflecting the beam of the transmitting antenna; also, the method enabled the depolarization of the reflected wave caused by each region of the reflecting surface to be studied (the results will be considered in Chapter 15).

13.3.2. *Interference of the Direct and Reflected Waves*

The numerous methods that have been used are all based on the same principle: if the field variations in space are analysed along the horizontal or the vertical, they can be shown to result from the superposition of two components:† a sinusoidal component resulting from the interference of the direct field E_0 with the specularly reflected field E_s, and a random component with more rapid variations resulting from the interference of the diffusely scattered field E_d with the sum E_0+E_s.

The methods of measurement consist in separating these two components, after which one may calculate

(a) the specular reflection coefficient by applying equation (3) to the sinusoidal component (as if there were no diffuse scattering);

(b) the reflection coefficient of the diffusely scattered component by applying the results of Fig. 12.15 to the random component (as if the specularly reflected field were constant and the resulting field were the sum of a Rayleigh-distributed field E_d and a constant field E_0+E_s).

Note. This analysis, performed by many authors, is not entirely correct: it is assumed that the specularly reflected field remains constant. Now it is sufficient to refer e.g. to Fig. 13.7 to see that it varies. Nevertheless, this manner of analysis is never in flagrant contradiction to experiment and we shall see that it gives results that are in agreement with theory (Chapters 14 and 15). This may be explained by referring to the analysis of fluctuations made in Part I (Chapter 7): the field E_s fluctuates as if it were the sum of a constant field E_{s0} and a variable field E_{sv} of mean value zero, the real and imaginary parts of which are each distributed normally with different standard deviations.

If the mean specular reflection coefficient is sufficiently small (this is often the case when diffuse scattering is also present), E_{sv} is distributed according to a law near to the Rayleigh distribution. Let us assume that

† This may easily be shown by a method similar to that of Section 12.6.2; relation (57) is generalized for the case where an additional component E_s is present.

this is the case; then instead of resolving the reflected field into a specular component E_s and a diffuse component E_d, it may also be resolved into

(a) a constant component given by the mean value of the specularly reflected field E_{s0},

(b) a variable component $E_v = E_{sv} + E_d$, which is Rayleigh-distributed.

Thus one again arrives the analysis of the reflected field as described above. It should also be pointed out that the components E_{sv} and E_d, which are distributed with very similar distributions, are difficult to distinguish if only ordinary reception equipment is available. They may be distinguished by means of a goniometer (Section 13.3.1), which confirms that they are of different physical origins. We shall also meet other methods of demonstrating their existence.† Thus the analysis of the reflected field by the goniometric method (Section 13.3.1) and by the interference method (Section 13.3.2) do not give exactly the same result: when both components are present, different values for the reflection coefficient of the diffusely scattered component are obtained; but this difference has no effect on either the order of magnitude or the general variation of the reflection coefficient. In view of the small number of experimental results and their inaccuracy, this difference is difficult to demonstrate by a simple field-strength measurement.

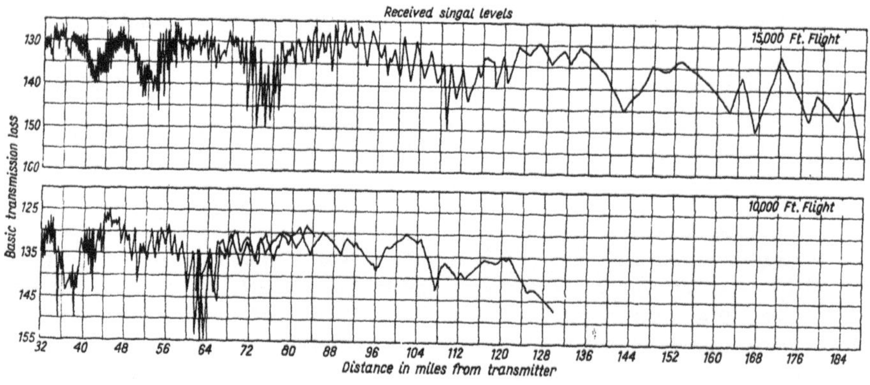

Fig. 13.8. Superposition of random and sinusoidal components [MacGavin and Maloney, 1959]. (*By permission of the authors.*)

To measure the reflection coefficients for E_s and E_d the sinusoidal and random components of the field variations must be separated from each other; various methods have been utilized to this end.

† Cf. the method of Straiton and Tolbert below.

Measurements by MacGavin and Maloney [1959]. MacGavin and Maloney [1959] have measured the field variations in an aircraft flying horizontally in the direction from the transmitter. Figure 13.8 gives an example of the obtained results: the two superposed components, sinusoidal and random, may be clearly recognized. They may be separated by Fourier analysis.

Measurements by Straiton and Tolbert [1956]. The separation of the two components becomes more difficult if the field is measured only at discrete points instead of continuously along a straight line.

To analyse reflection by the surface of a lake (Fig. 13.9) Straiton and Tolbert [1956] have used 10 horns as receiving antennae, spaced vertically at regular intervals. The authors first state that the *mean value* of the field

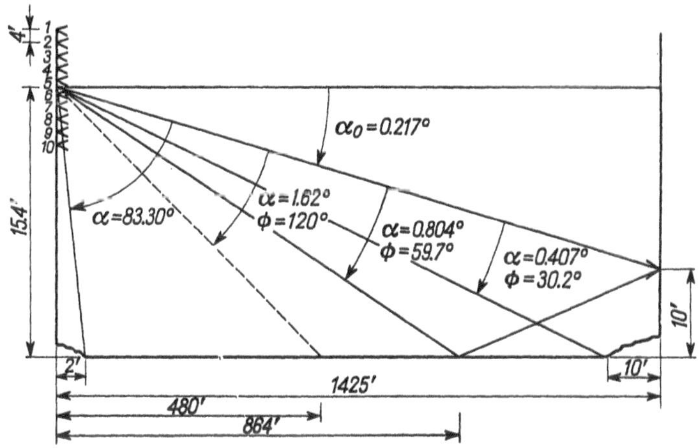

Fig. 13.9. Path profile used by Straiton and Tolbert [1956].
(*By permission of the I.R.E.*)

received by each horn is a sinusoidal function of the height (Fig. 13.10), within the accuracy of measurement. This is in agreement with the above analysis and the amplitude of the sinusoidal variation provides a measure of the specular reflection coefficient.

The authors then analyse the *instantaneous* fields received by the horn antennae; these are the result of the interference of three components: the direct wave E_0, the constant component of the reflected wave E_{s0} and the variable component of the reflected wave E_v. The first two are plane waves and E_0 and E_{s0} are known. It is assumed that the third component may be regarded as a plane wave within the space occupied by three neighbouring horn antennae. It is then shown that the amplitude E_v and the angle of arrival of the variable component may be measured on the basis of the fields received by three neighbouring horns.

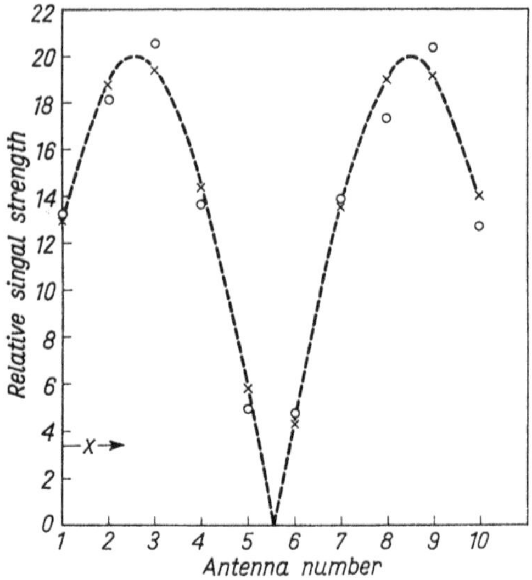

Fig. 13.10. Height gain curve for median data. Circles: normalized medians of signal strength fluctuation; crosses: theoretically fitted curve [Straiton and Tolbert, 1956]. *(By permission of the I.R.E.)*

Figure 13.11 gives an example of the values E_v measured by this method The results obtained from two groups of three different antennae are found to be in agreement.

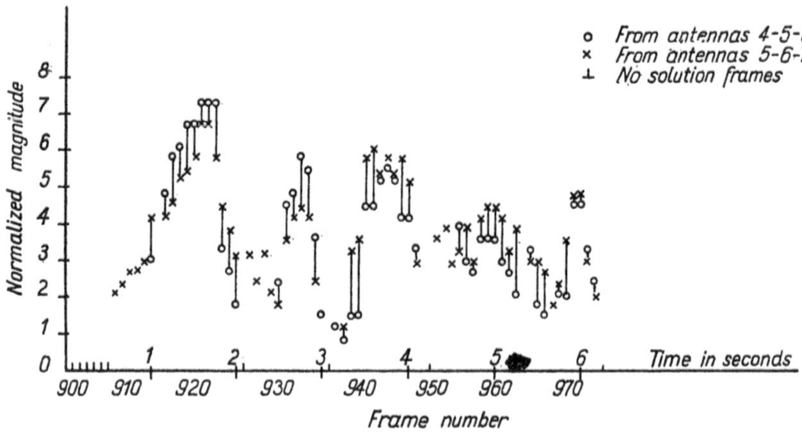

Fig. 13.11. Magnitude of E_v as a function of time [Straiton and Tolbert, 1956]. *(By permission of the I.R.E.)*

Figure 13.12 gives an example of the angles of arrival of the variable component as measured by this method: it may be seen that the field E_v at times comes from the same direction as the specularly reflected field, at times from another, quite different direction. This confirms our remarks on the decomposition of E_v as proposed above: $E_v = E_{sv}+E_d$. The similarity of Fig. 13.12 and 13.7 should be noted: very different methods, applied under different conditions, have led to similar results.

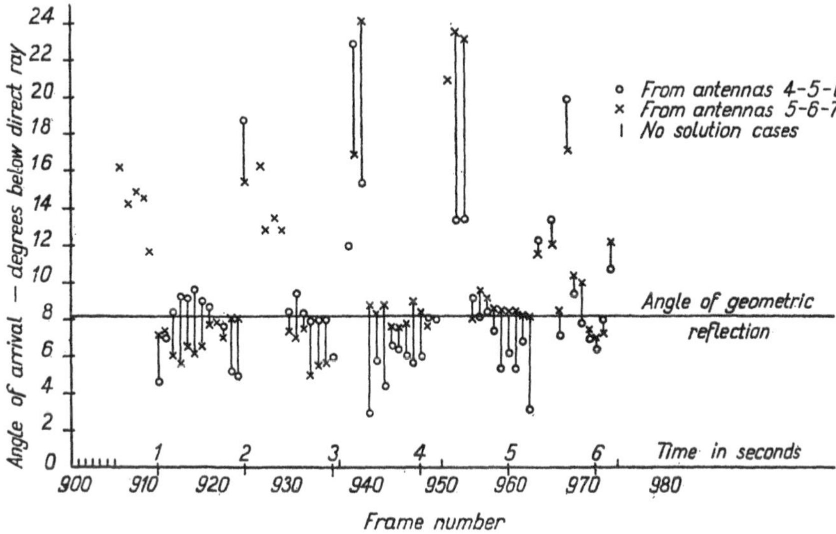

Fig. 13.12 Angle of arrival of variable component as a function of time [Straiton and Tolbert, 1956]. (*By permission of the I.R.E.*)

Measurement by Beard, Katz and Spetner [1956]. In the method proposed by these authors the two components of the field variations along the vertical may be separated on the basis of measurements at 3 equidistant points.

Let E_A, E_B, E_C be the fields measured at any instant at three points A, B, C such that $AB = BC$. Their mean square values are

$$\langle|E_A|^2\rangle = E_0^2+E_s^2+2E_0E_s\cos(\varphi+2\Delta)+\langle|E_d|^2\rangle$$

$$\langle|E_B|^2\rangle = E_0^2+E_s^2+2E_0E_s\cos(\varphi+\Delta)+\langle|E_d|^2\rangle$$

$$\langle|E_C|^2\rangle = E_0^2+E_s^2+2E_0E_s\cos\varphi+\langle|E_d|^2\rangle$$

where the phase shift Δ may be derived from the geometry of the link. Eliminating $\langle|E_d|^2\rangle$ from these three equations, we obtain

$$\langle|E_A|^2\rangle-\langle|E_C|^2\rangle = 2E_0E_s[\cos(\varphi+2\Delta)-\cos\varphi]$$

$$\langle|E_B|^2\rangle-\langle|E_C|^2\rangle = 2E_0E_s[\cos(\varphi+\Delta)-\cos\varphi]$$

From these two equations φ and the specular reflection coefficient $R_s = E_s/E_0$ may be obtained.

The reflection coefficient of the diffusely scattered field is derived from the variations of the received field in time: it is shown that if the diffusely scattered field is Rayleigh-distributed, its mean square value $\langle|E_d|^2\rangle$ equals twice the variance of the received field:

$$\langle|E_d|^2\rangle = 2\langle|E-\langle E\rangle|^2\rangle$$

This approximate results is valid only if $\langle|E_0|^2\rangle$ is very small. In general it is better to use the results of Fig. 12.15.

Measurements by Braude, Komarov and Ostrovsky [1958].† The authors endeavour to measure the reflection coefficient of the field scattered diffusely by the surface of the sea in the presence of the specularly reflected component. The method is based on the same assumptions as those made at the begininng of this section: the specular component E_s is assumed constant and its phase varies slowly. During a short interval of time the sum

$$E_1 = E_0+E_s$$

may be regarded as constant. The field E_1 interferes with the diffusely scattered field E_d, and the field is distributed with the distribution (56) of Chapter 12 (p. 279) with E_0 replacing E_1:

$$\mathrm{d}P\left\{\frac{E}{E_1} > \varepsilon\right\} = 2\frac{E_1}{E_d}\varepsilon\exp\left(-\frac{E_1^2+\varepsilon^2}{E_d^2}\right)I_0\left(2\frac{E_1}{E_d}\varepsilon\right)\mathrm{d}\varepsilon$$

where, for the sake of simplicity, E_1 and E_d stand for the mean square values of these quantities.

The mean values $\langle|E|\rangle$ and $\langle|E|^2\rangle$ derived from (6) equal

$$\langle|E|^2\rangle = E_1^2+E_d^2 = E_d^2(1+\eta^2)$$

$$\langle|E|\rangle = \sqrt{\pi}\cdot\frac{E_d}{2}\mathrm{e}^{-\eta^2/2}\left[(1+\eta^2)I_0\left(\frac{\eta^2}{2}\right)+\eta^2 I_1\left(\frac{\eta^2}{2}\right)\right]$$

where we have set

$$\eta = \frac{E_1}{E_d}$$

$\langle|E|\rangle$ and $\langle|E|^2\rangle$ may be obtained experimentally, and hence η and E_d may be derived. E_0 is determined by an analysis of the interference lobes in the vertical direction: denoting the maximum and minimum values of the field by E_M and E_m respectively,

$$E_0 = \frac{E_M+E_m}{2}$$

The ratio of E_d to E_0 then yields the reflection coefficient for the diffusely scattered field.

† Cf. also footnote † on p. 118.

All these different methods show the existence of a specularly reflected and a diffusely scattered component in ground reflections. The properties of the reflection coefficients R_s and R_d will be the subject of the next two chapters.

Annex to Chapter 13

CHARACTERISTICS OF SOME EXPERIMENTAL RADIO PATHS USED TO MEASURE THE REFLECTION COEFFICIENT OF THE EARTH'S SURFACE

SOME characteristics of experimental links used to measure the reflection coefficient of the earth's surface are given in the table below. The list does not in any way claim to be complete: only the links the results of which are used in Chapters 3, 4 and 5 are included in the table.

The data of the table have been compiled with a twofold purpose: firstly, to avoid overburdening the main text with detailed descriptions of experiments every time such an experiment is quoted, and secondly, the table may be useful for the reader who is confronted with a particular problem, for he may find a reference to an experiment performed under similar conditions to those that he requires. The original paper will then give all details of the experiment.

Abbreviations Used in the Table

Column "Object". The terms *specular*, *diffuse* refer to the purpose of the experiment, i.e. whether the specular or diffuse component was measured.

Column "Method of analysis". This column describes the method used to measure the reflection coefficient of the earth's surface (Chapter 13). The term *separation* refers to the method of separating the direct and reflected rays by means of directional antennae (Section 13.1.1, 13.2.1, 13.3.1); the terms *horizontal*, *vertical* refer to the method of measurement by interference of the direct and reflected waves using either the horizontal or the vertical variations, i.e. the resultant field variations either along the vertical or the horizontal directions.

Column "Data concerning the radio path":

d pathlength

h_1 height of transmitting antenna

h_2 height of receiving antenna

γ grazing angle (complement of angle of incidence) for the ray reflected in accordance with geometrical optics.

Experiments carried out by	Bibliography	Object	Method of analysis	Data concerning the radio path							
				f or λ	d (km)	h_1 (m)	h_2 (m)	γ	Polar.	Terrain	Location
Navy El. Lab. (U.S.A.)	Anderson *et al.* [1942]	Specular	Vertical (Radar)	44cm-2m	4.4			1°39′	V and H	Sea	near San Diego (U.S.A.)
Bell. Tel. Lab. (U.S.A.)	Sharpless [1944]	Specular	Vertical	3 cm	20			1°		Land	
U.S.Army Sign. Corps	Carroll [1946]	Specular	Vertical	6 cm	270	1750	1300	0,19°		Land	California
A.S.E. (U.S.A.)	Archer–Thomson *et al.* [1943]	Specular fluctuations	Separation	9 cm	0.21			11°	H and V	Calm sea	
R.L. (U.S.A.)	Rubenstein–Fishback [1943]	Diffuse & specular	Horizontal (airplane)	10 cm	0 to 80	30	180-2000	0.15°		Flat land little vegetation	near Orlando (U.S.A.)
R.L. (U.S.A.)	Fishback–Rubenstein [1944]	Specular	Horizontal (airplane)	3 cm 10 cm	8–95	37	150–1500	0.12 to 12°		Sea	Massachusets Bay (U.S.A.)
Gen. El. Co. (England)	Lamont and Watson [1946]	Specular	Horizontal (ship to shore)	5.81 cm 6.35 cm	1–10	5 and 14.5	19	0.14 to 2°		Sea	England
E.E.R.L. (U.S.A.)	Hamlin and Gordon [1946]	Specular phase measurements	Vertical	3 cm	43	0.60	0.60	0 to 0.175°	H	Land	Arizona Desert (U.S.A.)

Navy El. Lab.	Day–Trolese [1950]	Specular & diffuse Meteor-ol. meas-urements	Vertical	25, 63, 170, 520, 1000, 3300, 9375, 24000 MHz	43	0–60	0–60	0 to 0.175°	H	Land	Arizona Desert (U.S.A.)
A.S.E. G.E.C. (G.B.) A.S.E. G.E.C. (G.B.)	Megaw [1946] Megaw [1946] Megaw [1946]	Specular Specular & beyond-the-hori-zon Specular	Horizon-tal (ship-to-shore) ,, Separa-tion	9.2cm 9.2.cm 10.1cm 9.2cm	3 to 19 0.8 to 64 0.21	10.39 6	10 78	0.07 to 1° 0.08 to 6.6° 11°		Sea Sea Sea	English Channel Irish Sea
	Ford, Oliver [1946]	Specular Diffuse	Separa-tion	9cm	0.015	0 to 8	0 to 8	10 to 45°	H and V	Land†	
Sperry Gyr. Co. (U.S.A.)	Sherwood [1943] Sherwood–Ginzton [1955]	Specular Diffuse	Vertical	10cm	very short			4 to 28°	H and V	Land, calm, sea, lake, ice	New York (U.S.A.)
R.C.A. (N.J.) (U.S.A.)	Epstein *et al.* [1953]	Specular	Vertical	850MHz	0–74	0–250	12	var.	H	Land	New Jersey (U.S.A.)
Bell. Tel. Lab. and A.T.T. (U.S.A.)	Bullington [1954]	Specular diffuse	Vertical	4000MHz	42 and others	60	60	0.18		Land	Different paths (U.S.A.)

† Systematic study of different kinds of terrain.

Experiments carried out by	Bibliography	Object	Method of analysis	Data concerning the radio path							
				f or λ	d (km)	h_1 (m)	h_2 (m)	γ	Polar	Terrain	Location
N.B.S. (U.S.A.)	MacGavin & Maloney [1954]	Specular diffuse	Horizontal (aircraft)	1046MHz	0–320	850 2240 10,5	500–1400 500–2900 1250–2750	0–24°	H	Land	Colorado and Western Kansas (U.S.A.)
E.E.R.L. Univ. of Texas (U.S.A.)	La Grone, Tolbert [1952] MacKune, Smith [1953]		Vertical	0.86 3.2, 5.3, 9.0 cm	0.84 and 1.50					Sea	Gulf of Mexico
,,	Straiton *et al.* [1952]		Vertical	0.86, 3.2, 9.0 cm	0.43		50	about 10°		Calm, lake	Austin Lake (U.S.A.)
,,	La Grone, Straiton, Smith [1954]	Wave meas.†	Vertical	3 cm	25.6 26.4	25 78	56 56	0.2° 0.3°		Sea	Golden Gate San Francisco (U.S.A.)
,,	Straiton, Tolbert [1956]	Specular & diffuse	Vertical	0.86	0.43	1.5 and 3	variable (about 4.5)	about 1°	H	Calm, lake	Austin Lake (U.S.A.)
Vultee Aircr. Corp. U.S.A.	Beard, Katz, Spetner [1956]	Specular & diffuse	horizontal (aircraft)	3 cm	0–6.8	many values	62			Sea	San Clemente Island (Calif. U.S.A.)

† Cf. Jehn, Gerhardt, Metcalf, Prosser [1954].

E.E.R.L. Univ. of Texas (U.S.A.)	Beard, Katz. Spetner [1956]	Specular & diffuse	Horizontal (airplane)	3 cm	0–9	37.5	13.5		V	Sea	San Clemente Island (Calif. U.S.A.)
E.E.R.L. & Appl. Phys.Lab. (U.S.A.)	Beard, Katz [1957]	Wave meas.	Vertical†	3 cm	2.7 4.8	28–270 22.5–108	26 26	1.25–7° 0.6–1.7°		Sea	Golden Gate San Francisco (U.S.A.)
E.E.R.L. Univ. of Texas (U.S.A.)	Tomlinson, Straiton [1959]	Specular	Vertical	3 cm	(many paths)						Texas (U.S.A.)
U.S.S.R.	Braude, Komarov, Ostrovsky [1958]§	Specular & diffuse	Time variations Space var. (vertical)	3.2 cm	0.75	6	1, 7.5, 16	0.6, 1.1, 1.8°		Sea	U.S.S.R.
A.T.T. (U.S.A.)	Oxehufwud [1959]	Specular (Many reflection points)	Vertical	3950, 6310 and 11200 MHz	65 41 58 54	‡			H and V ,, ,, ,,	Land	Colorado
C.N.E.T. (France)	Spizzichino [1959], Boithias & Spizzichino [to be published]	Specular & diffuse Interferences Depolarization	Separation goniometry	3.2 cm	28.6	70	530	0.55°	H and V	Sea	Mediterranean

† Using the slope of a hill. ‡ Many reflection points, giving different values of h_1 and h_2. § Cf. also footnote † p. 118.

CHAPTER 14

EXPERIMENTAL INVESTIGATION OF THE SPECULAR REFLECTION COEFFICIENT OF THE EARTH

In Chapters 14 and 15 we shall consider the experimental values of the reflection coefficients of the earth obtained by the methods described in Chapter 13; we shall compare them with the theoretical results (Chapter 12) and attempt to draw practical conclusions.

14.1. Parameters Determining the Reflection Coefficient

Theoretically (Section 12.3), the specular reflection coefficient of the earth is the product of three factors:

$$R_s = \rho_s D R_0 \tag{1}$$

which we have successively considered: R_0, the reflection coefficient of a plane, smooth earth, in Section 11.1; D, the divergence coefficient, in Section 11.2; and ρ_s, the "scattering coefficient", in Section 12.3. The mean square value of the latter is given by†

$$\langle |\rho_s|^2 \rangle = e^{-(\Delta\Phi)^2} \tag{2}$$

with

$$\Delta\Phi = \frac{4\pi \Delta h \sin\gamma}{\lambda} \tag{3}$$

R_0 depends on the nature of the soil, on the polarization, on λ and on γ;‡ D depends on the effective radius of the earth,§ which varies with the meteorological situation; according to (2) and (3) ρ_s depends on the dimensions Δh of the irregularities of the earth's surface, on λ and on γ. The reflection coefficient R_s defined by (1) thus depends on all of these parameters and in a rather complex manner: for example, it depends on γ through the intermediate factors ρ_s, D and R_0, where D is an increasing, but R_0 a decreasing

† Relations (2) and (3) are identical with (5) and (6) of Chapter 12.

‡ Cf. Chapter 11, (1) and (2), or Figs. 11.1 and 11.2.

§ Cf. Chapter 11, equation (9).

function of γ. Theoretically, R_s may in fact be an increasing or a decreasing function of γ.†

However, the analysis of R_s simplifies if one notes that

(a) if $\gamma > 1°$, then almost invariably‡ $D = 1$; then R_s depends only on the nature of the terrain, the polarization, on λ, γ and Δh through the intermediaries ρ_s and R_0.

(b) if $\gamma < 1°$, then almost invariably§ $R_0 = 1$; then R_s will depend on λ, γ, Δh and the effective radius of the earth: meteorological conditions may thus have a considerable effect.

It is therefore convenient to distinguish these two cases in order to interpret the experimental results. In practice they correspond to two very different types of measurement: Case (a) is that of very short experimental links designed to measure the reflection coefficient of a small section of terrain; it is also sometimes the case of radars observing aircraft. Case (b) is that of practical links used for telecommunications.

14.2. Experimental Investigation of the Scattering Coefficient ρ_s

Before taking up the study of the reflection coefficient R_s given by (1), we shall consider the factor ρ_s associated with the irregularities of the earth's surface. We shall confront the theoretical relation (2) and (3) with the experimentally measured values of ρ_s.

When we speak of experimental values of ρ_s it should be understood that these are the values derived by (1) from the experimentally measured values of R_s and the calculated values of R_0 and D. The laws governing the variations of R_0 and D are now classic; they have been verified many times in the case of smooth surfaces and may be considered exactly valid.

Figure 14.1 shows the experimental values of $\langle|\rho_s|\rangle$ obtained by several authors‡‡ as a function of $\Delta\Phi$ calculated from (3) and the experimental data given by these authors. Although these authors have often drawn different

† An example of this is shown in Fig. 11.10, where R passes through a minimum when γ varies.

‡ According to (9) of Chapter 11, if $\gamma > 1°$, D may differ considerably from 1 only if r_1 or r_2 is of the order of at least 100 km. But for such values of r_1 and r_2, γ may be greater than 1° only if h_1 and h_2 are very large (aircraft flying at very high altitudes).

§ According to Figs. 12.1 and 12.2. However, the statement does not apply to vertical polarization above the sea if the wavelength is decimetric (UHF) or longer.

‡‡ Ford and Oliver [1946], Day and Trolese [1950], Megaw [1946], Archer-Thomson [1943], Bullington [1954], Sherwood and Ginzton [1955], Gough [1949, 1958], Straiton and Tolbert [1956] and Spizzichino [1959]. Most of the other authors indicated in the annex to Chapter 3 have not given sufficiently exact data on the nature of the terrain for Δh to be evaluated.

(if not contradictory) conclusions from their measurements, the entire set of all these measurements seems in good agreement with (2). This agree-

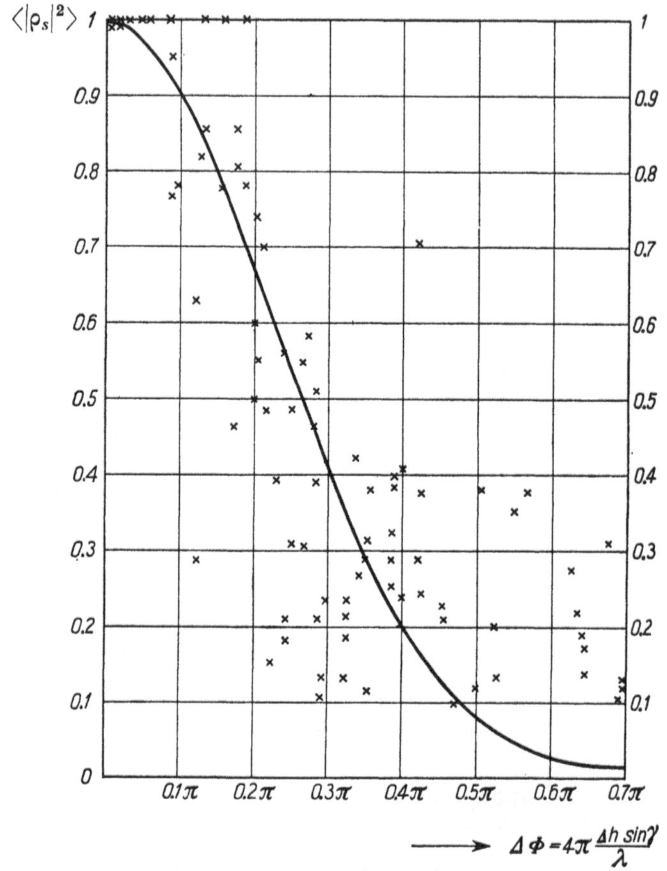

Fig. 14.1. Dependence of the scattering coefficient on the phase variance. The full curve plots the relation (2).

ment is all the more gratifying considering that the data are often given with very little precision:

1. The measurement of the reflection coefficient R_s is often inaccurate,† particularly near the values 0 and 1.‡ To this error one must add the inaccuracy of evaluating R_0.

† Cf. Section 13.1.2.

‡ It will be noted that a number of points in Fig. 14.1 corresponds to $\langle|\rho_s|\rangle = 1$. These points have been plotted in accordance with the available data, though in most cases they only give an order of magnitude.

2. The evaluation of Δh is even more inaccurate. We have defined Δh as the standard deviation of the normally distributed height h of the points of the earth's surface. Other definitions are met in certain papers. Some authors† use the range of h lying between the 1 and the 50% levels, Δh_{50-1}. For a normally distributed surface,

$$\Delta h_{50-1} = 2.33\,\Delta h,$$

Other authors use the difference between the 99 and 1% levels; for a normally distributed surface

$$\Delta h_{99-1} = 2\Delta h_{50-1} = 4.66\,\Delta h$$

Finally, some authors give the maximum dimension h_M of the irregularities. In this case we have set

$$h_M = \Delta h_{99-1} = 4.66\,\Delta h$$

A small number of points obtained in this way and plotted in Fig. 14.1 were in very good agreement with the theoretical relation (2). In the case of sea reflections, there are often only qualitative indications on the state of the sea; from these one may obtain a value of the heights of the waves from Table 12.1 (p. 248). The best agreement with the theoretical relation (2) is then obtained by assuming that this dimension represents $h_M = \Delta h_{99-1}$.‡

3. Finally, most of the points with abscissae $\Delta\Phi$ exceeding 0.4π very probably correspond either to diffuse scattering or to a combination of specular reflection and diffuse scattering (the authors sometimes give values of the reflection coefficient without making it clear to which of the two types of reflection it corresponds).

The relation (2) may be regarded as a very general one: the points in Fig. 14.1 were obtained by measurements over land and over the sea, in plane and hilly country, in different climates and in different countries, in the U.S.A, in Great Britain and the European continent. Moreover, they were obtained by different methods of measurement (Chapter 3).

Exception to the general rule. There is a series of values that does not agree with the general theoretical relation (2). They are all obtained in measurements made with vertical polarization and a large grazing angle (greater than 5°) and published by different authors.§

---

† In particular, this definition is used by Bullington.

‡ These results are not, however, plotted in Fig. 14.1.

§ Notably Fishback and Rubenstein [1944], Ford and Oliver [1946] and Sherwood and Ginzton [1955]. The points corresponding to $\langle|\rho_s|\rangle > 1$ have not been plotted in Fig. 14.1.

The values of $\langle|\rho_s|\rangle$ are in these cases abnormally high and may exceed unity.

The physical origin of these results becomes evident if one notes that these measurements have always been made with γ near the Brewster angle. If the measurements are sufficiently continuous for the dependence of $\langle|R_s|\rangle$ on γ to be traced as a curve, Fig. 14.2 is obtained: in the vicinity of the Brewster angle the minimum of the reflection coefficient is less pronounced in the case of irregular terrain than for a smooth surface.

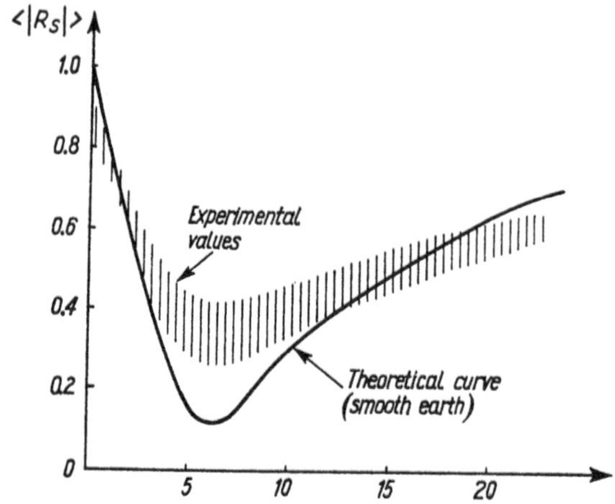

Fig. 14.2. Effect of surface roughness near the Brewster angle (vertical polarization).

Without claiming to provide a rigorous explanation of this phenomenon, it may easily be seen that the angle γ concluded by the incident ray and a wavy surface will not be the same at different points of this surface: it will vary about the value γ_0 which it would assume in the absence of the irregularities. If γ_0 is near the Brewster angle, the points where $\gamma = \gamma_0$ will reflect less energy than the others and (1) will no longer hold, since this relation states that the reflection coefficient R_s is proportional to $R_0(\gamma_0)$ as though all points on the earth's surface had the same reflection coefficient as for $\gamma = \gamma_0$.†

This special case is of little practical interest in the usual radio links, where γ is at most of the order of 1°, whereas the Brewster angle is greater than 10° for soil and greater than 4° for the sea if $\lambda < 1$ m. *In practical cases, relation* (2) *is in good agreement with experiment.*

† As stated in Section 12.3, relation (1) is valid only if R_0 is independent of γ or varies only slightly with γ.

14.3. Dependence of the Reflection Coefficient on Frequency and on the Dimensions of the Irregularities

In the following we shall no longer analyse the reflection coefficient R_s in its three factors ρ_s, D and R_0; the laws governing their variations are now well known and experimentally verified. We shall rather consider the *total* effect of some parameters on R_s and consider the practical consequences.

The frequency-dependence has been the subject of some experimental investigations described by Day and Trolese [1950], LaGrone and Tolbert [1952] and Oxehufwud [1959];† these measurements were carried out for practical purposes (radio relay lines, etc.). In these cases of practical radio links (other than radar), γ is at most of the order of 1°; theoretically, this corresponds to Case (b) of Section 14.1, where R_s depends on the frequency only through ρ_s, i.e. caused by the irregularities of the terrain.

The experiments quoted above confirm this effect of the irregularities which cause the reflection coefficient to decrease with growing frequency. Using these experiments with the additional aid of theory, the frequency range in which they play an important part may be given somewhat more accurately.

Sometimes the irregularities are classified as follows:‡

Irregularities of order 1: hills, valleys

Irregularities of order 2: houses, trees, etc.

Irregularities of order 3: grass, waves on a calm sea, etc.

For the usual values of γ (a few milliradians), one may define two frequencies for each of these types of irregularities bounding a frequency range within which the frequency considerably affects the reflection coefficient:

below a certain frequency f_1, $\Delta\Phi$ as given by (3) remains small and the reflection coefficient is that of a smooth earth;

above a certain frequency f_2, $\Delta\Phi$ is greater than $\pi/2$ and specular reflection disappears, leaving only diffuse scattering (which is only slightly frequency-dependent).

This frequency range, where the frequency has a pronounced effect on the reflection coefficient, is thus located

between 100 and 3000 Mc/s for irregularities of order 1,

between 1000 and 10,000 Mc/s for irregularities of order 2,

above 10,000 Mc/s for irregularities of order 3.

---

† Cf. Annex to Chapter 13 for details of these experiments.

‡ Cf., for example, Epstein and Peterson [1953].

14.4. Dependence on the Grazing Angle

Numerous experimental investigations have been made in the cases (a) and (b) distinguished in Section 14.1.

(a) *Measurements performed on very short experimental links* ($\gamma > 1°$, $D = 0$). In the experiments described by Ford and Oliver [1946] and Sherwood and Ginzton [1955],† a very short link was set up, the path length amounting to some tens of metres; the grazing angle γ could then be varied by altering the heights of the transmitter and receiver.

The general form of the obtained results is that of Figs. 14.2 and 14.3. The first of these has already been mentioned (Section 14.2), the second shows that with horizontal polarization, the irregularities of the terrain cause $\langle |R_s| \rangle$ to decrease with growing γ more rapidly than in the case of a smooth earth.

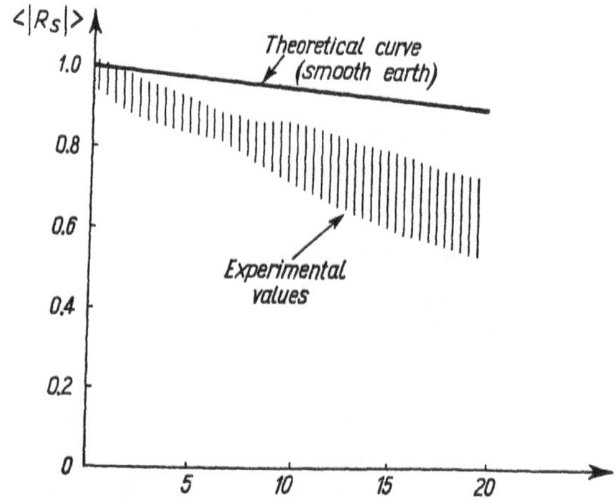

Fig. 14.3. Effect of surface roughness (horizontal polarization).

(b) *Measurements performed on links with pathlengths comparable to those used in telecommunications* ($\gamma < 1°$, $R_0 = -1$). The methods of measuring the reflected field by investigating the interference lobes enable the variations of the reflection coefficient in space, i.e. as a function of γ, to be studied if there is a large number of lobes. The authors who have supplied useful results of this kind include Epstein and Peterson [1953], Day and Trolese [1950], Hamlin and Gordon [1948], Tomlison and Straiton [1959], Straiton and Tolbert [1956],† who have used the method of analysing the field along the vertical, and Rubinstein and Fishback [1943], Day and Trolese [1950],

† Cf. Annex to Chapter 13 for details of these experiments.

Mac Gavin and Maloney [1954],† who analysed the field along the horizontal using an airborne receiver.

In theory, if $\gamma < 1°$, the reflection coefficient $|R_s|$ equals the product $D|\rho_s|$, where D is a growing‡ function of γ and $|\rho_s|$ a decreasing function of γ.§ One or the other of these two factors may dominate depending on the value of γ: it may be shown that as γ increases, $|R_s|$ will increase and then pass through a maximum.

In practice, the effect of D makes itself felt only in links approaching the limit of visibility, where γ is extremely small. As against this, the different experiments quoted above all demonstrate the effect of $|\rho_s|$: as γ increases, R_s decreases until specular reflection is completely masked by diffuse scattering. An example of this is given in Fig. 14.4, taken from MacGavin and

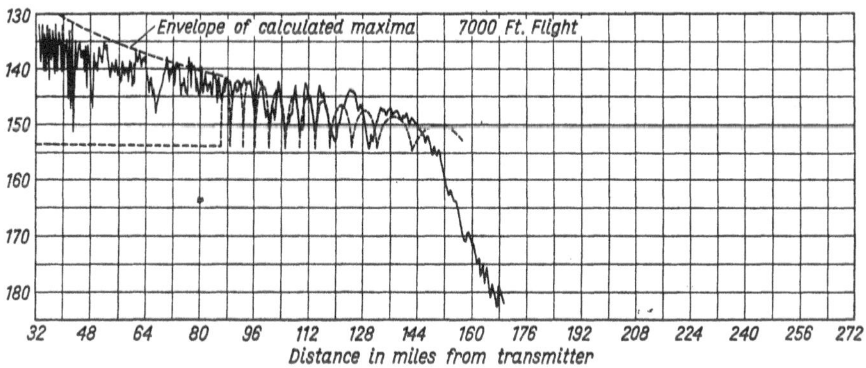

Fig. 14.4. Superposition of specular and diffuse components [MacGavin and Maloney, 1954]. (*By permission of the authors.*)

Maloney [1954]:† when the aircraft carrying the receiver approaches the transmitter, γ increases; it may be seen that at first the regular interference fringes of specular reflection appear, then their amplitude decreases and finally diffuse scattering dominates, the received field fluctuating in a random manner.‡‡

14.5. Dependence on Polarization

In the theoretical expression (1) for R_s, the polarization occurs only in R_0; thus R_s should depend on the polarization only if $\gamma > 1°$ (Case (a)

† Cf. footnote on p. 322.

‡ D is given by (1) of Chapter 11.

§ In accordance with (2) and (3).

‡‡ The same effect appears in Figs. 13.2 and 13.8.

of Section 14.1). We shall not return to this case which has already been considered (cf. Section 14.4, Figs. 14.2 and 14.3).

When $\gamma < 1°$ (Case (b) of Section 14.1), i.e. in links with the usual path length, some comparative measurements have been performed on horizontal and vertical polarization by the American Telephone and Telegraph Company (A.T.T.) ([Oxehufwud, 1959] at 6000 Mc/s over land) and by the C.N.E.T. (Paris) at 10,000 Mc/s over the sea:† the reflection coefficient hardly varied when the polarization was changed.

14.6. Dependence on Meteorologic Conditions

The meteorologic conditions may affect the reflection coefficient of the earth's surface in two ways;

(a) they affect the structure of the atmosphere, modifying the ray paths of electromagnetic waves;

(b) if the reflecting surface is that of the sea, they modify its reflection coefficient by altering the roughness of the surface.

Displacement of the Interference Fringes

Modifying the paths of the direct and reflected rays, a change in the structure of the atmosphere causes the phase of the reflected field to change and the interference fringes will be displaced. All authors who have observed interference fringes on sufficiently long radio links state that they are constantly being displaced.

Some authors (Hamlin and Gordon [1946], Tomlinson and Straiton [1959] and others) have studied this displacement and compared it with the results of meteorologic sondes. From measurements of the pressure, temperature and humidity the refractive index of the atmosphere along the vertical may be determined. It is assumed that the refractive index depends only on the height (stratified atmosphere) and then the paths of the direct and reflected rays may be calculated. The path difference between the two rays is $\Delta\delta = \Delta(n_R - n_D)$ where n_D is the mean value of the refractive index n along the direct path, and n_R the mean value along the path of the reflected ray. $\Delta(n_R - n_D)$ stands for the variation of $n_R - n_D$. From this the displacement ΔZ of the interference fringes observed along the vertical may be derived:

$$\Delta Z = -\frac{d^2}{2h_1}\Delta(n_R - n_D)$$

† The conditions of these experiments were those of Section 13.3.1.

Figure 14.5† gives an example of the comparison of the theoretical values ΔZ obtained in this manner with the experimental values obtained by tracking the displacements of a zero of the interference fringes. There

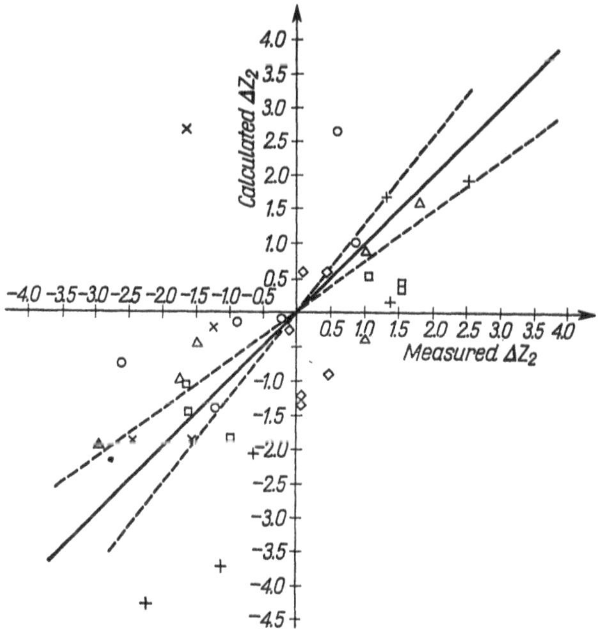

Fig. 14.5. Theoretical and experimental values of ΔZ [Tomlinson and Straiton, 1959]. *(By permission of the I.R.E.)*

is a clear correlation between these values, although in this example the situation is rather unfavourable: the path runs over hilly country, which may throw doubt on the hypothesis of a stratified atmosphere.

In practice one never attempts to predict the position of the interference fringes for any given instant. But it is interesting to know the distribution of the values of their displacement ΔZ because the distribution of the values of the received field may thus be derived. The problem may be solved in a satisfactory manner by assuming that the atmosphere is of the "standard" type and explaining the displacement of the interference fringes by the variations of the equivalent radius of the earth.

Although this model of the structure of the atmosphere describes the real atmosphere only very imperfectly, it leads to useful results: the ratio K to the equivalent of the real radius of the earth may be determined by

† Taken from Tomlinson and Straiton [1959].

radio methods: on the basis of the displacement of the interference fringes† and other methods‡ it has been shown that the value of K obtained for a given link is independent of frequency.§ The distribution of K for a given place and a given season may be determined experimentally‡ or one may simply assume that K varies between two limits, e.g. between 2/3 and infinity. From the distribution of K, the distribution of Δz may be obtained. The distribution of the values of the received field which one may thus derive, is, in general, in good agreement with experimental results.‡‡ However, this method does not always permit predicting the deepest fades measured experimentally.† (We shall return to this point in Chapter 16.)

Under the assumption of a "standard atmosphere" an interesting property of the displacement of the interference fringes may be demonstrated: considering the dependence of ΔZ on K it may be shown that for a given pathlength

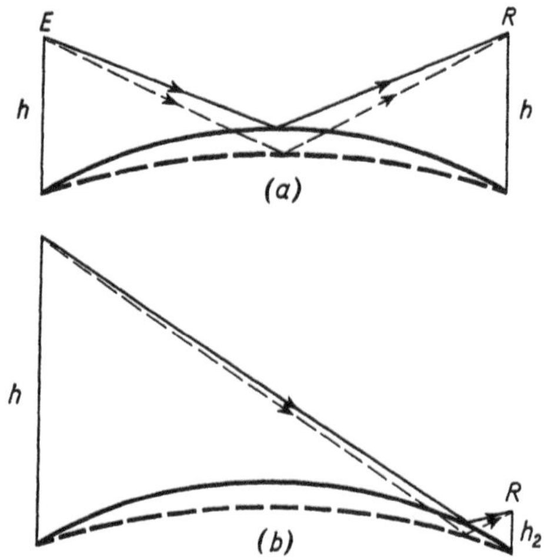

Fig. 14.6. Variation of the pathlength of the reflected ray with the effective radius of the earth. (a) Symmetrical link, (b) strongly assymetrical link.

of the link, the variations ΔZ are maximum if the point of reflection is located at the path midpoint (symmetrical link), whereas they may be very small if the link is very asymmetrical, i.e. if the transmitter and receiver heights h_1

† Oxehufwud [1951].

‡ Cf., for example, Misme: Le gradient équivalent; mésure directe et calcul théorique. *Annales des Téléc.* **15**, 99 (1960) and Gough [1955].

§ Oxehufwud [1951] reports a slight variation of K with the frequency, but it is too small to be of practical interest.

‡‡ Cf. Megaw [1946], Gough [1958], etc. We shall return to this point in Chapter 16.

† Cf. Bray and Corke [1952]; Bogle [1952]; Gough [1958].

and h_2 are very different. The reader may convince himself of this for example by chosing some numerical values and determining the variations of the optical paths with the aid of the nomograms in the Annex to Chapter 11. Intuitively one may accept that the phase of the reflected ray varies less rapidly with K when the point of reflection approaches one of the terminals of the link (Fig. 14.6).

Experiments have confirmed that meteorological conditions have no effect on the reflection from obstacles near the transmitter or receiver.

Variations of the Reflection Coefficient of the Earth's Surface

Meteorological conditions may affect not only the phase, but also the strength of the reflected ray by acting either on the structure of the atmosphere or on the surface of the sea.

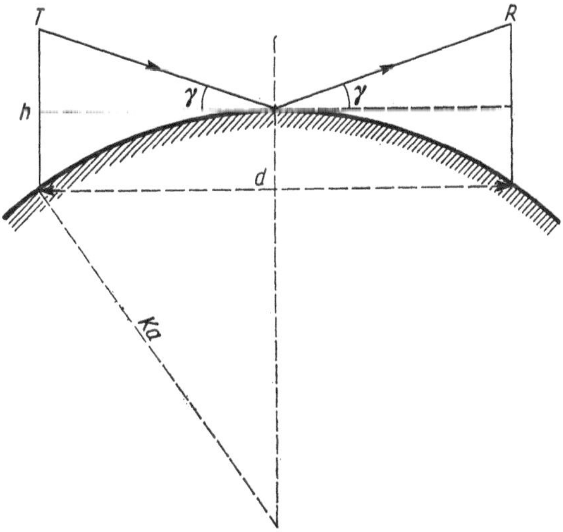

Fig. 14.7. Phase variation of the reflected ray with the effective radius of the earth.

In the former case theory shows that if the equivalent radius of the earth varies, it will also cause the grazing angle γ to vary. For simplicity, let us consider a symmetrical link ($h_1 = h_2 = h$). We obtain

$$\gamma = \frac{2h}{d} - \frac{d}{4Ka} \tag{4}$$

where d is the pathlength, and Ka the equivalent radius of the earth (cf. Fig. 14.7). If K varies from 1 to ∞, the variation of γ amounts to

$$\Delta\gamma - \frac{d}{4a} \tag{5}$$

This variation of γ entails a variation of the reflection coefficient R_s, which will be appreciable only if $\Delta\gamma/\gamma$ is not too small. In accordance with (4) and (5), we may write the condition $\Delta\gamma/\gamma \gg 1$ in the form

$$d^2 \geqslant 4ah$$

or

$$d \geqslant \sqrt{2} \cdot d_H$$

where $d_H = \sqrt{(2ah)}$ is the distance to the horizon. Thus *the reflection coefficient should vary only if the pathlength d is of the order of the distance to the horizon.*

If the link is asymmetrical, it may be shown that the absolute value of the reflection coefficient is, like the phase of the reflected ray, not appreciably affected by atmospheric influences.

Most of the numerous authors who have studied the time dependence of the interference lobes caused by specular reflection report the displacements of these lobes without noting a change in the reflection coefficient. However, a variation of the ground reflection coefficient which is clearly correlated with the variations of the refractive index, appears in the measurements described by Hamlin and Gordon [1946]. These experiments were performed in conditions† that are particularly favourable for this effect: $d = 43$ km, $d_H = 33$ km, and the link is symmetrical. On the contrary, MacGavin and Maloney [1958], whose measurements were performed up to the distance of the horizon on a strongly assymmetrical link (transmitter on the ground, receiver airborne), state that the reflection coefficient of the ground does not depend on the refractive-index profile of the air.

For oversea paths, the variations of the reflection coefficient are primarily due to the dimensions of the waves. As we have already indicated (Section 14.1), in this case, the values of the reflection coefficient given by (2) and (3) are in agreement with experimental values.

14.7. Very Irregular Terrain—Multiple Reflections

Up to the present we have assumed the surface of the earth to be near a mean plane with regard to which its irregularities may be defined. Let us now consider more irregular terrain: if, for example, there are two or three hills between the transmitter and receiver, they may not be considered as simple irregularities of the terrain and treated statistically as if there were a large number of them. In such a case we must consider each of them individually, investigate whether they include points of specular reflection and determine the reflection coefficient for each of these points.

† Cf. Annex to Chapter 13.

Theoretical methods. A method described by Norton [1952] consists in approximating the profile of the path by a simple curve of the second degree. The points of reflection, the dimensions of the Fresnel ellipse and the grazing angle γ may then be determined for each of these points.

In order to calculate the reflection coefficient of each of these points, the dimensions of the irregularities of the earth's surface must be known. The divergence coefficient may be determined from (9), Chapter 11, provided the radius of curvature of the profile is known.

It should be noted that this method is rarely efficient: no map can in general furnish sufficiently accurate data on the profile of the path.

On the other hand, if a reflection is produced by a hill with sufficiently great curvature at a small grazing angle (in which case the divergence coefficient D is small), the methods of geometrical optics become doubtful. It is then preferable to apply the theory of diffraction by a parabolic cylinder [Fok, 1945 and 1946] or the results of Norton's paper [1941] on the ground wave propagated over a spherical conducting surface.† Sometimes [Tomlinson and Straiton, 1959] the formulae for the diffraction by an infinitely thin conducting screen give correct account of the experimental results. The received power is then given by Fig. 14.8, the notation being evident from Fig. 14.9 (it has been assumed that the screen is equidistant from the transmitter and the receiver).

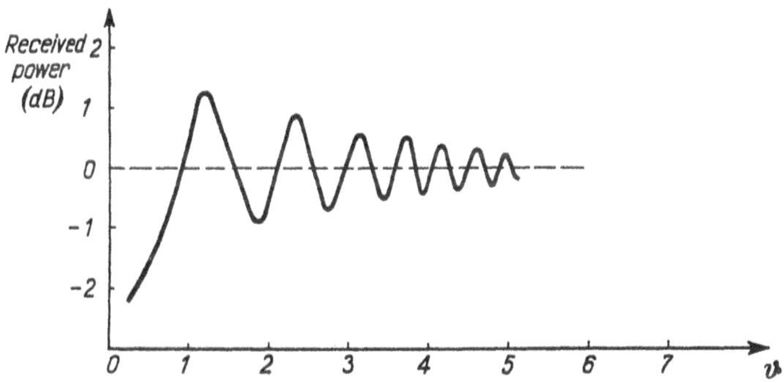

Fig. 14.8. Diffraction effect near the edge of a screen.

Experimental methods. The method of measuring the reflection coefficient of the earth's surface by analysing the interference fringes enables the point of reflection to be localized; if there are several points of reflection, the method will detect and localize these points and measure their reflection coefficients.

† In the same paper, Norton proposes a method of interpolation between the laws of geometrical optics and those governing the diffraction by a sphere.

In effect, the received field $E(z)$ measured along the vertical results from the superposition of the direct field

$$E = E_0 \exp\left[i\left(\omega t + \frac{2\pi z \sin\alpha}{\lambda}\right)\right]$$

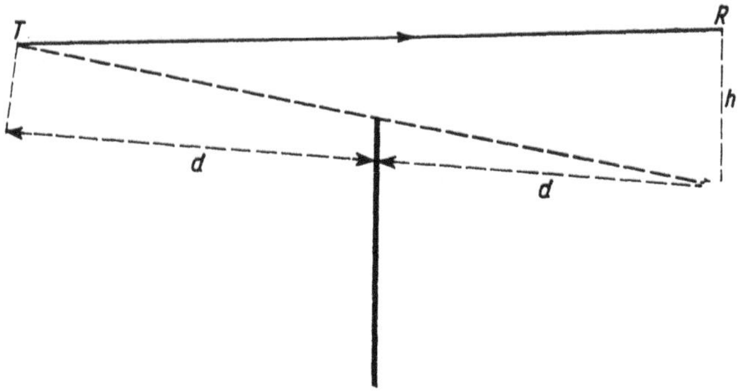

Fig. 14.9. Significance of the parameter $v = h/2\lambda d$ in Fig. 14.8.

and the fields E_i reflected by a series of points P_i on the earth's surface ($i = 1, 2, \ldots, n$):

$$E_i = R_i E_0 \exp\left[i\left(\omega t + \frac{2\pi z \sin\alpha_i + \beta_i}{\lambda}\right)\right]$$

where α_0 and α_i are the angles concluded by the direct and reflected rays with the horizontal; R_i stands for the reflection coefficient of the point P_i. The resultant fieldstrength $E(z)$ is then given by

$$\begin{aligned}|E(z)|^2 &= \left|E_0 + \sum_i E_i\right|^2 \\ &= E_0^2\left\{1 + \sum_i R_i^2 + 2\sum_i R_i \cos\frac{2\pi}{\lambda}[(\sin\alpha_i - \sin\alpha_0)z + \beta_i] + \right. \\ &\quad \left. + 2\sum_{i \neq j} R_i R_j \cos\frac{2\pi}{\lambda}[(\sin\alpha_i - \sin\alpha_j)z + (\beta_i - \beta_j)]\right\}\end{aligned}$$

If the reflection coefficients R_i are sufficiently small, the last term may be neglected; applying a Fourier analysis to the resultant field $|E(z)|$ one finds a series of sinusoidal components with periods

$$\frac{\lambda}{\sin\alpha_i - \sin\alpha_0} \qquad (i = 1, 2, \ldots, n)$$

and amplitudes $|R_i|$. Since α_0 is known, a measurement of the periods and amplitudes of these components will yield α_i and R_i. The values of α_i then enable the points P_i to be located by geometrical construction on the profile of the path (Fig. 14.10).

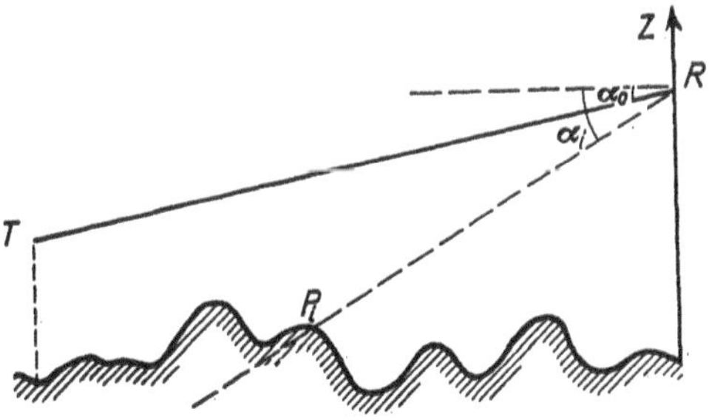

Fig. 14.10. Detection of reflecting points.

Tomlinson and Straiton [1959] give examples of applying this method to the study of experimental links over irregular terrain. The methods furnishes plausible results, always provided that the reflected field has the character of specular reflection.

Figure 14.11, taken from Oxehufwud [1959], shows an example of the variation of the field along the vertical above irregular terrain. Two sinusoidal components with very different periods may be recognized, the more rapidly

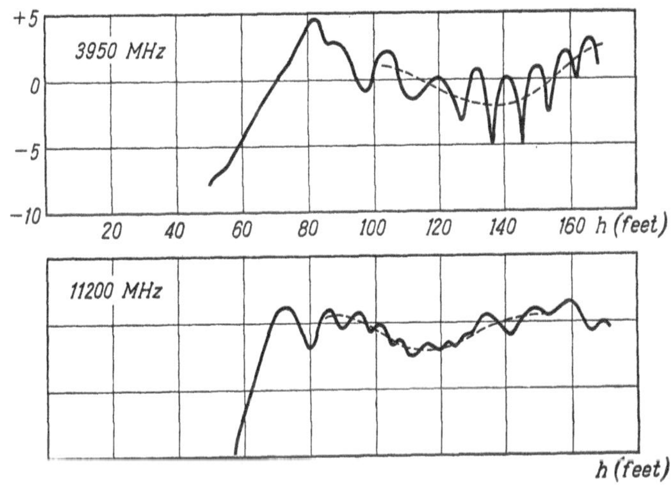

Fig. 14.11. Fieldstrength variations over irregular terrain [Oxehufwud, 1959]. (*By permission of the authors.*)

changing component originating at a point of reflection near the receiver. The amplitude of this latter component decreases substantially if a higher frequency is used.

14.8. Fluctuations of the Resultant Field

The fluctuations of the specularly reflected field have a very different character depending on whether the field is reflected by land or by the sea. In the former case these fluctuations are caused by deformations in the atmosphere and will appear only for sufficiently long paths. In the latter case, they are caused by the motion of the sea and appear for any pathlength.

It has been shown in Chapter 13 that the fluctuations of the reflected field are difficult to measure by the method of interference lobes. Practically all the interesting results in the literature are obtained by separation of the reflected ray with the aid of directional antennae. This method may only be used with short paths, so that there is little chance of measuring the fluctuations of the reflection coefficient of the soil: publications on this subject are rare.†

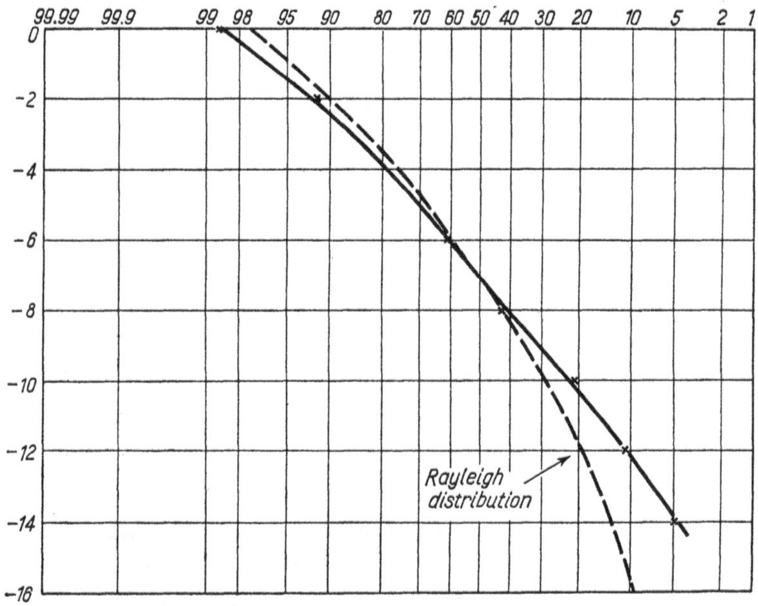

Fig. 14.12. Amplitude distribution of the reflected field.

On the other hand, there are some indications on the fluctuations of the field reflected by the sea.

† We have not found any; however, the experimental method used by Straiton and Tolbert (Section 13.3.2) brings out a variable component of the field coming from the direction given by geometrical optics, which may be interpreted as the variable part of the specularly reflected field. If this interpretation is accepted, this method should permit the fluctuations of the specularly reflected field to be studied.

According to Archer-Thomson, Brooke, Gold and Hoyle [1943], Megaw [1946] and Kerr [1951], the amplitude of these variations is very great: the reflection coefficient varies between 0 and a maximum *which may exceed unity*. This last point is important, for it implies that in interfering with the direct field, the reflected field may produce very deep fades when the reflection coefficient passes through the value 1. Experiments confirm that in the neighbourhood of a minimum of the field, the latter fluctuates with a very great amplitude; the fading depth may exceed 30 db.

Figure 14.12 shows the distribution of the values of the resultant field as measured in an experiment performed by the C.N.E.T. (Annex to Chapter 13). As explained in Part I of this monograph,† the variance of the amplitude of these fluctuations is found to be smaller than that of the Rayleigh distribution

The fading rate is of the order of c/s, but components with frequencies up to 50 c/s may by present.

† Cf. Part I, Section 7.2, p. 140.

CHAPTER 15

EXPERIMENTAL INVESTIGATION OF DIFFUSE SCATTERING FROM THE EARTH'S SURFACE

15.1 Survey of the Cases to be Considered

We have assumed that the reflection coefficient R_d of the diffusely scattered component is given by†

$$R_d = \rho_d R_0 \tag{1}$$

where R_0 is the reflection coefficient of a smooth earth and ρ_d a coefficient that depends only on the irregularities of the reflecting surface.

The theoretical investigation of Chapter 12 has demonstrated a case where the result is very simple (Section 12.4.3): ρ_d is constant for any geometry of the path, any irregularities of the earth's surface and for any wavelength. This case arises if the following conditions are satisfied:

(a) *The antennae are not too directional,* their beamwidth α should be greater than $2\beta_0$. If, on the other hand, the antennae are very directional, the expression for ρ_d is less simple (Section 12.4.4).

(b) *The "slope of the irregularities"* β_0 *is small.* If β_0 is large, certain assumptions of the derivation are no longer justified (Section 12.4.6), and the general relation (1) may no longer be valid; in addition, we have assumed the dimensions of the irregularities large compared with the wavelength. These conditions are not satisfied if the irregularities of the ground consist of vegetation.

(c) *The height of the irregularities is large compared with the space wavelength* (Rayleigh criterion). If this is not so, a part of the reflected energy is reflected specularly and the reflection coefficient of the diffusely scattered component will be smaller than calculated (Section 12.4.6).

(d) *The distribution of the irregularities is isotropical.* It is doubtful whether our results are still applicable if the irregularities are distributed along a priviliged direction (e.g. waves on the surface of the sea).

(e) *If the incident wave is vertically polarized, the angle of incidence is not near the Brewster angle.* We have, in effect, assumed that R_0 varies

† Equation (24) of Chapter 12.

very slightly with the grazing angle γ. We have made the same assumption for investigating specular reflection and found (Section 14.2) that in the neighbourhood of the Brewster angle it leads to results contradicting experience: we may expect that this will equally be true for diffuse scattering.

(f) *The transmitting and receiving antennae are beamed towards each other.*

We shall first consider the experimental results obtained when all of these conditions are satisfied (Section 15.2); fortunately, this is frequently the case for sufficiently long radio links, like those employed in practical applications. We shall then devote six sections (15.3 to 15.8) to experiments in which one of the conditions (a), (b), (c), (d), (e), (f) is not satisfied.

15.2. Results of Some Experiments Satisfying Conditions (a) to (f)

Three series of measurements have furnished a mass of important results on diffuse scattering on long paths. They are those described by Bullington [1954], Beard, Katz and Spetner [1956] and MacGavin and Maloney [1958]. They were performed in very different experimental conditions: the first at 4000 Mc/s over cultivated and inhabited terrain, the second at various frequencies over the sea, and the third at 1000 Mc/s over desert land.† Part of these measurements were performed on links between ground stations, another part employed links between a ground station and an aircraft.

In spite of these differences, the results obtained are very similar. Figure 15.1 shows the distribution of the values of $\rho_d = R_d/R_0$ measured in the course of various experiments. They are always rather widely scattered about a mean value which varies between 0.30 and 0.35. The majority of values is included between 0.2 and 0.4.

In the experiments described by MacGavin and Maloney [1958], the receiver is airborne. Taking measurements in horizontal flight and changing its altitude the authors were able to investigate the variations of ρ_d in space. The results of a large number of measurements were subjected to a statistical analysis to determine whether ρ_d varies with (1) the altitude h_2 of the receiver, (2) the grazing angle γ_s‡ (these two parameters define the position

† Cf. Annex to Chapter 13 for a description of these measurements.

‡ The notation γ_s emphasizes that this is the grazing angle as given by geometrical optics; it has a physical meaning only if the reflection is specular.

† A single series of measurements indicates a slight increase of ρ_d with growing γ_s: those made at 7000 ft altitude on the Cheyenne mountain path. It may be noted from

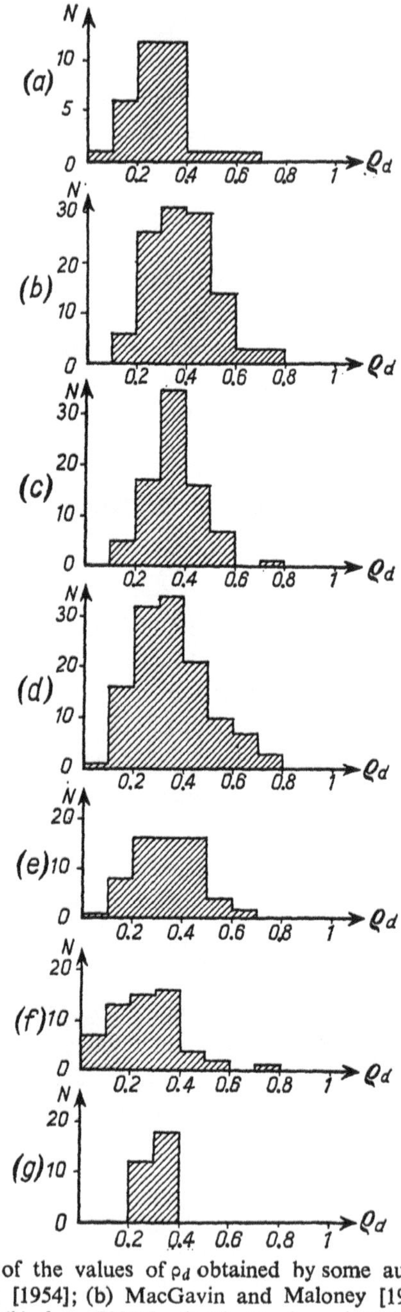

Fig. 15.1. Distribution of the values of ρ_d obtained by some authors (N is the number of cases). (a) Bullington [1954]; (b) MacGavin and Maloney [1958], Pikes Peak path, h_2 = 10,000 ft; (c) as in (b), h_2 = 7000 ft; d) MacGavin and Maloney [1958], Cheyenne Mountains path, h_2 = 15,000 ft; (e) as in (d), h_2 = 10,000 ft; (f) as in (d), h_2 = 7000 ft; (g) Beard, Katz and Spetner [1956].

of the receiver). The authors conclude that ρ_d is independent of γ_s and h_2.†

This result has been obtained independently by a different approach by Bullington [1954] and by Beard, Katz and Spetner [1956]. These authors investigated whether ρ_d depends on the expression

$$\Delta\Phi = \frac{4\pi \Delta h \sin \gamma_s}{\lambda}$$

which occurs in the Rayleigh criterion; they reached the conclusion that ρ_d is independent of $\Delta\Phi$.

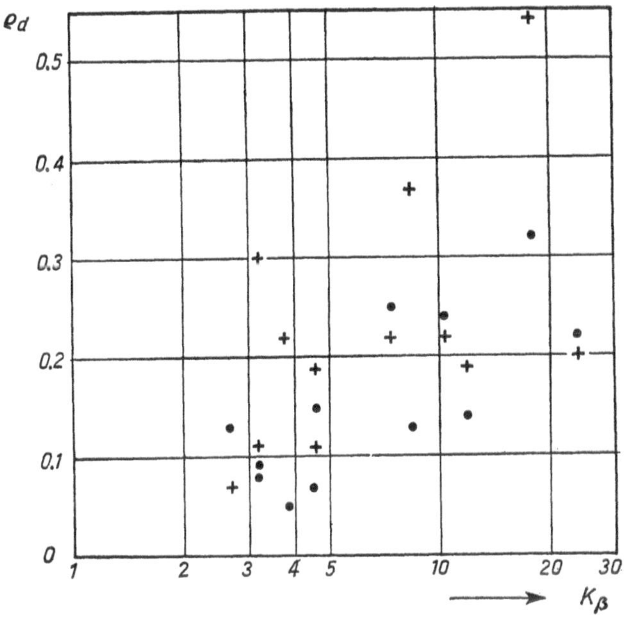

Fig. 15.2. Values of $\rho_d = R_d/R_0$ derived from the measurements by Ford and Oliver [1946] as a function of K_β. Circles: horizontal polarization, crosses: vertical polarization.

Moreover, since the mean values of ρ_d given by these different authors are very near in value, the experimental results are in good agreement with the theoretical results, according to which the mean value of ρ_d does not depend on γ_s, or h_2 or Δh or λ.†

Fig. 15.1 that on this path the values of ρ_d are particularly small. But it may also be noted (Fig. 14.4) that specular reflection is present on this path. This could explain these anomalies: this is the special case (c) of Section 15.1.

‡ Although we have quoted only three series of measurements, a number of isolated results obtained under very different conditions may be found in the literature; the values of ρ_d always lie between 0.2 and 0.4 [Rubenstein and Fishback 1943; Tomlinson and Straiton, 1959, etc.].

Variance of ρ_d. Figure 15.1 gives some idea of the scatter of the values found by different authors. The values given by Beard, Katz and Spetner [1956] are much less scattered than the others: it is possible that in the case of the surface of the sea, which lends itself much better to a statistical description, the variance of ρ_d is smaller than in the case of land.

For overland paths, MacGavin and Maloney [1958] have made a statistical study of the variance of ρ_d. The standard deviation $\Delta\rho_d$ of ρ_d from its mean value equals 0.15 for the entire set of their measurements, but $\Delta\rho_d$ is a decreasing function of the grazing angle γ_s: for $\gamma_s = 0$, $\Delta\rho_d$ is of the order of 0.2, whereas for $\gamma_s > 2°$, $\Delta\rho_d$ is at most of the order of 0.1.

15.3. Very Directional Antennae

A systematic study of the reflection coefficient of the ground was undertaken by Ford and Oliver [1946] using strongly beamed antennae to separate the reflected from the direct ray. Part of their results concern diffuse scattering and for this case the authors found that a portion of the earth's surface which was much larger than that illuminated by their antennae was often capable of reflecting waves: the reflection coefficient did not change appreciably when the antenna beams were deflected.

As predicted by theory, the authors obtained much smaller values of ρ_d than those in the preceding section. Apart from rare exceptions,† ρ_d varies between 0 and 0.25.

Some of the experiments by Ford and Oliver [1946] were performed on bare soil which was artificially made rough by sinusoidal undulations imitating the shape of waves on the sea.‡ Since the dimensions of the irregularities are exactly known, this experiment is particularly suited for comparisons with theoretical results.

Let us recall that if the antennae were not directional, the reflection coefficient and the dimension of the "glistening surface" would depend on the parameter§

$$K_\beta = \frac{\tan\gamma_s}{\tan\beta_0} \tag{2}$$

In the experiments by Ford and Oliver, the antennae are very directional and they do not always illuminate the entire "glistening surface"; hence

† They correspond to the reflection of a vertically polarized wave incident near the Brewster angle (cf. Section 15.7).

‡ Cf. Section 15.6.

§ Cf. equation (35), Section 12.4.3.

the reflection coefficient is diminished. The smaller K_β, the more extensive the "glistening surface"† and hence ρ_d will decrease.

From the data given by Ford and Oliver [1946], β_0 may be calculated, and since γ_s is known, K_β may be derived from (2). K_β is then found to vary between 2.7 and 24. According to theory, for $K_\beta = 2.7$ the "glistening surface" should occupy about half the length of the path; it is only partly illuminated by the antennae and ρ_d should be very small. On the other hand, for $K_\beta = 24$, the "glistening surface" is very small, and the antennae, which illuminate it entirely, may be considered non-directional: thus ρ_d, as in the preceding section, should be included between 0.2 and 0.4.

Figure 15.2 shows the experimental values of ρ_d gained by Ford and Oliver [1946] as a function of the parameter K_β.‡ Apart from a few explicable exceptions§ it will be seen that ρ_d varies in the way predicted by theory.

15.4. Effect of Vegetation

Measurements of the reflection coefficient of samples of soil covered by vegetation were taken by Ford and Oliver [1946] and Sherwood and Ginzton [1955] at wavelengths of 9 and 10 cm.

Vertical blades of grass reflect vertically polarized waves in a very selective manner: if the grass is sufficiently high, the wave is diffusely scattered and the reflection coefficient is small, of the order of 0.1. On the other hand, a horizontally polarized wave is reflected as if the ground were bare and either specular reflection or diffuse scattering may set in depending on the dimensions of the irregularities of the ground. For diffuse scattering, ρ_d may be of the order of 0.3 to 0.4 if the antennae are non-directional [Sherwood and Ginzton, 1955] or less if the antennae are very directional [Ford and Oliver, 1946].

If the vegetation is more dense and completely covers the ground, the reflection coefficient R_d is of the order of 0.1 regardless of the polarization [the expression (1) for R_d, which depends on the polarization, can no longer be regarded as valid].

---

† Cf. Fig. 12.7.

‡ All the values given by Ford and Oliver for different directions of the undulations of the ground have been included in this figure. To distinguish the values corresponding to diffuse scattering, the Rayleigh criterion in the form $\Delta\Phi \geqslant 0.5\pi$ has been applied (cf. Section 15.10).

§ Three abnormally high values of ρ_d at vertical polarization are probably due to the angle of incidence being near the Brewster angle.

15.5. Irregularities of the Ground Smaller Than Half the Space-wavelength: Diffuse Scattering Superimposed onto Strong Specular Reflection

In theory, ρ_d cannot reach its value $\rho_d = 0.35$ of Section 15.2 (15.1.c): it should equal zero for $\Delta h = 0$ and increase with Δh, approaching asymptotically the value 0.35 as shown in Fig. 15.3.†

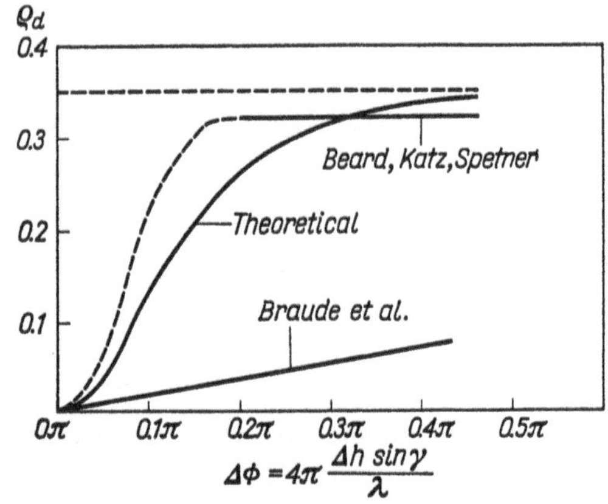

Fig. 15.3 Comparison of theoretical and exprimental data.

Beard, Katz and Spetner [1956] and Braude, Komarov and Ostrovsky [1958] have experimentally found the same result, but whilst the former find the asymptotic value of ρ_d practically attained for $(\Delta h \sin\gamma)/\lambda = 0.05$, the latter do not find it attained until $(\Delta h \sin\gamma)/\lambda$ exceeds a value of 0.01.

The mean curves obtained experimentally by these authors are shown in Fig. 15.3.

15.6. Irregularities Distributed along a Privileged Direction

Ford and Oliver [1946] have made a series of measurements of the reflection coefficient of artificially undulated terrain, imitating the shape of waves on the sea. Measurements were taken for different angles of incidence and different directions of the undulations: parallel to the radio path, perpendicular to it or inclined at an angle of 45°.

From their results it does not seem that the direction of the undulations has any appreciable effect on the reflection coefficient.

† The "theoretical" curve in Fig. 15.3 was obtained by assuming that (a) when all of the incident energy is diffusely scattered, $\rho_d = 0.35$ (the value found experimentally in Section 15.2), (b) the sum of the specularly reflected and diffusely scattered energy is constant.

15.7. Vertically Polarized Waves Incident near the Brewster Angle

The same experiments of Ford and Oliver give abnormally high values for ρ_d when the waves are vertically polarized and incident near the Brewster angle; some of these values may be seen in Fig. 15.2† and may be compared with the value of ρ_d for the corresponding points (same abscissa) for horizontal polarization.

We have already met this effect in connection with specular reflection (Section 14.2) and the same explanation remains valid. Equation (1) cannot be used under these conditions.

15.8. Deflection of Antenna Beams: Dimensions of the "Glistening Surface"

When very directional antennae are used to separate the reflected from the direct wave, it is interesting to deflect their beams to different parts of the reflecting surface to study the participation of each part of this surface in the process of scattering.

This experiment was performed by Ford and Oliver [1946] in the case of a very short link: in this case an extensive part of the path participated in diffuse scattering as predicted by theory.‡

A series of measurements was undertaken by the C.N.E.T. (Paris) using a fairly long path over the sea with the object of determining the region of the surface of the sea where diffuse scattering is produced. The experimental set-up was described in connection with the methods of measurement of the reflection coefficient (Section 13.3.1). Let us recall that it included the following:

A strongly beamed transmitting antenna (beamwidth 2° at 3 db). Swinging the beam downwards, the reflected ray may be isolated; deflecting it laterally, different portions of the surface of the sea may be illuminated (Fig. 15.4).

A very sensitive goniometer at the receiving end which enables the direction of arrival of the reflected ray to be analysed in both the horizontal and vertical directions.

In a first set of measurements, the transmitting antenna was deflected downwards, but along the axis of the link. By taking measurements of the angle of arrival of the reflected wave along the vertical direction, the goniometer enabled the participation of the points of the surface of the sea to

† These values of ρ_d correspond to values of R_0 smaller than 0.1; they are thus in addition liable to serious errors.

‡ Cf. Section 15.3.

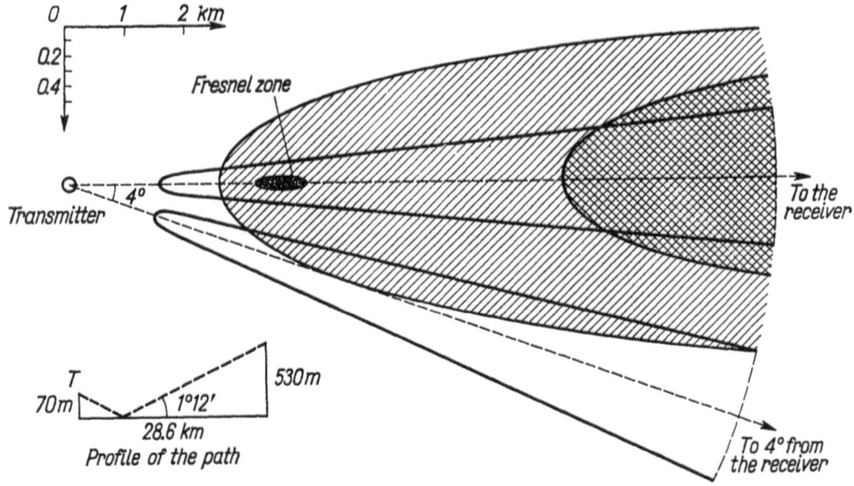

Fig. 15.4. Path geometry. The figure is a plan of the sea surface with the region illuminated by the transmitting antenna when it is either pointed towards the receiver or deflected by 4°. The region responsible for diffuse scattering is shaded, the more important sources of diffuse scattering lie in the doubly shaded area; the Fresnel zone, where the specular reflections were detected, is drawn in black.

be studied as a function of their abscissae (the abscissa axis coinciding with the axis of the link). An example of the results obtained in this way is shown in Fig. 13.7. In general, a set of points corresponding to specular reflection is obtained (the reflected energy coming from the first Fresnel zone), and another set of points corresponding to diffuse scattering: in Fig. 15.5 the points of these two sets lie in the shaded areas.

In addition to the separation of the speculary reflected and diffusely scattered components, this first set of measurements leads to two interesting results:

1. Unlike the specular component, the diffusely scattered component comes from an extensive region of the reflecting surface. In this experiment, it came from the entire length of the surface of the sea illuminated by the two antennae (with due regard to screening by natural obstacles).

2. The part of the surface of the sea that participates most intensely in diffuse scattering is that most distant from the point of specular reflection. This confirms our theoretical results (Fig. 12.7).

In a second set of measurement, the transmitting antenna was beamed in the same way, but at the receiving end the goniometer analysed the angle of arrival of the reflected field horizontally. These measurements also demonstrated the presence of a specularly reflected component coming from an almost fixed direction (in agreement with the small width of the Fresnel ellipse, cf. Fig. 15.4) and a diffusely scattered component coming from a much more extensive region (Fig. 15.4).

In a third set of measurements, the transmitting antenna was deflected by 4° horizontally, but always pointed downwards to eliminate the direct

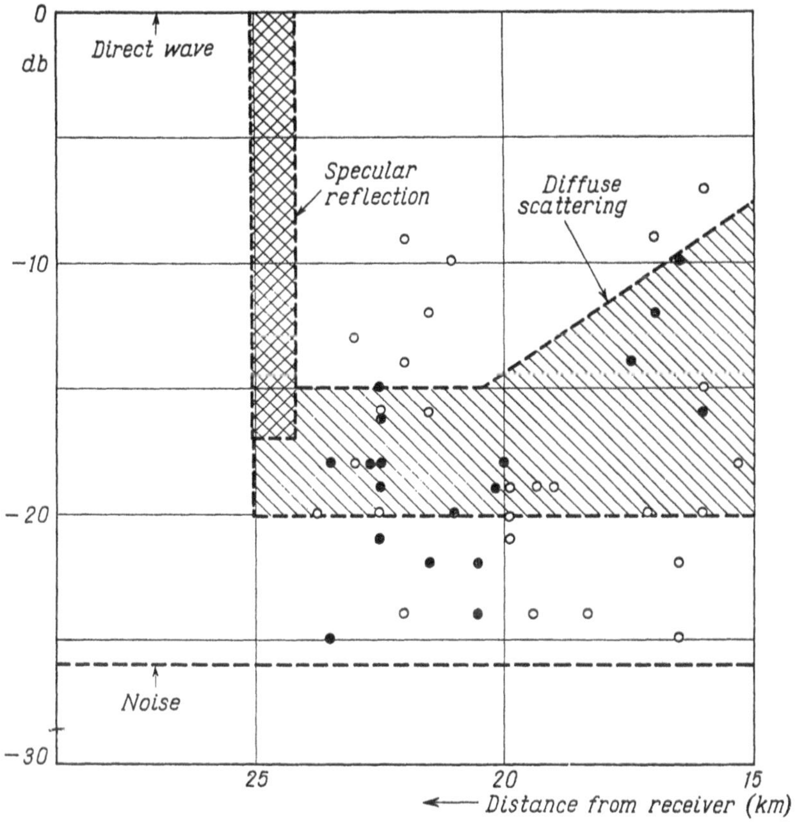

Fig. 15.5. The shaded areas correspond to the zones responsible for specular reflection and diffuse scattering when the transmitting antenna is beamed towards the receiver. The full and empty circles correspond to values measured when the transmitting antenna beam was deflected by 4°. Empty circles—calm sea (sea state 0 to 2); full circles—rough sea (sea state 3 to 5).

ray: in this manner it illuminated another part of the surface of the sea (Fig. 15.4). The goniometer analysed the angle of arrival of the reflected ray vertically.

An immediate result was the vanishing of the specular component, as was to be expected, since the first Fresnel ellipse was no longer illuminated by the transmitting antenna. A further result was the appearance of diffuse scattering which seemed to come from an extensive region of the surface of the sea, but the zone that is furthest removed from the transmitter does not overduly participate in the scattering. Figure 15.5 gives an example of the results that were obtained. It may be noted that the mean reflection coefficient for the diffusely scattered component is slightly smaller than in the case when the transmitting antenna is not deflected, but that the variance of the individual values is very large. The state of the sea does not seem to have a great effect on the value of the reflected field.

It may be concluded that the regions participating in the scattering have the shape shown in Fig. 15.4 (the shaded regions give only a qualitative indication).

A fourth set of measurements was devoted to the problem of depolarization.

15.9. Depolarization

The transmitted wave was first vertically polarized and the transmitting antenna was beamed towards the receiver. The receiving antenna was first vertically, then horizontally polarized. In the latter case, the received field was about 25 db below the field that was received when the antennae were identically polarized. The goniometer indicated that the cross-polarized component came from a different direction: thus this was a parasitic reflection. If the transmitted wave had contained a weak horizontally polarized component, the latter would have been *at least* 25 db below the vertical component.

In a further series of measurements, the transmitted wave was always polarized vertically, but the transmitting antenna was beamed towards the point of specular reflection on the surface of the sea. The field received in vertical polarization came from the point of specular reflection, the reflection coefficient being of the order of 0.6. The horizontally polarized component was 15 to 20 db below the vertical component, but the goniometer indicated that the wave arrived from a direction differing from that of the point of specular reflection.

This is in good agreement with theoretical predictions (Part I, Chapter 8): the specularly reflected field is not depolarized; on the other hand, the diffusely scattered field is strongly depolarized: in the above experiment the diffusely scattered field received at horizontal polarization seems to be of the same order (within a few db) as the scattered field that would have been received at vertical polarization if it had been possible to separate the diffusely scattered and the specularly reflected components.

The experiment was then repeated with interchanged polarizations and led to similar results.

15.10. Conclusions. The Rayleigh Criterion

Restricting our conclusions to practical links and excluding special effects appearing in short links (Brewster angle) and links with deflected antenna beams, we may, in summary, distinguish the following two cases:

Case I. Antennae not very directional, ground not covered by vegetation, no specular component. Then the value of ρ_d lies between 0.2 and 0.4; probable value: 0.35.

Case II. Some factor causes a decrease of the reflection coefficient of the diffusely scattered component: very narrow antenna beams, natural screening, ground covered with vegetation, or a part of the energy is reflected specularly: in this case ρ_d is smaller than 0.2 and may be negligible.

The Rayleigh Criterion

As we have seen† different authors give different expressions for the Rayleigh criterion, which is of the form

$$\Delta\Phi = 4\pi\frac{\Delta h \sin\gamma}{\lambda} < \Delta\Phi_0$$

where $\Delta\Phi_0$ is a constant that, depending on the value assessed by author, lies between $\pi/10$ and $\pi/2$. The diversity of these values points to the inaccuracy of the definition of Δh and to the diverse meanings that are given to this criterion.

If Δh is the standard deviation of the height of the points on the reflecting surface, and if by "Rayleigh criterion" we understand the condition for reflection of the specular type to be dominating, we may obtain an exact expression by writing

$$\rho_s > \rho_d$$

where ρ_s is given by (2) and (3) of Chapter 4. For the two cases distinguished above, we arrive at the following results.

Case I. Setting $\rho_d = 0.35$, we obtain

$$\Delta\Phi < 0.32\pi$$

or

$$\frac{\Delta h \sin\gamma}{\lambda} < 0.08$$

† Cf. Section 12.1.

Case II. Using the mean value $\rho_d = 0.1$, we obtain

$$\Delta\Phi < 0.5\pi$$

or

$$\frac{\Delta h \sin\gamma}{\lambda} < \frac{1}{8}$$

This last value was proposed by Rayleigh himself. The former value has not been proposed by any author before.

CHAPTER 16

SPECULAR REFLECTION: PRACTICAL APPLICATIONS TO LINE-OF-SIGHT PROPAGATION OF ULTRA-SHORT WAVES

16.1. Field Strength in Line-of-Sight Links. Origin of Fadings

In the preceding chapters we have considered certain radio links in which the interference between the direct wave and one or more ground-reflected waves causes variations in the resultant field. The analysis of these variations enabled the reflection coefficient of the earth's surface to be studied.

In practical applications the converse operation is performed: the laws governing the reflection coefficient of the earth are assumed to be known and are applied to the study of the resultant field in practical radio links, where the direct wave interferes with the reflected wave.

The analysis of the resultant field being given in the preceding chapters, we shall now consider its synthesis. This synthesis would present little difficulty if all radio links used in practice were as simple as those that are used for studying the reflection coefficient of the earth. In such a case we should choose links where the existence of a direct and a reflected wave is sufficient for explaining all the variations of the received field. However, in many practical links a number of other effects—especially of atmospheric origin—are superimposed onto the ground reflections and the results of propagation become more difficult to interpret. What are these other effects and what is their importance? This is the first question we shall pose.

We therefore start by enumerating the possible causes of the variations in fieldstrength in the case of a line-of-sight link.

(a) *Ground reflections (constant component).* The field that is reflected by the ground includes a constant, specularly reflected component.† When the vertical gradient of the atmospheric index of refraction varies, the phase of this constant component varies‡ and, when it is in phase opposition with the direct wave (Fig. 16.1), a fade will result.

---

† Cf. Section 13.3.2.

‡ Cf. Section 14.6.

Although these fades may be fairly short (a few seconds), in practical radio links they mostly last several minutes or sometimes even hours.† If the reflection coefficient of the earth's surface is near unity, these fades may be very deep: this is often so in the case of the sea.

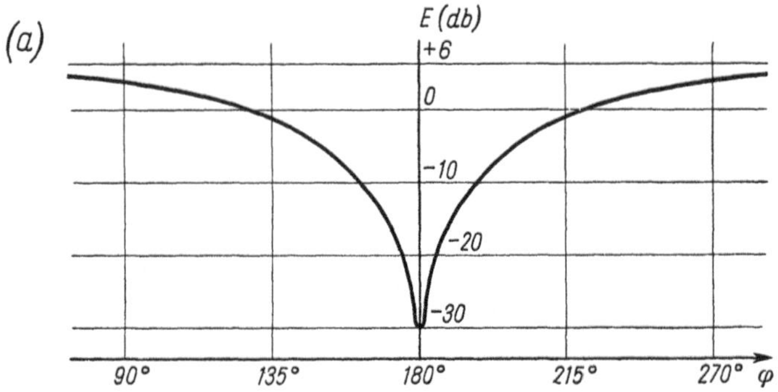

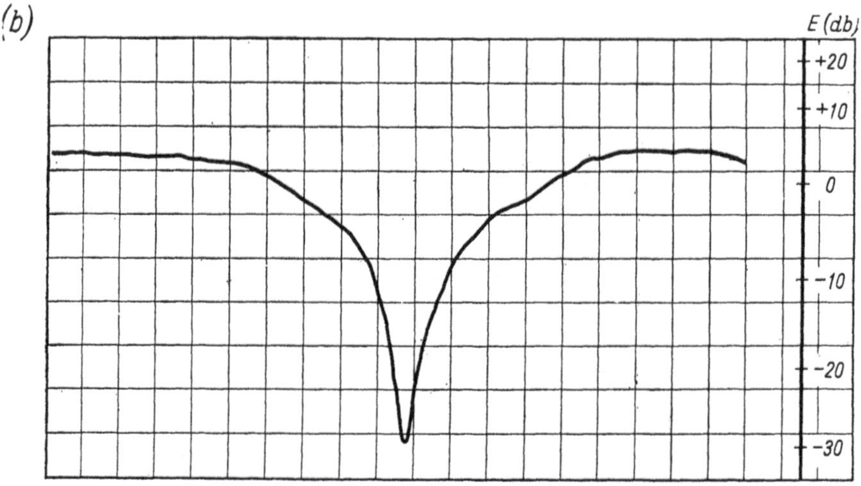

Fig. 16.1. (a) Dependence of the signal level on the phase shift φ between the direct and reflected waves; (b) recording of the signal level in time. (After Chavance [1952].)

This kind of fading is often important in short links with ample clearance from the earth's surface,‡ when the frequency is not too high. It is

† In practical radio links the frequency and the antenna heights are mostly chosen so that the receiver is located at one of the lowest interference fringes. Under these conditions the fringes move most slowly across the receiving point when atmospheric refraction causes their displacement.

‡ In long line-of-sight links the reflection coefficient is reduced, the divergence coefficient D being smaller than 1. This is particularly true for rough terrain, and in this case

characterized by its selectivity in space and frequency, which permits diversity reception to be employed.

(b) *Ground reflections* (*variable component*). The field reflected from the surface of the earth also has a variable component, which includes the variable part of the specularly reflected component and diffuse scattering.†

As has been shown theoretically‡ and verified experimentally by numerous authors,§ the diffusely scattered field, in interfering with the direct field, gives rise to very rapid fluctuations (of the order of one second). The variable component of the scattered field is particularly important in the case of the sea: theoretical investigations,‡‡ confirmed by experimental measurements† show that in interfering with the direct field, this variable component gives rise to rapid fluctuations.

If this variable component of the reflected field interferes only with the direct field, it gives rise to fadings with slight depth, for its amplitude is much smaller than that of the direct wave. But, if it interferes simultaneously with the direct wave and the constant component of the reflected field, and if these two fields are in phase opposition (minimum of an interference lobe), then the variable component of the reflected field may give rise to very deep fades (40 db and more).

These fades are very selective in space and frequency:‡ this selectivity may be utilized by employing diversity methods.

(c) *Multipath propagation of atmospheric origin*. For various reasons it is possible for several rays to be propagated along different paths and to arrive simultaneously at the receiver, where they will interfere:

---

the reflecting zones of the earth's surface may also be screened. (It will be noted that most of the links used for studying the reflection coefficient of the earth in Chapter 4 are short and have ample clearance.)

† Cf. Section 13.3.2.

‡ Cf. Section 12.6.

§ The rapidity of the fluctuations due to the diffusely scattered component has enabled several authors, such as Braude, Komarov and Ostrovsky [1958] to separate the diffusely scattered from the specularly reflected component (cf. Section 13.3.2).

‡‡ La Grone, Straiton and Smith [1955] and Beard and Katz [1957] have investigated the relation between the rate of the fluctuations of the received field and that of the deformation of the surface of the sea. The spectrum of the field fluctuations includes the basic component of the fluctuations of the surface of the sea, and some of its harmonics. The periods of these components are of the order of some seconds.

† Cf. Beard and Katz [1957].

‡ Cf. Section 12.6.

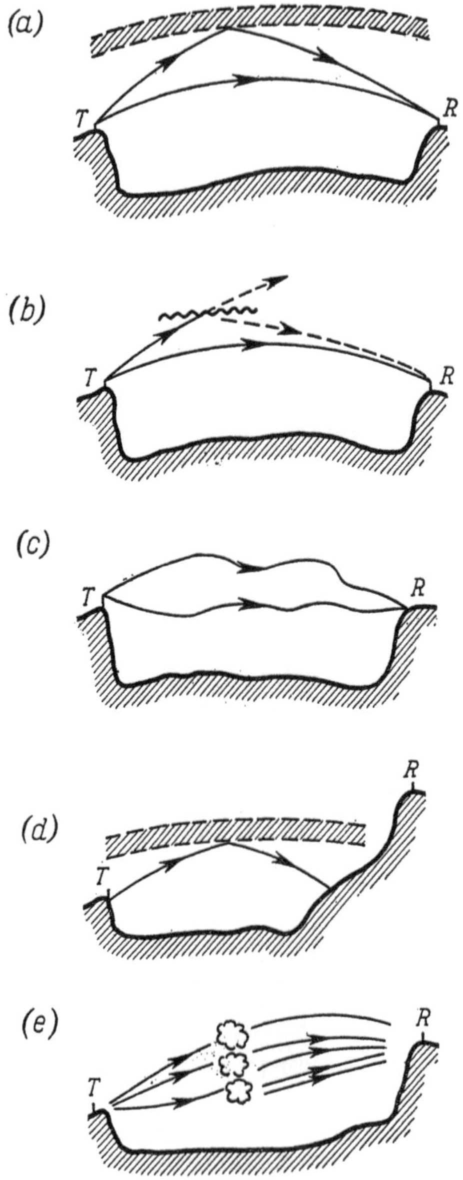

Fig. 16.2. Some causes of fading of atmospheric origin. (a) Reflection from an inversion layer, (b) partial reflection from a tropospheric sheet, (c) refraction in an inhomogeneous atmosphere, (d) reflection from a low inversion layer, (e) scattering by a turbulent atmosphere.

—a ray may be reflected by an inversion layer.† The interference lobes that are thus formed are displaced either as a result of atmospheric refraction or by a slight change in altitude of the inversion layer. In general, an inversion layer is fairly stable, it may last several hours. It may easily be demonstrated by an ordinary meteorological radio sonde (Fig. 16.2a).

—a ray may be partially reflected by an atmospheric "sheet".‡ A "sheet" is a very local discontinuity of the atmospheric index of refraction. It is a very brief phenomenon (the passage of a sheet lasts at most a few minutes), but more frequent than an inversion layer.§ It cannot be demonstrated by an ordinary meteorologic sonde, but may be observed with the aid of a refractometer‡‡ or a radar pointing vertically upwards.† Certain authors‡ are of the opinion that rapid fadings may be due to the interference of a ray reflected from a sheet with the direct ray§ (Fig. 16.2b).

—various phenomena of refraction in an inhomogeneous atmosphere may give rise to several rays striking the receiving antenna‡‡ (Fig. 16.3c). A special case of this phenomenon occurs in frontal discontinuity in the atmosphere, which has been investigated by Misme [1956].

The presence of several ray paths in the atmosphere has been demonstrated by using very directional antennae,† pulsed signals capable of distinguishing paths of different lengths‡ or (which theoretically amounts to the same thing) frequency-sweep techniques at the transmitter and receiver over a wide band of frequencies.§

The fading caused by multipath propagation of atmospheric origin is more frequent if the atmosphere is not turbulent (a stable atmosphere more

† Cabessa [1955], Thompson [1948], Millington [1949], Crawford and Jakes [1952], De Lange [1952], Chavance [1952].

‡ French original: *feuillet* (*Transl. Note*).

§ Cf. Section 19.1.

‡‡ Cf. Section 19.1 and Spizzichino [1960].

† Cf. Section 19.1 and Saxton [1960].

‡ According to Thomson [1948] atmospheric sheets may be responsible for fades shorter than 1 sec.

§ The atmospheric sheets form a rough surface (cf. Chapter 20) and above 3000 Mc/s we shall see that reflection by these sheets is both diffuse and partial; the reflection coefficient of a sheet is therefore very small and it seems that the sheets cannot play an important part in line-of-sight propagation.

‡‡ Gough [1958].

† Cf. Crawford and Jakes [1952].

‡ Cf. De Lange [1952]. This separation is particularly difficult becaute it necessitates very short pulses (of the order of 1 mμsec).

§ Crawford and Jakes [1952], Kaylor [1953].

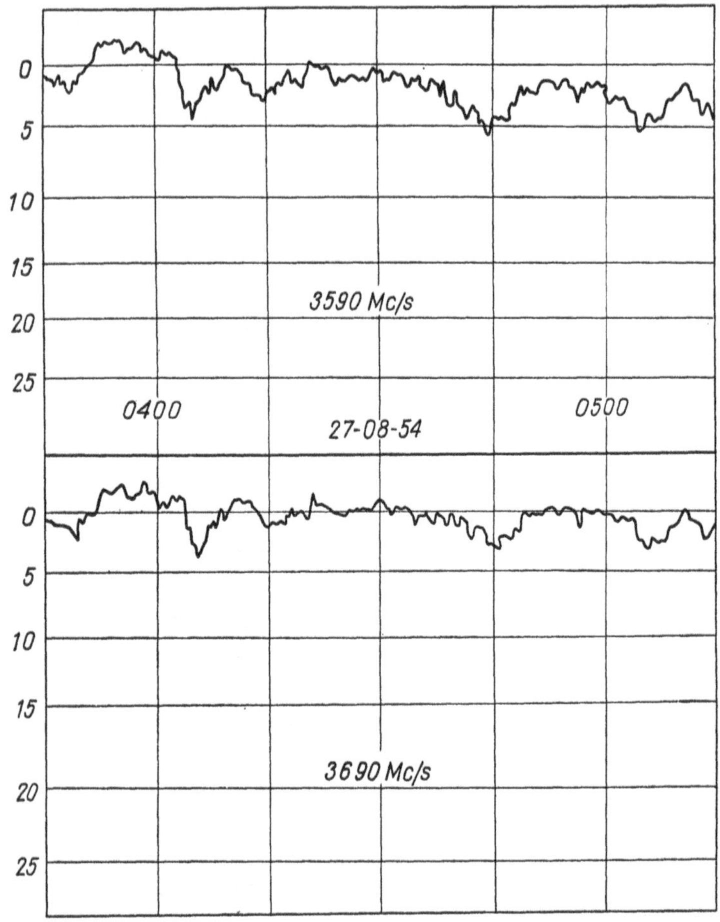

Fig. 16.3a. Different types of fluctuations on line-of-sight paths.

easily assumes a stratified structure: layers, sheets, etc.) and if it is rich in water vapour. Also, these fadings appear more often at night † and in summer. ‡

These fadings are most frequent in the case of long pathlengths and very short wavelengths. Like the fadings caused by ground reflections, the fading caused by reflections from layers and sheets (which, like the ground, are rough surfaces) contains slow and rapid components. The fading is selective in space and in frequency, and this selectivity may be utilized for diversity techniques.

† The strong cooling of the ground during summer nights is particularly favourable to the formation of inversion layers.

‡ Thompson [1948], Crawford and Jakes [1952] etc.

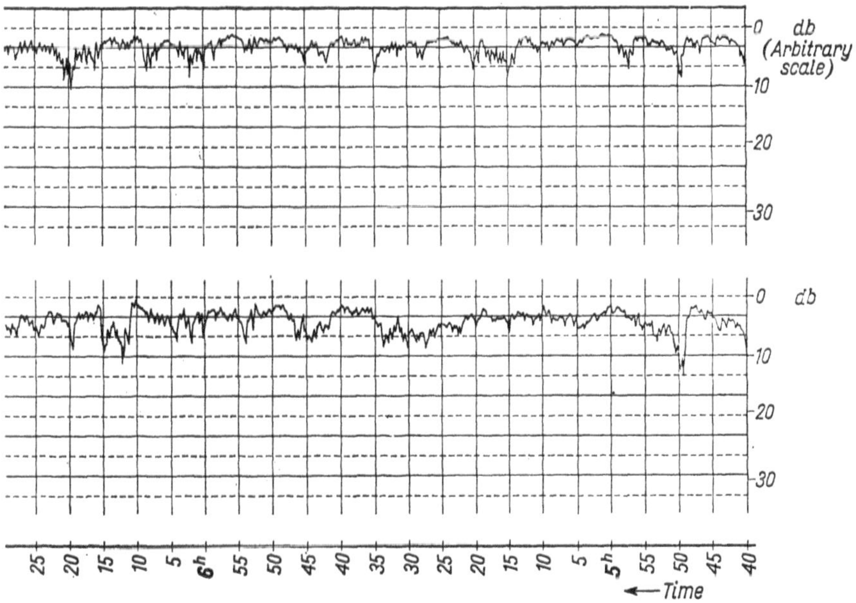

Fig. 16.3b. Different types of fluctuations on a line-of-sight path.

(d) *Fade-out (or attenuation) of the direct wave by atmospheric refraction.* The direct ray may be totally reflected by an inversion layer located at an intermediate altitude between that of the transmitter and that of the receiver† (Fig. 16.2d), or caught by a duct. The latter case is frequent in maritime paths at short wavelengths, where a duct is almost permanently present.‡

These fadings set in, like those of the preceding type and for the same reasons, in calm weather, especially during summer nights, and are most frequently observed on long paths or at short wavelengths.

They are characterized by their long duration (often several hours). They are sometimes selective in space, but never in frequency.§

(e) *Scattering by turbulent layers.* Turbulent layers in the atmosphere scatter radio waves; the received field then fluctuates rapidly (periods of the order of 1 sec), but the fluctuations only have a small amplitude (a few db).‡‡ This effect is more noticeable in the case of long paths (Fig. 16.2e).

† Cf. Rivet [1956], Crawford and Jakes [1952]. Rivet investigates more generally the fading due to all de-focusing effects in the atmosphere on the direct ray.

‡ This type of duct, demonstrated by Wust [1920] has been investigated by Kiely [1954] with the aid of a millimetre wave-link in which the transmitter was near the surface of the sea and the receiver at a greater height. Deep fades (up to 16 db) of long duration (up to several days) were observed.

§ Cf. Rivet [1956].

‡‡ Cf. Chavance [1952], Rivet [1956].

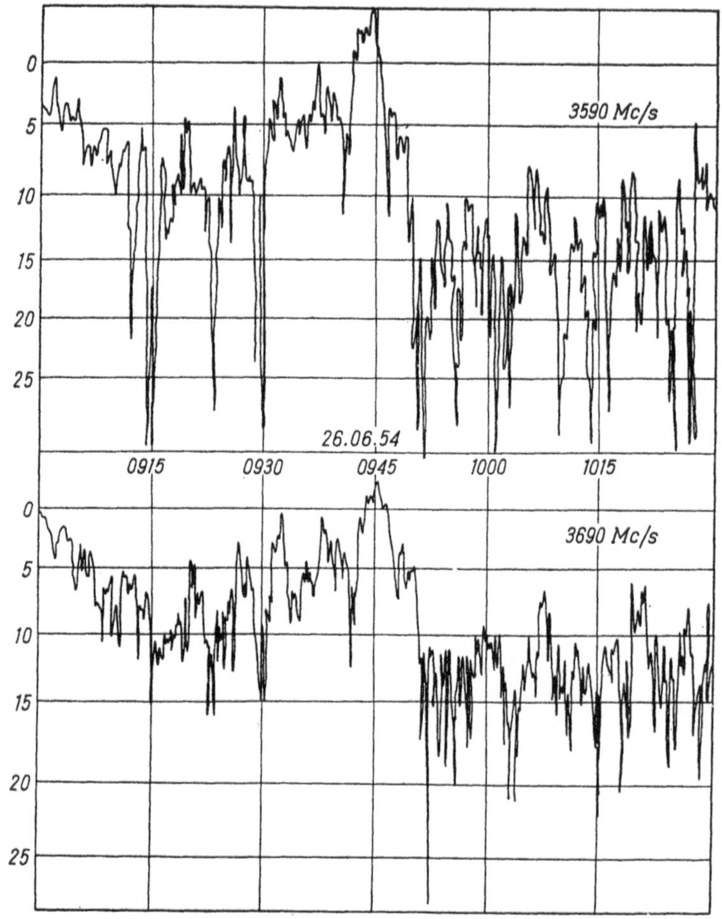

Fig. 16.3c. Different types of fluctuations on a line-of-sight path.

(f) *Absorption by precipitation, oxygen and water vapour in the atmosphere.* Rain, snow and hail are an important cause of absorption, especially above 10,000 Mc/s, as are also the absorption bands of oxygen and water-vapour;† this cause of fading is here given only for completeness, as it has no direct bearing on rough surfaces.

(g) *Diffraction by the earth.* This type of fading sets in if the line of sight is temporarily interrupted as a result of abnormal atmospheric refraction.‡

† Cf. Kerr [1951], Kiely [1956].

‡ Cf. Thompson [1948], Crawford and Jakes [1952], Chavance [1952].

If the first Fresnel ellipsoid is partially or totally screened,† the link will suffer from "diffraction fading". This type of fading is important in very long links‡ with small clearance. If the terminals are no longer above the radio horizon, the depth of the fading will increase with growing frequency. This effect often sets in if the atmosphere is stable and rich in water vapour (summer nights), like the fadings of types (c) and (d). Their duration is often very prolonged (several hours); their depths do not exceed 25–30 db at centimetre wavelengths on paths that are normally above the horizon.

16.2. Slow and Rapid Components of Fluctuations in Line-of-Sight Links

It would seem at first sight that by comparing the results given by some investigators of the question of fading, one may arrive at almost any conclusion: for example, according to the corresponding author, the fading depth may increase,§ decrease‡‡ or remain unaffected† when the frequency is increased. The most unfavourable season is summer,‡ spring§ or autumn‡‡ or else there is no seasonal variation.†

However, on examining the different types of fading more closely, some general conclusions may be drawn.

(1) The types of fading may be grouped into two classes: those due to the interference among several rays (a, b, c, e) and those where the direct field vanishes or is attenuated (d, f, g).

The fadings of these two groups have very different characteristics. For interference fading, the resultant field is greater or smaller than the free space field depending on whether the rays are in phase or in phase opposition.

---

† To estimate the orders of magnitude, we recall that if half of the first Fresnel ellipsoid is screened (line of sight just touching the obstacle), the theoretical attenuation with respect to free space propagation is 6 db in the case of an infinitely thin screen, and 15 db in the case of a perfectly spherical earth.

‡ It is easily verified that the longer the link, the more quickly the clearance diminishes when the refractive-index gradient decreases. Moreover, a short link is easily protected from diffraction fading by raising the antenna heights.

§ Millar and Byam [1950], Carassa [1955] and others.

‡‡ Chavance, Boithias and Blassel [1954], Day and Trolese [1950] and others.

† Megaw [1946].

‡ Kiely [1952], Crawford and Jakes [1952], Vecchiacci [1956] and others.

§ Vecchiacci [1956] in the case of one link; Gudmandsen and Larsen [1957] for certain links (the seasonal variations were not the same for receivers located at different heights.

‡‡ Vecchiacci [1956], Millar and Byam [1950] for certain links at certain frequencies.

† Megaw [1946], Smith-Rose and Stickland [1946], Oxehufwud [1959].

This is not so in the case of fadings caused by the attenuation of the direct ray. But these fadings may significantly affect the mean value only if they are fairly frequent, which will be in the following cases:

—if a semi-permanent inversion layer or duct is present, e.g. in a link using very short waves above the sea [Kiely, 1956b];

—if the wavelength is smaller than 3 cm, in which case atmospheric absorption may be important [Kiely, 1956a];

—if the link is very long (80–100 km) with the direct ray touching the horizon [Chavance, Boithias and Blassel, 1954; Thompson, 1948,† etc.].‡

For other links, the fadings of the second type will set in only for a short fraction of the time; it is therefore not surprising that according to most authors—in fact all except those we have just quoted—*the mean (or median) fieldstrength does not, within the errors of measurement, deviate from the free space level.*

(2) The fadings may be classified from another point of view, namely by their duration; this classification, however, differs little from the previous one.

In effect, *long* fades (several hours or even days) are due to the attenuation of the direct wave: by refraction (d), absorption (f) or diffraction (g).

The fadings due to the constant component of ground reflections (a) or to multipath propagation in the atmosphere (c) are usually of *medium* duration (several minutes, sometimes several hours).

The fadings caused by diffuse scattering from the earth's surface or from turbulent layers of the atmosphere appear as rapid scintillations (of the order of seconds).

Figure 16.3, taken from Thué [1955], shows examples of medium fadings (Fig. 16.3a), rapid fadings (Fig. 16.3b) and the superposition of the two (Fig. 16.3c).

This figure may be used to illustrate some more general remarks. When either type of fluctuation appears by itself (Fig. 16.3a or 16.3b), the fading depth is not very great; it rarely exceeds 3 or 4 db for the rapid scintillations§ or 20–25 db for the slower fading. But when both types are present simultaneously, very deep fades may result: during a minimum of the slow fading, the fields responsible for the rapid fluctuations (diffuse scattering by the earth or the atmosphere) are comparable to the field with which

† On the longest link (cf. Annex).

‡ Some data on the links quoted in this chapter will be found in the Annex.

§ This order of magnitude applies both to diffuse scattering by the surface of the sea [Beard, Katz and Spetner, 1956; Braude, Komarov and Ostrovsky, 1958] and to tropospheric scattering [Chavance, 1952; Kaylor, 1953, etc.].

they interfere, even though their amplitudes are small (Fig. 16.3c). The resulting fading depth may attain 40–45 db.† Even if only smaller values are attained, it may always be observed that

— the lower the mean level,‡ the greater the amplitude of the rapid fluctuations;§

— the deepest fades always result from the combination of slow and rapid fluctuations.‡‡

(3) Another general conclusion is illustrated by Fig. 16.3: comparing fields near in frequency, it is found that only the rapid fluctuations are very selective in frequency. Other authors have made similar observations† and have noted that the rapid fluctuations are selective not only in frequency, but also in space (diversity experiments). This may be explained in theory by examining succesively the causes of fading listed above: the shortest fades are those for which the diversity distance (which may easily be estimated) is shortest. But this also follows from a more general principle which we have already met in connection with ground reflections (Section 12.6) and which we shall meet again in connection with tropospheric scattering (Chapter 20): if a mechanism of propagation in the atmosphere gives rise to a field rapidly fluctuating in time, this field will also fluctuate rapidly in space and as a function of frequency, and vice versa.‡

16.3. Relative Importance of Specular Reflection and Other Components in Line-of-Sight Links

If the types of fading are classed not according to their effects, but according to the frequency with which they appear, it is found that

(1) *the fading due to specular reflection by the earth* appears especially on short links with ample clearance,§ at low frequencies;

---

† Cf. for example Kaylor [1953], Crawford and Jakes [1952], Chavance, Boithias and Blassel [1954].

‡ The mean value over a short interval of time (e.g. hourly mean).

§ Kiely [1956b], Kerr [1951], Vecchiacchi [1956], etc.

‡‡ Kaylor [1953], who analysed the fadings by means of frequency sweeps, reports that the deepest fades were the result of interference among at least six different components.

† Kaylor [1953], Carassa [1956]. Unpublished work by the C.N.E.T. (Transmission Dept.) has led to the same result.

‡ There is a single exception to this rule: the slow fading caused by defocusing is almost independent of frequency, but has a usable diversity distance [Rivet 1956].

§ The clearance of a link may be defined as the distance H between the direct ray TR and the point P of the earth's surface where the optical path TPR is minimum (cf. Fig. 16.15). For each of the links indicated in the annex, the value of H is given.

(2) *fading of atmospheric origin and diffraction fading* most frequently appears on long links at high frequencies. Diffraction fading appears only if the link has little clearance.

Only the fading due to diffuse scattering by the earth's surface is not covered by this classification.† In the present chapter, we shall assume that this kind of fading does not play an important part, which is equivalent to temporarily excluding short links at high frequencies; this subject will be dealt with in the next chapter.

In practical applications the problems presented by these two extreme types of link cannot be attacked in the same manner. In the first case, where a single effect dominates, it is best to analyse the cause of the fading directly (investigation of the point or points of reflection, of the interference fringes, etc.). The methods and results of Chapters 13 and 14 may be applied for this purpose. In the second case, the causes of the fading are in general too numerous to be easily analysed; but the fadings resulting from a large number of causes simultaneously may well be treated statistically.

We shall first consider the latter case.

16.4. Line-of-Sight Paths: Statistical Distribution of the Received Field

Figures 16.4 and 16.5 give some examples of the statistical distributions of the field as given by some authors.‡ They were taken over fairly long periods (several weeks to several months).§ Since the exact absolute value of the field is not always known, the ordinate gives the attenuation with respect to the median level. The data of the paths used for the curves in Figs. 16.4 and 16.5 are summarized in Tables 16.1 and 16.2.

To interpret these experimental results, we again refer to the two extreme types of link as mentioned in Section 16.3.

(1) *Long pathlengths at high frequencies in summer.* All the causes of atmospheric fading are favoured. Interference fading will play an important role and the field may be expected to be Rayleigh-distributed. However,

† These fadings occur most often in short links with ample clearance at high frequencies.

‡ A summary of the results obtained in a certain number of typical situations is given in Tables 16.1–2: pathlengths, frequencies, clearances, etc. (Some additional data on the links used for these measurements are given in the annex to this chapter.) Certain couples or sets of curves illustrating the effect of certain parameters such as frequency, pathlength, distance and variance have also been summarized for the same season and included in Figs. 16.4 and 16.5.

§ Except for Curves 5 and 6 of Table 16.2, which apply to a typical day.

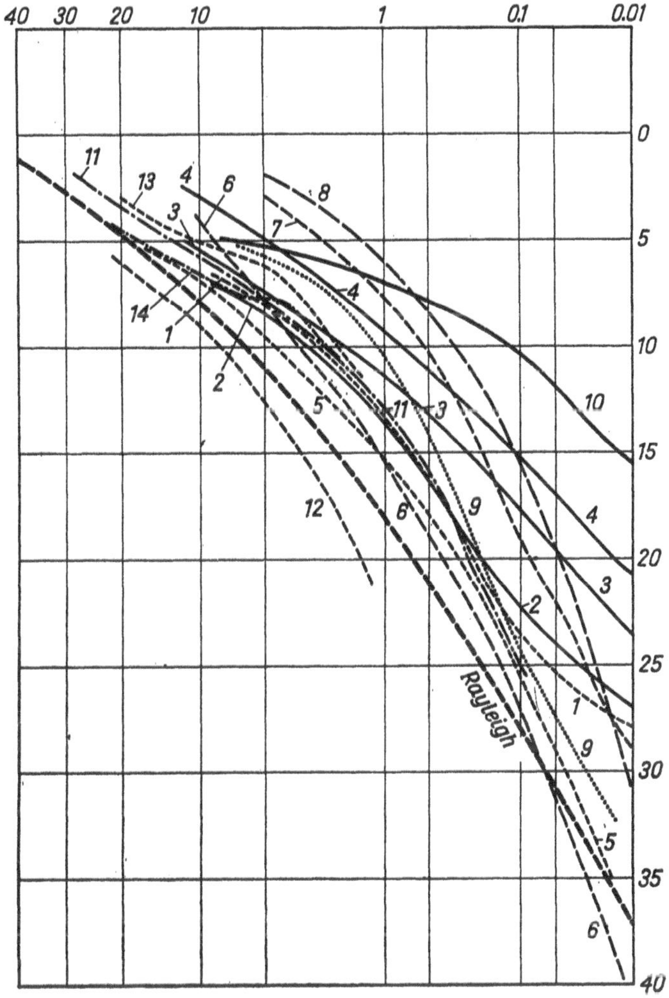

Fig. 16.4a. Amplitude distributions measured on overland paths in summer (cf. Table 16.1).

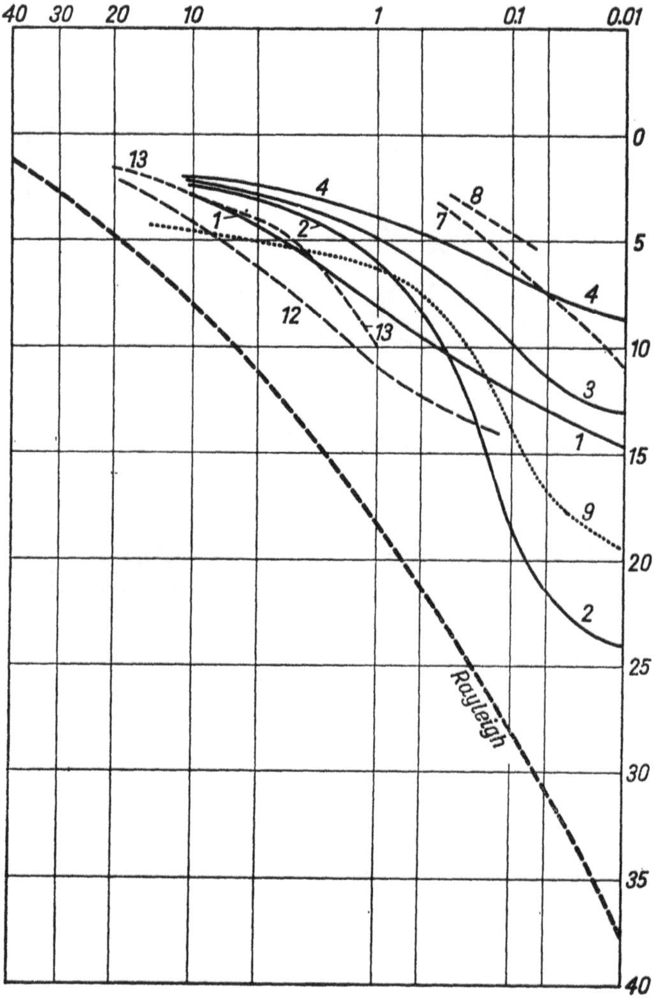

Fig. 16.4b. Amplitude distributions measured on overland paths in winter (cf. Table 16.1).

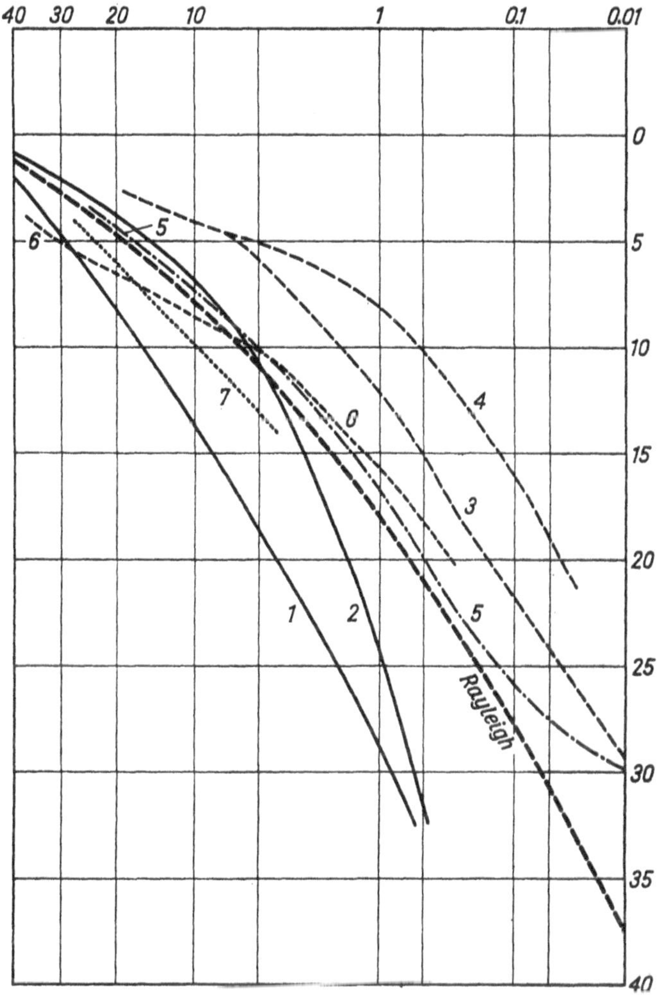

Fig. 16.5a. Amplitude distributions measured on oversea paths in summer (cf. Table 16.2).

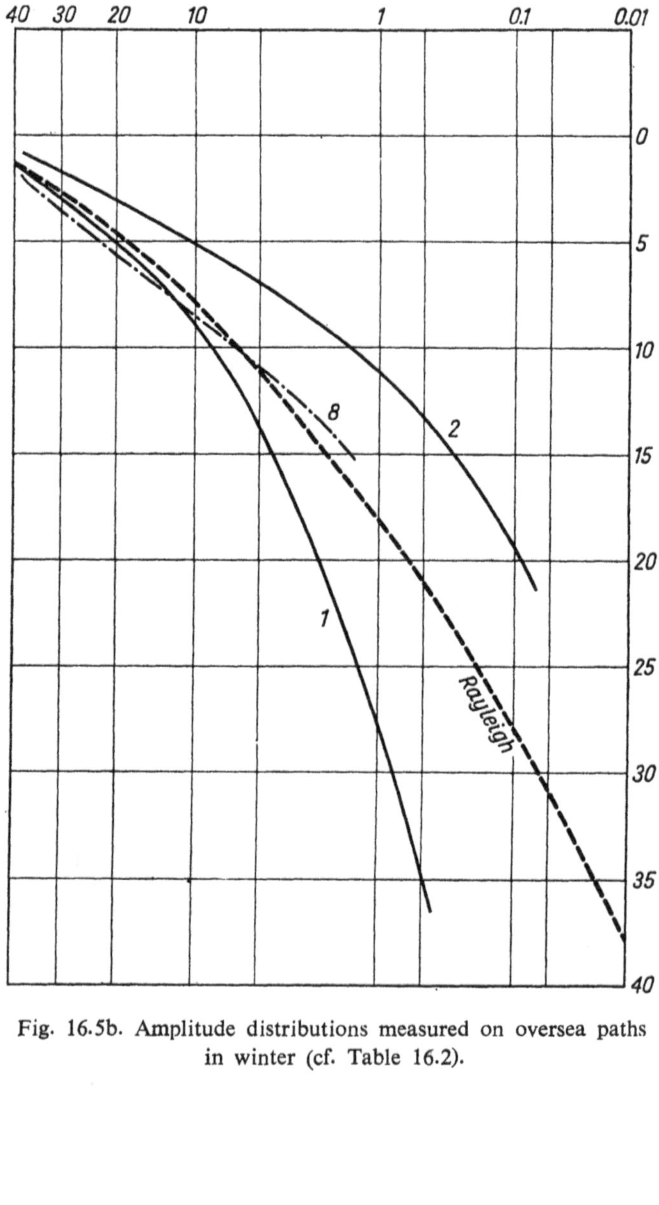

Fig. 16.5b. Amplitude distributions measured on oversea paths in winter (cf. Table 16.2).

TABLE 16.1—Overland Paths (Fig. 16.4)

No. of curve	Bibliographical reference	Frequency Mc/s	Distance km	Note
1	Millar and Byam [1950]	9550	65	
2	Millar and Byam [1950]	6325	65	
3	Millar and Byam [1950]	4150	65	
4	Millar and Byam [1950]	1850	65	
5	Kaylor [1953]	4000	48	June
6	Kaylor [1953]	4000	48	August
7	Crawford and Jakes [1952]	3950	27	
8	Crawford and Jakes [1952]	3950	35	
9	Chavance [1952]	3000	75	
10	Klein and Libois [1953]	300	160	In high mountains
11	Thué [1955]	4000	146	
12	Vecchiacchi [1956]	1000	189	
13	Vecchiacchi [1956]	1000	93.5	
14	Carassa [1956]	250	189	Coincides with Curve No. 12

TABLE 16.2—Oversea Paths (Fig. 16.5)

No. of curve	Bibliographical reference	Frequency Mc/s	Path-length km	Note
1	Chavance, Boithias and Blassel [1954], Rivet [1956]	1400	240	Small clearance
2	Blassel [1954], Rivet [1956]	3150	240	Small clearance
3	Gudmansen and Larsen [1957]	2000	82	September (curves 3–4)
4	Gudmansen and Larsen [1957]	2000	54	
5	Gudmansen and Larsen [1957]	2000	82	August—receiver not at the same height as before (curves 5–6)
6	Gudmansen and Larsen [1957]	2000	82	
7	Kiely [1956]	10,000	19.7	Semi-permanent duct
8	Cabessa [1955]	2000	90	Small clearance

for certain types of link, where the dominating type of fading is not caused by interference, this result will not apply.†

(2) *Short pathlengths at low frequencies.* The received field is the result of the direct field and the ground-reflected field. Attempts to predict its statistical distribution by the following method are often made:‡ it is assumed that the reflected field is limited to its constant component and that the atmosphere is of the "standard" type.§ The equivalent (effective) radius of the earth is

$$a' = Ka$$

where a is the true radius of the earth. For any value of K, the phase difference between the direct and reflected fields may be calculated and the resultant field may be derived. From an experimental or theoretical distribution of the values of K‡‡ the distribution of the resultant field may then be derived.

Figure 16.6 gives an example of such a distribution. It was calculated for a simple case† (Curve A), but the general aspect of the curve is characteristic for the distributions obtained in this way: in general, their variance will be smaller than that of the Rayleigh distribution, especially at the lower levels; if the reflection coefficient differs from unity, the resultant field can never fall below a certain minimum and, as in Fig. 16.6, the distribution curve must approach a horizontal asymptote.

(3) *Intermediate cases.* The distribution curve of the resultant field must be included between those of the above two extreme cases: the Rayleigh distribution should therefore be the most unfavourable of all distributions.

When these theoretical results are compared with experimental curves, it is first of all found that this last conclusion is rather well verified. In Fig. 16.4 (overland paths), the Rayleigh distribution is the most unfavourable. In Fig. 16.5 (oversea paths) this remains true if the curves No. 1 and 2 are

---

† These are the same types of link as listed in Section 16.2, where the median level may be considerably lower than the free-space value: very long paths at the limit of visibility, paths in the presence of a semi-permanent duct and millimetre-wave links where absorption is important.

‡ Bray and Corke [1952], Megaw [1946], Gough [1958].

§ The validity of these assumptions has been discussed in Section 14.6.

‡‡ E.g. Misme: Le gradient équivalent; mésure directe et calcul théorique. *Annales des Téléc.* **15**, 92–99 [1960], Gough [1955].

† This curve was constructed by allotting a definite value to the reflection coefficient of the earth ($R = 0.8$) and assuming that the phase of the reflected field may assume any possible value with equal probability.

excepted; however, the latter correspond to a very long path with small clearance (linking Corsica with a point on the French coast), where according to the authors the fadings are above all due to diffraction by the earth

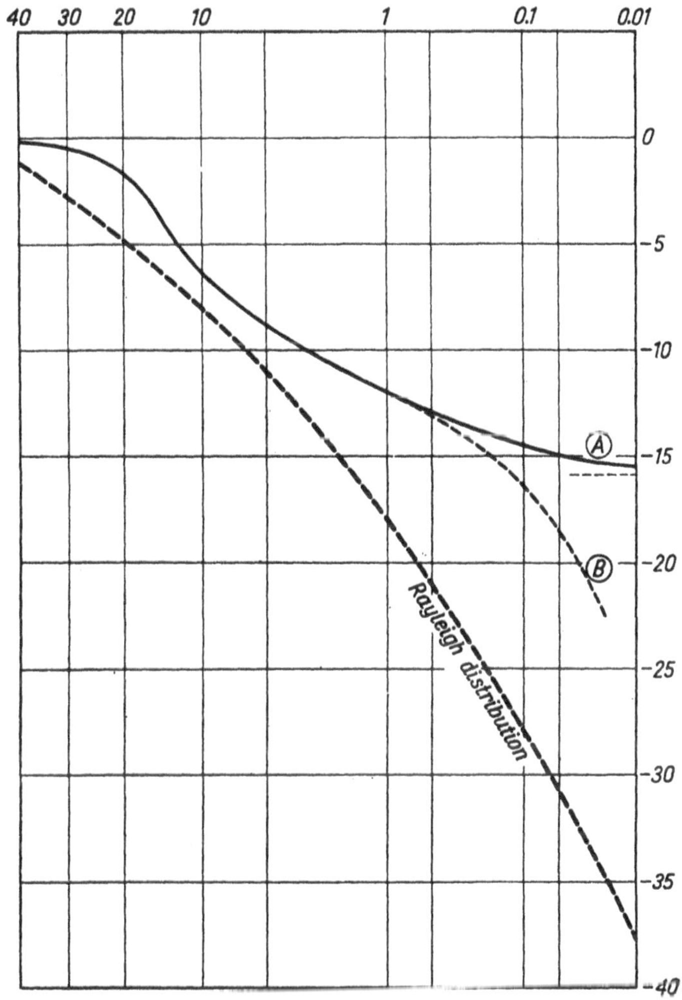

Fig. 16.6. Amplitude distribution of the field resulting from the interference between the direct and ground-reflected waves. A—theoretical curve (for $R = 0.8$), B—general shape of experimentally measured curves.

and to defocusing effects in the atmosphere. Thus this is one of the exceptional cases quoted above where interference fadings do not dominate; in this case there is no reason why the resultant field should be Rayleigh-distributed. We shall not here consider this kind of path, for which it is difficult to give a general rule.†

Examining the experimental results, a further conclusion confirming theoretical predictions may be drawn: it is particularly in summer that the measured distributions approach the Rayleigh distribution. This applies both to overland (Fig. 16.4) and oversea paths (Fig. 16.5).

Moreover, among the experimental distributions deviating least from the Rayleigh distribution, we find those corresponding to long paths at high frequencies (Fig. 16.4a, Curves No. 1, 2, 9, 12). On the other hand, among the curves deviating most from that of the Rayleigh distribution, one finds the shortest links (Fig. 16.4a, Curves No. 7 and 8) and the lowest frequencies (Fig. 16.4a, Curves No. 10 and 14), especially if the link is situated at a high altitude, where refraction effects are less important (Fig. 16.4a, Curve No. 10).

This point appears even more clearly if one compares the curves corresponding to the same pathlength but to different frequencies (Fig. 16.4 curves no. 1, 2, 3, 4, 12 and 14) or curves corresponding to the same frequency, but to different pathlengths (Fig. 16.4 Curves No. 7, 8, 12 and 13).

Thus of the three theoretical results stated at the beginning of this section, points (1) and (3), concerning the Rayleigh distribution, are in good agreement with experiment. There remains the second point concerning the distribution of the field resulting from interference of the direct and reflected rays.

The authors‡ who have attempted to predict the distribution by the above method have usually found that it agrees with the measured distribution only at relatively high levels, whereas at the lower levels the fadings are much deeper than predicted by the theoretical distribution (Fig. 16.6). In particular, the measured distribution curve falls below the minimum of the theoretical curve (i.e. below the asymptote of the latter). This result is not surprising: as we know, when the signal level is near its minimum, other phenomena with small amplitudes may interfere with this strongly attenuated field and cause large fluctuations (Section 16.2, point 2). Among these small-amplitude phenomena is the variable component of the specularly reflected field; even in the absence of any phenomena other than the specularly reflected field, the method is defective since it considers only the constant component of the reflected field, but as we have shown several times in

† Cabessa [1955] also found a distribution with greater variance than that of the Rayleigh distribution in the case of a link between two islands in the Agaean Sea with a pathlength of 120 km at the limit of visibility.

‡ Bray and Corke [1952], Bogle [1952], Gough [1958].

Chapters 13 and 14, the variable part of the specularly reflected field cannot be neglected. As we have seen, the coefficient of specular reflection may attain and exceed unity during its fluctuations; at the moment when the absolute value of the reflection coefficient equals unity, very deep fadings may result.

Although this method is very imperfect, it is sometimes useful, especially if a qualitative indication is needed; for example, for choosing the more favourable of two antenna sites, etc.† Sometimes it gives good account of experimental results even if the reflection coefficient of the earth is near unity (the variable component of the reflected field is then negligible). Finally, it may also be applied to investigations of diffraction fading.‡

Duration of Fades

The rapidity of the fluctuations (fading rate) depends strongly on the physical causes of the fading and varies over a wide range (Fig. 16.3). Some results obtained on oversea paths have been published by Wheeler and Mathwich [1955], Gudmansen and Larsen [1957], Kiely [1952] and others. The distribution of the duration of the fades below a given level is logarithmic-normal (i.e. the logarithm of the duration is normally distributed). The mean duration depends on the chosen level and on the frequency; it decreases with increasing frequency.

Seasonal Variations

Like all phenomena associated with the inhomogeneities of the atmospheric index of refraction, the fading of atmospheric origin is more frequent in summer; the fine variations of the atmospheric index of refraction are in effect largely due to the water vapour content of the atmosphere and it is during the warm season that the atmosphere attains its maximum absolute humidity. It is therefore not surprising that for most links, especially those with long pathlength and small clearance, summer is the season when the fieldstrength fluctuations attain their greatest amplitude.

This is borne out by the curves corresponding to summer or winter in Figs. 16.4 and 16.5. Figures 16.7 and 16.8, taken from Chavance, Boithias and Blassel [1954] show the evolution of the fading depth in the course of one year.

The amplitude of this seasonal variation is important. The attenuation exceeded in 1% of the time is usually greater than the free-space attenua-

† Bray and Corke [1952].

‡ Megaw [1946].

tion: it amounts to 15–18 db during a summer month (this figure is near to that corresponding to the Rayleigh distribution) and only 5–10 db during a winter month.

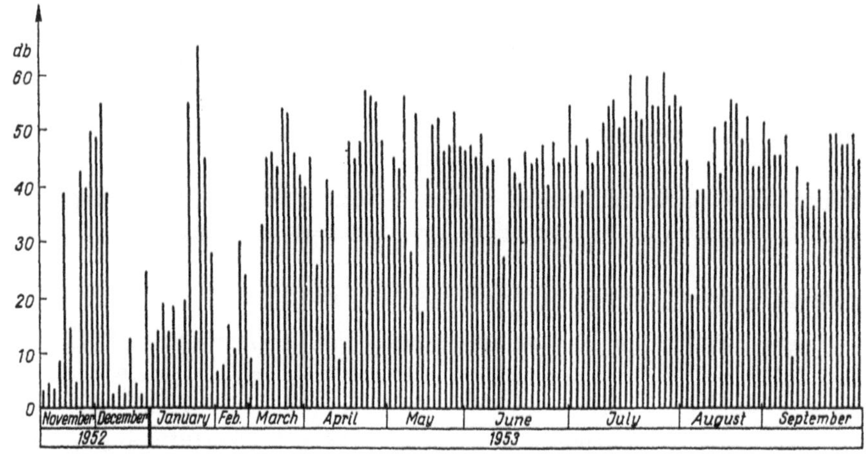

Fig. 16.7. Seasonal variations of the fading depth at 1400 Mc/s. [Chavance, Boithias and Blassel, 1954].

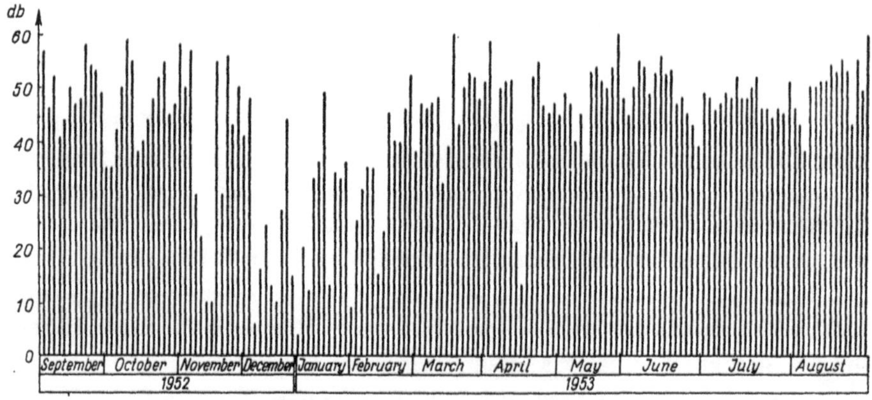

Fig. 16.8. Seasonal variation of the fading depth at 3150 Mc/s [Chavance, Boithias and Blassel, 1954].

On paths where specular reflection by the surface of the earth dominates, the amplitude of the seasonal variations is small or zero [Megaw, 1946, Oxehufwud, 1959, etc.]. In some cases it is found that the most unfavourable season occurs outside the summer months, e.g. in October.† The links on

† Vecchiacchi [1956], Millar and Byam [1950] for certain paths at certain frequencies.

which such seasonal variations have been observed are sometimes long, but usually have ample clearance, so that ground reflections may be an important cause of fading. It is hardly possible to draw general conclusions from these few and rare cases. Let us only note that care should be taken in considering paths where reflections by the earth's surface are the main cause of fading: changes in the effective radius of the earth will cause the interference fringes to move across the receiver and it is hardly possible to predict the moment at which the fading will be deepest.

Diurnal Variations

In examining the different causes of fading we have seen that all those of atmospheric origin should be maximum during the night, especially in summer. During the night the atmosphere is usually stratified, i.e. it contains inversion layers, sheets, etc. During the day, on the contrary, the atmosphere has a structure nearer to that of the "standard atmosphere".

If the fading is caused principally by specular reflection from the ground, the interference fringes are further displaced during the night, when atmospheric refraction changes. During the day they remain more or less constant. Furthermore, if the receiver is not placed near a minimum of the interference lobes in the standard atmosphere, it may be expected that deeper fading will set in during the night hours.

The deepest fading has practically always been observed during the night hours. Figure 16.9, taken from Thué [1955], provides an example of this: the fraction of the time during which the signal strength was below a certain value n (expressed in decibels with respect to the mean value of the field) is plotted as a function of the time of day (abscissa).

Figure 16.9 is typical of the diurnal variations of the field in summer. Similar variations are found in oversea paths;† they are also often found on short overland paths‡ and especially in long overland paths§ in temperate and desert climates.‡‡ The amplitude of these variations increases with growing frequency† and decreases in winter.‡

† Chavance, Boithias and Blassel [1954], Gudmansen and Larsen [1957], etc.

‡ Crawford and Jakes [1952] and Bray and Corke [1952] for overland and oversea paths respectively. On the other hand, Megaw [1946] finds no diurnal variation.

§ Vecchiacchi [1956], Carassa [1956] (overland paths); Chavance, Boithias and Blassel [1954] (oversea path); Thué [1955] (mountainous terrain).

‡‡ Day and Trolese [1950].

† Millar and Byam [1960], Carassa [1956].

‡ Vecchiacchi [1956], Carassa [1956], Millar and Byam [1950].

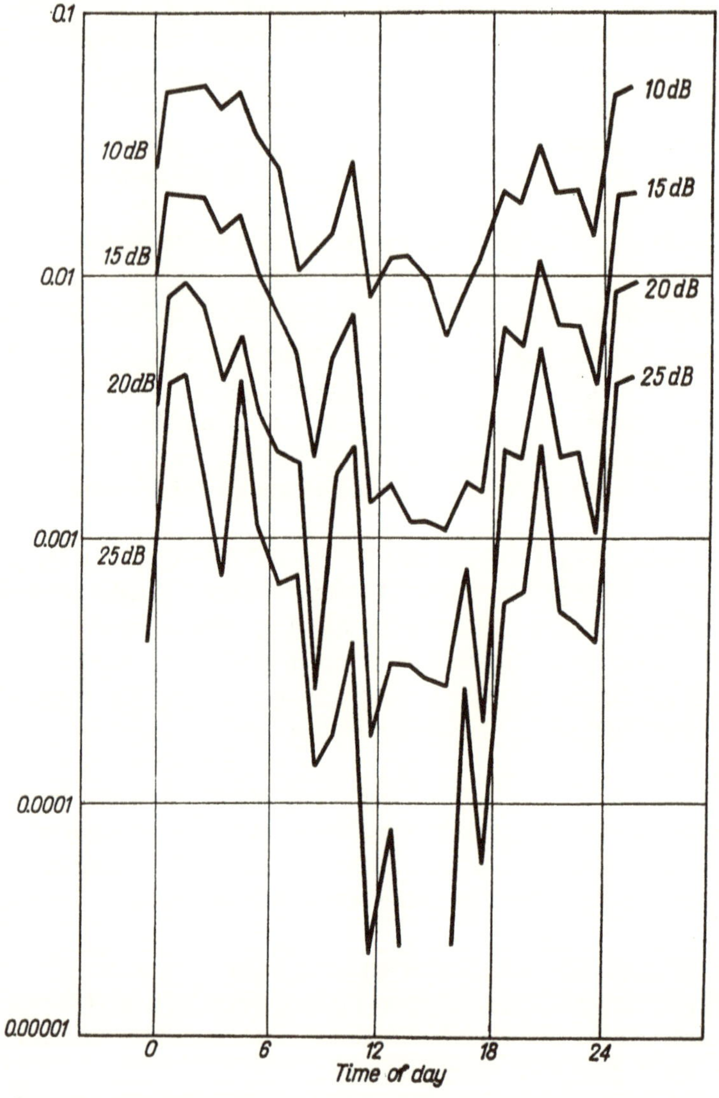

Fig. 16.9. Duration of fades. Ordinate: fraction of time during which the indicated attenuation is exceeded [Thué, 1955].

Frequency Dependence

The frequency dependence of the fadings varies widely in accordance with the dominating cause of the fluctuations, which may be of atmospheric origin or due to ground reflections.

If the cause of the fading is of atmospheric origin (long pathlength, little clearance, high frequency), then increased frequency will favour all effects of atmospheric refraction and reflection (other than partial reflections by atmospheric sheets). The amplitude of the fluctuations may therefore be expected to increase and their distribution should approach the Rayleigh distribution. If the distribution is already near the Rayleigh distribution, an increase in frequency should have little effect on the fading depth.

Curves No. 1, 2, 3 and 4 of Fig. 16.4 confirm these conclusions: the amplitude distribution first tends to the Rayleigh distribution as the frequency increases (Curves 4, 3, 2), then, in the neighbourhood of the Rayleigh curve, they become independent of frequency. Similar results are obtained by comparing curves No. 12 and 14 of Fig. 16.4 and No. 5 and 6 of Fig. 16.5.

Unfortunately the situation is not equally simple if the dominating type of fading is due to diffraction or due to the specular component of reflection by the earth's surface. We shall return to this point in Section 16.5.

Dependence on the Pathlength

If the fadings are of atmospheric origin, their distribution will tend to the Rayleigh distribution with growing pathlength, i.e. the variance of measured signal levels will increase and tend to a limit. This appears to be equally well confirmed for overland paths (Fig. 16.4, curves 7, 8, 12 and 13) and for oversea paths (Fig. 16.5, curves 3 and 4).†

If the fading is due to the specular component of the reflected wave, the opposite result may be obtained: the interference lobes become increasingly concentrated if the pathlength is decreased, so that the fading rate may increase; since the divergence factor increases, the fadings may at the same time become deeper. On the other hand, under other conditions the grazing angle increases when the pathlength is shortened, and if the surface roughness is sufficiently great (Rayleigh criterion), the reflection coefficient will decrease.‡

16.5. Fading due to Specular Reflection

Choice of the Receiver Site (Length of the Sections of a Radio Relay Line Network; Height of the Receiver)

Let us assume the path profile so smooth that to a first approximation it may be replaced by a plane so that there will be only a single point of reflec-

† However, as has already been pointed out, on certain very long oversea paths (longer than 100 km) with small clearence (curves 1 and 2 of Fig. 16.5, for example), where diffraction fading and defocusing effects are important, the variance of the received signal levels may exceed that of the Rayleigh distribution.

‡ Cf. Chapter 14.

tion. For a given position of the transmitter, the field will have a series of maxima and minima in space, the interference lobes presenting an aspect similar to that shown in Fig. 16.10.

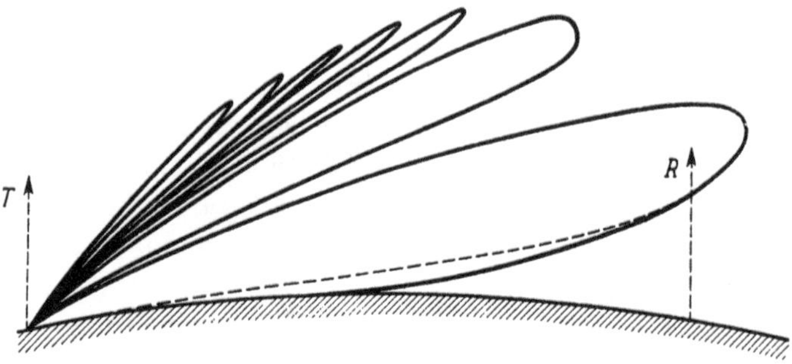

Fig. 16.10. Choice of the receiver location in the first interference lobe.

It is usually assumed that the most favourable position of the receiver is in the first maximum of the lobes, or—which reduces to the same thing—that only the first Fresnel ellipsoid is clear of the terrain. In this position the received field will be most stable when atmospheric conditions vary (this follows from the fact that the first Fresnel ellipsoid is the largest).

In this manner it is possible to choose either the lengths of the individual sections of a relay line network if the heights of the transmitter and receiver are more or less fixed, or the transmitter and receiver heights if the pathlength is given.

This result has been confirmed by many investigators. An illustration may be found in Bray and Corke [1952], where the fields received by two receivers placed at different heights are compared. The lower receiver is in the first interference lobe. By calculating the dependence of the received field at each of the receivers for variable K (ratio of effective to true radius of the earth), it may be verified that the field received by the lower receiver varies less rapidly when K changes. Experimentally it is also shown that the lower receiver receives the more stable fieldstrength.

When the receiver is placed in the first interference lobe, the stability of the field depends on the rate at which the lobes move when atmospheric conditions change. The results of Section 14.6 may be used here: the displacement of the lobes increases with increasing pathlength d (the displacement is proportional to d^2) and with increasing symmetry of the path ($h_1 = h_2$). In practice this means that:

(1) For long paths, the field may be subject to serious fading even if the receiver is placed in the first interference lobe.† (This is one more reason why the results of the present section should not be applied to long paths.)

(2) If one is free to choose the transmitter and receiver heights, they should be chosen as widely different as possible (asymmetrical link) while placing the receiver in the first interference lobe: under these conditions the received field may be very stable. This rule evidently applies only if the reflecting surface is not very rough in the neighbourhood of the point of reflection; it is principally useful for oversea paths.

If the path profile is very rough and there is a danger of several reflection points being present, they may be determined by one of the methods given in Section 14.7. In choosing the receiver site it is desirable to screen as many reflection points as possible. The presence of many reflected rays is a very unfavourable factor: even if each of them has only a small reflection coefficient, there is a risk of all the reflected rays being in phase opposition to the direct ray at certain moments and very deep fades may result.‡

Choice of Frequency and Antenna Directivity

On fairly short paths the stability of the received field may be considerably improved by eliminating the reflected ray.

A first possibility of eliminating this ray is to increase the frequency: if the irregularities of the earth's surface are sufficiently large, the reflection coefficient is a decreasing function of the frequency. This point has been considered in Chapter 14. In particular, in Section 14.3 the domain of frequencies in which the reflection coefficient decreases rapidly with growing frequency was given for each type of irregularity.

A second possibility is to use very strongly beamed antennae to eliminate the specularly reflected ray. This is possible only in the case of very short paths, where the grazing angle γ is at least of the order of one degree, and is more easily accomplished in the case of an asymmetrical link where $\psi = 2\gamma$ than in the case of a symmetrical link with $\psi = \gamma$ (ψ is the greatest angle concluded by the direct with the reflected ray, cf. Section 13.1.1). Finally, if the reflected ray is to be suppressed in this way, the frequency must be sufficiently high to avoid having to design antennae of prohibitive dimensions.

Figure 16.11§ gives an example of eliminating the field fluctuations by means of a very directional transmitting antenna on a short path.

† Chavance, Boithias and Blassel [1954].

‡ Oxehufwud [1959], Tomlinson and Straiton [1959] etc.

§ Taken from Chavance, Boithias and Blassel [1954]. Data of the corresponding link will be found in the annex to this chapter.

A third possibility of suppressing the reflected ray is to screen it by an obstacle, which is often possible. However, this obstacle must not itself be a reflector. It may be noted that reflections (or diffraction) by this obstacle

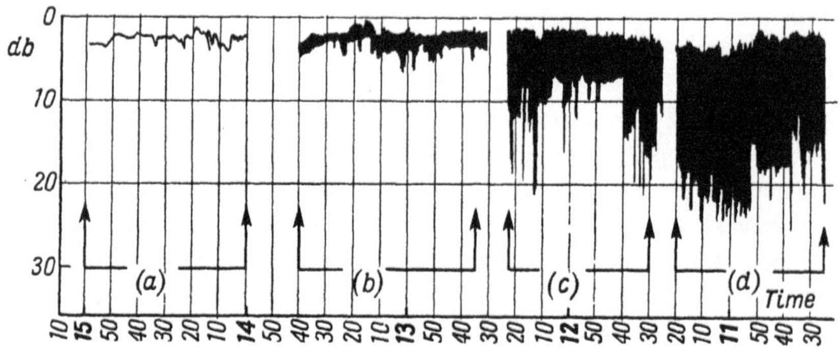

Fig. 16.11. Received field during a beam-swinging experiment. Transmitting antenna (parabolic mirror) beam deflected by (a) $+1°$, (b) $0°$, (c) $-1°$, (d) $-2°$ [Chavance, Boithias and Blassel, 1954].

are not troublesome if it is located near the transmitter or receiver. As we have seen (Section 14.6), the phase of the reflected ray is in this case not significantly affected by atmospheric conditions.

One may also use a screen which directly masks the first Fresnel ellipse† if the latter has reasonable dimensions.

If the first or the second method is employed, it is desirable to use the highest possible frequency and the greatest possible antenna directivities. Besides the technological difficulties one is in fact limited by

—the risk of deflecting the antenna beams by inhomogenities of the atmosphere if the beams are too narrow;

—the fading of atmospheric origin, which as we have seen, are enhanced by an increase in frequency (in particular the absorption by precipitation and the constituents of the atmosphere becomes considerable above 10 Gc/s).

If the third method (screening of the reflected ray by an obstacle) is employed, the choice of frequency is more delicate: for example, if the receiving antenna height is given, it may be desirable to use a higher frequency to keep the first Fresnel ellipse clear of the obstacle. But if the frequency is too high, the slightest modification of the form of the obstacle—building of a house, passing of a car, etc.—may have the most unexpected effects.‡

† Bussey [1951], Gough [1958].

‡ Cf., for example, Oxehufwud [1959].

Finally, if the reflected ray can be neither eliminated nor attenuated, a lower frequency should be used in order to decrease the effect of the displacement of the interference lobes on the stability of the field, because the lobes will have larger dimensions at lower frequencies.

Antenna-Beam Oreintation

The optimum antenna beam orientation is achieved when the beams are pointed towards each other or a little higher (if this may weaken the reflected ray without attenuating the direct field too strongly).

This orientation does not always coincide, at a given instant, with the strongest field. For example, at a time when the reflected field is in phase with the direct field, the orientation of the receiving antenna for which the maximum signal is received lies below the direction of the transmitter; when the reflected field is in phase opposition to the direct field, the receiving antenna should be pointed above the transmitter for maximum signal strength.†

16.6. Diversity Reception on Line-of-Sight Paths

It is well known that interference fadings are in general not synchronous at two receiving points separated by a sufficiently large distance. Similarly, if two links work between the same two points but at sufficiently separated frequencies, the interference fadings will not be synchronous. Thus the quality of reception may be improved by employing a system of space or frequency diversity.

As we have already seen (Section 16.2, point 2), the deepest fades on a line-of-sight path almost invariably result from the superposition of slow fading, which is not selective in space or frequency, and rapid fluctuations, which on the contrary are very selective. The use of diversity always leads to an improvement in quality of the transmission, especially if the slow fading (several hours or days) has a small amplitude.

In practice two questions arise:

(1) What improvement is achieved by diversity? To express this improvement, the values below which the signal falls during x per cent of the time with and without diversity are determined: the difference between these two signal levels is a measure of the improvement brought about by diversity reception. But this difference depends on x: in general it increases with decreasing x.

(2) What should be the distance between the two receiving points or the difference between the two frequencies for diversity to be effective?

† Cf., for example, Oxehufwud [1959].

The answer to this question varies according to the type of fading that dominates. In addition, the two types of paths encountered in Sections 16.4 and 16.5 should be distinguished:

(a) *Short paths with ample clearance at low frequencies.* If specular reflection by the earth's surface is the principal cause of fading, the best diversity effect is obtained

—by placing the two antennae on the same vertical, separated by a distance Δz corresponding to the distance between a maximum and a minimum of the interference lobes;†

$$\Delta z = \frac{\lambda d}{4h_1} \tag{1}$$

In this manner the field received by one of the antennae will be minimum when the field received by the other is maximum;

—by separating the two frequencies by

$$\Delta f = \frac{cd}{4h_1 h_2} \tag{2}$$

The field received by one antenna will then be minimum when it is maximum at the other.

In (1) and (2) it has been assumed that Δz and Δf remain practically constant when the effective radius of the earth $a' = Ka$ varies, i.e. that the interference lobes moving across the two antennae remain perfectly equidistant. This is not always true, especially in the neighbourhood of the lowest maximum.

A more general method consists in tracing the curve of fieldstrength variations for each of the two antennae (or the two frequencies) as a function of K.‡ Diversity reception will be effective if none of the minima received by the two antennae coincide.§

(b) *Long paths with small clearance at high frequencies.* There may be various causes of fading on paths of this kind, but they need not all be present simultaneously; depending on the cause of fading that dominates, diversity reception will be more or less effective and the diversity distance (or frequency separation) will vary.‡‡ We have already given an example

† Cf. Section 11.3.

‡ These curves are easily plotted point by point with the aid of the nomograms in the annex of Chapter 11.

§ Examples of this type of diversity, called "complementary diversity" are given by Bateman [1946].

‡‡ For example, Kaylor [1953] gives different values of the frequency separation Δf corresponding to the fading depth: for a fading depth of 20 db, $\Delta f = 160$ Mc/s; for a fading depth of 10 db, $\Delta f = 40$ Mc/s.

of this (Fig. 16.3). The efficiency of diversity reception may only be described statistically over a long period of time.

If the fieldstrengths are Rayleigh-distributed, it is most often found† that for a sufficiently large diversity distance (or frequency separation) the resulting signal strength is distributed with what is called the "Diversity Rayleigh Distribution". This distribution is derived theoretically by adding

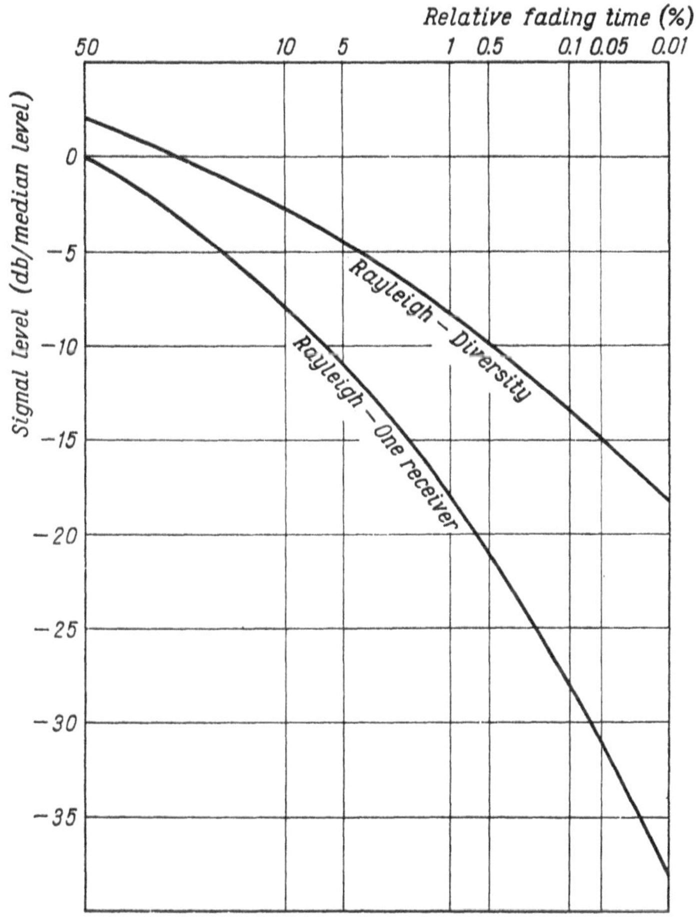

Fig. 16.12. Reduction of fading depth or fading time by diversity reception.

two statistically independent Rayleigh-distributed variables. The resulting distribution is shown in Fig. 16.12. The use of diversity reception should thus reduce the attenuation exceeded during 1% of the time by 10 db, and the attenuation exceeded in 0.1% of the time by 15 db.

† Gudmansen and Larsen [1957], Cabessa [1955].

The diversity distance is much more difficult to predict; it varies in time and depends on the cause of the fading. On oversea paths where the fading has many causes, one of which, interference of the direct wave with the wave reflected by the sea, may play an important part, Gudmansen and Larsen [1957] have obtained the best results by choosing the diversity distance equal to the distance between a maximum and a minimum of the interference lobes. Similar conclusions were reached by Chavance, Boithias and Blassel [1954].

16.7. Radar Coverage

The results of the investigation of line-of-sight propagation at ultra-short wavelengths may be applied to the study of radar coverage. As in the preceding sections, the following cases should be distinguished.

(a) propagation to short distances from a site with ample clearance where fluctuations are principally caused by the interference of the direct wave with the ground-reflected wave; this is the case of a radar observing an aircraft at a short distance.

(b) long-range propagation from a site with small clearance with very many causes of fading; this is the case of a radar observing a target at long range near the surface of the earth.

Radar Observing an Aircraft

Let the site and position of the radar be given; as the target moves, the fieldstrength striking it depends on the reflection coefficient R, which is a function of the grazing angle γ, and on the phase shift between the direct and reflected rays, which equals

$$\Delta\Phi = \frac{4\pi h_1 h_2}{\lambda d}$$

in the plane-earth approximation. $\Delta\Phi$ depends on the position of the receiver through the intermediate ratio

$$\frac{h_2}{d} \approx \tan\gamma$$

Since R and $\Delta\Phi$ practically depend only on γ, the fieldstrength E is of the form

$$E = E_0 f(\gamma)$$

where E_0 is the free-space field and γ is practically equal to the elevation angle of the target as seen by the radar.

The curve in Fig. 16.13 maps the function $f(\gamma)$ (apart from a constant factor) in polar coordinates; it may be used to determine the value of $f(\gamma)$ (radius vector) as a function of γ (polar angle), or the values of d and h_2

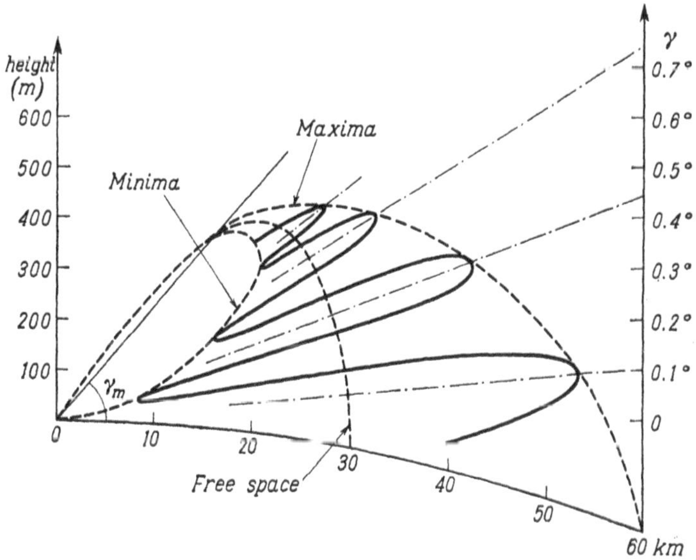

Fig. 16.13. Radar coverage diagram. $h_1 = 100\,\lambda$, $\Delta h = 7.5\,\lambda$.

(abscissa and ordinate). Particular, but typical numerical values of h_1/λ and $\Delta h/\lambda$ and also a typical antenna pattern (curve marked "Free Space") have been chosen. The polarization is assumed horizontal.

As γ increases, the field oscillates between minimum and maximum values lying on envelope curves (Fig. 16.13). Their shape depends on the ratio $\Delta h/\lambda$ where Δh is the height of the irregularities of the earth's surface (Chapter 4). Those drawn in Fig. 16.13 correspond, for example at $\lambda = 3$ cm, to $\Delta h = 22$ cm (sea state 1 to 2).

If γ exceeds a certain value γ_m, specular reflection disappears and the field E becomes more stable (the fluctuations due to diffuse scattering have a smaller amplitude, especially if the radar beam is very narrow †). γ_m may be given by the Rayleigh criterion‡

$$\frac{\Delta h \sin \gamma_m}{\lambda} = \frac{1}{8}$$

† Cf. Sections 12.4.4 and 12.4.5.

‡ Values of γ_m above the sea in terms of λ and the state of the sea have been given by Court [1955].

Since the free-space field E_0 is inversely proportional to the distance, the curve in Fig. 16.13 is also a curve along which the resultant field E remains constant, and hence also a curve of constant fieldstrength backscattered by the target towards the radar. Thus *apart from a constant factor it is the curve limiting the coverage of a radar.*

As the interference lobes change their position in time, the interception of a target is doubtful in the entire zone between the two envelope curves.

Radar Observing a Target on the Earth's Surface at Long Range

Besides the mean range of a radar, which is derived from its power and the mean transmission loss,† one may also define its "minimum range", i.e. its range realized e.g. in 90, 99 or 99.9% of the time. In general it is desirable to place the "minimum range" of a radar between the last maximum of the field and the horizon. Under these conditions the Rayleigh distribution is often satisfactory for use in calculations.

An example of such a calculation is given by Kiely [1952]: to calculate the range of a 3 cm radar above the sea with a transmitting antenna height of 20 m, the author uses an experimentally measured signal-amplitude distribution (which is close to the Rayleigh distribution, cf. Fig. 16.5). In this example, the range realized in 99% of the time amounts to 40% o the mean range.

16.8. Communication by Specular Reflection (without Visibility)

Sometimes it happens, especially in mountainous terrain, that the line of direct visibility is obstructed by some obstacle; a radio link may then be accomplished by reflection from a mountain slope. In links that have been established in this way the mountain slope rarely has the correct orientation to reflect the waves to the receiver according to the laws of geometrical optics; in this case the link is accomplished by diffuse scattering, which will be the subject of Chapter 17.

However, sometimes it happens that the reflection is specular.‡ But even in this case the measured values of the reflection coefficient are small, mostly below −20 db.§ These small values of the reflection coefficient are due to two causes: the reflecting surface is rough, and its dimensions are usually smaller than the first Fresnel ellipse (in most cases where the re-

† Norton and Omberg [1947], Kiely [1952], Court [1955].

‡ Poleschinsky [1959].

§ Klein and Dufour [1948], Klein [1952].

flection is specular, the reflecting surface does not cover the entire mountain slope, but only a small area, a rock, etc.).

The power received by reflection equals

$$P = P_0 \rho_s \frac{\delta^2}{\lambda^2}\left(\frac{d_1+d_2}{d_1 d_2}\right)^2 \sin^2\gamma \tag{3}$$

where d_1 and d_2 are the distances from the reflecting surface to the transmitter and receiver respectively, P_0 is the free-space power received at a distance of d_1+d_2, S the active area of the reflecting surface, γ the grazing angle (Fig. 16.14a) and ρ_s the scattering coefficient defined in Chapter 12.

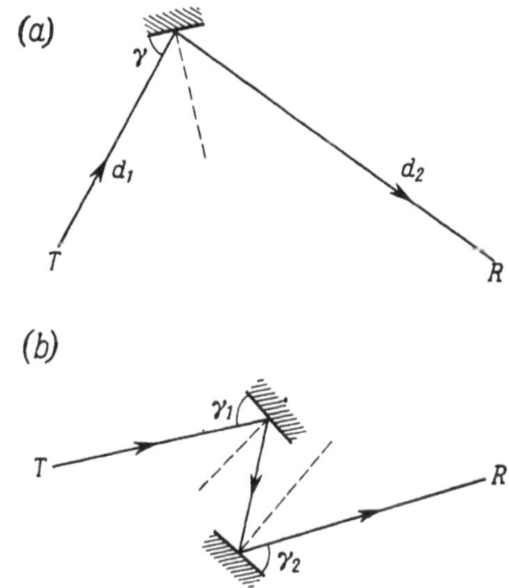

Fig. 16.14. (a) Communication via a passive reflector, (b) reflectors mounted in "periscope" fashion.

One of the principal dangers of this mode of propagation is the risk of distortion by multipath interference. If δ is the greatest path difference between two interfering rays, the largest bandwidth that can be transmitted without distortion is of the order of

$$\Delta f = \frac{c}{2\delta}$$

where c is the velocity of light in free space; for practical purposes this may be written as

$$\Delta f = 150\,\frac{1}{\delta}$$

where f is in Mc/s and δ in metres.

Artificial Reflectors

Certain links using artificial reflections have been put into operation.† In this case the field is again given by (3) with $\rho_s = 1$. It will be seen that if the reflected power is to be sufficiently large without using reflectors with prohibitive dimensions, the following conditions must be fulfilled:

—d_1 and d_2 should be small (short path) and if possible unequal;

—the wavelength λ should be small; but if it is too small, the tolerance limits of the surface finish will be narrow and the technological difficulties will mount;

—the grazing angle γ should be large (Fig. 16.14a); but if this is accomplished by removing the reflector far from the axis of the link, d_1 and d_2 will at the same time increase. There is an optimum value of γ: for example, if $d_1 = d_2$ are fixed and S is constant, it is easily shown that the reflected field will be maximum for $\gamma = \pi/4$.

However, Chaux and Dascotte [1950] have given examples of links where γ is increased without increasing the distance by making use of two reflectors mounted in "periscope" fashion (Fig. 16.14b).

Annex to Chapter 16

Characteristics of Some Experimental Paths

The characteristics of some experimental line-of-sight paths on ultra-short waves are listed in Tables 16.3 and 16.4. This list does not claim to be complete: it is restricted to the paths the results of which have been used in Sections 16.1 to 16.6.

We have grouped all these data in this list to evade overburdening the text of the above chapter and to facilitate reference for the reader who is confronted with a particular problem and may find the pertinent bibliographical reference corresponding to analogical conditions.

In Tables 16.3 and 16.4, f denotes frequency and d the pathlength. For oversea paths, h_1 and h_2, the transmitter and receiver heights above the reflecting surface are identical with their altitudes h_1' and h_2'. For overland paths, h_1 and h_2 are difficult to define if the path profile is very irregular and *unless otherwise stated*, only the altitudes h_1' and h_2' are given.

† A general study of paths with artificial reflectors has been undertaken by Andrieux [1956]; Chaux and Dascotte [1950] give examples of installations with artificial reflectors in Morocco; Magnuski and Koch [1953] describe a path employing an artificial reflector in the U.S.A.

The clearance H has been defined (Section 16.3) as the distance from the direct ray TR to the point P of the reflecting surface where the optical path TPR is minimum (Fig. 16.15).

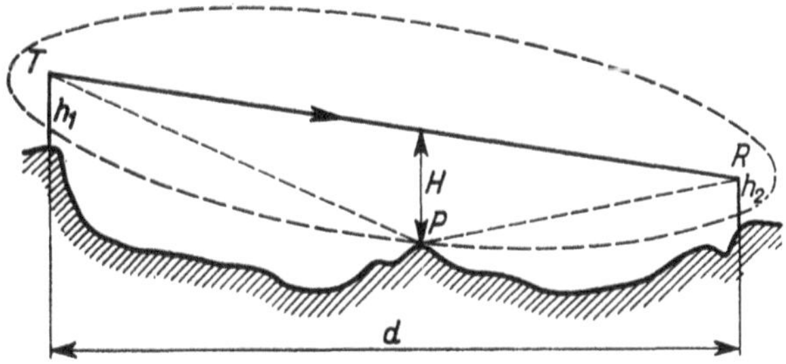

Fig. 16.15. Definition of the clearance H.

Finally, the paths given in the annex to Chapter 13 should be added to the present list.

TABLE 16.3—Overland Paths

Location	Bibliography	f Mc/s	d km	h_1' m	h_2' m	Clearance	Test Period	Particular tests
England—Hilly terrain near London	Megaw [1946]	3000	29,43, 58 and 82	85 to 180	225	(diff. values)	February	Two polarizations
England—East Anglia	Megaw [1946]	3000, 3400 and 10,000	46, 78 and 96	51–100	25–185		August	
U.S.A. Low terrain near New York	Thompson [1948]	3300 and 4000	33 43 60	30 100 70 above the ground	70 30 100 above the ground	20 to 30 m	One year	Space diversity
U.S.A. Low terrain near New York	Millar and Byam [1950]	1850, 4150 6325 9550	65	200	120	30 m	One year	

Location	Bibliography	f Mc/s	d km	h'_1 m	h'_2 m	Clearance	Test period	Particular tests
U.S.A., Arizona Desert	Day and Trolese [1950	25, 63, 170, 520, 1000, 3300, 9375, 24,000	43 and 74	0–60 above the ground	0–60 above the ground	Variable	Winter	Meteorological measurements Variable height
North of France Plains and hilly terrain	Chavance [1952]	3150 and 9350	76			Almost none	One year	
England, near London	Bray and Corke [1952]	3930	45	27 above the ground	35 above the ground	Small (trees)	May	
U.S.A., New Jersey	Crawford and Jakes [1952] Crawford and Jakes [1952] De Lange [1952]	25,000 3950±225 4000	27 35	 115	 195	20 m 85 m	Three years Summer and autumn	Antenna beam sweep Frequency sweep Very short pulses } study of multipath propagation
Between France and Switzerland over high mountains	Klein and Libois [1953]	300 and 3000	160	592	1609	Very low	Summer	
U.S.A., Iowa	Kaylor [1953]	3750–4190	48				July, August	Frequency sweep

Location	Bibliography	f Mc/s	d km	h'_1 m	h'_2 m	Clear-ance	Test period	Particular tests
U.S.A., New Jersey. Various kinds of terrain	Epstein *et al.* [1953]	850	0–74	0–250	12	Var-ia ble	Summer	
U.S.A., Colorado (Cheyenne Mountains)	Bean [1954]	1046	80,110	840 above the ground	13 above the ground	Large	One year	Meteorological measurements
France, between the Massif Central and the Alps, over the Rhone Valley	Thué [1955]	3590 and 3690	146	1910	1230	400 m	Summer	
	Thué [1955]	3790 and 3690	98	1910	650	120 m	Summer	
Greece, Mountains	Cabessa [1955]	2000	90	1700	1000		Summer	Diversity
		2000	71	2000	200			
North of Italy, over the Po Valley	Vecchiacchi [1956]	1000	189	165	2165	1st Fresnel zone	One year	
	Carassa [1956]	1000	196	1320	582	Large		
		500	83.5	861	135	Large		
			88	415	90			
		250	189	165	2165	1st Fresnel zone		
U.S.A., Colorado and Wyoming	Oxehufwud [1959]	4000	70	2500	1500		July–August and April–May	Variable height—Study of multiple reflections. Two polarizations
		6310 and 11,200	43	2300	2200			
			63	2100	2200			
			59	2300	2100			

TABLE 16.4—Oversea Paths

Location	Bibliography	f Mc/s	d km	h_1 m	h_2 m	Clearance	Test period	Particular tests
English Channel	Megaw [1946]	3400	3–19†	10 and 39	10	Variable	September	Measurement of sea reflection coefficient
English Channel		3400	98–102†	24 and 150	24–220	Variable	March	Depolarization
Irish Sea		3400 and 10,000	165 and 180	350, 1050	300	Variable	June	
Irish Sea	Bray and Corke [1952]	2000	65	250	20 120	Large	May–October	
Pacific Ocean, near San Francisco (Golden Gate)	La Grone, Straiton and Smith [1955] Brooks, Smith, Du Bose [1954] Beard, Katz [1957]	10,000	5	26	15	Large		Fluctuation spectrum studies Fluctuation spectrum studies
Across Loch Goil (Scotland)	Kiely [1954] Kiely [1956a]	40,000	1.6	5	30	Large		Ducts
Near Portsmouth (England)	Kiely [1956b] Kiely [1952]	9375	19	18.3	18.3	1st Fresnel zone	January to July	Meteorological measurements

† Shore-to-ship paths.

Location	Bibliography	f Mc/s	d km	h'_1 m	h'_2 m	Clearance	Test period	Particular tests
Between Continental France and Corsica	Chavance, Boithias and Blassel [1955] Rivet [1956]	1400 3150 and 435	240	1136	790	180 m. (= many Fresnel zones at 1.4 and 3.1 GHz)	One year	Frequency and space diversity Horizontal space diversity Depolarization Meteorological measurements
Mediterranean Sea	Chavance, Boithias and Blassel [1955] Spizzichino [1959]	9400	28.9	70	530	Very large	Various parts of the year	Separation of the direct and the reflected ray-goniometry
Greece	Cabessa [1955]	2000	90 120	221 575	1070 810		Summer	Diversity
Denmark	Gudmansen and Larsen [1957]	1800 and 4700	54 82	57.68 148	76 to 155	One Fresnel zone or more	Spring and Summer	Diversity

† Shore-to-ship paths.

CHAPTER 17

DIFFUSE SCATTERING: I. PRACTICAL APPLICATIONS

17.1. Line-of-Sight Paths

The received fieldstrength on line-of-sight paths was considered in Chapter 6, but we had to exclude the case where diffuse scattering by the earth's surface is the principal cause of fading.† This will be the case if

(a) the path is very short, so that atmospheric effects do not play an important part;

(b) the earth's surface is rough in the sense of the Rayleigh criterion, so that specular reflection is less important than diffuse scattering (Section 15.10).

If these two conditions are fulfilled the results of Chapter 15 may be applied: the reflection coefficient of the diffusely scattered component is of the order of 0.35, unless it is reduced by some unfavourable factor, e.g. very directional antennae, obstacles screening the reflecting surface or presence of vegetation.‡

Thus diffuse scattering will play a dominant role on short paths with ample clearance at high frequencies, i.e.

—paths at millimetre wavelengths above land or above the sea in the absence of a quasi-permanent duct; §

—paths at centimetre wavelengths in hilly terrain or in cities;

—paths with high grazing angles, e.g. links between an aircraft and the ground,‡‡ radars observing aircraft† etc.

The fluctuations of the field at the receiving point which result from the interference of the diffusely scattered field with the direct wave, have been considered in Section 12.6.1‡ and experimental data have been given in Chapter 15. Figure 17.1 shows the theoretical distribution of the resultant

† Cf. Section 16.3.

‡ Cf. Section 15.10.

§ Kiely [1954, 1956a].

‡‡ MacGavin and Maloney [1958].

† Diffuse scattering is the dominating component if $\gamma > \gamma_m$ (cf. Section 16.7).

‡ The distribution of the resultant field is given by (56) of Section 12.6.1.

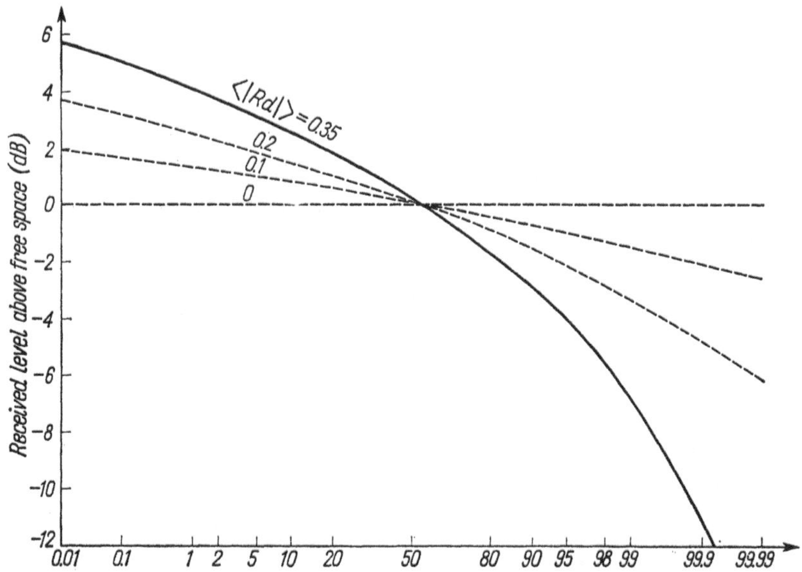

Fig. 17.1. Amplitude distributions of the received field on a line-of-sight path where the fluctuations are caused only by diffuse scattering with a mean reflection coefficient $\langle|R_d|\rangle$.

field with $\langle|R_d|\rangle$ set equal to several values up to its maximum value of 0.35. It will be seen that these fluctuations have a much smaller amplitude than those which were met in Chapter 16 (specular reflection).

We recall that the amplitude of the fluctuations may be much greater if in addition to diffuse scattering a specular component is also present (Section 16.2).

To protect the link from fadings due to diffuse scattering, one may make use of the fact that the scattered radiation originates at two zones A and B near the transmitter and receiver (Fig. 17.2a; cf. Section 12.4.3). It is sufficient to use antennae with sufficiently narrow beams to prevent these zones being illuminated, or to make use of natural obstacles to screen them.†

Another effect of diffuse scattering is the generation of interference with respect to other radio links (Fig. 17.2b): even if a very directional antenna illuminates the receiver R' very feebly, the earth's surface may scatter waves towards it.

† These two solutions are facilitated by the fact that diffuse scattering is the dominating component at high frequencies. Moreover, since the scattered radiation comes from the foot of the transmitter (and receiver), its separation from the direct wave is much easier than that of the specularly reflected ray. The antenna gain necessary to eliminate the scattered component may be determined from Section 12.4.4.

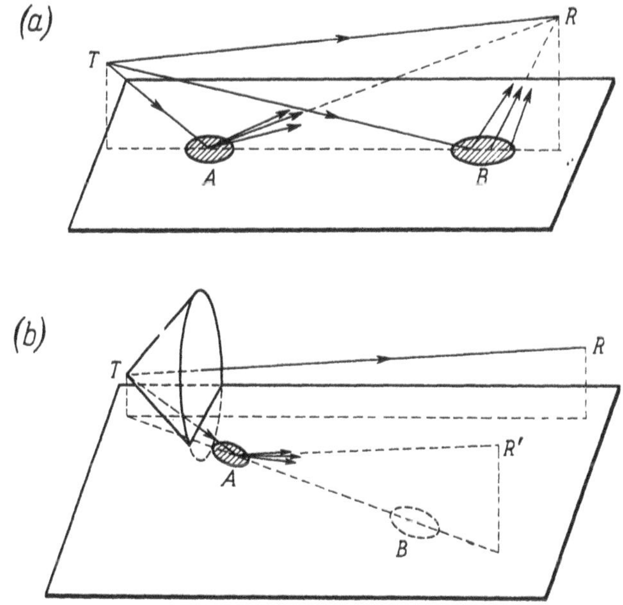

Fig. 17.2. Effects of diffuse scattering: (a) fluctuations caused by interference with the direct wave, (b) interference between different paths.

Theoretically the scattered radiation may have its origin in two zones, A and B, but only A, in the neighourhood of the transmitter, is in fact, partially illuminated by the transmitting antenna (Fig. 17.2b). Thus the interference arises in the neighbourhood of the transmitter and the effect is the same as if the transmitting antenna pattern were deformed.

Figure 17.3, taken from Kirby, Taff and Moore [1952] shows some examples of this apparent deformation of an antenna pattern above different types of irregular terrain. The transmitting antenna was mounted at a height of 30 m above the ground in an inhabited area and was successively pointed in different directions. For each position of the transmitting antenna the receiver, mounted on a truck, was moved along an approximately circular path about the transmitting point at a distance of 0.4 miles (640 m). The frequency was 460 Mc/s. The deformation of the antenna pattern depended above all on the type of irregularities of the ground in front of the transmitting antenna. In general, it appeared as a widening of the principal lobe and an increase of the secondary lobes. The last diagram of Fig. 17.3 was obtained by pointing the transmitting antenna towards large buildings and the appearance of strong secondary lobes is due to specular reflection from these buildings.

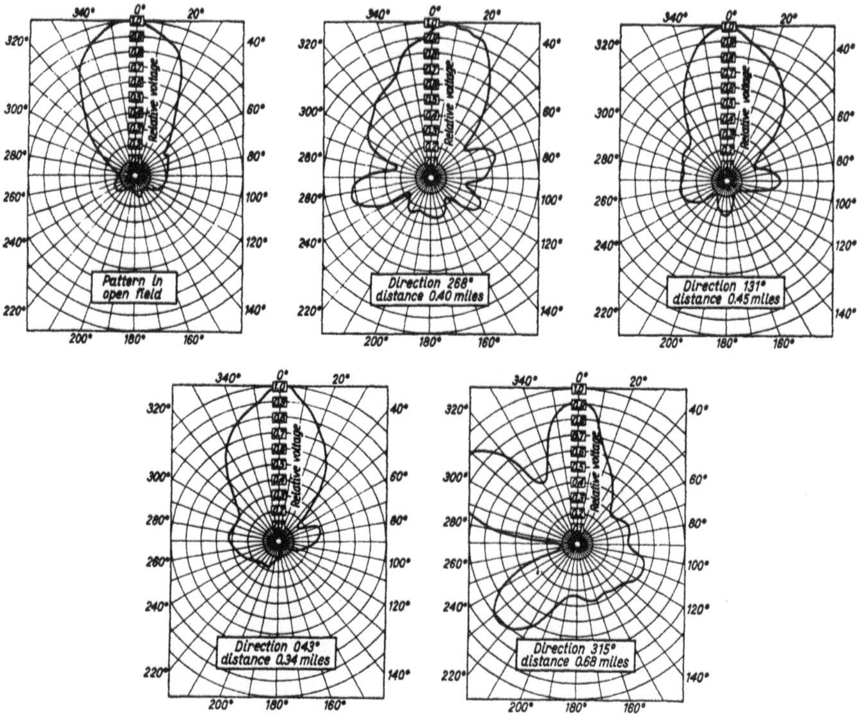

Fig. 17.3. Antenna patterns of an antenna in free space (a) and above irregular terrain: small buildings (b,c,d) and large buildings (e). [Kirby, Taff and Moore, 1953]. (*By permission of the I.R.E.*)

In connection with the measurements of the reflection coefficient of the diffusely scattered component (Section 15.8) we have already given an example of the field scattered out of the transmitting antenna beam by the surface of the sea.

Finally, the strong depolarization of the diffusely scattered component† is an additional factor in the mutual interference of radio links.

The back-scatter of waves by the terrain and the surface of the sea is a special case of this interference. In view of the importance of this problem, we shall devote a separate Chapter (Chapter 18) to it.

17.2. Communication by Diffuse Ground Reflections without Direct Visibility

In Section 16.8 we considered links between two points without direct visibility by means of specular reflection from a mountain slope. This type

† Cf. Section 12.7 and 15.9.

of link is rather exceptional, since the mountain slope will rarely be oriented in the necessary direction to reflect the waves between the two given points. On the other hand, diffuse scattering by mountain slopes plays an important part in the propagation through mountainous terrain for all frequencies above some tens of megacycles.

The theorctical value of the scattered field is given by (15) of Chapter 12; in accordance with Section 12.4.1 we have

$$\langle |R|^2 \rangle = \frac{1}{4\pi}\left(\frac{r_1+r_2}{r_1 r_2}\right)^2 \frac{S}{\tan^2\beta+\tan^2\beta_0} \tag{1}$$

where the reflecting surface S is assumed sufficiently small for r_1, r_2 and β to be reasonably constant over S. Comparing (1) with (3) of Chapter 16, it may be shown that the reflection coefficient of the mountain slope is the same as that of a specularly reflecting plane, smooth sheet, the area Σ of which is given by

$$(\Sigma \sin\gamma)^2 = \frac{\lambda^2 S}{4\pi(\tan^2\beta+\tan^2\beta_0)}$$

This kind of propagation has been studied by Klein and Dufour [1948] and by Poleschinsky [1959]. The former have studied the propagation of waves at 50 Mc/s in the Swiss Alps, the latter investigated propagation at 160 Mc/s in the Austrian Alps. In both cases, the attenuation with respect to free space varies between 25 and 40 db. The values of Σ given by Klein and Dufour vary between 1000 and 4000 m². At 50 Mc/s, propagation by scattering quickly becomes the dominating mode of propagation as soon as the attenuation caused by diffraction exceeds 20 db; the waves scattered by different mountain slopes arrive with comparable orders of magnitude and mutually interfere, producing a distorted signal.† When the attenuation caused by diffraction axceeds 40 db, the resultant field is usually solely caused by scattering by the mountain slopes and multipath interference often makes the link unusable.

As against the case of reflection by horizontal ground, the reflection coefficient is greater when the polarization is vertical.

17.3. Scattering of Solar Radiation by the Earth—Albedo

The solar radiation striking the earth is almost entirely contained in the wave band from 0.3 to 4μ (microns). For such small wavelengths the earth always behaves like a very rough surface: the reflection of solar radiation by the earth is always diffuse.

† The analysis of the distortions in Section 16.8 may also be applied to the case of diffuse scattering

The laws of the reflection of solar radiation by the earth are of interest in the study of exchanges of energy between the earth and different layers of its atmosphere. It is also necessary to explain the variations of temperature of the earth's surface, which also must be considered in investigating the scattering of light by the atmosphere.

To measure the solar radiation one may use a detector of visible or infrared light, pointed downwards and placed at a low altitude, e.g. on a tower or a building. The radiation thus measured is reflected by the small portion of the earth's surface in the immediate neighbourhood of the detector. In this way, the reflection of solar radiation has been studied for various samples of ground, e.g. snow, rock, cultivated land, etc.

In meteorology, large-scale phenomena are of greater interest. The corresponding measurements are therefore taken at a higher altitude by using balloons, rockets or artificial earth satellites; the measured radiation then comes from a more extensive part of the earth's surface. But the received radiation is then also more complex, for a detector pointed downwards will, in addition to the radiation reflected by the earth, receive

—the (very intense) radiation reflected by the clouds;

—the radiation scattered by the molecules and dust particles of the atmosphere;†

—the primary radiation of the earth at wavelengths exceeding 4μ, which may be separated by means of a filter.

Experiments made with the aid of a balloon at an altitude of 25,000 m, described by Murcray, Brooks, Sible and Westdal [1959], provide an example of the separation of these different components. The detector was mounted in a reflector telescope, the observation beam of which could, by means of a mobile mirror, scan all directions below the horizontal plane. In this way the terrestrial radiation could be eliminated and the radiation reflected by the earth could be distinguished from that scattered by atmospheric particles.

Before giving some experimental results, we shall consider the quantities that can effectively be measured. Depending on the employed method of measurement, the reflection coefficient of the earth or the albedo coefficient of the earth may be measured.

Reflection coefficient of the earth. This is the ratio of the reflected to the

† This radiation may be recognized on the basis of its directivity. The molecules of the atmosphere back-scatter a large part of the incident radiation (Rayleigh scattering), whereas dust particles have a forward-scattering characteristic, if their dimensions are of the same order as the wavelength. In both cases the radiation is particularly intense when the sun is near the horizon.

incident field and we shall assume that, as in the preceding chapters, it may be written in the form

$$R = \rho R_0 \tag{2}$$

where R_0 is the reflection coefficient of a smooth earth and ρ a coefficient smaller than 1. One may then define a reflection coefficient R_t assuming the receiving antenna to be isotropic (or insignificantly directional, cf. Section 12.4.4); if the receiving antenna beam is very narrow, R is smaller than R_t—in the limit as the solid angle $d\Omega$ of the beam tends to zero, the reflection coefficieint may be defined with respect to a unit solid angle, $dR/d\Omega$.

Since detectors of light waves do not in fact measure fieldstrength but power, one is usually more interested in the "power reflection coefficients"† R^2, R_0^2, $d(R^2)/d\Omega$.

In the experiments by Murcray *et al.* quoted above, the quantity $d(R^2)/d\Omega$ was in fact measured.

Albedo coefficient of the earth. This is the ratio of the energy flux incident onto a portion of the earth's surface to the energy flux reflected by this surface.‡ It is the albedo coefficient, and not the reflection coefficient, which is important in problems of energy exchange. To measure it, the energy flux intercepted by a horizontal surface must be measured.

Let us assume that the conditions under which (2) is valid are fulfilled; § then at any point of the earth's surface, a fraction $1-R_0^2$ of the incident energy is absorbed by the earth and the rest is reflected. If R_0 is, as we have assumed, the same for any point on the earth's surface, *the albedo coefficient of the earth equals* R_0^2.‡‡ It is thus independent of the form of the irregularities of the earth's surface and depends only on its material nature. If the ground is not very absorbing, the albedo coefficient should be near 1.

---

† Since phase considerations are irrelevant here, we denote the absolute values of the reflection coefficients for simplicity by R, R_0, etc.

‡ In general the *albedo coefficient* at an altitude h is defined as the ratio of the energy flux leaving the part of the terrestrial atmosphere below altitude h to the energy flux penetrating it. Thus this definition includes not only the direct solar radiation and that reflected by the earth's surface, but also the radiation scattered by the atmosphere etc. The albedo coefficient therefore depends on the altitude at which it is measured. The *albedo coefficient of the earth* is that measured at the surface of the earth.

§ This assumes that (a) the radius of curvature of the earth's surface is at any point much larger than the wavelength, (b) R_0 does not vary appreciably with the grazing angle γ.

‡‡ The radius of curvature of the sea surface is always much greater than the wavelength, and the slope β_0 of the irregularities is smaller than γ.

Experimental Data

The experimentally measured values of the albedo coefficient range from 0.85 (fresh snow) to 0.07 (cultivated land, moist earth).† If the grazing angle γ, i.e. the elevation of the sun, is not too small, the albedo coefficient does not vary much; according to Montieth [1961], for γ between 60° and 210°, it varies from 0.16 to 0.19 in the case of bare soil and from 0.23 to 0.28 for ground covered with vegetation. Finally, it does not depend appreciably on the wavelength: the spectral composition of the radiation scattered by the earth is very near to that of the incident radiation.

The radiation reflected by the surface of the sea seems to come from the point of specular reflection. However, there is no doubt that the reflection is diffuse, the heights of the waves being much larger than the space wavelength. The explanation may be found by noting that the surface of the sea corresponds to the model of a rough surface considered in Section 12.4.5.‡ When the slope β_0 of the irregularities of the reflecting surface is small with respect to γ, the "glistening surface", where the scattered energy originates, is a small area encircling the specular point of reflection.

On the other hand, the reflection by dry ground is much less directional. For example, in the experiments by Murcray *et al.* [1959] described above the only very directional component of the received radiation arrives at an almost horizontal direction; it disappears when the sun stands sufficiently high above the horizon and may be interpreted as scattering by atmospheric dust particles. The rest of the received radiation, which is interpreted as radiation reflected by the earth, comes almost uniformly from the entire surface of the earth, which behaves almost as an isotropic scatterer.

This result is hardly surprising: at radio frequencies, the surface of the earth may be considered as a gently undulated surface and the slope β_0 of its irregularities is small. However, at wavelengths of the order of microns, each grain of sand constitutes an irregularity; β_0 is large, and in addition, the radius of curvature of the irregularities may be comparable to the wavelength. For these two reasons, the scattering is not directional.

Isotropic Scattering

Isotropic scattering is a particularly simple case of diffuse scattering in general and is often used in optics. It is the limiting case which is ap-

---

† Rouleau, J. and Trochon, R.: *Météorologie Génerale.* vol. I, 1952, vol. II, 1958, Gauthier-Villars, Paris.

‡ The radii of curvature of the surface of the sea are always much larger than the wavelength and the slope of the irregularities is smaller than β_0 (cf. Section 12.4.6).

proached if β_0 and γ tend to $\pi/2$ or if the irregularities of the reflecting surface are very large compared with the wavelength.

If the scattering is isotropic, the energy flux scattered by an element of surface dS into a solid angle dΩ is given by

$$\mathrm{d}W = W_1 \,\mathrm{d}S \sin\gamma \sin\delta \,\mathrm{d}\Omega$$

where γ and δ are evident from Fig. 17.4, and where W_1 is called the *radiance* of the reflecting surface. Setting the total scattered energy flux equal to the product of the incident flux W_0 and the albedo coefficient R_0^2, we obtain

$$W_1 = \frac{1}{\pi} R_0^2 W_0$$

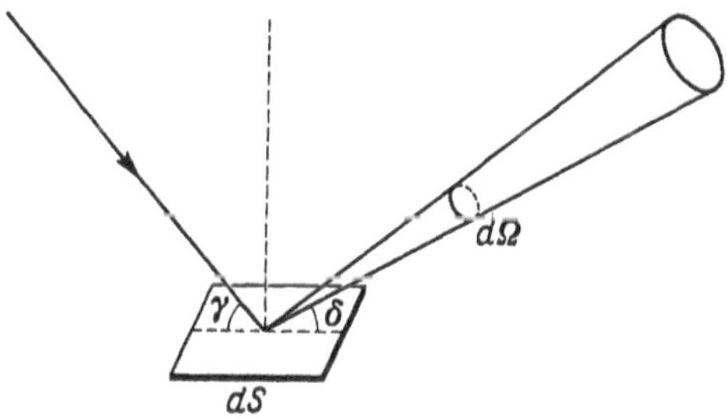

Fig. 17.4. Definition of γ, δ and dΩ.

The field scattered by an element dS into any given direction depends only on δ: hence it is symmetrical about a vertical axis. The field scattered by the entire terrestrial surface at any given point will therefore also be rotationally symmetrical. The Poynting vector is therefore vertical and its flux across a unit area equals

$$W_r = R_i^2 p_0^2 \tag{3}$$

where R_i is the reflection coefficient of the diffusely scattered component for an isotropic antenna (defined above) and p_0 is the strength of the incident Poynting vector. Now by definition of the albedo coefficient R_0^2, W_r is related to the incident flux W_0 by

$$W_r = R_0^2 W_0 = R_0^2 p^2 \sin\gamma \tag{4}$$

From (3) and (4) we obtain the reflection coefficient of the diffusely scattered component:

$$R_i^2 = R_0^2 \sin\gamma$$

In conclusion we may say that the experimental results may be quite well explained on the basis of two models of scattering by rough surfaces: "isotropic scattering" for the scattering by dry ground, and the model of Section 12.5.4 for scattering by the surface of the sea.

CHAPTER 18

DIFFUSE SCATTERING: II. APPLICATION TO BACK-SCATTERING BY THE EARTH'S SURFACE

BACK-SCATTERING by the terrain or by the surface of the sea implies that a part of the incident energy is scattered back to the transmitter. In radar, it appears as a diffuse echo coming from an extensive area of the ground or the sea surface, which is superimposed on other echoes and may considerably impair the sensitivity of the radar.

With the possible exception of vertical incidence, back-scatter is always a diffuse, not a specular reflection.

18.1. THE BACK-SCATTERING CROSS-SECTION σ^0

The amount of back-scatter could be defined by the reflection coefficient of diffuse scattering R_d of the earth's surface in the direction of the transmitter. In practice it is more convenient to proceed in analogy with scattering problems and to define the back-scattering cross-section $\mathrm{d}\sigma$ of an element of the earth's surface $\mathrm{d}S$, introducing

$$\sigma^0 = \frac{\mathrm{d}\sigma}{\mathrm{d}S}$$

Applying the definition of a scattering cross-section it is easily shown that the power scattered by an element of the earth's surface $\mathrm{d}S$ and received in the neighbourhood of the transmitter equals

$$\mathrm{d}P_r = P_t \frac{g^2\lambda^2}{(4\pi)^3 r^4} \sigma^0 \mathrm{d}S \tag{1}$$

where P_t is the transmitted power (assuming continuous-wave transmission), r is the distance of the surface element $\mathrm{d}S$ from the transmitter, g the gain of the radar antenna, and λ the wavelength.

$\mathrm{d}P_r$ may also be written in the form

$$\mathrm{d}P_r = P_t \left(\frac{\lambda}{8\pi r}\right)^2 \mathrm{d}(R_d^2) \tag{2}$$

where $d(R_d^2)$ is the reflection coefficient for diffuse scattering by the element dS.† From (1) and (2) the relation between σ^0 and the reflection coefficient of the diffusely scattered component may be obtained.

From (1), the power received by a pulsed radar may be obtained:‡

$$P_r = P_t \frac{g^2\lambda^2}{(4\pi)^3 h^2} \int_0^{\pi/2} f^4(\gamma)\Phi(\gamma)\sigma^0(\gamma)\sin\gamma\, p\left(\frac{h}{\sin\gamma}\right) d\gamma \tag{3}$$

where P_t is the peak value of the transmitted power, h the height of the transmitting antenna above the surface of the earth, γ the grazing angle, g the maximum gain of the radar antenna, whereas $f^2(\gamma)g$ is its gain in the direction making an angle γ with the horizontal, $\Phi(\gamma)$ is the area of the plane at an inclination γ intercepted by the radar beam, and $p(r)$ is proportional to the absolute value of the Poynting vector of the incident wave at a distance r from the radar; we assume $p(r) = 1$ during the propagation of a pulse and $p(r) = 0$ during the rest of the time.

Relation (3) simplifies in two cases:‡

—*if* γ *is small*, we obtain

$$P_r = P_t \frac{g^2\lambda^2}{(4\pi r)^3} f^4(\gamma)\Phi \frac{\tau c}{2} \sigma^0(\gamma) \tag{4}$$

where $r = h/\sin\gamma$ is the distance from the radar to the corresponding part of the reflecting surface, τ is the pulse duration, c is the velocity of light in free space, and Φ is practically independent of γ: $\Phi(\gamma) \approx \Phi(0)$.

—*if* γ *is large* (near 90°), we obtain

$$P_r = P_t \frac{g^2\lambda^2}{(4\pi h)^2} \Theta\Phi\sin\gamma\, \sigma^0(\gamma) \tag{5}$$

where Θ is the area of the vertical plane intercepted by the radar beam.

Experimental Investigations of Back-Scatter

Using one of the relations (1), (3), (4) or (5), the value of σ^0 may be derived from the measured value of the received power. These "experimental" values of σ^0 will be considered in the rest of this chapter.

The numerous parameters contained in (1), (3), (4) and (5) are not all known with great accuracy, and the power received by a radar is often difficult to evaluate. Also, it is wise to be cautious in comparing the absolute values of σ^0 obtained in different experiments. On the other hand, the varia-

† The reflection coefficient $d(R_d^2)$ is defined as the ratio of the back-scattered power to the power that would be received on a path of equal length $2r$.

‡ Cf., for example, Kerr [1951], pp. 483 ff.

tions of σ^0 in the course of the same experiment may be investigated more accurately, and especially its dependence on the following parameters:

—the grazing angle γ (or angle of incidence θ),†

—the wavelength λ,

—the polarization of the incident wave,

—the nature of the terrain in the case of back-scattering by the ground, and the state of the sea, wind speed and direction in the case of back-scattering by the sea.

Experience shows that both for dry ground and the sea, the dependence of σ^0 on γ (or on θ) is of the general form shown in Fig. 18.1. This means

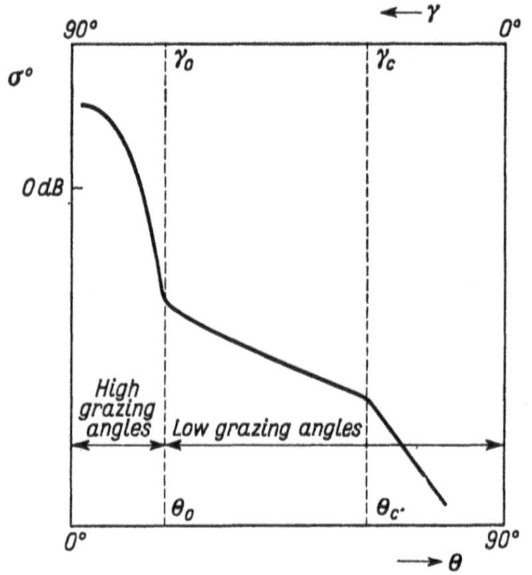

Fig. 18.1. General shape of the dependence of σ^0 on the grazing angle γ (or the angle of incidence θ). γ_0 is generally of the order of some tens of degrees, γ_c is of the order of one degree.

† We assume that the scattering cross-section σ^0 depends only on γ and not on the distance r from the radar or on its height h above the ground. This assumption is theoretically justifiable if the wave incident onto any part of the reflecting surface may be considered plane, so that it is then uniquely defined by γ. This assumption, which is implicitly contained in (1), (2) and (3) has been the subject of various experimental verifications: e.g. Kerr [1951, p. 490]; Katzin [1957], who performed measurements at different heights h, shows that σ^0 depends only on the ratio r/h. However, this is only true if the state of the sea, which also affects σ^0, is independent of r; this is why the experimental verification cannot be carried out by radars located on the coast, since the state of the sea varies rapidly with increasing distance from the shore. The experiments described by Kerr [1951] and Katzin [1957] were performed on board of an aircraft, the radar being pointed towards a fixed point on the sea.

that two domains of the variation of σ^0 as a function of γ should be distinguished:

Large values of γ $(\gamma > \gamma_0)$: σ^0 increases rapidly with growing γ, i.e. the scattering is very directional.

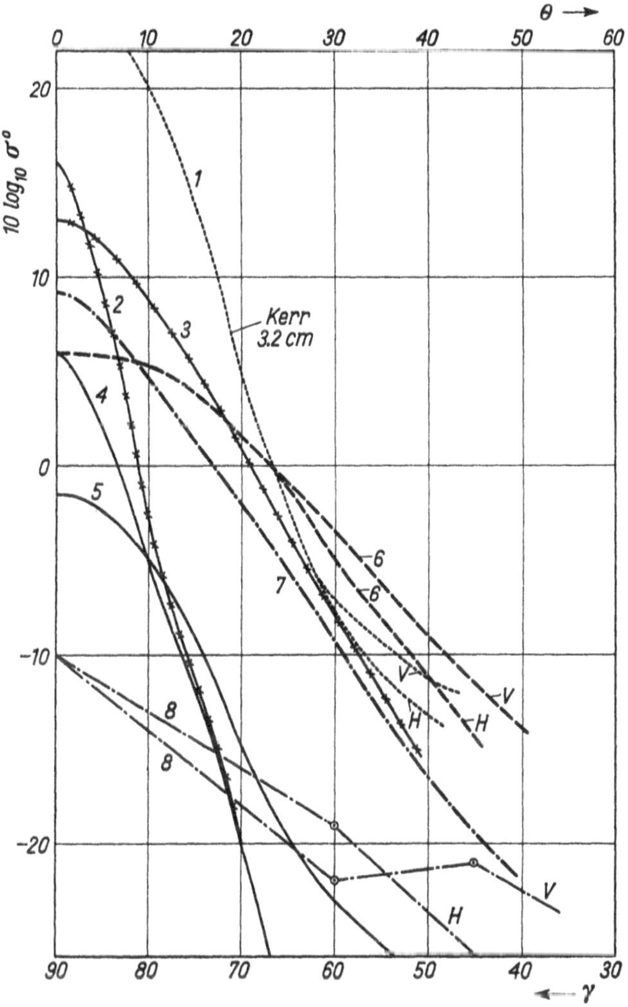

Fig. 18.2. Experimentally measured curves of $\sigma^0(\gamma)$ (cf. Table 18.2).

Small values of γ $(\gamma < \gamma_0)$: σ^0 varies very slowly with growing γ, i.e. the scattering is not very directional. However, σ^0 decreases rapidly for $\gamma < \gamma_c$ (γ_0 is of the order of some tens of degrees, γ_c of the order of one degree).

Experimental data show that σ^0 has very different properties in these two domains: for example, σ^0 depends very little on the frequency and the polarization for large values of γ, but for small values of γ, it depends on them very strongly.

We shall successively consider the properties of σ^0 in these two domains.

18.2. Back-Scatter by the Sea at Large Grazing Angles

Experimental Data

Table 18.1 lists the experiments the results of which we shall use in the following.

Figure 18.2 shows some experimental results in the form of curves giving the dependence of σ^0 on γ (or on the angle of incidence $\theta = \pi/2-\gamma$). The

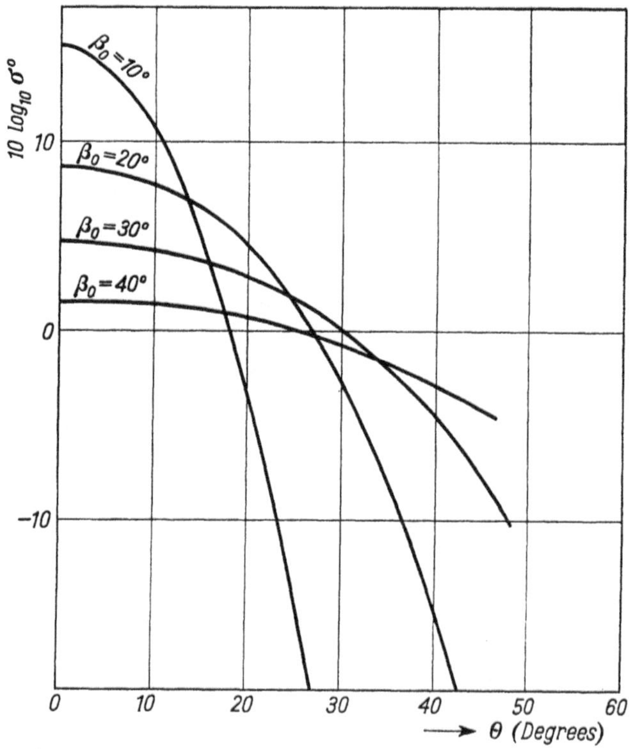

Fig. 18.3. Theoretical curves of $\sigma^0(\gamma)$ for a "discontinuous" (uniformly distributed) surface.

conditions under which these results were obtained are given in Table 18.2.

Some general conclusions may be drawn from these measurements, in particular from the curves of Fig. 18.2.

(1) The growth of σ^0 with increasing γ is the more rapid the more agitated

TABLE 18.1—Back-scatter from the Sea at High Grazing Angles

Reference	Wavelength or frequency	γ	Polarization	Method of observation	Sea and wind measurements	Other measurements
Cowan [1946] Kerr [1951]	3 cm	5–90°	H.V.	Pulses Airborne	Visual estimate of the sea state	
Maclusky and Davies [1945] Kerr [1951]	3 cm	10–90°	H.	Pulses Airborne	Visual estimate of the sea state	
Wiltse *et al.* [1957]	9.6, 15, 24, 26, 35, 38, and 48.7 Gc/s	15–90°	H.V. circular, cross polarizations	Ship C.W. (Doppler system)	Photographic observations of the sea	Doppler effect Comparison of upwind, downwind and crosswind measurements
MacDonald [1956]	1–2 and 9.3 Gc/s	1–90°	H.V.	Pulses	Wind and sea state measurements	Comparison of upwind, downwind and crosswind measurements
Grant and Yapplee [1957]	8.6 mm, 1.25 and 3.2 cm	10–90°	V.	C.W. (Doppler system)	Wind measurements	
Campbell [1958]	3.4 cm	40–90°	H.V. cross polarization	Airborne	Wind and sea state measurements	

TABLE 18.2

No. of curve	Bibliographic reference	Wavelength or frequency	Windspeed
1	Kerr [1951], Cowan [1946]	3 cm	12 m.p.h.
2	Grant and Yapplee [1957]	8 mm	5 m.p.h.
3	Grant and Yapplee [1957]	8 mm	15–20 m.p.h.
4	Grant and Yapplee [1957]	3 cm	5 m.p.h.
5	Grant and Yapplee [1957]	3 cm	15–20 m.p.h.
6	Wiltse *et al.* [1957]	1.25 cm	26 m.p.h.
7	Wiltse *et al.* [1957]	1.25 cm	4–10 m.p.h.
8	MacDonald [1956]	1250 Mc/s	34 m.p.h.

the sea (i.e. the stronger the wind). But for $\theta = 0$, the value of $\sigma^0(0)$ is the higher the calmer the sea (cf. curves no. 2 and 3; 4 and 5; 6 and 7).

(2) At the higher frequencies (10–50 Gc/s), σ^0 is almost independent of frequency for any γ [Wiltse *et al.* 1957]. But the curves given by MacDonald [1956] at 1250 Mc/s give much lower values of σ^0.

(3) For $\theta < 30°$, σ^0 is practically independent of polarization—horizontal, vertical or circular (Fig. 18.3, curves No. 1, 6 and 8).

(4) For reception of the cross-polarized component (transmitted wave polarized horizontally, reception of vertical component, or *vice versa*), the value of σ^0 is about 10 to 15 db below the value measured for identical polarizations [Wiltse *et al.* 1957, MacDonald, 1956].

(5) σ^0 is slightly larger for the upwind direction than for other directions.

Interpretation of Results; Theoretical Models

All of these results may be well explained by considering the surface of the sea to consist of undulations much larger than the wavelength, which, to a first approximation, are isotropic. The results of Chapter 12 may therefore be used; (19) and (20) of Chapter 12 then yield for the case of back-scatter

$$\mathrm{d}P_r = P_t \frac{g^2\lambda^2}{(4\pi)^3 r^4} G(\theta, \beta_0)\,\mathrm{d}S \tag{6}$$

where $\mathrm{d}P_r$ P_t, g, λ and r have the same meaning as in (1) of the present chapter, and where $G(\theta, \beta_0)$ is a function of the form

$$G(\theta, \beta) = \frac{1}{\tan^2\beta_0} G_1\left(\frac{\tan\theta}{\tan\beta_0}\right) \tag{7}$$

with θ the angle of incidence. The form of the function G depends on the chosen model of a rough surface: $G(\theta, \beta_0)$ is proportional to the probability that at a given point the slope of the reflecting surface is β, and β_0 is proportional to the standard deviation of β.

Comparing (1) and (6) we obtain

$$\sigma^0(\theta) = G(\theta, \beta_0) \tag{8}$$

We have so far only considered the effect of reflections from the earth's surface on radio links where the grazing angle was always very small. In that case it was not possible to derive the form of the function $G(\theta, \beta_0)$ from the experimental data. In Chapter 12, some "models" of the function G were compared and found to give very similar results.

However, in the case of back-scatter, the dependence of σ^0 on θ may be experimentally measured and Fig. 18.2 shows some of the resultant curves. These results should therefore enable us to choose the function $G(\theta, \beta_0)$, correctly representing the surface of the sea, *experimentally.*

Figure 18.3 shows the theoretical dependence of σ^0 on θ for a particular model: the normally distributed surface, studied in Part I, Chapter 5 and in Part II, Chapter 12, σ^0 is given by:

$$\sigma^0(\theta) = G(\theta, \beta_0) = \cot^2\beta_0 \exp\left(-\frac{\tan^2\theta}{\tan^2\beta_0}\right) \tag{9}$$

To account for the properties (1), (2), (3), (4), (5) of σ^0 listed above, any model may be used. But to use the correct model, we must consider

—direct measurements of the surface of the sea,

—the shape of the experimentally measured curves.

Experimental Data on the Surface of the Sea

The surface of the sea has been the subject of many theoretical and experimental investigations.† Schooley [1954] and Cox and Munk [1954] have measured the distribution of the slopes β of the surface of the sea by optical methods. According to Cox and Munk [1954] this distribution is to a first approximation normal (as in the normally-distributed model of surface we have just considered).‡ The standard deviation of this distribution is a function of the windspeed W; the mean values found by these authors have been plotted in Fig. 18.6.

In reality, the surface of the sea is not isotropic: its slopes are slightly higher in the upwind direction [Cox and Munk, 1954; Schooley, 1954; MacDonald, 1956].

Reduction of the Experimental Curves

We shall now attempt to identify the experimental curves $\sigma^0(\theta)$ of Fig. 18.2 as a set of curves corresponding to a particular theoretical function

† Cf. Kerr [1951], p. 486. Cf. also footnote † on p. 118.

‡ Cf. Part I, Appendix D (p. 193).

$G(\theta, \beta_0)$, e.g. to the set of Fig. 18.3. To do this, we shall compare the curves of Fig. 18.2 with the set of curves given by

$$\sigma^0(\theta) = \mu G(\theta, \beta_0) \tag{10}$$

where $G(\theta, \beta_0)$ is of the form (5) and μ is a coefficient equal to 1 if all of the incident energy is diffusely scattered, and smaller than 1 if part of this energy

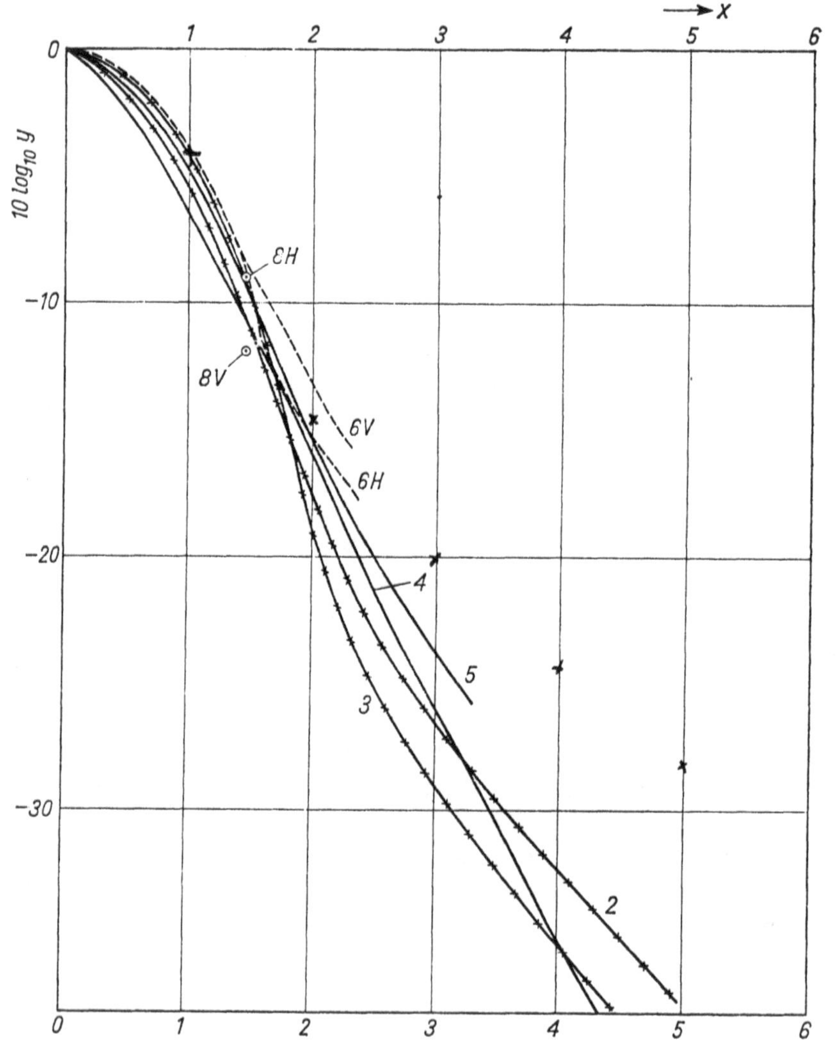

Fig. 18.4. Reduced curves with $y = \sigma^0(\theta)/\sigma^0(0)$ and $X = \tan\theta/\tan\beta_0$. Within the errors of measurement, the experimental curves coincide. The number of the curves are the same as in Fig. 18.2; their origin is given in Table 18.2.

is absorbed ($|R_0| < 1$) or specularly reflected. If this is so, then by (7) and (8),

$$\sigma^0(\theta) = \frac{\mu}{\tan^2 \beta_0} G_1\left(\frac{\tan\theta}{\tan\beta_0}\right)$$

and on introducing the transformation of variables

$$x = \frac{\tan\theta}{\tan\beta_0}$$
$$y = \frac{\sigma^0(\theta)}{\sigma^0(0)} = \frac{1}{G_1(0)} G_1\left(\frac{\tan\theta}{\tan\beta_0}\right) \tag{11}$$

all the experimental curves should coincide.

Figure 18.4 shows that it is in fact possible to find a transformation of the form (11) to make the experimental curves coincide reasonably well. Figure 18.5 shows that these "reduced curves" correspond quite well to

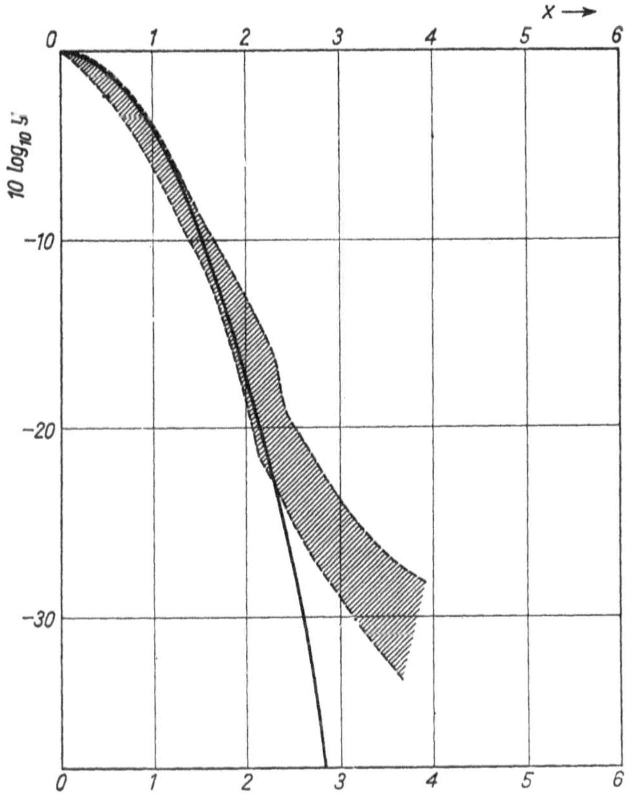

Fig. 18.5. Comparison of reduced experimental curves with the theoretical model of Fig. 18.5 (normally distributed slopes). The broken curves are the envelopes of the reduced experimental curves of Fig. 18.6, the full curve is the reduced curve obtained by applying the same transformation (10) to the theoretical curves of Fig. 18.5.

those obtained from the theoretical model of normally distributed slopes of the facets of the sea surface [cf. (9) and Fig 1.8.3]. These results are thus in good agreement with those optical measurements that also indicate a normal distribution of slopes.

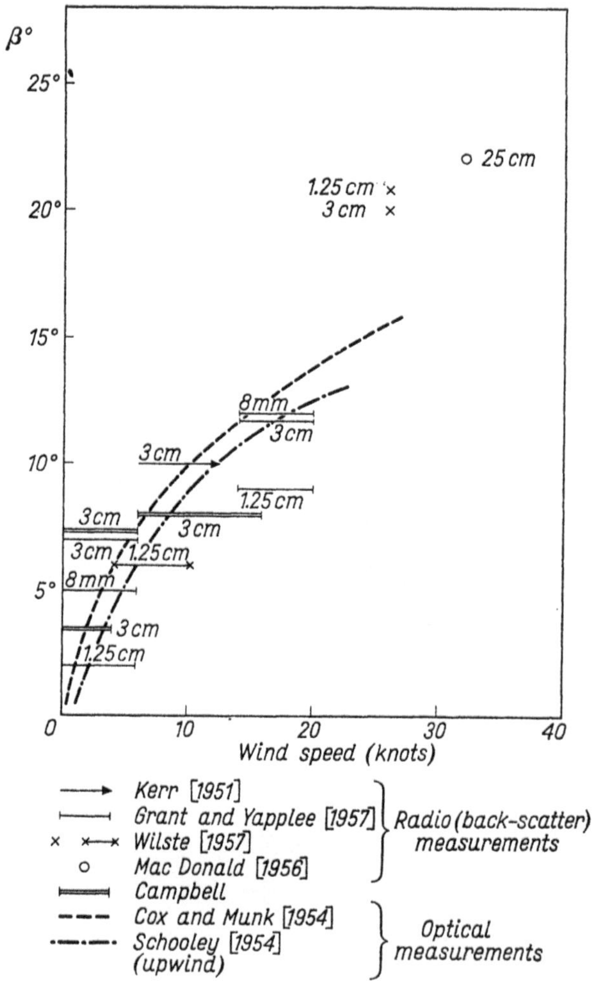

Fig. 18.6. Experimental values of β_0 as a function of the windspeed.

To obtain the superposition of the experimental curves, particular values of β_0 and μ had to be chosen.† The superposition has in fact been successfully

† The choice of $G_1(\theta)$ is arbitrary, but we have set $G_1(0) = 1$ to permit comparison with the theoretical "reduced curve" given by (8), for which $G_1(0)$ has this value.

tried on a larger number of curves than could be drawn in Figs. 18.2 and 18.4, and in this way a set of *experimental* values of β_0 and μ was obtained.

The dependence of some experimental values of β_0 on the windspeed obtained in this way has been plotted in Fig. 18.6. Considering the inaccuracy of these values, they may be regarded as being in good agreement with the curves derived from optical measurements. The distribution of the facet slopes β of the surface of the sea derived from measurements of back-scatter and from optical methods are thus in good agreement: in both cases the distribution is normal and the standard deviation β_0 obeys the same law. It will be noted that the experimental values of β_0 plotted in Fig. 18.6 seem to be independent of the frequency.

Some experimental values of μ have been plotted in Fig. 18.7 as a function of λ. Unfortunately these values of μ are very inaccurate (the error may be of the order of ± 5 db).† However, it appears that $10 \log_{10} \mu$ is of the order 0 in the millimetre wave band and much smaller in the decimetre

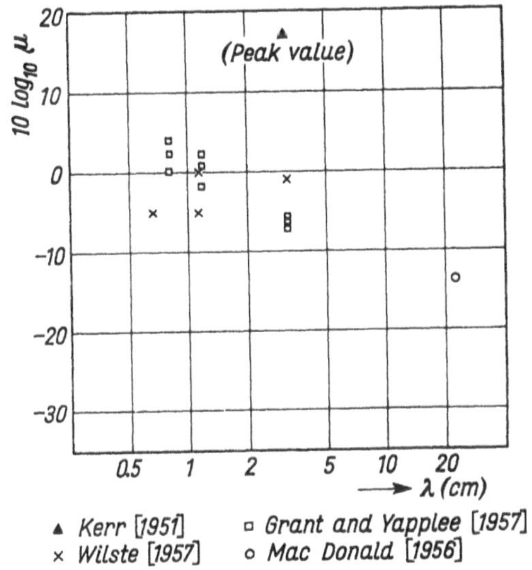

Fig. 18.7. Experimental values of μ as a function of the wavelength.

(UHF) band; this may be explained by noting that in the decimetre wave band part of the incident energy is reflected specularly. According to some data it seems that μ is independent of the state of the sea and the windspeed at the shorter wavelengths ($\lambda \leqslant 3$ cm).

† The measurement of the mean value of the field received by a pulse radar is necessarily always inaccurate and caution should be used in comparing the absolute values of σ^0 reported by different investigators.

Conclusion

It seems that the experimental results may be well accounted for by assuming σ^0 to be given by (10) with $G(\theta, \beta_0)$ given by (9); in the millimetre wave band, $\mu = 1$ and σ^0 is given by Fig. 18.3, the dependence of β_0 on the wind-speed being given by Fig. 18.7; in the decimetre wave band, σ^0 is reduced by a factor μ, which may be lower than -10 db; MacDonald [1956].

18.3. Back-Scatter by Terrain at Large Grazing Angles

Measurements of the back-scatter by terrain by means of airborne radar have been reported by Grant and Yapplee [1957] at wavelengths of 8.6 mm, 1.25 cm and 3.2 cm, by Campbell [1958] at 3.4 cm, Edison *et al.* [1960] at 415 and 3800 Mc/s, and by some others. The investigation of back-scatter by terrain is much less simple than that of back-scatter by the sea, for

—the types of terrain are very varied and it is very risky trying to represent them by a single model;

—terrain is less easily described statistically than the surface of the sea; the variance of the results is very large (sometimes exceeding 10 db).

Some conclusions may be drawn from the above experimental investigations.

(1) In the measurements at low frequencies [Edison, 1960] over terrain which is not entirely covered by vegetation (desert, agricultural areas, inhabited areas), a rapid decrease of σ^0 with increasing θ is observed as in the case of the sea. Considering the inaccuracy of the experimental results, it is always possible to make the measured curve coincide with the theoretical curve $\sigma^0(\theta)$ corresponding to any model of a rough surface within the errors of measurement.† The experimental curves given by Edison [1960] would coincide fairly well with the curves in Fig. 18.3 for β_0 between 10° and 20° according to the type of terrain and for an attenuation factor μ of the order one (0 db) at 3800 Mc/s, and of the order of -5 to -10 db at 415 Mc/s.

(2) In the measurements made over wooded terrain almost isotropic scattering is observed, i.e.

$$\sigma^0(\theta) = \sigma^0(0)\cos^2\theta$$

The values of $\sigma^0(0)$ are very variable from one experiment to another (-15 to -30 db at 3 cm) and increase with growing frequency [Grant and Yapplee, 1957].

---

† Many other models may be found in the literature; within the errors of measurement, they are always in agreement with the experimental values.

At 3 cm or shorter wavelengths, terrain covered with grass may also behave as an isotropic scatterer. The values of $\sigma^0(0)$ are of the same order for wooded regions.

(3) In the case of moist soil partially covered by grass, the dependence of σ^0 on θ often has the aspect of a superposition of curves of type (1) and (2): rapid decrease for small values of θ; then, for larger values of θ, slow decrease as in the case of an isotropic scatterer.

(4) In general, σ^0 increases with the frequency† [Grant and Yapplee, 1957; Edison, 1960].

(5) There is no significant difference between the values of σ^0 measured with horizontal and vertical polarizations. The cross-polarized component is from 3 to 10 db below the component of the same polarization [Campbell, 1958].

18.4. Back-Scatter by the Sea at Small Grazing Angles

A detailed study of this problem has been made by Goldstein [Kerr, 1951]. This analysis remains valid and fairly complete to the present day; the experiments performed since that time in general confirm Goldstein's conclusions. However, the measurements carried out by the N.R.L. [Katzin, 1955, 1956, 1957; MacDonald, 1956; Schooley, 1956] have made many points more exact and have enabled the theory of this problem to be understood more fully.

In the present account we shall only briefly state the points that the reader may find more fully treated in Goldstein's work, and rather concentrate on more recent results.

Dependence on the Grazing Angle

Experience shows that if a radar located at a height h measures the energy scattered by the surface of the sea at a distance r, the value of σ^0 derived from P_r and relation (4) depends only on the ratio h/r, i.e. on the grazing angle γ‡ [Kerr, 1951; Katzin, 1957]. The dependence of σ^0 on γ has the general form of Fig. 18.1. More precisely, the dependence of $\log \sigma^0$ on $\log \gamma$ may be represented by two straight segments intersecting at $\gamma = \gamma_c$ [Kerr, 1951; Katzin, 1957]. They may be defined by four parameters:

—γ_c, which is of the order of one degree;

---

† Grant and Yapplee [1957] find σ^0 minimum for $\lambda = 1.25$ cm over moist ground, but attribute this to the absorption by water-vapour.

‡ Cf. Section 18.1, footnote (p. 400).

—the slopes n_1 and n_2 of the straight lines for $\gamma > \gamma_c$ and $\gamma < \gamma_c$ respectively. The variation of σ^0 in these two intervals is thus given by

$$\sigma^0 \propto \gamma^{n_1} \quad \text{and} \quad \sigma^0 \propto \gamma^{n_2}$$

In general, n_1 varies between 0 and 1.5, and n_2 between 2 and 4;

—the values of σ^0 for a particular value of γ, e.g. $\gamma(1°)$, which generally lies between -40 and -80 db.

However, to arrive at more accurate results than these orders of magnitude, the dependence of these four parameters, viz. γ_c, n_1, n_2, $\sigma^0(\gamma_0)$ on the state of the sea, the polarization and the frequency must be considered.

Dependence on the State of the Sea

The value of σ^0 will in general increase for any γ as the sea becomes more rough.

In a set of experiments described by Davies and MacFarlane [1946],† the height of the waves Δh was measured from a submarine, while measurements were taken by a radar for values of γ between 0.5 and 5° at 3.2 and 1.25 cm. As Δh increases, a rapid increase of σ^0 is observed, until for large values of Δh a kind of "saturation" sets in: the quantity σ^0/γ then remains constant for any γ and Δh. At 10 cm this saturation effect seems to appear only for much higher values of Δh.‡

In another set of experiments described by Goldstein,§ measurements were taken at a wavelength of 10 cm for different states of the sea; Δh was not measured, but the value of $\sigma^0(1°)$ may be considered as an indication of the state of the sea. These experiments, performed with an airborne radar, enable the curve of the dependence of σ^0 on γ to be drawn for $\gamma < 5°$; this curve may then be approximated by two straight segments of slopes n_1 and n_2 (Fig. 18.1). It is then found that as the sea becomes more rough, i.e. as $\sigma^0(1°)$ increases,

—σ^0 increases for all values of γ;

—γ_c decreases ($\gamma_c = 4°$ for the calmest, and 1° for the roughest sea);

—n_1 and n_2 vary only slightly: n_1 remains in the range of 0 to 1.3, and n_2 between 2 and 3.8, but these variations do not seem to be correlated with the state of the sea.

The dependence of σ^0 on the state of the sea thus has the form schematically indicated in Fig. 18.8.

† Cf. also Kerr [1951], p. 513.

‡ Kerr [1951], p. 513.

§ Kerr [1951], pp. 508 ff.

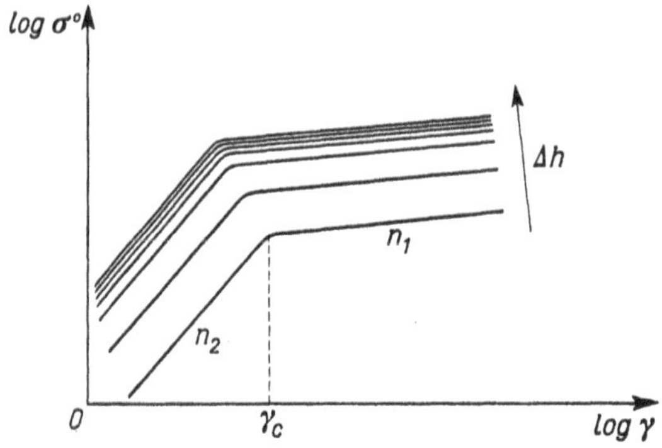

Fig. 18.8. Effect of the state of the sea. Δh is the height of the waves.

Other measurements carried out by the N.R.L. have shown that there is a good correlation between the values of σ^0 and the windspeed W [Schooley, 1956; Katzin, 1957]; σ^0 may be proportional to W or W^2. The same experiments have also shown that the value of σ^0 changes with the direction of the wind [MacDonald, 1956]. If σ_u^0 denotes the upwind value of σ^0, σ_d^0 the downwind value, and σ_c^0 the crosswind value, then

$$\sigma_u^0 > \sigma_d^0 > \sigma_c^0$$

for vertical polarization, and

$$\sigma_u^0 > \sigma_c^0 > \sigma_d^0$$

for horizontal polarization. The differences between σ_u^0, σ_c^0 and σ_d^0 are of the order of 0 to 5 db.

Dependence on Frequency

To study the effect of frequency on σ^0, one generally compares the experimental values of σ^0 for two values of λ. It is convenient to assume that σ^0 is proportional to λ^k and to express the result in terms of the value of k.

Goldstein reports several experiments where the values of σ^0 were compared either for $\lambda = 1.25$ and 3.2 cm, or for $\lambda = 3.2$ and 9 cm. The values obtained for k are very variable; they depend on the state of the sea and vary from -3 to -4 for a calm sea and from 0 to 1 for a rough sea. They also depend on the polarization: k is larger for horizontal polarization. Finally, they depend on γ: the knee (change of slope) appearing for $\gamma = \gamma_c$, where γ_c is of the order of 1 to 4° for $\lambda = 10$ cm, does not appear or appears for much smaller values of γ for $\lambda = 3.2$ or 1.25 cm. Thus γ_c *is an increasing function of frequency*. However, in the experiments at 1.25 and 3.2 cm where this knee does not appear ($\gamma > \gamma_c$), the values of k are independent of γ. This is equivalent to saying that n_1 *is independent of frequency.*

Katzin [1957] describes a series of experiments made at 3.2, 9 and 24 cm wavelength by the N.R.L. According to Katzin the curves of the dependence of σ^0 on γ obtained at different frequencies may be made to coincide by a transformation of coordinates with the ratio γ/λ as abscissa. This means that

—n_1 and n_2 are independent of frequency,

—γ_c is proportional to λ.

These results confirm and implement those obtained by Goldstein.

Dependence on Polarization

This problem has been studied in detail by Goldstein and further precision is due to the measurements by the N.R.L. [MacDonald, 1956].

The polarization does not have the same effect for any state of the sea or at any wavelength. For a *calm sea*, σ^0 is larger for horizontal polarization than for vertical polarization; the ratio σ^0_V/σ^0_H increases with decreasing frequency: for $\lambda = 9$ cm, the ratio σ^0_V/σ^0_H is of the order of 15 to 25 db; for $\lambda = 3$ cm, σ^0_V/σ^0_H does not exceed 10 db; for $\lambda = 1.25$ cm there is no appreciable difference between the two polarizations. For a *rough sea*, σ^0_V/σ^0_H varies between 0 and 5 db for $\lambda = 9$ cm, and between -5 and 0 db for $\lambda = 3.2$ cm.

Theoretical Interpretation

The properties of back-scatter are very different at high and at low grazing angles. We have seen that the properties of back-scatter, at high grazing angles could be satisfactorily explained by assuming that the waves on the surface of the sea are much larger than the wavelength. However, back-scattering by the sea at low grazing angles, which is much less directional (Fig. 18.1), has the property of diffuse scattering by a surface with irregularities which are much smaller than the wavelength. How can the simultaneous presence of these two effects be explained?

The surface of the sea consists of both large undulations, the waves (sea, swell), and small undulations, ripple (Fig. 18.9a). The majority of authors writing on the subject of back-scatter represent the surface of the sea by reflecting facets; the sea should therefore be represented by a set of large facets, representing the waves, which are themselves decomposable into smaller facets representing ripple (Fig. 18.9b). It is assumed that the large and small facets reflect independently as though they were separated. Thus the surface of the sea may be replaced by two fictitious surfaces (Fig. 18.9c), one of which consists of large facets and gives rise to very directional back-scatter, whereas the other consists of small facets and gives rise to back-scatter of small directivity, which will dominate at low grazing angles.

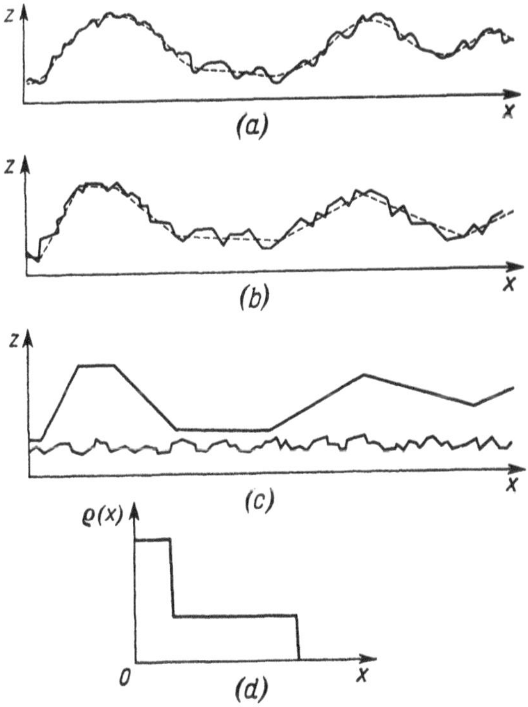

Fig. 18.9. The sea surface (a) and the models by which it may be represented: decomposition into facets (b), separation of large and small facets (c), Voge's model (d).

This resolution of the back-scattered radiation into two components may be justified in a more rigorous manner by applying a paper by Voge,† in which the author considers the scattering of waves by a surface of the type shown in Fig. 18.9a. Instead of decomposing the surface into facets, it is defined by the autocorrelation function $\rho(x)$ of the heights $z(x)$ of its points; to respect the existence of large and small undulations, the function $\rho(x)$ is introduced in the form of Fig. 18.9d. The author then shows rigorously that the scattered radiation contains two components: the one called "quasi-specular" is scattered as if the surface consisted only of large undulations; the other, called "diffuse", is scattered as if it consisted only of small undulations.

† Cf. 4th part of Du Castel, Misme, Spizzichino and Voge [1960]. Voge's study concerns the scattering by tropospheric sheets, but the model of the reflecting surface may also represent the surface of the sea.

On the basis of the hypothesis of a surface that is decomposable into small facets, some tentative assumptions have been made to explain the properties of back-scatter at low grazing angles theoretically. Katzin [1957] makes the following assumptions:

—The surface of the sea is decomposable into facets that for purposes of calculation may be regarded as plane disks;

—The distribution of the slopes of these facets is that obtained by Cox and Munk [1954] experimentally by optical methods. As we have seen, this distribution is to a first approximation normal, the dependence of its standard deviation on the windspeed W being given by Fig. 18.6. To a second approximation, it is slightly asymmetrical and depends on the direction of the wind.

—The distribution of the dimensions of these facets remains arbitrary, but if it is given, σ^0 is a given function of λ. Setting

$$\sigma \propto \lambda^{-1}$$

(the mean experimentally measured dependence), the distribution of the dimensions of the facets is fixed.

Under these assumptions, Katzin [1957] obtains the result

$$\sigma^0 \approx 9 \times 10^{-3} N_0 \lambda^{-1} W^{3/4} (1 \pm 2 \times 10^{-3} W)$$

where N_0 is a constant (experimental value: 7.2×10^{-5}) and the double sign corresponds to the upwind (+) or downwind (−) directions.

This theoretical result accounts very approximately for the principal properties of σ^0 for $\gamma > \gamma_c$. But it underestimates the effect of the wind direction and does not bring out the interdependence of the effects of different parameters: we have seen, for example, that the dependence of σ^0 on the frequency is very different for different states of the sea and different polarizations; that the effect of the wind direction may be reversed by a change in polarization, etc. All this leads one to believe that to account for *all* experimental results, a much more complex theory than that by Katzin would have to be worked out.†

The rapid decrease of σ^0 with γ for $\gamma < \gamma_c$ may be explained by assuming that the wave incident onto a scatterer results from the interference of the direct wave with the waves scattered by the rest of the sea surface.‡ Let

---

‡ Katzin's theory is qased on many simplifying assumptions, in particular, the distributions of the slopes and dimensions of the facets are assumed independent, shadowin effects are neglected, etc.

† Katzin [1957 .

H be the height of a scatterer above the mean level of the sea surface; it is then easily shown that if

$$\gamma > \gamma_c \quad \text{with} \quad \gamma_c = \frac{\lambda}{4H}$$

then the scatterer is in an interference zone where the resultant field passes through a series of maxima and minima: the entire set of scatterers will then receive a field which on the average equals the direct field and the effect of the field scattered by the sea surface may be neglected.

On the other hand, if $\gamma < \gamma_c$ defined above, there will always be destructive interference between the direct field and that scattered by the sea surface. The resultant field is essentially proportional to $1/r^2$. If the back-scattered power is proportional to γ^{n_1} in the absence of the multiply scattered field, it will be proportional to γ^{n_1+4} in the presence of destructive interference.

The principal conclusions reached in this way are in good agreement with experimental conclusions:

—γ_c is proportional to λ,

—$n_2 = n_1 + 4$ essentially corresponds to the values obtained experimentally [Katzin, 1956, 1957].

The increased effect of the polarization for $\gamma < \gamma_c$ and the frequently spiky aspect of the back-scattered echo may also be explained in this way.

CHAPTER 19

SCATTERING BY ATMOSPHERIC SHEETS

It is now well established that the lower atmosphere (troposphere) has a stratified structure: its humidity, temperature, etc., and hence its index of refraction undergo sharp changes across horizontal surfaces called "sheets".† An electromagnetic wave incident onto a sheet is partially reflected. Reflections by sheets play an important role in the propagation of ultra-short waves beyond the horizon.

This chapter begins with an account of the scattering of waves by tropospheric sheets.‡ We shall later see that other atmospheric layers (in the lower ionosphere and the exosphere) probably also have a structure of sheets, and that certain results obtained for tropospheric sheets may be extended to ionospheric and exospheric sheets.

19.1. Some Data on the Structure of the Troposphere

The problem of the structure of the troposphere has been approached from a theoretical point of view by Misme.§

For partial reflections to take place in the troposphere, sharp changes of the index of refraction are necessary. An acceptable order of magnitude of the variation of the index of refraction Δn at heights up to several kilometres is obtained from the formula

$$\Delta N \approx 0.3\Delta P - \Delta T + 7\Delta r \tag{1}$$

where $N = 10^6\Delta n$, ΔP is the variation in pressure in millibars, ΔT the variation of the temperature in degrees Kelvin (or Centigrade) and Δr the variation of the mixing ratio, defined as the mass of water vapour (in grams) mixed with 1 kg of dry air.

To study the inhomogeneities of the index ΔN of the atmosphere one

---

† French original: *feuillets* (*Transl. Note*).

‡ In this chapter we shall review the principal theoretical and experimental results obtained by a team of French research workers of the C.N.E.T. (Paris). The reader may find a more detailed account in the series of papers of Du Castel, Misme, Spizzichino and Voge [1960].

§ Cf. Parts I and II of Du Castel, Misme, Spizzichino and Voge [1960].

must therefore consider the variations of pressure, humidity and temperature in the atmosphere. The pressure varies regularly with the altitude so that for local inhomogeneities the term with ΔP is practically negligible in (1).

On the other hand the variations of humidity play an important part in the fluctuations of ΔN. The water vapour contained in the atmosphere is supplied by the earth: after evaporation it will tend to diffuse in a vertical upward movement so as to distribute itself over the entire atmosphere. If the atmosphere is calm, this diffusion proceeds very slowly under the influence of gravity, water vapour being lighter than dry air. However, if the atmosphere is turbulent, this diffusion is facilitated and proceeds about 10,000 times faster. From this it may be concluded that *in a turbulent region of the atmosphere, the mixing ratio varies little about a constant mean value. In a stable region, however, r may have a very strong vertical gradient.*

The temperature decreases with the height, but this decrease is not regular. Apart from rare exceptions, the vertical gradient of the temperature does not exceed a value of approximately 10°/km at any altitude; in fact it follows from the equations of thermodynamics that if the gradient exceeds this value—called "adiabatic gradient"—the atmosphere becomes unstable and the resulting turbulence tends to diminish the temperature gradient until it is again equal to the adiabatic gradient. On the other hand, if the temperature gradient is smaller than the adiabatic gradient, the atmosphere may remain stable.

Thus *in a turbulent region of the atmosphere the temperature decreases with growing altitude by* 10°C/km; *in a stable region, the decrease is always smaller than this value.*

From what has been said about the variations of humidity and temperature and from (1) it thus follows that

—in a turbulent zone, the vertical gradient of the index of refraction varies only slightly about a constant value which is found equal to

$$\operatorname{grad} N = 20\ N\text{-units/km}$$

—in a stable zone, the vertical gradient of the index of refraction always exceeds this value:

$$\operatorname{grad} N > 20\ N\text{-units/km}$$

Hence by sounding the atmosphere to measure the variations of r, T or N, it may be ascertained whether there are stable or turbulent zones in the atmosphere.

Atmospheric Soundings

There are two ways of sounding the atmosphere directly:

(1) All over the world, the various meteorological services take radio sonde observations by means of balloons and determine the values of pressure, temperature and humidity at different altitudes. Unfortunately the instruments of a radio sonde only provide values which are strongly averaged over space and time and the small inhomogeneities of the atmosphere escape their observation.

(2) The variations of the index of refraction may be measured directly by a *refractometer*. This instrument has the advantage of having virtually no time constant so that it may provide data on the fine variations of the index of refraction. Its disadvantage is that it does not separately indicate the variations of pressure, temperature and humidity, which are indispensable for a study of the physics of the atmosphere.

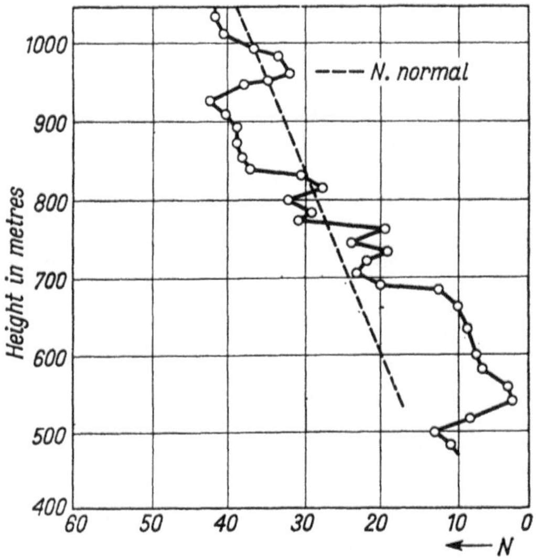

Fig. 19.1. Example of an atmospheric sounding by means of a radio sonde on a captive balloon (above the Mediterranean, 12 Dec. 1957). Fine fluctuations of the refractive index escape this type of observation. But, in the above case, important stable layers may be distinguished from turbulent layers in the regions where the vertical gradient of N has its "normal" value 20 N-units/km.

Figures 19.1 and 19.2 give some examples of the results of measurement by the two types of instruments. It is exceptional for very thick stable layers, characterized by a high refractive-index gradient, to occupy an appreciable fraction of the atmosphere. In general, the soundings show that the greater part of the atmosphere is occupied by turbulent layers (where grad

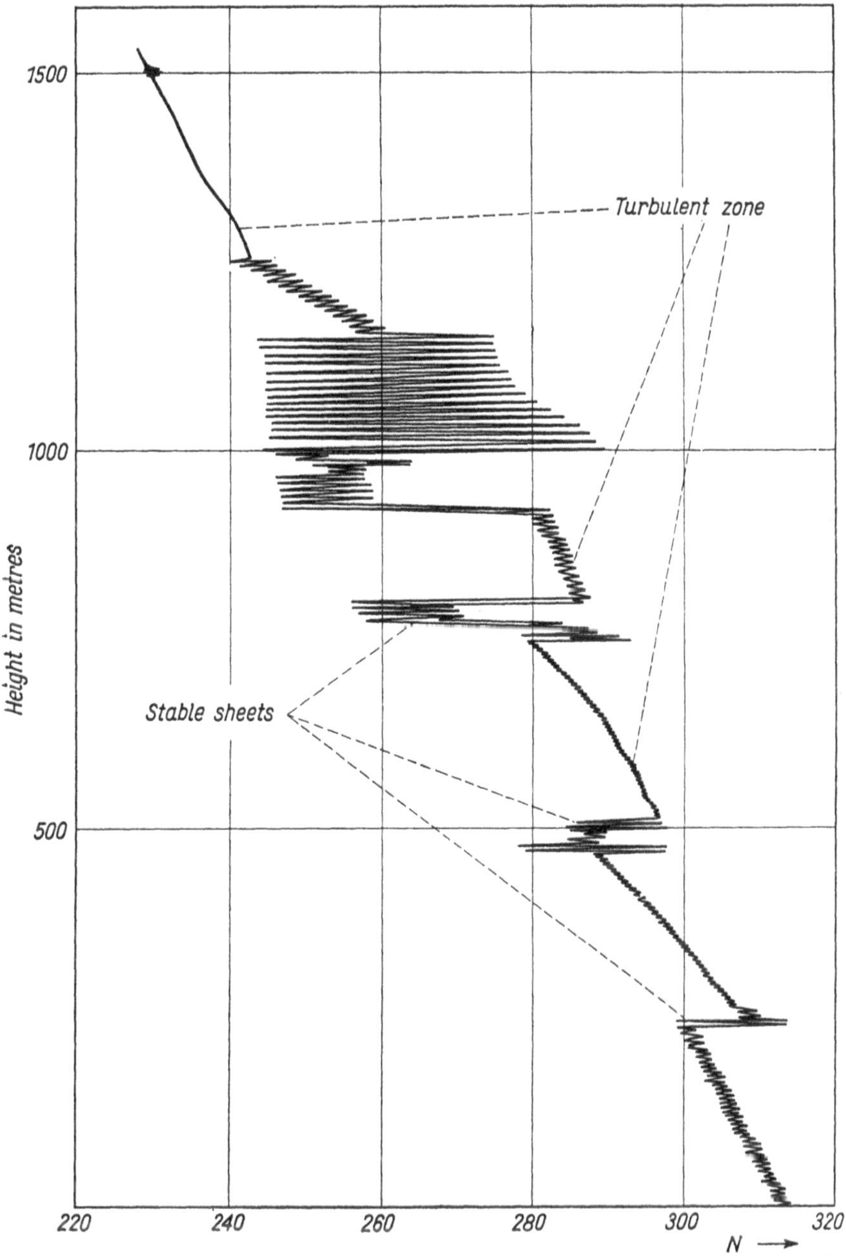

Fig. 19.2. Example of a sounding by means of an airborne refractometer (region of Tours, France, June 1957). Information on the fine fluctuations of the refractive index may be obtained in this way, but unlike the preceding method, these recordings are taken during an almost horizontal flight and certain rapid fluctuations recorded in the neighbourhood of stable sheets are probably due to the fact that a stable sheet, which is of an irregular shape, may be traversed several times during a horizontal flight.

$N = 20\,N$-units/km) within which there is a certain number of *stable sheets* of small thickness (Fig. 19.3). These *sheets,* which seem to exist most of the time at various altitudes, are characterized by variations of the refractive

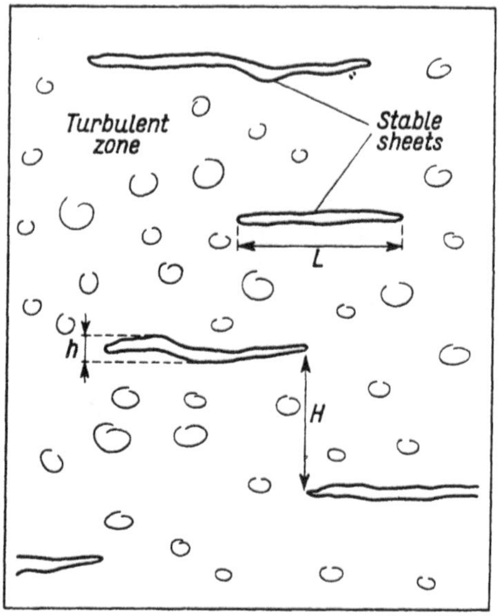

Fig. 19.3. Structure of the troposphere.

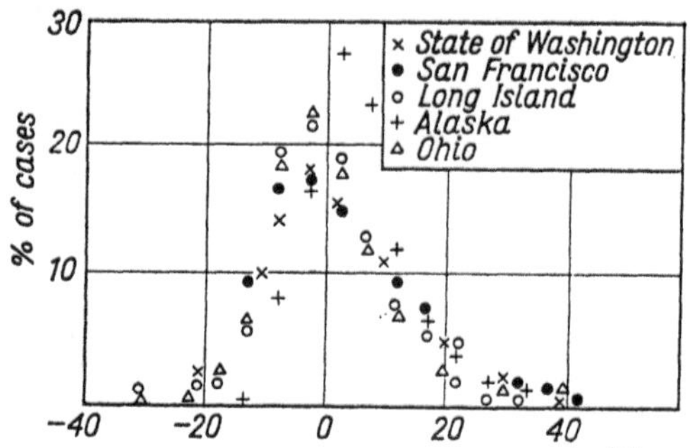

Fig. 19.4. Distribution of the values of the discontinuity ΔN in different regions (adapted from Smyth and Anderson [1957]).

index of the order of ten N-units over their thickness which rarely exceeds some tens of metres (Fig. 19.4).

The measurments taken by a refractometer carried by a climbing aircraft (Fig. 19.2) provide another item of information: if the aircraft climbs at a small angle with respect to the horizontal, the measurements show that it traverses the same stable sheet several times (it may be seen from Fig. 19.3 that when the aircraft traverses a stable sheet, N runs through its range of values several times, attaining its extreme values at the top and bottom of the sheet). This shows that the sheets have undulations: they are rough surfaces.

Radar Observations

Experiments carried out by the D.S.I.R. Radio Research Station (Slough, England) and described by Saxton [1960] have shown that these tropospheric sheets may be observed by radar.

(a)

(b)

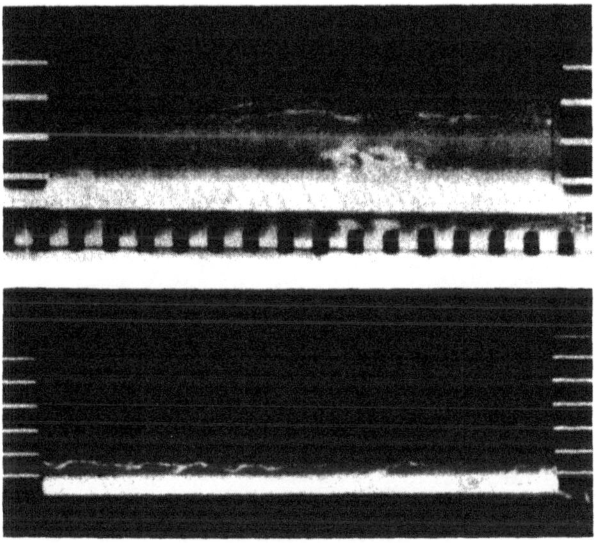

Fig. 19.5. Vertical soundings by a 10 cm radar. Height-time records with height markers at 1 km intervals. (a) 12.30–12.45 hours, 10.6.1959, (b) 12.00–12.15 hours, 30.4.1959 [Saxton, 1960]. (*By permission of the authors.*)

A radar at 10 cm wavelength pointed vertically upwards with very short pulse durations (0.1 μsec) was used. Its resolving power was thus 30 m. Figure 19.5 shows some examples of the echoes that were obtained. They are returned by very thin sheets the thickness of which is at times smaller and at times greater than the resolving power of the radar (30 m). These experiments have again demonstrated the irregular form of the sheets and have enabled their dimensions to be studied.

Characteristics of the Sheets

Some characteristic orders of magnitude of tropospheric sheets have been derived from these various experiments (Fig. 19.6).

The variation of the index of refraction ΔN across a sheet amounts to some tens of N-units.

The thickness e of a sheet is of the order of several metres.

Its length L is of the order of some 10 km.

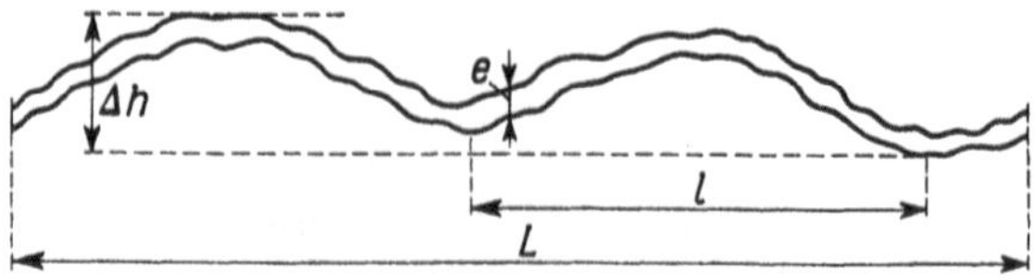

Fig. 19.6. Characteristics of a tropospheric sheet.

The sheets have large undulations of several kilometres length (l) and some tens of metres in height (Δh); smaller deformations due to turbulence are probably superimposed on them.

The sheets are present most of the time. A balloon sonde climbing vertically passes through an average of 2 or 3 sheets per kilometre.

19.2. Theory of Reflection by Tropospheric Sheets

Reflection by a Perfectly Plane Sheet

The "sheets" are thus zones of small thickness in which the index of refraction undergoes a sharp change. Can such a zone reflect electomagnetic waves?

In classical optics it is assumed that a partial reflection of light waves takes place at a *surface of discontinuity* between two media with different indices of refraction n, whereas a *continuous variation* in a medium does not produce a reflection of light. It may therefore be queried whether a sharp change of the index of refraction of the kind met in the "stable sheets" should be regarded as a discontinuity.

In general, the energy scattered or reflected by a volume V in which the index of refraction n is inhomogeneous, i.e.

$$n = n_0 + \Delta n(x, y, z)$$

(z axis vertical), is given by an expression of the form

$$E \propto E_0 \int_V \Delta n(x, y, z) \exp\left[\frac{2\pi i \delta(x, y, z)}{\lambda}\right] dx dy dz \tag{2}$$

where $\delta(x, y, z) = TP + PR$ is the optical path of a ray emitted by the transmitter and propagated to the receiver R after being scattered at a point $P(x, y, z)$ of the volume V. The dimensions of the volume V are assumed small with respect to the distance from the transmitter T or receiver R to any point of V. E_0 denotes the free-space field.

Using (2) to calculate the field reflected by a perfectly plane sheet located between the horizontal planes $z = Z_1$ and $z = Z_2$, we obtain the expression for the reflection coefficient R_0 of the sheet † in the form

$$R_0 \propto \int_{Z_1}^{Z_2} \Delta n(z) \exp\left(\frac{4\pi i \gamma z}{\lambda}\right) dz \tag{3}$$

where γ is the grazing angle of the incident ray, which is assumed to be small.

Equation (3) may be written as ‡

$$R_0 = \frac{\gamma^2}{2}\left\{\left[n + i\frac{\lambda}{4\pi\gamma}\frac{dn}{dz} - \left(\frac{\lambda}{4\pi\gamma}\right)^2 \frac{d^2n}{dz^2} - i\left(\frac{\lambda}{4\pi\gamma}\right)^3 \frac{d^3n}{dz^3} + \dots\right] e^{4\pi i\gamma z/\lambda}\right\}_{z=Z_1}^{z=Z_2} \tag{4}$$

This expression shows that R_0 may assume high values if n or one of its derivatives dn/dz, d^2n/dz^2, ..., etc., varies rapidly in the interval $[Z_1, Z_2]$ which is sufficiently narrow for the exponential $\exp(4\pi i\gamma z/\lambda)$ to retain approximately the same phase for $z = Z_1$ and $z = Z_2$. This condition will be assumed satisfied for

$$Z_2 - Z_1 = \frac{1}{4}\frac{\lambda}{\gamma} \tag{5}$$

We may thus conclude that *a reflection will be produced at any level where the index of refraction or one of its derivatives pass through a vertical discontinuity, denoting by "discontinuity" an important change taking place over an interval*

$$\varepsilon < \frac{\lambda}{4\gamma} \tag{6}$$

where λ/γ is the space wavelength. We shall call ε the *thickness* of the discontinuity.

Application to Tropospheric Radio Links

In order that a sheet may reflect the waves transmitted by a transmitter T towards a receiver R, it must be within the volume common to both antenna beams (Fig. 19.7). As we shall see, it must be near the bottom of this common

† To simplify the notation, we write, in this chapter, R_0 for the *absolute value* of the reflection coefficient.

‡ Equation (4) is obtained by reiterated integration by parts of the integral (3).

volume; the grazing angle γ is therefore near the angle γ_0, which equals half the angle concluded by the tangents to the earth at T and R:

$$\gamma \approx \gamma_0 = \frac{d}{2a} \tag{7}$$

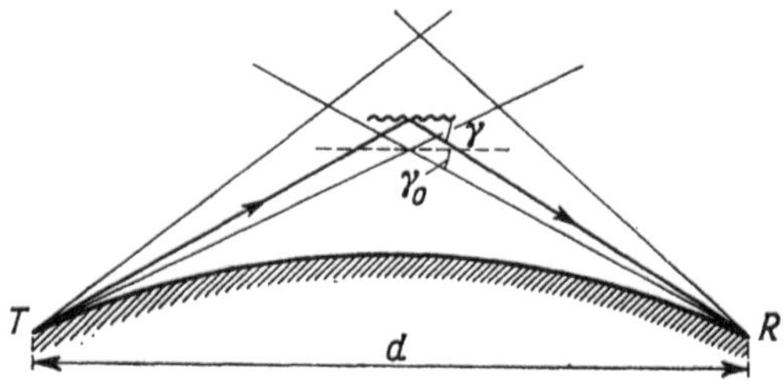

Fig. 19.7. Reflection by a sheet in a tropospheric radio link.

where d is the pathlength and a the radius of the earth.† In the usual tropospheric links, γ as given by (7) is of the order of hundredths of a radian, and λ is at most equal to 1 m; the right side of (6) is therefore at most of the order of a few metres. The thickness of a sheet (a few metres) is therefore

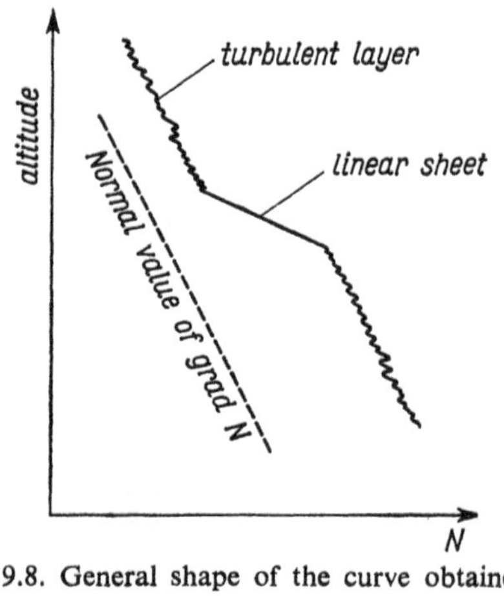

Fig. 19.8. General shape of the curve obtained by a refractometer in traversing a linear sheet.

often too large for the variation of n across the sheet to produce a reflection. But soundings show that the variation of n in a sheet often has the "linear" aspect of Fig. 19.8, and one is led to believe that partial reflections may be due to the discontinuities of the first derivative dn/dz at the top and bottom

† The symbol a is here used to denote the effective radius of the earth as obtained by respecting atmospheric refraction.

of the sheet, the "thicknesses" ε of which are always small compared with $\lambda/4\gamma$. We shall for the moment retain this assumption.

If $\varepsilon < \lambda/8\gamma$, the variation of the exponential factor in (4) may be neglected in the interval ε, so that we may write

$$R_0 \approx \frac{\lambda}{8\pi\gamma^3}\left[\frac{dn}{dz}(Z_2) - \frac{dn}{dz}(Z_1)\right] \tag{8a}$$

If, on the other hand, $\varepsilon > \lambda/8\gamma$, it may be shown that (4) leads to expressions for R_0 of the form

$$R_0 \approx \frac{\lambda}{8\pi\gamma^3}\left[\frac{dn}{dz}(Z_2) - \frac{dn}{dz}(Z_1)\right] F\left(\frac{\gamma\varepsilon}{\lambda}\right) \tag{8b}$$

where $F(\gamma\varepsilon/\lambda)$ is a function which depends on the form of the function $n(z)$. Figure 19.9 gives examples of such functions.

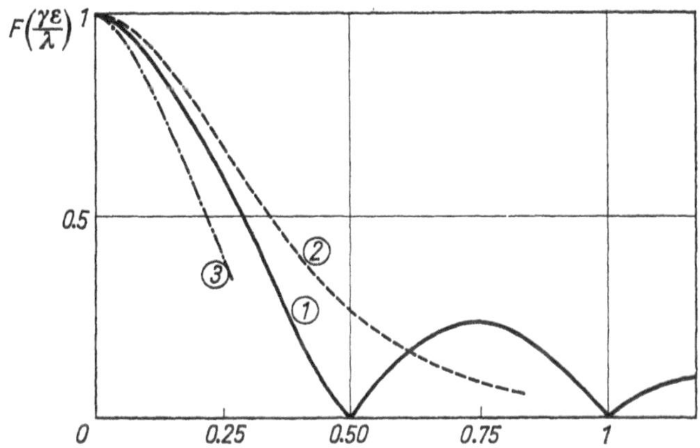

Fig. 19.9. Values of the function $F(\varepsilon\gamma/\lambda)$ determining the reflection coefficient of a layer, for different functions grad $N(z)$ in the interval of discontinuity (z_1, z_2, ε). Curve 1: grad N is a linear function of z; curve 2: grad N is a quadratic function of z with d^2N/dz^2 continuous; Curve 3: $d^2N/dz^2 = \cosh^{-2}(z/\varepsilon)$. As may be seen, the three curves lie close to each other for $\varepsilon\gamma/\lambda < 1/4$.

In practice, the experimental values of ε seem to be generally of the order of 1 metre. Hence

(1) In the UHF (decimetre) band and for the usual pathlengths (up to 300 km) it is certain that $\lambda/8\gamma$ is greater than 1 m; R_0 is given by (8a).

(2) In the SHF (centimetre) band or for very long paths, R_0 is given by (8b). In the usual cases, $\lambda/\gamma > 3$ m, and if it is assumed that $\varepsilon \approx 1$ m, the corresponding domain is given by the left part of Fig. 19.9, where the different functions $F(\gamma\varepsilon/\lambda)$ assume comparable values.

Tropospheric Sheets as Rough Surfaces

Equation (8a) and (8b) give the reflection coefficient of a sheet assumed to be perfectly plane. Applying (2), it may be shown rigorously that the reflection coefficient of a rough sheet is proportional to R_0:†

$$R = \rho R_0 \tag{9}$$

where ρ is the value of R for $R_0 = 1$; it is the reflection coefficient of a perfectly conducting surface of the same form as the sheet.

The expressions for ρ have been derived in Part I of this monograph.‡§ It has been shown that the secondary radiation consists of two components: specular reflection and diffuse scattering.

Specular Reflection by a Sheet

According to (9), the reflection coefficient of the specularly reflected component is given by

$$R_s = \rho_s R_0 \tag{10}$$

where ρ_s^2 would be equal to

$$\exp\left[-\left(\frac{4\pi\gamma\Delta h}{\lambda}\right)^2\right]$$

if the sheet covered the entire Fresnel ellipse.

In reality, only part of the Fresnel ellipse lies inside the sheet. The axes A and B of the first Fresnel ellipse are given by‡‡

$$A = \sqrt{(\lambda d)}; \quad B = \frac{\sqrt{(\lambda d)}}{\gamma}$$

In the usual case of tropospheric links, i.e. d from 200 to 400 km, λ between 10 cm and 3 m, and the length L of the sheet some 10 km,† we have

$$A < L < B$$

so that only a fraction L/B of the first Fresnel ellipse lies within the area of the sheet. Hence

$$\rho_s^2 = \left(\frac{L}{B}\right)^2 \exp\left[-\left(\frac{4\pi\gamma\Delta h}{\lambda}\right)^2\right]$$

† In this chapter, R, R_0 and ρ stand for the absolute values of the mean reflection coefficients.

‡ Cf. Part I, Section 5.3.

§ The above definition of ρ is not identical with, or even equivalent to that given in Part I, Section 3.1; however, this difference in no way affects the further treatment of this chapter (*Transl. Note*).

‡‡ Cf. Annex to Chapter 12.

† Cf. Section 19.1

or

$$\rho_s^2 = \frac{L^2\gamma^2}{\lambda d} \exp\left[-\left(\frac{4\pi\gamma\Delta h}{\lambda}\right)^2\right] \tag{11}$$

Diffuse Scattering by a Sheet

The reflection coefficient of the diffusely scattered component is according to (9)†

$$R_d = \rho_d R_0 \tag{12}$$

where ρ_d is given by (23) of Chapter 12, which when applied to a tropospheric link, becomes‡

$$\rho_d^2 = \frac{4}{\pi d^2} \int_\Sigma G(\beta, \beta_0)\, \mathrm{d}S$$

where Σ is the surface of the sheet. If $G = G_0$ as given by (18) of Chapter 12, we obtain

$$\rho_d^2 = \frac{4S}{\pi d^2 \beta_0^2} \tag{13}$$

where S is the area of that portion of the sheet for which

$$\beta < \beta_0$$

where β is the angle concluded by the bisector of the angle TPR and the vertical at a point P of the sheet.

Physically, the area S is that part of the sheet where there can be reflecting facets sufficiently inclined to reflect a wave from the transmitter to the receiver: this is the "*useful area*" which effectively participates in the reflection.

A point P near the centre 0 of the path, is given by

$$\tan \beta = \frac{2x}{\gamma d}$$

where the x axis is horizontal and perpendicular to the path and the origin is at the path midpoint 0. The area S is then found to be limited by the length L of the sheet in the direction of the path; in the direction of the x axis its length is

$$X = \gamma d \tan \beta_0 \approx \gamma d \beta_0 \tag{14}$$

† R_d and ρ_d denote the absolute values of the mean reflection coefficients in this chapter.

‡ To apply (23) of Chapter 12 to a tropospheric link, it may be assumed that (1) $r_1 = r_2 = d/2$ and $r = d$, where d is the pathlength; (2) the sheet is entirely contained within the volume common to both antenna beams (i.e. its dimensions are much smaller than those of the common region). In this case $g_1 = g_{10}$ and $g_2 = g_{20}$.

Substituting for S in (13), we obtain

$$\rho d^2 = \frac{4}{\pi} \frac{\gamma L}{d\beta_0} \tag{15}$$

19.3. Tropospheric Scatter Propagation—the Mean Field beyond the Horizon

Since the atmosphere contains stable sheets capable of reflecting waves, and turbulent layers capable of scattering them, the field on tropospheric beyond-the-horizon paths consists of three components: specular reflection, diffuse reflection and turbulent scattering.† We shall first compare the properties of these three components.‡

Orders of Magnitude of the Three Components

It is difficult to compare the reflected and scattered fields: the expressions for the reflection coefficient and for the scattering cross-section§ contain quantities that vary over a wide range. Nevertheless, it appears that

(1) Specular reflection by a sheet will set in only if

$$\Delta h < \frac{\lambda}{2\gamma}$$

(Rayleigh criterion). But if this condition is satisfied, it may be shown that specular reflection will dominate over other phenomena (in effect, all of the reflected energy is concentrated in a narrow beam). Specular reflection is characterized by a particularly high signal level.

Whenever there is a sheet of thickness Δh in the volume common to both antenna beams, the received field is due to specular reflection, provided that the frequency is smaller than a cut-off frequency f_c defined by

$$f_c = 2\gamma\Delta h \tag{16}$$

† To distinguish diffuse scattering by tropospheric sheets (i.e. by a rough surface) from the scattering by turbulent layers (i.e. by an inhomogeneous volume), we shall in the rest of this chapter designate the former by "diffuse reflection" rather than by "diffuse scattering" as we have done so far, reserving the expression "scattering" for the propagation mechanism associated with atmospheric turbulence (*Transl. Note*).

‡ For this comparison we shall use a very simplified theory of turbulent scattering to clarify the fundamental differences between these three components.

§ Cf. Booker, H. G. and Gordon, W. E.: A Theory of radio scattering in the troposphere. *Proc. I.R.E.* **38**, 401–412 (1950); Staras, H. and Wheelon, A. D.: Theoretical research on tropospheric scatter propagation. *Trans. I.R.E.* AP-7, 80–87 (1959).

(2) Diffuse reflection and scattering give rise to fieldstrengths of the same order.† It thus seems that in the absence of specular reflection, one or the other of these two effects will dominate in accordance with the characteristics of the path (distance, frequency, etc.) and the climatic and meteorologic conditions.

Comparison of Frequency Dependences

The specularly reflected field decreases rapidly with the frequency f and vanishes for $f > f_c$. Considering the usual orders of magnitude of Δh (Sec-

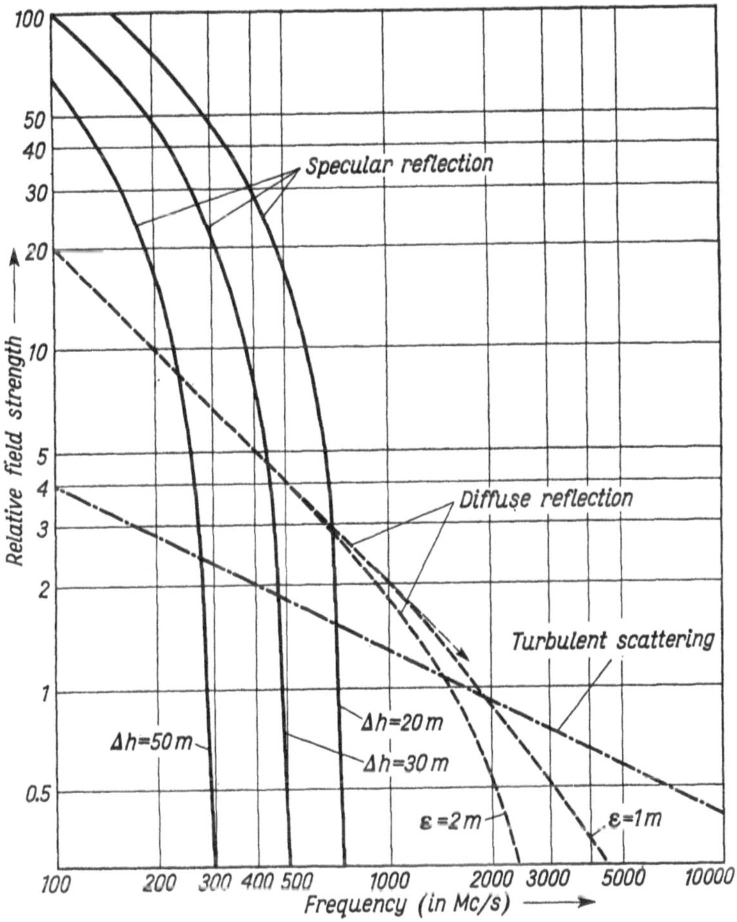

Fig. 19.10. Comparison of the frequency dependences of turbulent scattering, diffuse reflection and specular reflection. The values of the three components include an unspecified constant factor.

† Cf. Part 4 of Du Castel, Misme, Spizzichino and Voge [1960].

tion 19.1) and of γ given by (7), f_c generally lies in the UHF (decimetre) band.

The reflection coefficient of the diffuse component R_d is given by (12), with R_0 given by (8a) or (8b) and ρ_d by (15). In the VHF and UHF bands, R_d is in accordance with (8a) proportional to f^{-1}. In the SHF band, R_d decreases according to (8b) more rapidly with f than f^{-1}.

Opinions differ on the frequency dependence of the turbulently scattered field. We shall retain a mean frequency dependence and set the scattered field proportional to $f^{1/2}$.

Figure 19.10 shows the frequency dependence of the three components: *As the frequency increases, the specularly reflected field decreases more quickly than that of diffuse reflection, which in turn decreases more quickly than the scattered field.*

Comparison of the Dependence on Pathlength

The specularly reflected field given by (10) and (11) decreases rapidly as the pathlength d of the link is increased. It may also be noted that the longer the pathlength d, the lower the cut-off frequency f_d given by (16): specular reflection should therefore set in more often on the shorter paths.

The field due to diffuse reflection is given by (12) and (15) and we shall assume that R_0 is given by (8a). The ratio of the power reflected by a sheet to the free-space power varies as $\gamma^{-6}L$, where γ is given by (7) and varies as d; it may be shown† that the length L of a sheet increases with the height of the common volume.

To calculate the total energy reflected by the total number of stable sheets, some assumption regarding their distribution in space must be made. If they are assumed to be uniformly distributed over all heights, the mean energy P_{rd} reflected by the sheets may be obtained by integrating over the common volume. One then obtains

$$\frac{P_{rd}}{P_0} \propto d^{-5}L$$

For the diffusely reflected power the average dependence may be assumed as

$$\frac{P_d}{P_0} \propto d^{-5}$$

Since L increases with d, P_{rd} decreases less quickly than P_d as d is increased. *Diffuse reflection should be the more frequent the longer the pathlength of the link.*

† Cf. Complement to Part 2 of [Du Castel, Misme, Spizzichino and Voge, 1960].

19.4. Tropospheric Scatter Propagation—Fieldstrength Fluctuations, Bandwidth Limitations, Diversity Effects

Before considering the fieldstrength fluctuations, we must consider two preliminary problems.

Simultaneous Reflection by Several Sheets

As mentioned in Section 19.1, the simultaneous occurrence of tropospheric sheets is observed; on the average, 2 or 3 per kilometre are found on a vertical run.

The power reflected by a single sheet is a rapidly decreasing function of γ and hence of the height; if several sheets are present, the lowest of them will have the greatest reflection coefficient; the reflection coefficients of the others will decrease with their altitude.

This is illustrated by Tables 19.1 and 19.2, which give the power reflected by sheets numbered successively I, II, III... from the lowest sheet upwards. The values of power are relative and normalized so that the power reflected by the first sheet equals unity. It is assumed that the sheets are similar in dimensions and equidistant.

Tables 19.1 and 19.2 show that even in the most adverse case the contribution of the lowest sheet represents about half of the total reflected power; in most cases it contributes almost all of the reflected power.†

Table 19.1—Specular Reflection

d km	H m	Relative power reflected by successive sheets				
		I	II	III	IV	V
170	500	1	0.16	0.037	0.011	0.004
340	500	1	0.37	0.16	0.075	0.037
170	250	1	0.37	0.16	0.075	0.037
340	250	1	0.60	0.37	0.24	0.16

Table 19.2—Diffuse Reflection

d km	H m	Relative power reflected by successive sheets						
		I	II	III	IV	V	VI	VII
170	500	1	0.21	0.06	0.02	0.009	0.004	0.002
340	500	1	0.43	0.21	0.11	0.06	0.03	0.02
170	250	1	0.43	0.21	0.11	0.06	0.03	0.02
340	250	1	0.65	0.43	0.30	0.21	0.15	0.11

† This is even better demonstrated if one considers the dissimilarity of the sheets present at the same moment in the troposphere, instead of assuming that they are all of the same kind as we have done here.

In the following we shall assume, unless otherwise indicated, that *the reflected energy comes from a single sheet.*

The Useful Volume

By the term "useful volume" we shall denote the part of the atmosphere effectively taking part in the transmission of waves in a tropospheric scatter link. Reflection or scattering, as the case may be, takes place in this volume.

It is interesting to compare the dimensions of the "useful volume" for the different propagation mechanisms. We denote its width perpendicular to the path by X, its length in the direction of the path by Y and its height by Z.

For specular reflection by a sheet, the "useful volume" is the part of the sheet lying inside the first Fresnel ellipse. The length L of a sheet is generally larger than the minor axis of the Fresnel ellipse and smaller than its major axis (Section 19.2). Thus $X = \sqrt{(\lambda d)}$, $Y = L$, $Z = \Delta h$ (height of the irregularities of the sheet). For the usual type of path, X is of the order of hundreds of metres and Z of the order of tens of metres (Section 19.1).

For diffuse reflection by a sheet, the "useful volume" is the part of the sheet inside the "useful surface" defined above (Section 19.2). Hence by (14), $X = \gamma d \tan \beta_0$; $Y = L$; $Z = \Delta h$. X is of the order of hundreds of metres to 1 km.

For isotropic scattering by atmospheric turbulence, it may be shown† that the dimensions of the useful volume are the following: X of the order of $\gamma d/2$, Z of the order of $\gamma d/10$. X and Z are of the order of several kilometres.

Thus X and Z are much larger in the case of scattering than in the case of either specular of diffuse reflection: the dimensions of the "useful volume" are viewed from the terminals of the link under the angles $2X/d$ and $2Z/d$, which are of the order of a few degrees for scattering and of the order of a few tenths of a degree for reflection by sheets. This means that the dimensions of the useful zone of scattering may be observed with the usual antennae, whereas the useful zone of reflections appears as a brilliant point (rarely as several brilliant points) without any width even for the most directional antennae.

Experimental Evidence for the Coexistence of Reflection and Scattering

Let us note in passing that several investigators have experimentally studied the dimensions of this "useful volume" by means of antennae with

† Cf., for example, GORDON, W. E.: Radio scattering in the troposphere. *Proc. I.R.E.* **43**, 23–28 (1955).

very narrow beams that could more or less rapidly sweep through several degrees in a scanning movement.

The reader may be referred to the papers by Trolese [1955], Chisholm [1956, 1957], Waterman [1958], Hoggs and Kummer [1959] and Crawford [1959].

In general this type of experiment shows that the received field originates at times in a fairly large zone, at times at a single point in space; in the latter case the antenna pattern is the same as that in free space. Figure 19.11 shows some results published by Crawford and his colleagues; these observations were made on the same path on different days.

The fact that the received field on the same path comes at times from a very narrow, and at times from a very extensive zone cannot be explained

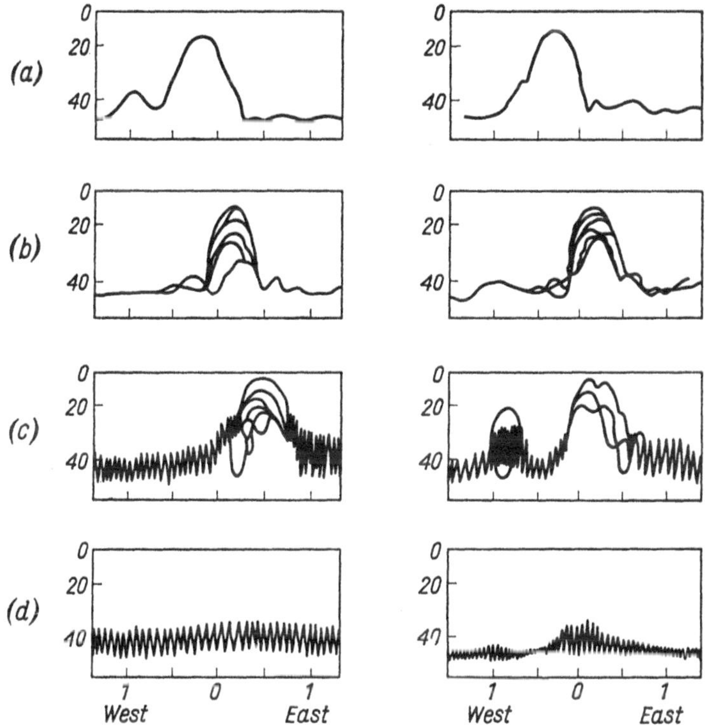

Fig. 19.11. Dependence of the received field (ordinate) on the angle of arrival (abscissa) at 4000 Mc/s. Measured with antennae 10 m in diameter. To explore an angle of 3°, the antennae turn about their axes with a period of 15 sec. (a) Free space; (b), (c), (d) field received on a 280 km path. We interpret (b) as a reflection phenomenon, (d) as scattering, (c) as a superposition of the two. (Adapted from Crawford *et al.* [1959]; *by permission of the American Telephone and Telegraph Company.*)

solely by reflections or solely by scattering: it must be accepted that reflections and scattering may coexist.

Connection between Fieldstrength Fluctuations and "Useful Volume"

The connection between the dimensions of the "useful volume" and the rapidity of the fluctuations of the field scattered by the troposphere may be established by a course of reasoning which we shall only sketch in outline:

(1) "Fieldstrength fluctuations" should be understood to mean rapid variations simultaneously

—in time (short fades of the order of seconds),

—in space (possibility of space diversity).

—as a function of frequency (bandwidth limitations and possibility of frequency diversity).

(2) A turbulent zone may theoretically be decomposed into elementary scatterers ("blobs"), and a sheet into small plane elementary mirrors. It is assumed that the field scattered by one of these scatterers or mirrors does not fluctuate as rapidly in time, space or frequency as the experimentally observed resultant field.

The fluctuations of the field are thus caused by the *interference* among the waves scattered from several elementary scatterers or mirrors. Since each of these elementary fields varies much more slowly than the resultant field, the elementary fields may be considered constant in time, space and frequency when studying the fluctuations of the resultant field.

(3) The field resulting from the interference of several sources varies the more rapidly in time, space and as a function of frequency, the further the sources are spaced from each other, i.e. the greater the "useful volume" containing them.

When two sources interfere, the interference lobes are the more closely crowded the longer the distance between the sources. It is also well known that the resulting field is a sinusoidal function of frequency and that with increasing distance between the sources the field will pass through a series of maxima and minima with increasing rapidity. Similarly, it may be shown that if the two sources move with random velocities (correlated or not), the resulting field will fluctuate at an increasing rate as the distance between the sources is increased.

These results may be generalized† for any number of sources distributed over the "useful volume". We shall simply state the general result here:

† Cf. Part 5 of [Du Castel, Misme, Spizzichino and Voge, 1960].

The greater the "useful volume" the more rapidly the tropospheric field will fluctuate in time, space and as a function of frequency.

Hence in general, *the scattered field is characterized by much more rapid fluctuations than the field reflected by a sheet.*

We shall investigate this point more closely by considering the fieldstrength fluctuations in space, time and as a function of frequency.

Fluctuations in Space—Diversity Distance

Whatever the propagation mechanism, the fluctuations of the reflected or scattered field are due to the interference of waves re-radiated by a large number of sources (elementary mirrors or elementary scatterers) occupying the useful volume. It may be shown† that the fieldstrength fluctuations result from the superposition of the fluctuations that would be obtained by the interference of two sources at a time. For simplicity we shall seek to determine only the most rapid component of these fluctuations, i.e. we shall require two points P_1 and P_2 of the "useful volume" such that the paths passing through P_1 and P_2 produce the fastest possible fluctuations.

The points P_1 and P_2 corresponding to the smallest horizontal diversity distance are two points on the same horizontal straight line perpendicular to the path separated by a distance equal to the width X of the "useful volume".

The diversity distance Δ obtained by making these two points interfere is

$$\Delta = \lambda \frac{d}{2X}$$

Using the above expressions for X, we obtain

—*for specular reflection*:

$$\Delta = \tfrac{1}{2}\sqrt{(\lambda d)} \tag{17}$$

Δ is of the order of several hundreds of metres;

—*for diffuse reflection*:

$$\Delta = \frac{\lambda}{2\gamma \tan \beta_0} \tag{18}$$

Δ is of the order of several hundreds of wavelengths for a path of 300 km;

---

† We restrict the following to a short summary indicating the orders of magnitude of the characteristics of the fluctuations. A more detailed treatment will be found in Part 5 of [Du Castel, Misme, Spizzichino and Voge, 1960].

—*for turbulent scattering*:

$$\Delta \approx \frac{\lambda}{\gamma} \tag{19}$$

Δ equals 50λ for a pathlength of 300 km.†

Thus the reflected field varies much more slowly in space than the scattered field. In practice, when the two antennae are separated by the diversity distance that can usually be realized, a diversity effect will only be obtained if the field is caused by turbulent scattering.

Fluctuations in Time

Following the same method as that used to study the fluctuations in space, we assume that the field produced by each elementary scatterer or elementary mirror varies slowly in time: the rapid fluctuations of the resultant field are due to the interference between the fields produced by these elementary sources, which are moving with respect to each other.

This movement may be of two kinds:

(1) *High-altitude winds.* The "useful volume" illuminates the space in front of it in a non-uniform manner: the energy reflected or scattered by it is concentrated in a series of narrow beams, the width of which, at the receiving point, equals the diversity distance.

A wind displacing the "useful volume" at a velocity v will cause these beams to move across the receiving point at a velocity $2v$. It will thus produce fluctuations in time; their rate, expressed as the number of positive crossings N of the median level per unit time (the "fading rate") will be of the order of

$$N \approx \frac{2v}{\Delta}$$

(2) *Turbulent movement.* In this case the different scatterers (or mirrors) causing the interference move with random velocities, but as has been indicated, it may be shown that the fading rate increases with the size of the "useful volume", and hence with decreasing Δ. For a first approximation the fading rate is given by an expression of the form‡

$$N \approx \frac{v'}{\Delta}$$

† Equation (19) is obtained by setting $X = \gamma d$, which applies only if the useful volume is limited by scattering effects. It may be shown that if the diameters of the antennae are greater than Δ given by (19), the useful volume is limited by the beamwidths of the antennae, so that the diversity distance becomes equal to the antenna diameter, i.e. there will be a diversity effect between two antennae placed side by side (cf. Spizzichino [1960]).

‡ Cf. Part 5 of [Du Castel, Misme, Spizzichino and Voge, 1960].

To these two primary causes of fluctuations in time a third cause must be added.

(3) *Atmospheric refraction.* The narrow beams in which the energy reflected or scattered by the "useful volume" is concentrated, may be slightly deflected along their paths by inhomogeneities in the atmosphere. Let us recall that the angular widths of these beams amount to only some tenths of a milliradian ($\Delta = 20$ m and $d = 200$ km give an angular width of 0.2 mR); much larger deflections than this value are often observed in the atmosphere.

In this case N is inversely proportional to Δ. We set

$$N = \frac{2v+v_t}{\Delta} \tag{20}$$

where v is the transversal component of the wind and v_t is a term associated with the effects (2) and (3), i.e. with atmospheric turbulence.

As we shall see in Section 19.5, it is experimentally confirmed that N is related to Δ and that the effect of the wind may be directly studied by observing the motion of the scattered beams across several receiving points lying on a straight line. These observations show that the term "wind" $2v/\Delta$ in (20) is the dominating term slightly less than 50% of the time. We shall therefore assume that on the average the terms "wind" $2v/\Delta$ and "turbulence" v_t/Δ are of the same order.

To determine the order of N, we substitute a mean value for v in (20): $v = 50$ km/h, and we assume that $v_t = 2v$. The value of Δ is then given by (17), (18) or (19) in accordance with the propagation mechanism. We thus obtain

—*for specular reflection*:

$$N = \frac{1600}{\sqrt{\lambda d}} \tag{21}$$

—*for diffuse reflection*:

$$N = 6600\,\frac{\gamma \tan\beta_0}{\lambda} \tag{22}$$

—*for turbulent scattering*:

$$N = 3300\,\frac{\gamma}{\lambda} \tag{23}$$

In these expressions N is the fading rate, i.e. the number of positive crossings of the median level per minute, the lengths d and λ are in metres, and γ is in radians. For a 300 km path at 500 Mc/s with $\beta_0 = 2°$, we find

—for specular reflection $N = 4$

—for diffuse reflection $N = 9$

—for turbulent scattering $N = 110$.

As was to be expected, the reflected field fluctuates much more slowly in time than the scattered field. If it is assumed that one or the other of these two components alternatively dominates, the values of N given by (21) or (22) and by (23) represent respectively the minimum and maximum of the fading rate N that is to be expected on a tropospheric link.

If we wish to determine the dependence of the mean value of N on the frequency and the pathlength, not only the variations of the maximum and minimum of the fading rate N should be considered. The manner in which the reflected and scattered field vary should be investigated.

Dependence on the Pathlength

Let us assume that the frequency is sufficiently high or the pathlength sufficiently long for specular reflection to be exceptional. The maximum and minimum of the fading rate N, given by (22) and (23) respectively, are both proportional to d. However, since the reflected field, which corresponds to the smaller values of N, decreases with the pathlength less quickly than the scattered field (Section 19.3), the mean value of N is a decreasing function of the pathlength (cf. Section 19.5).

Frequency dependence. Under the same assumptions as above, the minimum and maximum of N, given by (22) and (23), are both proportional to f. But the mean value of N may increase more quickly than f: as the frequency is increased, the scattered field decreases less quickly than the reflected field; at the highest frequencies the scattered field will dominate and N will attain its maximum value. At the lowest frequencies, on the contrary, N will always attain its minimum value.

Fluctuations in Frequency—Bandwidth Limitations

The method of analysis is the same as that used for fluctuations in space. If the optical path difference corresponding to two extreme points of the "useful volume" is δ, the bandwidth is limited to

$$\Delta f = \frac{c}{\delta} \tag{24}$$

The value of δ for these two extreme points is

$$\delta = 2\gamma Z + \frac{X^2}{d} \tag{25}$$

For diffuse reflection, X and Z are small; the second-order term X^2/d may be neglected in (25). According to (24) and (25)

$$\Delta f = \frac{c}{2\gamma \Delta h} \tag{26}$$

where Δh is the thickness of the sheet. For $d = 270$ km and $\Delta h = 50$ m we obtain $\Delta f = 150$ Mc/s.

For turbulent scattering, the values of X and Z should be substituted in (24) and (25). At the beginning of this section values of X and Z for turbulent scattering were given under the assumption that the "useful volume" is not limited by the antenna beams. The curves of Fig. 19.12 give the approximate dependence of Δf on d and the aperture angle α of the antenna beams; for very small values of α, the "useful volume" may be limited by the antenna beams and this has been respected in the figure.

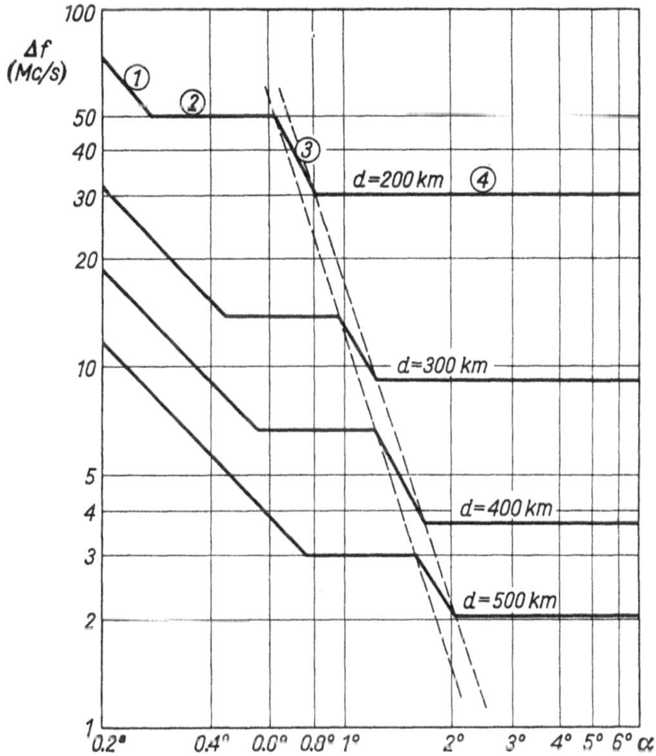

Fig. 19.12. Dependence of the bandwidth on the angular aperture of the antenna beams and the distance. As the beamwidth increases, the bandwidth depends first on the vertical dimension of the "useful volume" (1), then is independent of it (2). The bandwidth then depends on the horizontal dimensions of the "useful volume" (3) and then becomes independent of it (4).

The values of Δf for the case of turbulent scattering as given by Fig. 19.12 are always much smaller than those corresponding to reflection. Like the fluctuations in space and time, the fluctuations as a function of frequency are much more rapid in the case of turbulent scattering than in the case of reflection.

The Statistical Distribution of Instantaneous Amplitudes

The study of the *rapidity* of the fieldstrength fluctuations (the fading rate) enabled us to confront turbulent scattering with reflection (diffuse or specular) and hence to demonstrate the existence of these two components. The study of the ***amplitude distribution*** of these fluctuations will enable us to confront specular reflection with the other two components.

Diffuse reflection and scattering are phenomena with incoherent phase; hence the instantaneous values of the fieldstrength amplitude will be Rayleigh-distributed. On the other hand, in the case of specular reflection, the fieldstrength amplitude is distributed with a different distribution the variance of which is smaller than that of the Rayleigh distribution (cf. Part I, Chapter 7).

The existence of such distributions has been demonstrated experimentally. Crawford [1959] observed that they appear only if the fieldstrength fluctuates very slowly, which confirms our hypothesis, according to which they may be explained only by a phenomenon of *reflection*.

These distributions appear particularly at the lower frequencies. On paths of 200 to 300 km length they are observed

during 90% of the time at 80 Mc/s
during 35% of the time at 460 Mc/s (Crawford [1959] and unpublished work by the C.N.E.T)
during 20% of the time at 850 Mc/s (unpublished work by the C.N.E.T.)
during 3% of the time at 4000 Mc/s [Crawford 1959]

Since a sheet of thickness Δh has a cut-off wavelength $\lambda_0 = 2\gamma\Delta h$, the frequency of occurrence of sheets with a certain thickness may be deduced from the above data:

sheets of thickness $\Delta h <$ 100 m are present during 90% of the time
sheets of thickness $\Delta h <$ 16 m are present during 35% of the time
sheets of thickness $\Delta h <$ 10 m are present during 20% of the time
sheets of thickness $\Delta h <$ 2 m are present during 3% of the time

19.5. Some Experimental Verifications

We shall not attempt to give a detailed and systematic survey of the experimental results obtained on tropospheric paths; we shall only illustrate the theoretical results given in the preceding section by a few examples.

Since the study of the fieldstrength fluctuations allow its different components to be analysed, we start by some experimental data concerning these fluctuations.

Fluctuations in Space and Time

Figure 19.13 illustrates the fact that the rapidity of the fluctuations in space is related to their rapidity in time. Similar results were found on comparing the rapidity of the fluctuations in time and in frequency.

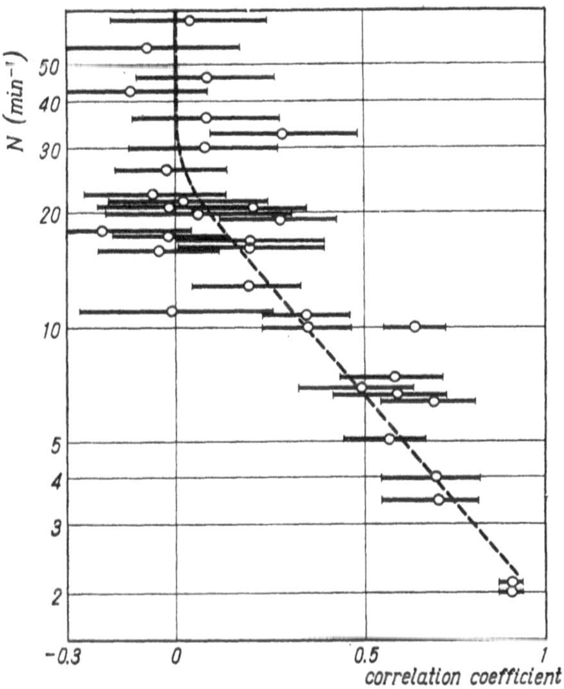

Fig. 19.13. To show that the rapidity of fieldstrength fluctuations in time is related to their rapidity in space, the above data were plotted. Abscissa: correlation coefficient of the instantaneous fields received at two points 20 m apart (respecting the inevitable uncertainty of the value of the correlation coefficient); ordinate: fading rate per minute (number of positive crossings of the median level).

The relation between the fluctuations in time and in space being thus established, we may check the assumptions that we have made concerning the mechanisms that may link these two quantities (Section 19.4).

Effect of the wind. Figure 19.14 gives a typical example of the fluctuations due to winds at high altitudes: the same fluctuations in space move across the two receiving points separated by a distance of 20 m. Figure 19.15 shows

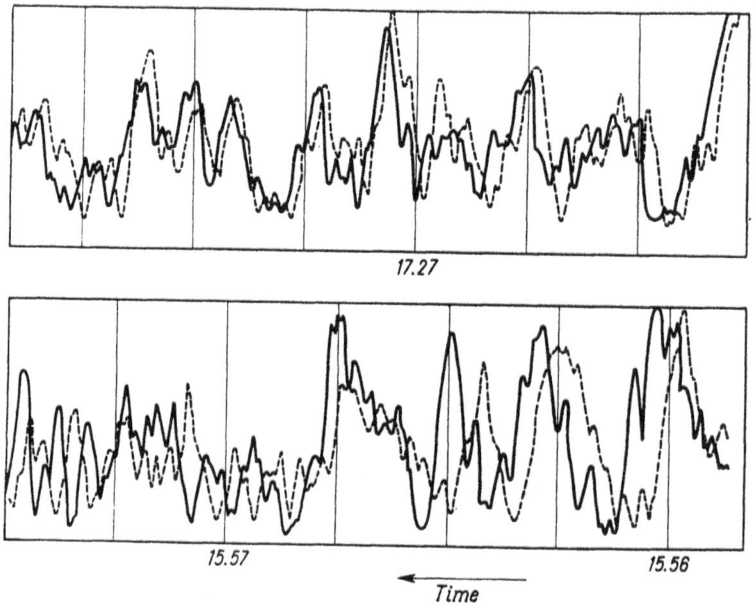

Fig. 19.14. Demonstration of wind on a tropospheric path. The recordings of the fieldstrength received at two points separated by 20 m have been plotted in the same figure (April and May 1959).

the same effect from another aspect: the correlation coefficient of the fields received by two neighbouring receivers passes through a maximum for a finite (non-zero) value of the time shift. This effect, more or less pronounced, sets in during about 30–50% of the time.

Effect of atmospheric turbulence. According to (20), the fading rate N depends on one hand on the windspeed v, on the other on a term v_T associated with atmospheric turbulence. The correlation between N and the windspeed has often been demonstrated experimentally (cf., for example, Crawford [1959]). Figure 19.16 shows that there is a broad correlation between the fading rate N and a parameter s defining the stability of the atmosphere. This parameter, which defines the stability of an atmospheric layer between the altitudes H_1 and H_2 equals the work necessary to raise 1 g of air from H_1 to H_2.† In Fig. 19.16, the quantity s has been calculated for the altitudes $H_1 = 500$ m and $H_2 = 2000$ m.

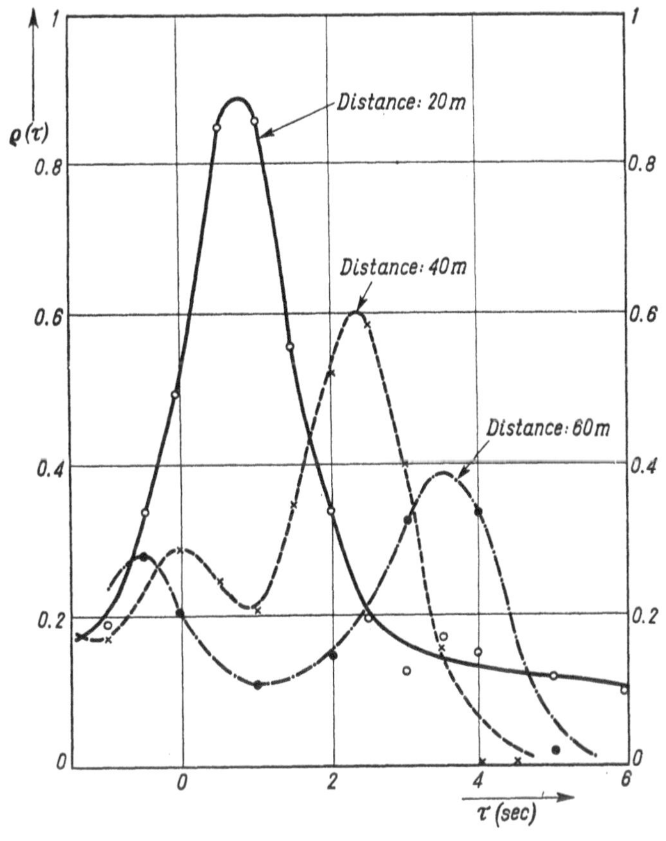

Fig. 19.15. Demonstration of wind on a tropospheric path. The dependence of the cross-correlation coefficient $\rho(\tau)$ between the fields received at two points on the time interval τ for different spacings of the two receiving antennae.

Figure 19.17 shows that the correlation between N and s becomes worse if s is determined for a thicker layer of the atmosphere, viz. a layer included between the heights of 500 and 4000 m; this confirms that the "useful volume" should, for $d = 230$ km, have a thickness of at most 1500 m.

† It has been shown by Misme how this parameter may be calculated from the radio sonde data measured by the meteorological services. Cf. MISME, P.: The correlation between the electric field at a great distance and a new radio-meteorological parameter *Trans. I.R.E.* AP-6, 289–292 (1958).

Mean Values of the Reflected and Scattered Fields

As we have seen, the largest portion of the atmosphere is turbulent and hence causes scattering. Reflections are due to the passing of stable sheets, which occupy only a small volume, but transmit an important amount of energy towards the receiver.

To explain these processes, one may use the following scheme: for given conditions of temperature and humidity (in the course of the same day or the same month), the mean scattered field, which essentially always comes from the same volume, varies little in time, whereas the reflected field appears only if stable sheets are present. Thus *the cases when the reflected component of the field dominates should correspond to the highest values of the mean field.*

This provides the explanation of the different phenomena observed experimentally.

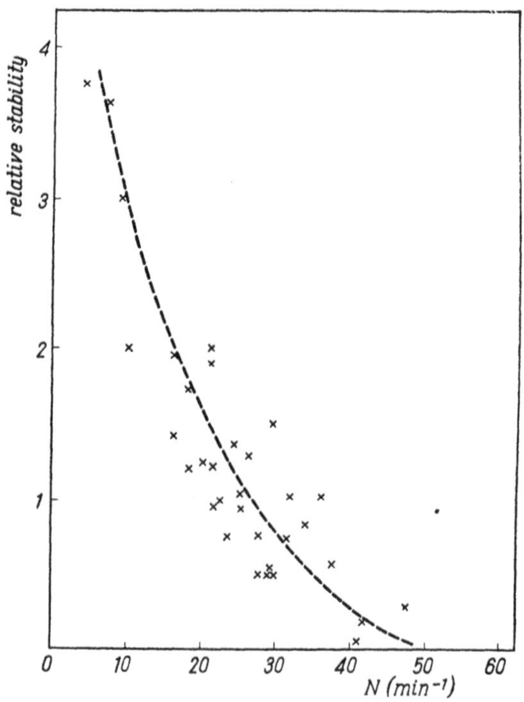

Fig. 19.16. Correlation between the stability of the atmosphere and the number of positive crossings of the median level (fading rate). The stability applies to the layer between $h_1 = 500$ m and $h_2 = 2000$ m; $f = 850$ Mc/s, $d = 230$ km.

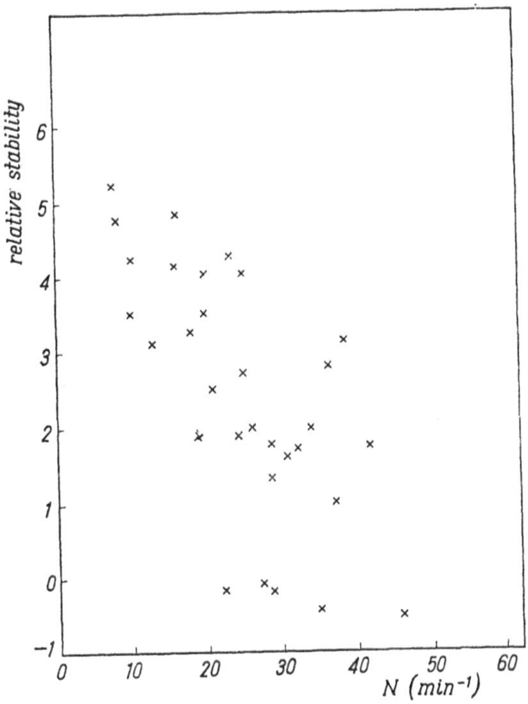

Fig. 19.17. Correlation between the stability of the atmosphere and the fading rate. The stability applies to the layer between $h_1 = 500$ m and $h_2 = 4000$ m, otherwise the conditions are the same as in Fig. 19.16. As may be seen, the correlation is less good than for $h_1 = 500$ m, $h_2 = 2000$ m (Fig. 19.16), which indicates that the layers responsible for tropospheric propagation are entirely included between the altitudes of 500 and 2000 m.

Fading rate and mean level. Depending on whether the reflected or the scattered component dominates, the fading rate N is low (units per minute) or very high (tens per minute). Also it is often observed that a sudden drop of the fading rate is accompanied by an enhancement of the received signal level.

Figure 19.18 shows that if the fields received in the course of the same month are compared, the most rapidly fluctuating fields correspond to the lowest *signal levels*. But, as we can see on the same figure, 19.18, this result is no longer valid over a longer period: if the satistical evaluation is prolonged over one year, the opposite result is obtained: during the warm months, the signal level is high, but in this period, the fading rate is high. Indeed, the scheme we have just used to explain the variations of the received level (the scattered field varies little in time, the high values of the received field

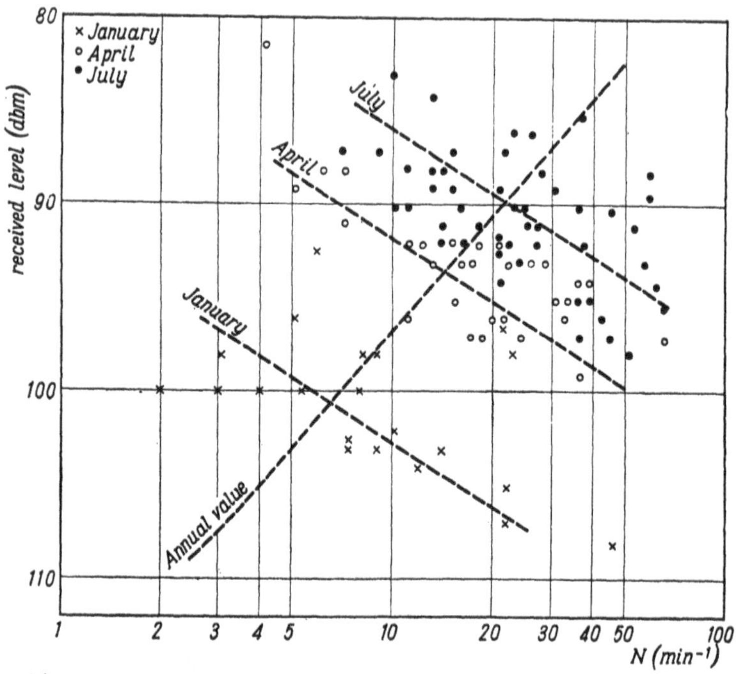

Fig. 19.18. Correlation between the fading rate and the signal level.

appear only when stable sheets are present), is not valid over a very long period. The enhancement of the received signal during the warm months cannot be explained by the presence of sheets: during the warm months atmospheric turbulence increases, stable sheets become more rare and the field will, on the average, fluctuate more rapidly (cf. Fig. 19.19). But at the same time the atmosphere becomes richer in water vapour and its vertical gradient of the refractive index increases; stronger atmospheric refraction diminishes the scattering angle γ and hence leads to a higher level of the received signal.

Diurnal variation. The atmosphere is in general more stable during the night than during the day, i.e. stable sheets will be more numerous and more important during the night. In the UHF band this results in a very characteristic diurnal effect: the mean signal level is higher at night than during the day. Figure 19.20 shows that this diurnal effect is, in France, maximum in summer: evidently during this period atmospheric turbulence during the day is particularly important so that the contrast between night and day is more conspicuous.

However, as the frequency is increased, the diurnal effect diminishes

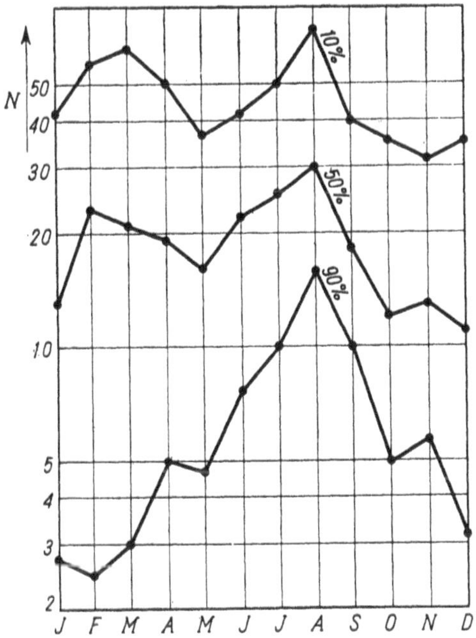

Fig. 19.19. Seasonal variation of the fading rate. The fading rate N (ordinate) exceeded during 10, 50 and 90% of the time is plotted for each month of the year (abscissa) for a 230 km path at 470 Mc/s (Palaiseau–Ligueil).

and vanishes towards 2000 Mc/s; it may even be reversed in the SHF (centimetre) band.† This may be explained by noting that the reflected field decreases rapidly with growing frequency; at the highest frequencies, practically only the scattered field will remain.

Dependence on the pathlength. The decrease of the mean field with growing frequency slows down for very long pathlengths. This may be explained by the fact that the (diffusely) reflected field, which decreases with the distance less rapidly than the scattered field (Section 19.3), is very often the dominating component at large distances. This is also confirmed by the fact that the fading rate decreases with the pathlength (Fig. 19.21). (Let us recall that if the field far beyond the horizon were due to a single phenomenon, only reflection or only scattering, the fading rate would on the contrary increase with increasing distance.)

Conclusions. We have restricted our summary in this section to giving some examples of experimental results that can only be explained by assuming that the resultant field is due to several propagation mechanisms.

† Trolese [1955].

Some practical conclusions may be drawn from this decomposition. Thus it is often useful to perform some propagation tests before establishing a tropospheric link; these tests are often carried out in a short period, e.g.

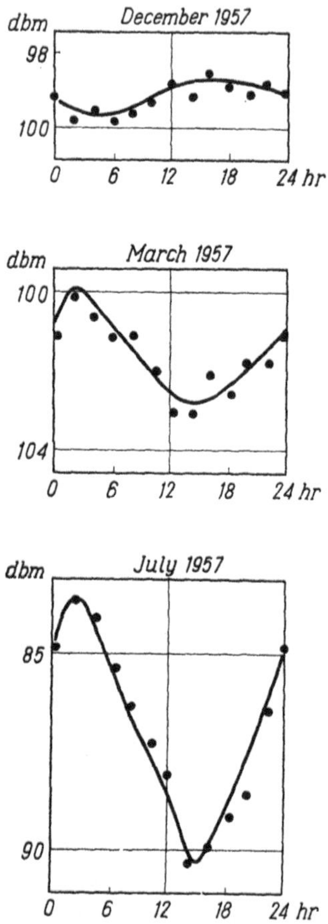

Fig. 19.20. Diurnal fieldstrength variation. The mean hourly fieldstrength is plotted against the time of day for the different days of the same month (Palaiseau–Ligueil path, 230 km, 470 Mc/s).

a month or two, and before extrapolating the obtained results to a much longer period, it is often advisable to make sure that these tests have not been carried out during an unfavourable period, or on the contrary, during too favourable a period. What we have called "specular reflection" is a source of abnormally high fieldstrengths; on the contrary, turbulent scattering results in the lowest fieldstrengths. We have seen that a number of indices may be utilized to ascertain whether, for example, one of these two extreme cases applies, e.g. the fading rate or the diurnal effect for a given climate.

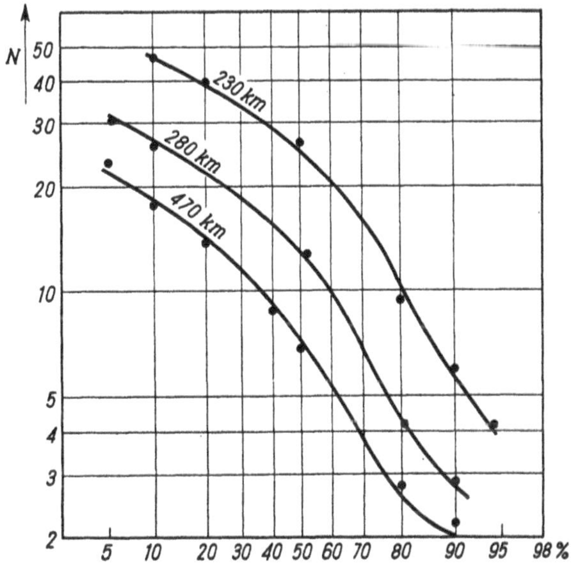

Fig. 19.21. The fading-rate distribution for different paths: Palaiseau–Ligueil (230 km), Palaiseau–Poitiers (280 km), Palaiseau–Arcachon (470 km), The curves give the percentage of the time during which the fading rate N exceeded the value indicated on the ordinate.

Similarly, when diveristy or bandwidth tests are carried out, the same indices may be used to make sure that the conditions of the test are not exceptionally favourable or adverse.

In general, our study shows that the diversity of propagation mechanisms entails large variations of the characteristics of the field: for a given radio link, the transmission loss, the bandwidth and the diversity distance vary over wide ranges. We have tried to show that these variations are not completely random: they are inter-related and correlated with certain tropospheric phenomena.

19.6. Ionospheric and Exospheric Sheets

The concept of a *sheet*, i.e. a rough surface along which the index of refraction changes abruptly, may be extended to the higher atmosphere.

The existence of thin ionized layers, the "sporadic E" layers, have often been observed in the lower ionosphere. Radio contact may be established when reflections by these layers set in; they may be studied in the same way as those caused by tropospheric sheets.

Reflections by the sporadic E layer may also be observed with the aid of an ionospheric sounding station. It will then be noted that

—below a frequency f_0 (occlusion frequency), the sporadic E layer gives rise to total reflections: no echo from altitudes above that of the sporadic E layer is observed;

—above a certain limiting frequency f_1, which is greater than f_0, the echoes from the sporadic E layer disappear.

The frequency f_0 is the plasma frequency of the sporadic E layer. It is given by

$$f_0^2 = \frac{e^2}{\pi^2 m \varepsilon_0} N \tag{27}$$

where m is the electron mass and e its charge, and N is the electron density.

At frequencies exceeding f_0, the received echoes may be explained by partial reflections from small inhomogeneities of the refractive index n of the atmosphere; in this case the same theoretical results as those for tropospheric sheets (Section 19.2) may be applied. The reflections by the sporadic E layer are diffuse: ionospheric soundings and rocket measurements have shown that the thickness of these layers is large compared with the wavelengths used by ionospheric sounding stations (decametric band). Their reflection coefficient R_d is given by (9): $R_d = \rho_d R_0$, where ρ_d is given by (15), and where R_0 depends on the inhomogeneities δn of the index of refraction n, given by

$$n^2 = 1 - \frac{f_0^2}{f^2} \tag{28}$$

According to (27) and (28), a variation δN of the electron density corresponds to a variation in refractive index δn with

$$\delta n = \frac{1.5 \times 10^{-5}}{f^2} \delta N \tag{29}$$

where f is in megacycles per second. Thus R_0 is a rapidly decreasing function of f for the following two reasons:

—the discontinuities of the refractive index δn are proportional to f^{-2},

—in addition, as in the case of tropospheric sheets, R_0 is a decreasing function of f if the thickness ε of the irregularities is of the order of the wavelength.

The rapid decrease of the reflection coefficient of the sporadic E layer with the frequency may explain the existence of a limiting frequency f_1, above which a given sounding station no longer observes any echoes.

Finally, the existence of sheets in the exosphere may be considered as probable. These sheets would be aligned parallel to the lines of force of the terrestrial magnetic field and could be responsible for the guided propagation along these lines of force.† But the characteristics of these sheets are as yet not well enough known to study the role of their irregularities.

† Voge, J.: Propagation guidée le long d'un feuillet atmosphérique ou (plus particulièrement) exosphérique. *Ann. d. Télécomm.* **16**, 288–295 (1961) and **17**, 34–43 (1962).

CHAPTER 20

REFLECTION OF RADIO WAVES BY THE MOON AND THE PLANETS

20.1. Problems Posed by the Reflection of Radio Waves by the Moon

Anyone may observe that the moon reflects visible light like a very rough surface: when it is entirely illuminated by the sun, the whole lunar disk scatters the sunlight toward the earth. It has been shown by measurements in the visible and infrared ranges of the spectrum† that the surface of the moon behaves practically like an isotropic scatterer. A surface will behave like an isotropic scatterer if its irregularities are of the order of the wavelength of the incident radiation; hence the surface of the moon is very rough on a scale of the order of microns.

Investigations of the reflection of radio waves by the moon have provided information on the irregularites of the lunar surface on a larger scale, on the scale which might be called "human". The first observations of the lunar surface by means of radio waves were carried out simultaneously by Bay [1946] in Hungary and by a team from the U.S. Army Signal Corps‡ at wavelengths of the order of 3 m. The first experiments of this type were all performed with the aid of radars; the low level of the received signal and the insufficient quality of the equipment at that time made it necessary to use very long pulses,§ so that the effects of different parts of the lunar surface could not be separated.

Thus only global information was available for studying the linar surface: the field reflected by the entire surface of the moon, and its fluctuations in time. The latter consist of two components:

(1) Very slow and regular fluctuations that may be explained by the rotation of the polarization plane caused by the Faraday effect during the propagation of the wave through the ionosphere [Murray and Hargreaves, 1954].

† Cf. Markov, *Astron. J. U.S.S.R.* **25**, 172–179 (1948); Pettit and Nicholson, *Astrophys. J.* **71**, 102–135 (1930).

‡ Mofensen [1946], De Witt and Stodola [1949].

§ A list of experiments and their principal characteristics will be found in the Annex to this chapter.

This rotation has been demonstrated by receiving the moon-reflected wave by means of two antennae polarized at right angles [Browne *et al.*, 1956, Blevis and Chapman, 1960, etc.]. These fadings vanish if circularly polarized antennae are employed [Evans, 1961; Ingalls, 1961]. Their period is of the order of half an hour at low frequencies and increases with growing frequency.† Their amplitude becomes negligible at 300 Mc/s [Straiton and Tolbert, 1960].

(2) Much more rapid fluctuations (of the order of one second), which are irregular and can easily be distinguished from the above fluctuations; they are associated with the irregularities of the lunar surface. Kerr and Shain [1949, 1951] attribute them to the relative rotation of the moon with respect to the earth, the lunar *libration.* Browne *et al.* [1956] have shown that there is a relation between an assumed structure of the irregularities of the lunar surface and the rapidity of the fluctuations. On the basis of an experimental investigation of the fluctuations of the received field it was established by Evans [1957] that at sufficiently high frequencies (120 Mc/s) the reflection by the moon is very directional and that the reflected energy comes from an area 10 times smaller than that of the lunar disk.

This result was in contradiction with previous theoretical work [De Witt and Stodola, 1949; Kerr and Shain, 1951]: in an analogy with light waves it was at first assumed that the moon behaves like an isotropic scatterer at all wavelengths. However, if one had proceeded in an analogy with the back-scattering of waves by the surface of the earth,‡ this would have led to the conclusion that, on the contrary, reflection by the moon should be very directional at radio frequencies.

The rapid progress of radio equipment in the last few years has enabled the results obtained by studying the time fluctuations of the moon-reflected signal to be confirmed by other methods. In particular, the reflections caused by different parts of the lunar surface have been separated by using radars with very short pulse durations (several μ sec). Numerous investigations have been performed, notably at Jodrell Bank§ and Malvern‡‡ (England) and by the Naval Research Laboratory in the U.S.A.†

† Browne *et al.* [1956] report a period of about half an hour at 120 Mc/s. At 488 Mc/s, Blevis and Chapman [1960] have observed a period of 6 hr.

‡ Back-scattering by the surface of the earth in the absence of vegetation is treated in Section 18.3.

§ Cf. Browne *et al.* [1956], Evans [1957], Evans, Evans and Thomson [1959]. The principal characteristics of these and the following experiments will be found in the Annex to this chapter.

‡‡ Hey and Hughes [1959], Hughes [1960].

† Yapplee *et al.* [1958, 1959], Trexler [1958].

At the same time the frequency band used in these experiments was extended to the higher frequencies: with a lower limit of 20 Mc/s set by ionospheric reflections and disturbances, the frequency band used for obtaining reflections from the moon now extends to 3000 Mc/s. At the lowest frequencies (20 Mc/s), the moon is an isotropic scatterer, from which it is deduced that the undulations of the lunar surface are of the order of at least some tens of metres. But the moon does not seem to contain undulations on a smaller scale, for at frequencies above about 100 Mc/s diffuse scattering by the moon is on the contrary very directional.

Besides the mean directivity of the reflection by the moon, the use of radars with very short pulse durations has enabled the problem of the fine structure of the lunar surface to be approached. Certain authors† have the impression that the reflected field is due to a small number of very directional reflectors scattered over the lunar surface: the response to a very narrow pulse seems in effect to be composed of a discrete series of spikes. However, this interpretation is very controversial.‡

Finally, another field of application of reflections by the moon is the determination of its material (chemical) nature. A very good knowledge of the irregularities of the lunar surface and their effect on the reflection coefficient is necessary; if this is known, the value of R_0 for a perfectly smooth moon may be derived from the measured value of the reflection coefficient R. Since R_0 depends only on the chemical nature of the moon soil, an investigation of the frequency dependence of R_0 should yield its electrical characteristics [Senior and Siegel, 1960]. However, owing to the inaccuracy of the present results it is hardly possible yet to draw definite conclusions on this subject.

20.2. Reflections by the Rough Surface of the Moon—Theory

We have listed the problems posed by the reflection of radio waves by the moon; we shall now examine more closely the problems directly connected with the roughness of its surface.

We assume that any ionospheric effects are suppressed (e.g. by the use of circular polarization). At any instant t, the moon presents a scattering cross-section $\sigma(t)$, so that

$$P_r = P_t \frac{g^2\lambda^2}{(4\pi)^3 r^4} \sigma(t) \tag{1}$$

† Yapplee [1958], Trexler [1958], Senior and Siegel [1958].

‡ Hughes [1960].

where P_r and P_t are the received and transmitted powers respectively, g is the antenna gain, and r is the distance from the antennae to the moon (the variations of this distance are neglected). As in the case of back-scattering by the earth's surface (Chapter 18), a scattering cross-section per unit surface may be defined; if its mean value is σ^0, then

$$\langle\sigma(t)\rangle = \int_S \sigma^0 \, dS \tag{2}$$

where the integration is extended over the part of the lunar surface which is struck by the incident wave. If the undulations of this surface are large compared with the wavelength, then σ^0 is a function of the form†

$$\sigma^0 = R_0^2 G(\theta, \beta_0) = \frac{R_0^2}{\tan^2\beta_0} G_1\left(\frac{\tan\theta}{\tan\beta_0}\right) \tag{3}$$

where R_0 is the reflection coefficient of a perfectly smooth lunar surface, β_0 the mean inclination of the undulations of the surface with respect to the mean spherical surface, θ the angle of incidence, i.e. the angle concluded by the incident ray with the normal to the mean spherical surface at a given point, and $G_1(x)$ is a decreasing function of x which vanishes for $x \gg 1$.

In the particular case of a normally distributed surface, (3) becomes‡

$$\sigma^0 = \frac{R_0^2}{\tan^2\beta_0} \exp\left(-\frac{\tan^2\theta}{\tan^2\beta_0}\right) \tag{4}$$

Response to a Long Pulse

Let us assume that the duration of a pulse transmitted by a radar towards the moon is sufficiently long to illuminate the entire visible part of the lunar surface simultaneously.§ In this case the integration (2) must be extended to the entire visible surface, so that with regard to (3) we obtain

$$\langle\sigma\rangle = R_0^2 \frac{\pi a^2}{\tan^2\beta_0} \int_0^{\pi/2} G_1\left(\frac{\tan\theta}{\tan\beta_0}\right) 2\sin\theta \, d\theta$$

If in addition it is assumed that β_0 and θ are small, $\langle\sigma\rangle$ is independent of β_0 and

$$\langle\sigma\rangle = R_0^2 \pi a^2 \tag{5}$$

† Cf. Chapter 18, equation (10).

‡ Cf. Chapter 18, equation (9).

§ The radius of the moon is $a = 1.74 \times 10^6$ m; the time necessary for an electromagnetic wave to travel through this distance equals 5.8 msec. For the entire visible surface of the moon to be illuminated by a radar pulse, its duration must be 11.6 msec.

as if the moon were perfectly smooth. (If β_0 is not small, $\langle\sigma\rangle$ may be slightly greater than this value.)

In the course of time, σ fluctuates about its mean value. Since the scattering is diffuse, its values are Rayleigh-distributed. To investigate the rapidity of the fluctuations, it is assumed that the moon revolves about an axis passing through its centre with an angular velocity l (libration).

Let dS be an element of the lunar surface in the neighbourhood of a point P defined by its coordinates x, y, z in the coordinate system shown in Fig. 20.1; Ox is parallel to the direction of the incident rays, Oz is the axis of rotation of the moon and O is the point on the moon that is nearest

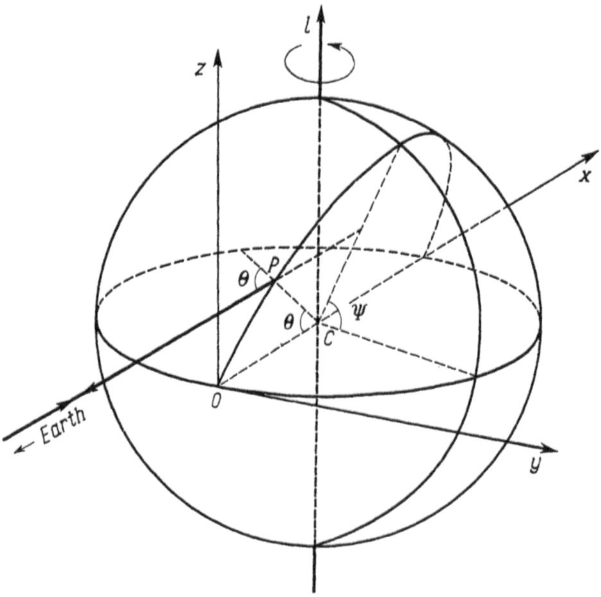

Fig. 20.1. Geometry of moon reflections.

to the earth. Let the element dS be sufficiently small for the amplitude of the field $|dE|$ reflected by this element to be constant during a short interval of time. During this interval, its phase variations are determined by the variations of the optical path δ of the ray reflected at P. In the vicinity of the instant $t = 0$ we have

$$\delta(t) \approx \delta(0) + 2lyt$$

The field reflected by the entire surface of the moon is

$$E(t) = \int_S \frac{dE(0)}{dS} \exp\left(\frac{4\pi ilyt}{\lambda}\right) dS$$

which may be written as

$$E(t) = \int_{-a}^{a} \varepsilon(y) e^{4\pi i l y t/\lambda} dy \tag{6}$$

with

$$\varepsilon(y) = \int_{-\sqrt{(a^2-y^2)}}^{\sqrt{(a^2+y^2)}} \frac{dE(0)}{dS} \cos\theta \, dz$$

where a is the radius of the moon.

Equation (6) shows that $E(t)$ is the Fourier transform of $\varepsilon(y)$.† We may thus obtain its power spectrum by determining the autocorrelation function $\gamma(Y)$ of $\varepsilon(y)$. For values of y much larger than the horizontal dimensions of the irregularities of the lunar surface we obtain

$$\gamma(Y) = \frac{1}{\pi a^2 R_0^2} \int_{-a}^{a} \sqrt{[F(y) F(y+Y)]} \, dy \tag{7}$$

where $F(y)$ stands for the function

$$F(y) = \int_{-\sqrt{(a^2-y^2)}}^{\sqrt{(a^2+y^2)}} \sigma^0 \cos\theta \, dz \tag{8}$$

Thus $F(y)\,dy$ is the energy reflected by the points on the moon with coordinates between y and $y+dy$.

The power density $S(F)$ of the signal $E(t)$ is proportional to $\gamma(\lambda F/2l)$.

Let us apply these results to a normally distributed surface with σ^0 given by (4). Using (7) and (8) we obtain

$$\gamma(Y) = \exp\left(-\frac{Y^2}{4a^2 \tan^2 \beta_0}\right)$$

so that

$$S(F) \propto \exp\left(-\frac{F^2}{F_0^2}\right) \tag{9}$$

with

$$F_0 = \frac{2al}{\lambda} \tan \beta_0 \tag{10}$$

Equation (10) shows that the rapidity F_0 of the fluctuations of the field scattered by the moon depends on the velocity of libration l and on β_0, i.e. on the directivity of the lunar reflections.

† The integration limits $-a$ and $+a$ may be replaced by $-\infty$ and $+\infty$, since $\varepsilon(y)$ practically vanishes for $y \geqslant a$.

Applying the Fourier transformation to $S(F)$ as given by (9), the autocorrelation function $\Gamma(t)$ of the received signal $E(t)$ is obtained:

$$\Gamma(t) = e^{-2\pi^2 F_0^2 t^2} \tag{11}$$

where F_0 is given by (10).

By a similar method the variations of the field at a given instant as a function of frequency may be examined. This investigation is of interest from two points of view: on one hand, the maximum bandwidth for communication via moon reflections may be determined in this way; on the other hand, we shall see that the fluctuations of the field as a function of frequency are, like the fluctuations in time, connected with the assumptions of the structure of the lunar surface. This provides a second method for checking these assumptions. Let $\mathrm{d}S$ be an element of the lunar surface, sufficiently small for the amplitude of the reflected field $\mathrm{d}E$ to be independent of frequency, and let it cause phase variations $\Delta\varphi$ when the frequency varies by a small amount Δf about the mean frequency f; in the coordinate system of Fig. 20.1 we obtain

$$\Delta\varphi = \frac{4\pi x \Delta f}{c}$$

The field reflected by the entire surface of the moon is

$$E(\Delta f) = \int_S \frac{\mathrm{d}E(f)}{\mathrm{d}S} e^{4\pi i x \Delta f/c} \mathrm{d}S$$

which may be written as

$$E(\Delta f) = \int_0^a \eta(x) e^{4\pi i x \Delta f/c} \, \mathrm{d}x \tag{12}$$

with

$$\eta(x) = \int_0^{2\pi} \frac{\mathrm{d}E(f)}{\mathrm{d}S} a \, \mathrm{d}\psi$$

where the meaning of ψ is evident from Fig. 20.1.

Equation (12) shows that $E(\Delta f)$ is the Fourier transform of $\eta(x)$, and the power spectrum of $E(\Delta f)$ may be determined by calculating the autocorrelation function $\beta(x)$ of $\eta(x)$. If X is sufficiently large with respect to the dimensions of the irregularities of the lunar surface, we obtain

$$\beta(X) = \frac{2}{aR_0^2} \int_0^a \sqrt{[\sigma^0(x)\sigma^0(x+X)]} \mathrm{d}x \tag{13}$$

The power spectrum $s(\nu)$ of the signal $E(\Delta f)$ is proportional to $\beta(c\nu/2)$.

In the case of a normally distributed surface σ^0 is given by (4). From (13) we obtain

$$\beta(X) = \exp\left(-\frac{X}{a\tan^2\beta_0}\right)$$

so that

$$s(\nu) \propto e^{-\nu/2T} \tag{14}$$

with

$$T = \frac{a}{c}\tan^2\beta_0 \tag{15}$$

It may be seen that the quantity $1/T$, which represents the maximum bandwidth transmitted by moon reflection, strongly depends on β_0, i.e. on the directivity of the moon reflection. T also represents the minimum pulse duration that may be used without distorting the pulse.

The autocorrelation function $c(\Delta f)$ of the signal $E(\Delta f)$ is obtained by applying the Fourier transformation to $s(\nu)$:

$$c(\Delta F) = \frac{1}{1+(4\pi T\Delta f)^2} \tag{16}$$

All of the above results concerning long pulse durations will evidently also apply to the case of a continuous wave.

Response to a Short Pulse

Let us assume that a radar transmits pulses of sufficiently short duration for only a small part of the lunar surface to be illuminated at a time: then it will be possible to determine which part of the moon is responsible for the reflected energy and the directivity of the reflection may be derived.

Let us assume the lunar surface to be normally distributed. At any instant t, the back-scattered field is given by (1), (2) and (4). The integration in (2) is extended over the area of the lunar surface illuminated by a pulse of duration τ at the instant t; the pulse travels with the velocity of light. If the leading edge of the pulse reaches the moon at the instant $t = 0$, the mean resposen to the impulse will be†

$$\left.\begin{aligned} \langle\sigma(t)\rangle &= \sigma_\infty(1-e^{-t/T}) && \text{for} \quad t < \tau \\ \langle\sigma(t)\rangle &= \sigma_\infty(1-e^{-\tau/T})e^{-(t-\tau)/T} && \text{for} \quad t > \tau \end{aligned}\right\} \tag{17}$$

† In the derivation of these expressions β_0 and θ were assumed small. Under these assumptions (17) is a very good approximation except for very large values of t: for $t > 2a/c$, $\langle\sigma\rangle$ should vanish, whereas (17) yields a finite value of $\langle\sigma\rangle$. But the received signal s is in practice only detectable for $t \ll 2a/c$; for these values of t, (17) is correct.

where $\sigma_\infty = \pi a^2 R_0^2$ is the value of $\langle\sigma\rangle$ for a very long pulse and T is given by (15). An indication of the curves $\langle\sigma(t)\rangle$ is given in Fig. 20.2.

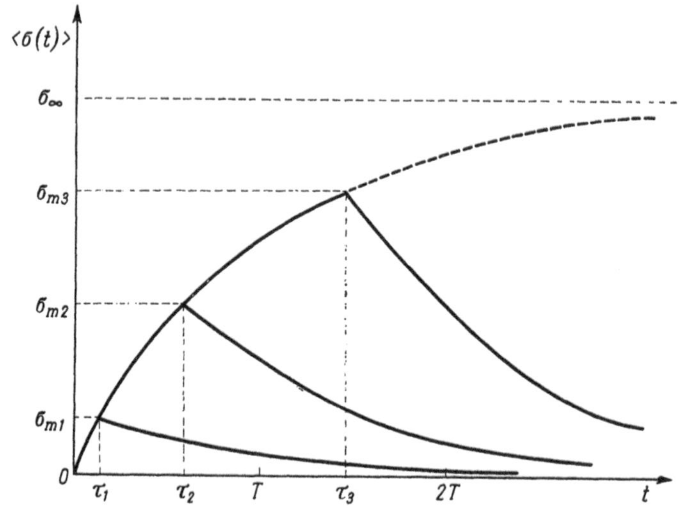

Fig. 20.2. Response to pulses of increasing durations τ_1, τ_2, τ_3.

The variation of $\langle\sigma(t)\rangle$ may be experimentally directly verified. It may also be verified indirectly by measuring the dependence of the maximum σ_m of $\langle\sigma(t)\rangle$ on τ. From (17) we have (Fig. 20.2)

$$\sigma_m = \sigma_\infty(1-e^{-\tau/T}) \tag{18}$$

20.3. Reflections by the Rough Surface of the Moon—Experimental Investigations

The preceding theoretical investigation shows that there are several experimentally accessible parameters that are related to the directivity of the back-scattering by the moon. From the experimental values of these parameters it should be possible to find the "model" of a rough surface which accounts best for the reflections from the moon; in particular, it should be possible to find out whether the normally distributed model, which has the advantage of being simple, is an adequate approximation and in this case it should be possible to determine the mean slope β_0 of the irregularities of the lunar surface.

Response to a Long Pulse

The experiments performed with long pulses or with continuous waves are the oldest and most numerous. A brief list of some of them will be found in the annex to this chapter.

The mean value of the back-scattered field, in accordance with (5), depends more on R_0, i.e. on the chemical nature of the lunar surface, than on the structure of its irregularities; we shall therefore not consider it here.†

An investigation of the rapid fluctuations of the reflected field (after eliminating the slow fluctuations caused by the Faraday effect) confirm that the reflection is diffuse: the amplitude of the scattered field is Rayleigh-distributed (Fig. 20.3).

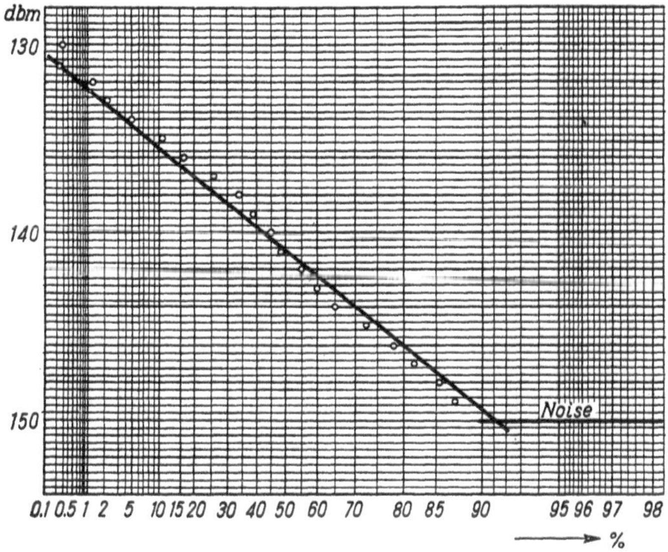

Fig. 20.3. Amplitude distribution of the reflected field (experiments performed between the U.S.A. and France).

As first verified by Kerr and Shain [1951], the rapidity of the fluctuations is proportional to the velocity of the moon's libration l. At the frequency used by these investigators—20 Mc/s—the values obtained indicate that the entire surface of the lunar disk participates in the reflection. Since then, these experiments have been repeated on higher frequencies (cf. Annex): the rapidity of the fluctuations is always proportional to l, but the constant of proportionality is smaller than at 20 Mc/s. Only a small part of the lunar surface participates in the reflection.

For example, Fig. 20.4 shows some values of the autocorrelation function $\Gamma(t)$ of the received signal as a function of l for a constant value of the time interval t. These values have been obtained at 120 Mc/s by Browne *et al.*

† Senior and Siegel [1960].

[1956]. They are compared with the theoretical dependence (11) which was obtained for the case of a normally distributed surface. Good agreement is obtained for $\beta_0 = 11°\pm 2°$.

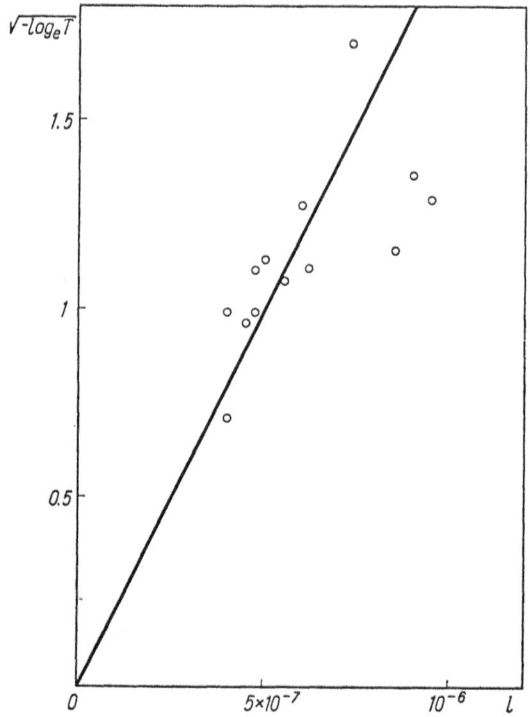

Fig. 20.4. The correlation coefficient $\Gamma(t)$ between the strengths of two successive echoes separated by a constant time interval t, as a function of the libration l of the moon: comparison of the theoretical curve with the experimental values given by Browne [1956].

Similar measurements have been reported by Evans [1957] at 120 Mc/s. In this case, either the dependence of $\Gamma(t)$ on l for constant t (equal to the interval between two successive pulses), or the dependence of $\Gamma(t)$ on t at a given time (l constant) was measured; in the latter case the interval between two successive pulses was variable. The experimental values of $\Gamma(t)$ are fairly closely represented by the function

$$\Gamma(t) = \exp\left[-1.3\left(\frac{2a}{\lambda}lt\right)^2\right]$$

which coincides with (11), with due regard to (10), obtained under the assumption of a normally distributed surface, provided we set $\beta_0 = 14°$.

The same experiment repeated at 100 Mc/s yielded

$$\Gamma(t) = \exp\left[-0.8\left(\frac{2a}{\lambda}lt\right)^2\right]$$

which corresponds to $\beta_0 = 11°$ [Evans, Evans *et al.* 1959].

Another method of studying the structure of the lunar surface consists, as has already been pointed out, in comparing the fields received at the same time on different frequencies. Experiments using simultaneously two frequencies have been performed by Ingalls [1961] at 440 Mc/s. Some experimental values of the correlation coefficient $c(\Delta f)$ of the fields received

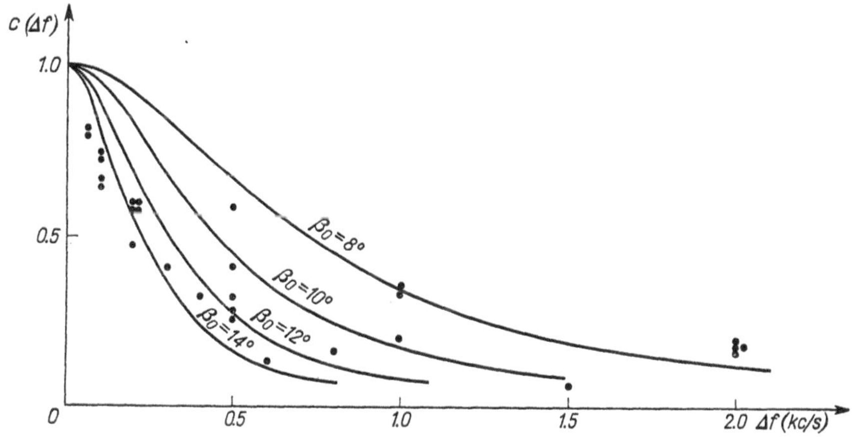

Fig. 20.5. Correlation coefficient of the fields reflected at two frequencies separated by Δf. Comparison of the theoretical curves and the experimental data given by Ingalls [1961].

on two frequencies separated by Δf are shown in Fig. 20.5. These values are compared with the theoretical curves derived from (16) corresponding to a normally distributed surface. Considering the inevitable inaccuracies of the experimental results, the latter are compatible with the theoretical curves for values of β_0 between about 11° and 15°.

Evans *et al.* [1961] describe experiments in which the distortion of amplitude-modulated waves was measured. From a theoretical point of view these experiments are equivalent to the preceding ones; the two phenomena are related, but their relation is unfortunately not simple. Nevertheless, Evans' experiments demonstrate that the maximum bandwidth that may be transmitted without distortion at 162 Mc/s is of the order of 1 kc/s, which means that the autocorrelation interval of the received signal is also of the order of 1 kc/s; this is compatible with the values of β_0 suggested above.

Response to a Short Pulse

The response to very short pulses (some microseconds) has been investigated by several authors (cf. Table 20.2 of the Annex to this chapter).

Figure 20.6 shows the dependence of the received power between the instant t_0 of the arrival of the leading edge and the instant $t_0+\Delta t$ on the time interval Δt. The experimental values given by Trexler [1958] at 198 Mc/s are compared with the theoretical exponential dependence derived from (17)

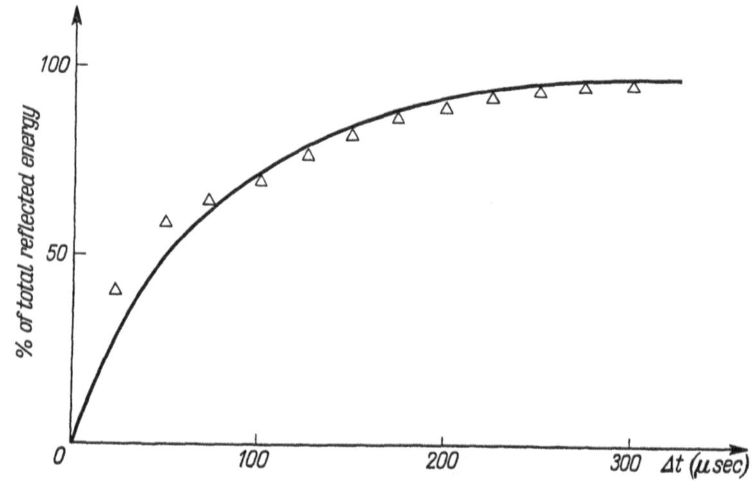

Fig. 20.6. Response to a very narrow pulse. Fraction of the total reflected energy received between the instant of arrival of the leading edge of the pulse t_0 and the instant $t_0+\Delta t$. Comparison of the theoretical curve with the experimental results given by Trexler [1958].

assuming $\tau \ll T$, i.e. neglecting the time during which the echo rises. The agreement is very good if the time constant T of this exponential dependence is set equal to a value between 80 and 100 μsec, which corresponds to $\beta_0 = 7\text{–}8°$.

In the set of experiments, the dependence of the maximum signal strength of the received echo on the pulse duration τ was measured. In Fig. 20.7 these experimental values are compared with the theoretical dependence (18); the best agreement is obtained for values of β_0 slightly higher than those in the preceding experiments, viz. $\beta_0 = 8\text{–}12°$.

Finally, these results may be compared with those obtained on higher frequencies. On the basis of the response to very short pulses at 3000 Mc/s, Hughes [1950] attempted to establish the dependence of the scattering cross-section σ^0 per unit lunar surface on the angle of incidence θ. The function

$$\sigma^0(\theta) = e^{-10\theta}$$

accounts for the experimental results for $3° < \theta < 14°$. In this interval this dependence deviates slightly (less than 2 db) from the "normal" dependence (4) for $\beta_0 = 8$–$9°$.

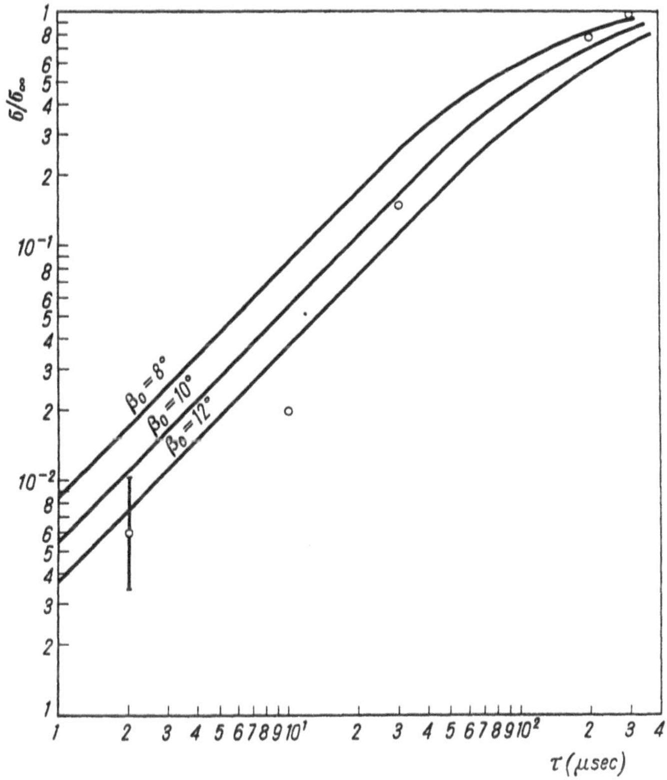

Fig. 20.7. Dependence of the echo strength on the pulse duration τ. Comparison of the theoretical curve with the experimental results given by Trexler [1958].

In conclusion, it may be said that the dependencies obtained by different experimental methods and on different frequencies are very close to each other. In particular, the angle β_0 defining the directivity of the moon reflections seems to a first approximation independent of the frequency between 100 and 3000 Mc/s.† This confirms the assumption that the lunar surface contains undulations which are large with respect to the wavelength.

† The variations of β_0 that are found in the above results and which seem to depend much more on the experimental conditions than on the frequency, may be partly due to a slight discrepancy between the real and the Gaussian law ruling $\sigma^0(\theta)$; the former yields slightly larger values of β_0 for small values of θ than for large values of θ.

Spiky Structure of the Echo

Investigating the lunar surface by means of radars with very narrow pulses, certain authors [Yapplee, 1958; Trexler, 1958, etc.] had the impression that the echo does not decrease regularly by fluctuating about a mean value, but that it contained a series of spikes. Subsequently the idea developed that reflections from the moon are the result of a small number of specular reflections from favourably oriented regions of the lunar surface. The idea assumes that the surface of the moon is very gently undulated.

This interpretation later became very controversial.† The original objection was that the spikes did not represent a constant value, but that, on the contrary, they fluctuated in a random manner in the course of time. But as we have already seen several times,‡ a specularly reflected field may fluctuate in time. Another objection was the absence of any correlation between the intensity of the same spike received at different points on the earth.§

On the other hand, the hypothesis of a lunar surface sufficiently gently undulated to present only a small number of points of specular reflection leads to a very large and improbable value of the mean radius of curvature of this surface‡‡ and in addition contradicts the fact that the reflection by the moon at 20 Mc/s is not directional.

For all these reasons it seems very unlikely that the reflection of radio waves by the entire surface of the moon may be reduced to a small number of specular reflections. However, there is no reason why there should not be certain gently undulated regions (small β_0) on the inhomogeneous surface of the moon, which scatter the incident waves in a very directional manner with an increased reflection coefficient, even though this will be a case of diffuse scattering.

Conclusions on the Structure of the Lunar Surface

In conclusion we may say that the irregularities of the surface of the moon may only be either much greater than the wavelength between 100 and 3000 Mc/s (at least tens of wavelengths), or much smaller than the

† Cf. for example, Hughes [1960].

‡ Cf. Chapters 7 and 14.

§ Cf. the discussion following Straiton and Tolbert [1960].

‡‡ It may be shown [Spizzichino, 1959] that the mean number $\langle n \rangle$ of points of specular reflection presented by a surface of area S is related to the mean radius of curvature R_c by the relation $\langle n \rangle = S/\langle R_c \rangle^2 \beta_0^2$. Applying this relation to the surface of the moon, we find that for $\langle n \rangle$ to be of the order of some tens, $\langle R_c \rangle$ must be of the order of some hundreds of kilometers.

wavelength (in the sense of the Rayleigh criterion). The small inhomogeneities can thus be only smaller than about 1 cm. The rock constituting the surface of the moon is thought to be entirely covered with dust which assumes the shape of the large undulations of the terrain.

Practical Applications

Moon reflections can be used to establish radio communication between two distant points on the globe. Without going into the details of the problems posed by such radio links,† we may note that the principal characteristics of the propagation involved in such a link have already been given above:

—the mean reflected field is given by (1) and (5);

—its amplitude distribution is the Rayleigh distribution;

—the rapidity of the fadings F_0 is given by (10); it is proportional to the frequency;

—the bandwidth that may be transmitted without distortion is the reciprocal value of T given by (15). According to (15), its value is approximately 0.5 to 1 kc/s (this value has been confirmed experimentally) and does not depend either on the frequency or on any other characteristic of the link, for which it constitutes a severe limitation.

20.4. Reflection of Radio Waves from Venus and from the Sun

The observation of the sun and the nearest planets of the solar system by means of radar, though not impossible, is outside the limits of presently available equipment.‡ Several successful attempts of detecting venus by radar were undertaken during 1959.§ But it was not until March 1961 that the Jet Propulsion Laboratory (U.S.A.) was able to set up sufficiently

† Cf. Emma, T.: Successful moon bounce satellite communication. *Electronics* **33**, 31 (1960); Hahn, P. M.: Availability of moon for global communications. *Trans. I.R.E.* SET—7, 57–59 (1961); Radford, W. H.: Moon relay communications. *I.R.E. Intern. Conv.* Rec. Pt. 5, 277–283 (March 1960); Senior, T. B. A., Siegel, K. M. and Weil, H.: The influences of radar reflection characteristics of the Moon on specifications for Earth–Moon–Earth communication systems. *I.R.E. Wescon. Conv.* Rec. Pt. 1 (August 1958); Sulzer, P. G., Montgomery, G. F. and Gerks, I. H.: A UHF moon relay. *Proc. I.R.E.* **40**, 361 (1952); Webster, P. A.: Long distance communication via the moon. *J. Brit. I.R.E.* **22**, 257–264 (1961); Yeh, L. P.: Use of the moon or satellite relays for global communications. *Elec. Eng.* **79**, 1043–1048 (1960); Yeh, L. P.: Communicating in Space. *Electron. Industr.* **18**, 94–99 (1959).

‡ Kerr [1959].

§ Price *et al.* [1959]; Evans and Taylor [1959].

powerful equipment (at 2388 Mc/s) to carry out a long series of systematic measurements.†

During these experiments, not only the mean strength of the echo was measured, but also its characteristics were studied, such as the widening of its frequency spectrum by the Doppler effect (which is of the same order as the frequency F_0 of the fluctations of the received signal in time).

The investigation of the fluctuations of the field reflected by venus poses essentially the same problems as the field reflected by the moon: if one excludes the possibility of a moving surface on venus such as that of a sea, the fluctations are due to a libration movement. Their rapidity F_0 is thus given by (10), F_0 being proportional to the product lt an β_0. But in the case of the moon, its rotation (libration) l was known, so that from F_0 the mean slope β_0 of its irregularities could be derived; in the case of venus, however, neither its velocity, nor its axis of rotation is known. The measurements by the Jet Propulsion Laboratory have indicated that the value of F_0 is very low, of the order of a few c/s. Should this be explained by a small velocity of rotation l or by assuming that the surface of venus is fairly smooth?

There are several ways of resolving this ambiguity.

(1) The libration l, the *apparent* rotational motion of venus with respect to an observer on the earth, consists of two components: that due to the rotation of venus, l_1, is unknown but constant in time; that due to the relative motion of venus with respect to the observer,‡ $l_0(t)$, is variable, but may be calculated. Theoretically it is therefore sufficient to measure F_0, which is proportional to $[l_0(t)+l_1] \tan \beta_0$, for two different values of t in order to derive the values of l_1 and β_0. To accomplish this, the spectrum of the echo in two as widely as possible separated points of venus' orbit must be measured.

Unfortunately the term $l_0(t)$ is smaller than l_1 and the method is very inaccurate.

(2) Another method consists in measuring the depolarization of the field reflected by Venus. As has been shown in Chapter 8, the depolarization increases with the surface roughness, so that its measurement gives some indication of the value of β_0.

The depolarization measured in the course of the experiments carried out by the Jet Propulsion Laboratory was of the order of 11 db, so that it is comparable to that obtained in the reflections from the moon.

---

† Victor, Stevens and Golomb [1961].

‡ Although this is a translation with respect to the Galilean axes, it contains a rotational component with respect to the axes determined by the straight line joining the observer and the centre of venus.

(3) As we have seen in considering reflections from the moon, the measurement of the rapidity of the time fluctuations of the received echo is not the only method of deriving the slope β_0 of the irregularities. Even if the application of narrow pulses is excluded for the moment (this is difficult in view of the low level of the received signal), the instantaneous values of the fields at two close frequencies could be measured. Their correlation coefficient given by (15) and (16) depends only on β_0. In this expression, $\tan \beta_0$ occurs in the fourth power, so that the corresponding measurements could be fairly accurate.

It does not seem that this method has as yet been attempted.

Finally, the sun may also be the object of radar measurements. Besides certain adverse factors such as its great distance, the high noise level caused by the sun itself in the receiver, the wide spectrum of the echo that is to be expected, the absorption by the gas in the solar atmosphere which will further attenuate the reflected field, etc., a favourable factor consists in the sun's large scattering cross-section. A more detailed consideration of the possibilities of such observations will be found in Kerr [1959] and Blattner [1960].

Annex to Chapter 20

Tables of Data on Experiments with Moon Reflections Quoted in Chapter 20

Table 20.1—Radar Contacts with the Moon Using Long Pulses or Continuous Waves

Country, Research Institution	Bibliographical reference	Frequency Mc/s	Pulse duration	Polarization	Object of investigation
Hungary	Bay [1946]	120			
U.S.A., U.S. Army Signal Corps	Mofensen [1946] De Witt and Stodola [1949]	111.5	0.25 sec		Intensity of echo
Australia	Kerr, Shain and Higgins [1949] Kerr and Shain [1951]	20	0.25 sec 2.25 sec		Libration fluctuations Ionospheric influences
United Kingdom, Jodrell Bank, Univ. of Manchester	Browne *et al.* [1956]	120	30 msec	H	Libration fluctuations, Faraday effect
	Evans [1957]	120	2 and 10 msec, 20 msec (double pulse)	H	Libration fluctuations, Faraday effect
	Evans, Evans *et al.* [1959] Evans *et al,* [1961]	100 162	25 msec CW, AM	circular circular	Libration fluctuations Frequency dependence of fluctuations
Canada, near Ottawa	Blevis and Chapman [1960]	488	CW	H and V	Faraday effect, libration fluctuations, depolarization of scattered signal
U.S.A., M.I.T.; and Canada, Def. Res. Telec. Est., Ottawa	Ingalls [1961]	440	CW, 2 frequen.	circular	Frequency dependence of fluctuations
France, C.N.E.T.; U.S.A., Alpha-Corpor., Texas		8505	CW	circular	Libration fluctuations

TABLE 20.2—Narrow-pulse Radar Contacts with the Moon

Country, Research Institution	Bibliographical reference	Frequency Mc/s	Pulse duration μsec	Polarization	Object of investigation
U.S.A. Naval Research Laboratory	Trexler [1958]	198	12		Structure of the lunar surface
	Yapplee *et al.* [1958] Yapplee *et al.* [1959]	2860	2		Do.; exact measurement of the distance Earth–Moon
United Kingdom, Radar Res., Est., Malvern	Hey and Hughes [1959] Hughes [1960]	3000	5		Structure of the lunar surface
Contact between Radar Res. Est., Malvern, England, and E.E.R.L., Texas, U.S.A.	Straiton and Tolbert [1961]	3000	5	variable	Do.; investigation of depolarization

CHAPTER 21

CONCLUSION

CLOSE analogies are to be found in the study of the scattering of electromagnetic waves from different kinds of rough surfaces, whatever they may be—terrain, sea, atmospheric sheets or distant planets. The same problems are often met and similar methods are used to solve them. Having read the recent literature on reflections by atmospheric sheets or by the moon, for example, the authors have frequently applied the methods to reflections by the earth's surface.

Among the general conclusions which follow from the study of these different applications, we may first of all note the necessity of clearly distinguishing several components of the scattered field: specular reflection and diffuse scattering, the latter having two components in its turn—scattering by a surface with undulations much larger than the wavelength (this kind of scattering is very directional) and almost isotropic scattering by irregularities with dimensions of the same order as the wavelength. This last component has not been investigated in great detail because it does not play such an important part as the other components at radio frequencies; we have met it almost only in the case of back-scattering at low grazing angles (Chapter 18). But it may play a very important role in the infrared and visible parts of the electromagnetic spectrum (Chapter 17). It is significant that the necessity of distinguishing several components of the scattered field has appeared in the theoretical investigation (Part I of this book) as well as in the applications.

It may also be noted that in all these applications, the general method of studying the fluctuations of the scattered field is the same. The dimensions of the "glistening surface", i.e. the portion of the reflecting surface effectively responsible for the scattered field, are first determined; the smaller this area, the more directional the scattering. It is then shown that the rapidity of the fluctuations of the scattered field depends on the dimensions of the "glistening surface": the larger the latter, the more rapid the fluctuations. This remains true regardless of whether fluctuations in time, space or as a function of frequency are considered.

This method shows that the more directional the scattering, the more slowly the field fluctuates. The method has been applied to scattering by the earth's surface (Chapter 12) as well as to atmospheric sheets (Chapter 19) and the surface of the moon (Chapter 20). But the conclusions appear more general. Thus when we investigated the propagation of very short waves on line-of-sight paths (Chapter 15) and on beyond-the-horizon paths (Chapter 19), we noted that apart from reflections other mechanisms of propagation also intervened. In each case, the least directional mechanisms gave rise to the most rapid fieldstrength fluctuations in space, time and frequency—these three aspects always being closely linked.

Finally, we may note that the analogy between the different applications may be extended even further: the model which we have used to investigate back-scattering by the earth's surface (a normally distributed surface, the slopes of which are for most practical purposes also normally distributed†) may equally well be used to explain the reflections by the moon, and according to the most recent measurements it seems that this model may also be applied to the surface of venus.

† Cf. Part I, Appendix D.

BIBLIOGRAPHY

(The paper is in the language of the title, unless otherwise stated)

ABILD, B.: Beitrag zur Ultrakurzwellen-Ausbreitung über See. *Techn. Hausmitt. NWD-Rundf.* **8**, 103-108 (1956).

AKSENOV, V. I. (Аксенов, В. И.): On the scattering of electromagnetic waves from sinusoidal and trochoidal surfaces of finite conductivity (in Russian). *Radiotekhn. i Elektr.* **3**, 459 (1958).

AKSENOV, V. I. (Аксенов, В. И.): Experimental investigation of the scattering of electromagnetic waves by periodically rough surfaces (in Russian) *Radiotekhn. i Elektron.* **5**, 782–795 (1960).

AKSENOV, V. I. (Аксенов, В. И.): Application of the Kirchhoff approximation to the problem of the scattering from periodically rough surfaces of finite conductivity (in Russian). *Radiotekhn. i Elektr.* **6**, 347–354 (1961).

ALPERT, Y. L., GINZBURG, V. L. and FEINBERG, E. L. (Альперт, Я. Л., Гинзбург, В. Л., and Фейнберг, Е. Л): *The Propagation of Radio Waves* (in Russian). Gostekhizdat, Moscow, 1953.

AMENT, W. S.: Toward a theory of reflection by a rough surface. *Proc. I.R.E.* **41**, 142–146 (1953).

AMENT, W. S.: Application of a Wiener-Hopf technique to certain diffraction problems. Naval Research Laboratory Report No. 4334, 10 May 1954.

AMENT, W. S.: Forward and back-scattering by certain rough surfaces. *Trans. I.R.E.* **AP-4**, 369–373 (1956).

AMENT, W. S.: Reciprocity and scattering by certain rough surfaces. *Trans. I.R.E.* **AP-8** 167–174 (1960).

AMENT, W. S. and KATZIN, M.: Signal fluctuations in long-range overwater propagation. *Trans. I.R.E.* **CS-4**, 118–122 (1957).

AMES, L. A., NEWMAN, P. and ROGERS, T. F.: VHF tropospheric over-water measurements far beyond the radio horizon. *Proc. I.R.E.* **43**, 1369–1373 (1955).

ANATOLSKY, M. L. (Анатольский, М. Л.): The reflection of waves from a rough, perfectly reflecting surface (in Russian). *Dokl. Akad. Nauk SSSR* **62**, 203 (1948).

ANDERSON, L., SMYTH, J. B., ABBOTT, F. R. and REVEILLE, R.: Radar wave propagation N.R.S.L. Report No. W-P-2, 30 Nov. 1942.

ANDRIEUX, G.: Reflecteurs passifs pour faisceaux hertziens. *L'Onde Electrique,* **36**, 57 (1956).

ANDRONOW, A. and LEONTOWITSCH, M.: Zur Theorie der molekularen Lichtzerstreuung an Flüssigkeitsoberflächen. *Z. Physik* **38**, 485 (1926).

ARCHER-THOMSON, H., GROOKE, N., GOLD, T. and HOYLE, F.: Preliminary report on the reflection of 9 cm radiation at the surface of the sea. A.S.E.Report No. M-542 (1943).

ARTMAN, K.: On the theory of anomalous reflection of periodic structures (in German). *Z. Physik* **119**, 529 (1942).

BACHYNSKI, M. P.: Microwave propagation over rough surfaces. *RCA Rev.* **20**, 308–335 (1959).

BASS, F. G. (Басс, Ф. Г.): Boundary conditions for the mean electromagnetic field on a surface with random irregularities and fluctuations of impedance (in Russian). *Izv. Vyssh. Zav. Radiofizika* **3**, 72–78 (1960).

BASS, F. G. (Басс, Ф. Г.): On the theory of combinational scattering of waves by a rough surface (in Russian). *Izv. Vyssh. Zav. Radiofiz.* **4**, 58–66 (1961).

BASS, F. G. (Басс, Ф. Г.): The propagation of radio waves over a statistically rough surface. *Izv. Vyssh Zav., Radiofiz.* **4**, 476–483 (1961).

BASS, F. G. and BOCHAROV, V. G. (Басс, Ф. Г., Бочаров, В. Г.): On the theory of scattering of electromagnetic waves from a statistically rough surface (in Russian). *Radiotekhn. i Elektr.* **3**, 180–185 (1958).

BATEMAN, R.: Elimination of interference type fading at microwave frequencies with spaced antennas. *Proc. I.R.E.* **34**, 662 (1946).

BAUERMEISTER, KNÖPFEL: UKW Feldstärkevoraussage in gebirgigem Gelände. *Techn. Hausmitt. NWDR* **4**, 67–73 (1952).

BAY, Z.: Reflection of microwaves from the moon: *Acta Phys. Hungar.* **1**. 1–22 (1946).

BAZER, J.: Multiple scattering in one dimension. Inst. Math. Sci., New York University, Report EM 122 (1959).

BEAN, B. R.: Prolonged space wave fadeouts at 1046 Mc observed in Cheyenne Mountain propagation program, *Proc. I.R.E.* **42**, 848 (1954).

BEARD, C. I.: Coherent and incoherent scattering of microwaves from the ocean. *Trans. I.R.E.* **AP-9**, 470–483 (1961).

BEARD, C. I. and KATZ, I.: The dependence of microwave radio signal spectra on ocean roughness and wave spectra. Appl. Phys. Lab. Johns Hopkins Univ. Report, CF-2456 (1956).

BEARD, C. I. and KATZ, L.: The dependence of microwave radio signal spectra on ocean roughness and wave spectra. *Trans. I.R.E.* **AP-5**, 183–191 (1957).

BEARD, C. I., KATZ, I. and SPETNER, L. M.: Phenomenological model of microwave reflection from the ocean. *Trans. I.R.E.* **AP-4**, 162–167, 182–184 (1956).

BECKMANN, P.: On the problem of the scattering of very short waves from rough surfaces (in Czech). *Slabopr. Obz.* **14**, 302–309 (1953).

BECKMANN, P.: The reflection of electromagnetic waves and synthesis of media. *Acta Technica ČSAV* **1**, 251–267 (1956).

BECKMANN, P.: A new approach to the problem of reflection from a rough surface. *Acta Techn. ČSAV* **2**, 311–355 (1957).

BECKMANN, P.: The probability distribution of the vector sum of n unit vectors with arbitrary phase distributions. *Acta Techn. ČSAV* **4**, 323–335 (1959).

BECKMANN, P.: A generalized Rayleigh distribution and its application to tropospheric propagation. Electromagnetic wave propagation (Symposium, Liège 1958), Academic Press, London 1960, p. 445–449.

BECKMANN, P.: The scattering of waves from a periodic surface. *Czechosl. J. Phys.* (1961a)

BECKMANN, P.: The depolarization of electromagnetic waves scattered from rough surfaces. *Acta Techn. ČSAV* **6**, (1961b).

BECKMANN, P. and SCHMELOVSKY, K. H.: Über ein bei Schwunderscheinungs-untersuchungen auftretendes Integral. *Wiss. Z. Hochsch.* Ilmenau **4**, 167–171 (1958).

BERNING, J. A.: Performance of diffraction gratings with random errors in the position of the grating grooves. *J. Opt. Soc. Amer.* **47**, 339 (1957).

Bespalov, V. I. and Gaponov, A. V.: The effect of small irregularities on the propagation of electromagnetic waves in periodic structures (in Russian). *Radiotekhn. i Elektr.* **1,** 722–748 (1956).

Biggi, J., Du Castel, F., Simon, J. C., Spizzichino, A., Voge, J.: Etude expérimentale de la structure fréquentielle instantanée de la propagation en milieu hétérogène. *Annales des Télécom.* **15,** 274–276 (1960).

Biot, M. A.: Reflection on a rough surface from an acoustic point source. *J. Acoust. Soc. Amer.* **29,** 1193–1200 (1957a).

Biot, M. A.: Some new aspects of the reflection of electromagnetic waves on a rough surface. *J. Appl. Phys.* **28,** 1455–1463 (1957b).

Biot, M. A.: On the reflection of electromagnetic waves on a rough surface. *J. Appl. Phys.* **29,** 998 (1958).

Blake, L. V.: Reflection of radio waves from a rough sea. *Proc. I.R.E.* **38,** 201–304 (1950).

Blattner, D. J.: Radio relaying by reflection from the sun. *Trans. I.R.E. C.S.* **8,** 169–172 (1960).

Blevis, B. C. and Chapman, J. H.: Characteristics of 488 Mc/s radio signal reflected from the moon. *J. Res. NBS,* **64,** 331–334 (1960).

Bogle, A. G.: Some aspects of microwave fading on an optical path over sea. *Proc. I.E.E.*, Part III, **99,** 236 (1952).

Bocharov, V. G. and Bass, F. G. (Бочаров, В. Г., Басс, Ф. Г.): On the scattering of electromagnetic waves by a statistically rough surface (in Russian). *Radiotekhn. i Elektr.* **3,** 577–578 (1958).

Booker, H. G., Ratcliffe, J. A. and Shinn, D. H.: Diffraction from an irregular screen with application to ionospheric problems. *Phil. Trans. Roy. Soc. London*, Ser. A, **292,** 579 (1950).

Bouix, M.: Study of the electromagnetic field in the vicinity of a reflecting surface (in French). *Compt. Rend. Acad. Sci.* **240,** 1763–1765 (1955).

Bowhill, S. A.: Statistics of a radio wave diffracted by a random ionosphere. *J. Res. NBS* **65D,** 275–292 (1961).

Braude, S. Ya. (Брауде, С. Я.): On the propagation of ultra-short waves in mountainous terrain (in Ukrainian). *Ukr. Fiz. Zhurn.* **R3,** 246–253 (1958a).

Braude, S. Ya. (Брауде, С. Я.): On the distribution of scattering elements in the propagation of radio waves over the rough surface of the sea (in Russian). *Izv. Vyssh. Zav. Radiofiz.* **1,** 25–29 (1958b).

Braude, S. Ya. (Брауде, С. Я.): The Fresnel coefficients for a rough surface (in Russian), *Izv. Vyssh. Zav. Radiofiz.* **2,** 691-696 (1959).

Braude, S. Ya. and Komarov, N. N. (Брауде, С. Я., Комаров, Н. Н.): Generalized curves of the Fresnel coefficients for horizontal and vertical polarization (in Russian). *Radiotekhnika* **6,** 100–106 (1959).

Braude, S. Ya., Komarov, N. N. and Ostrovsky, I. E. (Брауде, С. Я., Комаров, Н. Н., Островский, И. Е.): On the statistical character of the scattering of centimetre radio waves by the rough surface of the sea (in Russian). *Radiotekhn. i Elektr.* **3,** 172–179 (1958).

Bray, W. J. and Corke, R. L.: A technique for 4000 Mc/s propagation testing for radio-relay systems. *Proc. I.E.E.*, Part III A, **99,** 281 (1952).

Brekhovskikh, L. M. (Бреховских, Л. М.): Diffraction of sound waves by a rough surface (in Russian). *Dokl. Akad. Nauk SSSR* **79,** 585 (1951a)

BREKHOVSKIKH, L. M. (Бреховских, Л. М.): Diffraction of electromagnetic waves by a rough surface (in Russian). *Dokl. Akad. Nauk SSSR* **81,** 1023 (1951b).

BREKHOVSKIKH, L. M. (Бреховских, Л. М.): The diffraction of waves by a rough surface, part I (in Russian). *Zh. Eksper. i Teor. Fiz.* **23,** 275–289 (1952a).

BREKHOVSKIKH, L. M. (Бреховских, Л. М.): The diffraction of waves by a rough surface, part II (in Russian). *Zh. Eksper. i Teor. Fiz.* **23,** 289–304 (1952b).

BREKHOVSKIKH, L. M. (Бреховских, Л. М.): The reflection of bounded wave beams and pulses (in Russian). *Usp. Fiz. Nauk* **50,** 539–576 (1953).

BRIGGS, B. H.: Roughness of the moon as a radar reflector. *Nature* **187,** 490 (1960).

BROWN, G. G., EPSTEIN, J. and PETERSON, D. W.: Comparative propagation measurements; television transmitters at 67.25, 288, 510 and 910 Mc/S. *RCA Review,* **7,** 171–201 (1948).

BROWN, R. H. and LOVELL A. C. B.: *The Exploration of Space by Radio.* Chapman Hall, London (1957).

BROWNE, I. C. and EVANS, J. V.: The moon as a scatterer of radio waves. *Radio Astronomy* University Press, Cambridge 1957, pp. 408–409.

BROWNE, I. C., EVANS, J. V., HARGREAVES, J. K. and MURRAY, W. A. S.: Radio echoes from the moon. *Proc. Phys. Soc. London* **69,** 901–920 (1956).

BUCKINGHAM, A. D. and STEPHEN, M. J.: A theory of depolarization of light scattered by a dense medium. *Trans. Faraday Soc.* **53,** 884 (1957).

BULLINGTON, K.: Radio propagation at frequencies above 30 Mc/S. *Proc. I.R.E.,* **35,** 1122 (1947).

BULLINGTON, K.: Radio propagation variations at V.H.F. and V.H.F. *Proc. I.R.E.* **38,** 27 (1950).

BULLINGTON, K.: Reflection coefficients of irregular terrain. *Proc. I.R.E.* **42,** 1258–1262 (1954).

BURKE, J. E. and TWERSKY, V.: On scattering of waves by the grating of elliptical cylinders. Sylvania Electronic Defense Laboratories, Report EDL-E44 (1960).

BURROWS, C. R. and ATWOOD S. S.: *Radio Wave Propagation.* Acad. Press (1949).

BUSSEY, H. E.: Suppressing microwaves by zonal screens. *Tele-Tech.* **10,** 45 (1951).

CABESSA, R.: Realisation de liaisons par faisceaux hertziens de qualité sur des trajets maritimes en Grèce. *L'Onde Electrique,* **35,** 714 (1955).

CAMPBELL J. P.: Backscattering characteristics of land and sea at X band. *Transactions of the 1959 Symposium on Radar Return, Univ. of New Mexico* (11–12 May 1959).

CARASSA, F.: Prove di propagazione con le frequenze di 250, 500, 1000 MHz. *Alta Frequenza,* **25,** 378 (1956).

CARLSON, J. F. and HEINS, A. E.: The reflection of an electromagnetic plane wave by an infinite set of plates. *Quart. Appl. Math.* **4,** 313–329 (1946) and **5,** 82–88 (1947).

CARROLL, T. J.: Complementary diversity reception of microwaves. O. C. Sig. O. Report No. RPS-4 (1946).

CHAUX, R. and DASCOTTE, J.: Les relais passifs de la liaison Afourer-Bin-el-Ouidane. *Ann. Radioélectricité,* **5,** 220–229 (1950).

CHAVANCE, P.: Etude de propagation des ondes centimétriques dans le Nord de la France. *Annales des Télécomm.* **7,** 254–261 (1952).

CHAVANCE, P., BOITHIAS, L. and BLASSEL, P.: Etude de propagation d'ondes centimétriques dans le région méditerranéenne. *Annales des Télécomm.* **9,** 158–185 (1954).

CHISHOLM, J. H.: Recherches expérimentales sur la diffusion angulaire et sur les possibilités d'utilisation de la propagation troposphérique. *L'Onde Electrique* **37,** 427–434 (1957).

CHU, C. M. and CHURCHILL, S. W.: Multiple scattering by randomly distributed obstacles —method of solution. *Trans. I.R.E.* **AP-4,** 142 (1956).

COCKROFT, A. L., DAVIES, H. and SMITH, R. A.: A quantitative study of sea returns at centimeter wavelengths for moderately small angles of elevation. *Proc. Phys. Soc.* **58,** 717 (1946).

COURT, G.: Determination of the reflection coefficient of the sea for radar coverage calculation by an optical analogy method. *Proc. I.E.E.,* Pt.B, **102,** 728–830 (1955).

COWAN, E. W.: X-Band sea return measurements. R. L. Report No 870, 19 Jan. (1946).

COX, C. and MUNK W.: Measurement of the roughness of the sea surface from photographs of the sun's glitter. *J. Opt. Soc. Amer.* **44,** 838–850 (1959a).

COX, C. and MUNK, W.: Statistics of the sea surface derived from sun glitter. *J. Marine Res.* **13,** 198–227 (1954b).

CRAWFORD, A. B., HOGGS, D. C. and KUMMER, W. H.: Studies in tropospheric propagation beyond the horizon. *Bell Syst. T. J.* **38,** 1067–1178 (1959).

CRAWFORD, A. B. and JAKES, W. C.: Selective fadings of microwaves. *Bell Syst. T. J.* **31,** 68 (1952).

CROMBIE, D. D.: Doppler spectrum of sea-echo at 13.56 Mc/s. *Nature* **175,** 681 (1955).

DANIELS, F. B.: Radar determination of the surface structure of the moon. U. S. Army signal research and Dev. Lab., Fort Monmouth (N. J.) Technical Report 2110 (Apr. 1960).

DAVIES, H.: The reflection of electromagnetic waves from a rough surface. *Proc. I.E.E.,* Pt. III, **101,** 209–214 (1954). Discussion on above paper, *Proc. I.E.E.,* Pt. III, **102,** 148 (1955).

DAVIES, H. and MACFARLANE, G. G.: Radar Echoes from the sea surface of one centimetre wavelength. *Proc. Phys. Soc.* **58,** 717 (1946).

DAY, J. P. and TROLESE, L. G.: Propagation of short radio waves over desert terrain. *Proc. I.R.E.* **38,** 165–175 (1950).

DE, A. C.: Reflection of microwaves from earth's surface. *Indian J. Meteorol. Geophys.* **11,** 45–49 (1960).

DE LANGE, O. E.: Propagation studies at microwave frequencies by means of very short pulses. *Bell Syst. T. J.,* **31,** 91 (1952).

DERYUGIN, L. N. (Дерюгин, Л. Н.): Equations for the coefficients of reflection of waves from a periodically rough surface (in Russian). *Dokl. Akad. Nauk SSSR* **87,** 913–916 (1952).

DERYUGIN, L. N. (Дерюгин, Л. Н.): On the theory of diffraction by a reflecting grid (in Russian). *Dokl. Akad. Nauk SSSR* **93,** 1003–1007 (1953).

DERYUGIN, L. N. (Дерюгин, Л. Н.): On the surface resonance of a reflecting grid (in Russian). *Dokl. Akad. Nauk SSSR* **94,** 203–206 (1954).

DERYUGIN, L. N. (Дерюгин, Л. Н.): The reflection of a laterally polarized plane wave from a surface of rectangular corrugations (in Russian). *Radiotekhn.* **15,** No. 2, 15–26 (1960a).

DERYUGIN, L. N. (Дерюгин, Л. Н.): The reflection of a longitudinally polarized plane wave from a surface of rectangular corrugations (in Russian). *Radiotekhn.* **15,** No. 5, 9–16 (1960b).

DERYUGIN, L. N. and FRIDMAN, G.Kh. (Дерюгин, Л. Н., Фридман, Г. Х.): Resonance curves of double resonance at a diffraction grating (in Russian). *Dokl. Akad. Nauk SSSR* **111**, 1209–1211 (1956).

DERYUGIN, L. N. and MYAKISHEV, B. Ya. (Дерюгин, Л.Н., Мякишев, Б. Я.): A diffractional antenna with lateral excitation (in Russian). USSR patent No. 110721, Sept. 14th, 1956.

DE WITT, J. H. and STODOLA, E. K.: Detection of radio signals reflected by the moon, *Proc. I.R.E.* **37**, 229–242 (1949).

DINGER, H. E., GARNER, W. E., HAMILTON, D. H. and TEACHMAN, A. E.: Investigation of long-distance overwater tropospheric propagation at 400 Mc. *Proc. I.R.E.* **46**, 1501–1510 (1958).

DU CASTEL, F.: Note sur les phénomènes de propagation troposphérique. *Ann. Télécomm.* **15**, 137–142 (1960). Part VII of Du Castel, Misme, Spizzichino and Voge (1960).

DU CASTEL, F.: *Propagation Troposphérique et Faisceaux Hertziens Transhorizon.* Editions Chiron, Paris (1961).

DU CASTEL, F., MISME, P., SPIZZICHINO, A. and VOGE, J.: Réflexions partielles dans l'atmosphère et propagation à grande distance. Editions de la revue d'optique, Paris, 1960; reprinted from *Ann. Télécomm.* **13**, No. 7–8, 9–10 (1958), **14**, No. 1–2 (1959); No. **15**, 1–2, 5–6 (1960).

DU CASTEL F., SPIZZICHINO A.: Réflexion en milieu inhomogène. (Réflexions partielles dans l'atmosphère et propagation à grande distance, 3ème partie.) *Ann. Télécomm.* **14**, 33–40 (1959).

DURKEE, A. L.: Results of microwave propagation tests on a 40 mile overland path. *Proc. I.R.E.* **36**, 197 (1948).

DYMOVICH, N. D. (Дымович, Н. Д.): Propagation of radio waves in large cities (in Russian). *Elektrosvyaz* **12**, 26–33 (1958).

DYMOVICH, N. D. (Дымович, Н. Д.): Propagation of metre and decimetre waves over the rough surface of the earth. *Elektrosvyaz* **11**, 26–33 (1957).

ECKART, C.: The scattering of sound from the sea surface. *J. Acoust. Soc. Amer.* **25**, 566–570 (1953).

EGLI, J. J.: Radio propagation above 40 Mc over irregular terrain. *Proc. I.R.E.* **45**, 1383–1391 (1957).

EDISON, A. R., MOORE, R. K. and WARNER, B. D.: Radar terrain return measured at near-vertical incidence. *Trans. I.R.E.* **AP-8**, 246–254 (1960).

ELLIOT, R.: On the theory of corrugated plane surfaces. *Trans. I.R.E.* **AP-2**, 71–81 (1954).

EPSTEIN, J. and PETERSON, D. W.: An experimental study of wave propagation at 850 Mc/S. *Proc. I.R.E.* **41**, 595 (1953).

EVANS, J. V.: The scattering of radio waves by the moon. *Proc. Phys. Soc. London* **70**, 1105–1112 (1957).

EVANS, J. V., EVANS, S. and THOMSON, J. H.: The rapid fading of the moon echoes at 100 Mc/S. Paris Symp. on Radio-Astronomy. Stanford Univ. Press (1959).

EVANS, J. V.: Bandwidth of a moon communication circuit. *Brit. J. Appl. Phys.* **12**, 406–409 (1961).

EVANS, J. V. and TAYLOT, G. N.: Radio echo observations of Venus. *Nature* **184**, 1258–1259 (1959).

FEINBERG, Ye. L. (Фейнберг, Е. Л.): *The Propagation of Radio Waves along the Surface of the Earth* (in Russian). Izd. Akad. Nauk SSSR, Moscow (1961).

FEINSTEIN, J.: Some stochastic problems in wave propagation. Part I. *Trans. I.R.E.* **AP-2,** 23–30 (1954).

FELDMAN, C. B.: Optical behaviour of the ground for short radio waves. *Proc. I.R.E.* **21,** 764–801 (1933).

FELSEN, L. B.: The scattering of electromagnetic waves by a corrugated sheet. *Canad. J. Phys.* **37,** 1565 (1959).

FENSLER, W. E., SENIOR, T. B. A. and SIEGEL, K. M.: Exploring the depth of the surface layer of the moon from a radar space observatory. *Aero Space Eng.* **18,** 38 (1959).

FIELDER, G.: *Structure of the Moon's Surface.* Pergamon Press, Oxford (1961).

FISHBACK, W. T. and RUBENSTEIN, P. J.: Further measurements of 3 and 10 cm reflection coefficients of sea water at small grazing angles. R.L. Report No. 568, 17 May 1954.

(FOK, V. A.) (ФОК, В. А.): The current distribution induced by a plane wave on the surface of a conductor (in Russian). *Zhurn. Eksp. Teor. Fiz.* **15,** 693–702 (1945).

FOK, V. A. (ФОК, В. А.): The field of a plane wave near the surface of a conducting body (in Russian). *Izv. Akad. Nauk SSSR, Ser. Fiz.* **10,** 171–186 (1946).

FOLDY, L. L.: The multiple scattering of waves. *Phys. Rev.* **67,** 107 (1945).

FORD, L. H. and OLIVER, R.: An experimental investigation of the reflection and absorption of radiation of 9 cm wavelength. *Proc. Phys. Soc.* **58,** 265 (1946).

FRICKER, S. J., INGALLS, R. P. and MASON, W. C.: Computation and measurement of the fading rate of moon reflected V.H.F. signals. *J. Res. NBS,* Part D, **64,** 455–465 (1960).

FRICKER, S. J., INGALLS, R. P., MASON, W. C. and STONE, M. L.: U.H.F. moon reflections. Presented at the U.R.S.I.–I.R.E. Meeting, Washington D.C. (1958).

FRICKER, S. J., INGALLS, R. P., MASON, W. C., STONE, M. L. and SWIFT, D. W.: Computation and measurement of the fading rate of moon-reflected UHF signals. *J. Res. NBS* **64-D.** 455–465 (1960).

FRIIS, H. T., CRAWFORD, A. B. and HOGG, D. C.: A reflection theory for propagation beyond the horizon. *Bell Syst. T. J.* **38,** 627–644 (1957).

FURUTSU, K.: Wave propagation over an irregular terrain, part I. *J. Radio Res. Labs. (Japan),* **4,** 135–153 (1957).

GLENTZER, K. V.: 450 Mc. coverage tests at Chicago. *Trans. I.R.E.* **V.C-6,** 20–32 (1956).

GOLDSTEIN, H.: The frequency dependence of radar echoes from the surface of the sea. *Phys. Rev.* **69,** 695 (1946a).

GOLDSTEIN, H.: Frequency dependence of the properties of sea echo. *Phys. Rev.* **70,** 938–946 (1946b).

GORELIK, G. S. (Горелик, Г. С.): The influence of correlation between the velocities of scatterers on the statistical properties of the scattered radiation (in Russian). *Radiotekhn. i Elektron.* **2,** 1227–1233 (1957).

GOUGH, M. W.: V.H.F. and U.H.F. propagation within the optical range. *Marconi Rev.* **12,** 121 (1949).

GOUGH, M. W.: Some features of V.H.F. tropospheric propagation. *Proc. Instn. Elect. Engrs.* **102 B,** 43 (1955).

GOUGH, M. W.: Microwave line of sight propagation. *Elec. Eng.* **30,** 237–247 (1958).

GRANT, C. R. and YAPPLEE, B. S.: Back-scattering from water and land at centimeter and millimeter wavelengths. *Proc. I.R.E.* **45,** 976–982 (1957).

GREENWOOD, J. A. and DURAND, D.: The distribution of length and components of the sum of *n* random unit vectors. *Ann. Math. Stat.* **26,** 233–246 (1955).

GROSSKOPF, J.: Propagation of a surface wave over curved and stratified ground (in German). *Hochfrequ. u. Elektroak.* **58,** 163–171 (1941).

GROSSKOPF, J.: Beitrag zur Ausbreitung der dm- und cm-Wellen über See. Fernmeldetechnisches Zentralamt der deutschen Bundespost, Technischer Bericht 5561 (1960).

GROSSKOPF, J.: Reflexion an rauhem Boden. Fernmeldetechnisches Zentralamt der deutschen Bundespost, Technischer Bericht No. 5570 (1961).

GUDMANDSEN, P. and LARSEN, B. F.: Statistical data for microwave propagation measurements on two overseas paths in Denmark. *Trans. I.R.E.* **AP-5,** 255 (1957).

HAMLIN, E. W. and GORDON, W. E.: Comparison of calculated and measured phase difference at 3.2 cm wavelength. *Proc. I.R.E.* **36,** 1218 (1948).

HARGREAVES, J. K.: Radio observations of the lunar surface. *Proc. Phys. Soc.* **73,** 536–537 (1959).

HATCHER, R. D. and ROHRBAUCH, J. H.: Theory of the echelette grating. Part I. *J. Opt. Soc. Amer.* **46,** 104 (1956): Part II. *J. Opt. Soc. Amer.* **48,** 704 (1958).

HATCHER, R. D., ROHRBAUCH, J. H., PRINE, C. and ZOELLNER, W. G.: Theory of the echelette grating. Part III. *J. Opt. Soc. Amer.* **48,** 410 (1958).

HAYRE, H. S.: Radar scattering cross-section applied to moon return. *Proc. I.R.E.* **49,** 1433 (1961).

HAYRE, H. S. and MOORE, R. K.: Theoretical scattering coefficient for near vertical incidence from contour maps. *J. Res. N.B.S.* D. Radio Propagation, **65-D,** 427–432 (1961).

HEAPS, H. S.: Reflection of plane waves of sound from a sinusoidal surface. *J. Appl. Phys.* **28,** 815–818 (1957).

HEINS, A. E. and BALDWIN, G. L.: On the diffraction of a plane wave by an infinite plane grating. *Math. Scand.* **2,** 103 (1954).

HEY, J. S., PARSONS, J. J. and JACKSON, F.: Reflection of centimetre electromagnetic waves over ground and diffraction effects with wire netting screens. *Proc. Phys. Soc.* **59,** 847 (1947).

HEY, J. S. and HUGHES, V. A.: Paris U.R.S.I. Symposium on radioastronomy, Stanford Univ. Press (1959).

HIATT, R. E., SENIOR, T. B. A. and WESTON, W. H.: A study of surface roughness and its effect on the backscattering cross section of spheres. *Proc. I.R.E.* **48,** 2008–2016 (1960).

HOFFMAN, W. C.: Back-scatter from perfectly conducting doubly-trochoidal and doubly-sinusoidal surfaces. *Trans. I.R.E.* **AP-3,** 96 (1955a).

HOFFMAN, W. C.: Scattering of electromagnetic waves from a rough surface. *Quart. Appl. Math.* **13,** 291 (1955b).

HOYT, R. S.: Probability functions for the modulus and angle of the normal complex variate. *Bell Syst. Tech. J.* **26,** 318–359 (1947).

HUFFORD, G. A.: An integral equation approach to the problem of wave propagation over an irregular surface. *Quart. Appl. Math.* **9,** 4 (1952).

HUGHES, V. A.: Roughness of the moon as a radar reflector. *Nature* **186,** No. 4728, 873–874 (1960).

INGALLS, R. P., BIRD, L. E. and DAY, J. W. B.: Band pass meas. of a lunar reflection. *Proc. I.R.E.* **49,** 631–632 (1961).

ISAKOVICH, M. A. (Исакович, M. A.): The scattering of waves from a statistically rough surface (in Russian). *Zhurn. Eksp. Teor. Fiz.* **23,** 305–314 (1952).

JELONEK, Z., FITCH, E. and CHALK, J. H. H.: Diversity reception. *Wireless Engr.* **24,** 54 (1947).

JONES, J. L. and BARTON, L. E.: Acoustic characteristics of a lake bottom. *J. Acoust. Soc. Amer.* **30,** 142 (1958).

KAYDANOVSKY, N. L. and SALOMONOVICH, A. E. (Кайдановский Н. Л. and Саломонобич А. Е.): Determining the characteristics of the surface of the moon from the observations by radio-telescopes with high resolving power (in Russian). *Izv. Vyssh. Zav. Radiofiz.* **4,** 40–43 (1961).

KALININ, A. I. (Калинин, А. И.): Approximate methods for calculating the field strength of very short waves with respect to the effect of the terrain relief (in Russian). *Radiotekhnika* **12,** No. 4, 13–23 (1957).

KARP, S.: Diffraction by an infinite grating of cylinders. Research Report EM-85, Inst. Math. Sci., New York University (1958).

KARP, S. and RADLOW, J.: On resonance in infinite gratings of cylinders. *Trans. I.R.E.* **AP-4,** 654 (1956).

KATSENELENBAUM, B. Z. (Каценеленбаум, Б. З.): On the problem of normal incidence of a plane electromagnetic wave onto a periodic boundary between two dielectrics (in Russian). *Radiotekhn. i Elektr.* **5,** 1929–1932 (1960).

KATZ, I. and SPETNER, L. M.: Polarization and depression-angle dependence of radar terrain return. *J. Res. N.B.S.* **64-D,** 483–486 (1960).

KATZIN, M.: Back-scattering from the sea surface. I. R.E. Convention Record, Part I, 72–77 (1955).

KATZIN, M.: Recent developments in the theory of sea clutter. I.R.E. Convention Record, Part I, 19, (1956).

KATZIN, M.: On the mechanism of sea clutter. *Proc. I.R.E.* **45,** 44–54 (1957).

KAY, I. and SILVERMAN, R. A.: Multiple scattering by a random stack of dielectric slabs. *Nuovo Cimento* **9,** 626–645 (1958).

KAYLOR, R. L.: A statistical study of selective fading of super high frequency radio signal. *Bell Syst. T. J.* **32,** 1187 (1953).

KELLER, J. B.: A geometrical theory of diffraction. *Calculus of Variations and Its Applications,* ed. L. M. Graves, McGraw-Hill, New-York, 1958.

KENNAUCH, E. M. and COSGRIFF, R. L.: The use of impulse response in electromagnetic problems. *I.R.E. Nat. Conv. Rec.* **6,** 72–77 (1958).

KERR, D. E.: Propagation of very short waves (part 2). *Electronics* **21,** 118 (1948).

KERR, D. E.: *The Propagation of Short Radio Waves.* MIT Radiation Lab. series No. 13 McGraw-Hill, New York 1951.

KERR, F. J.: Radio echoes from sun, moon and planets. *Encycl. of Phys.* **52,** 449–464 (1959).

KERR, F. J., SHAIN, C. A. and HIGGINS, C. S.: Moon echoes and penetration in the ionosphere. *Nature* **163,** 310–313 (1949).

KERR, F. J. and SHAIN, C. A.: Moon echoes and transmissions through the ionosphere. *Proc. I.R.E.* **39,** 230–242 (1951).

KHIZHNYAK, N. A. (Хижняк, Н. А): Artificial anisotropic dielectrics formed by two-dimensional lattices of infinite strips and rods (in Russian). *Zh. Tekh. Fiz.* **29,** 604–614 (1959).

KIELY, D. G.: Measurements of the reflection coefficient of water at $\lambda = 8.7$ mm. *Proc. Phys. Soc.* **63,** 46–49 (1950).

KIELY, D. G.: Some measurements of fading at a wavelength of 8 mm over a very short sea path. *J. Brit. I.R.E.* **14,** 89–92 (1954).

KIELY, D. G.: Quelques problèmes posés par la propagation des ondes de 8 mm et de 3 cm ac dessus de la mer et à travers la pluie. *Annales des Télécomm.* 1ère partie: **11**, 233–244 (1956a); 2ème partie: **11**, 267–279 (1956b).

KIELY, D. G. and CARTER, W. R.: An experimental study of fading propagation at 3 cm wavelength over a sea path. *Proc. I.E.E.*, Pt. III (1952).

KIRBY, R. S.: Correlation in VHF propagation over irregular terrain. *Trans. I.R.E.* **AP-4,** 78–85 (1956).

KIRBY, R. S., DOUGHERTY, H. T. and McQUATE, P. L.: VHF propagation measurements in the Rocky Mountain region. *Trans. I.R.E.* PGVC, July 1956.

KIRBY, R. S., TAFF, J. M. and MOORE, H. S.: Measurments of the effect of irregular terrain on VHF and UHF directive antenna patterns. *Trans. I.R.E.* AP, 167–178 (1952).

KLEIN, W.: Wave propagation studies in the Alps over 15 cm, 90 cm and 2 m wave radio relays. *Brown–Boveri Rev.* **36,** 387 (1949).

KLEIN, W.: Probleme der drahtlosen Übertragung bei mobilen Telephonanschlüssen. *Bull. Tech. P.T.T. Berne* **30,** 331–347 (1952).

KLEIN, W. and DUFOUR, J.: Rundspruchversuche mit frequenz-modulierten Ultrakurzwellen. *Bull. Tech. P.T.T. Berne* **30,** 331–347 (1952).

KLEIN, W. and LIBOIS, L. J.: Essais de transmissions par faisceau hertzien sur un long parcours en visibilité optique entre la France et la Suisse. Résultats d'essais de propagation. *L'Onde Electrique* **33,** 664 (1953).

KOVALEV, A. A. and POZDNYAK, S. I. (Ковалев А. А., Поздняк С. И.): The scattering of electromagnetic waves by statistically rough surfaces with finite conductivity (in Russian). *Radiotekhnika* **16**, 31–36 (1961).

KÜHN, U.: Ein Beitrag zur Kenntnis der Ausbreitungsbedingungen in den Bändern I, II, III und IV. *Radio u. Ferns.* **6,** 439–442 (1957a).

KÜHN, U.: Die Ausbreitung ultrakurzer Wellen im Gelände. *Deutsche Post* **2,** 215–220 (1957b).

KÜHN, U.: Vergleichende Untersuchungen der Ausbreitungsbedingungen in den Bändern I, II, III und IV. *Radio u. Ferns.* **6,** 459 (1957c).

KÜHN, U.: Ausbreitungsuntersuchungen über unterschiedlichem Gelände in den Frequenzbändern I, II und III. Techn. Mitt. BRF 1–7 (1958a), 41–43 (1958b).

KÜHN, U.: Die Ausbreitung von Dezimeterwellen im Gelände. *Radio u. Ferns.* **7,** 346–351 (1958c).

KÜHN, U.: Investigation of radio wave propagation conditions over irregular terrain in bands, I, II, III, IV and V. Doc. Inf. Bull. OIR February 1959.

KÜHN, U.: Ausbreitung von VHF and UHF in unregelmässigem Gelände. *Electromagnetic Wave Propagation*, ed. by Désirant and Michiels. Academic Press, London–New–York 1960, pp. 615–630.

LABRUM, N. R.: The scattering of radio waves by meteorological particles. *J. Appl. Phys.* **23,** 1324–1330 (1952).

LACASCE, E. O.: Note on the backscattering of sound from the sea surface. *J. Acoust. Soc. Amer.* **30,** 578–580 (1958).

LACASCE, E. O. and TAMARKIN, P.: Underwater sound reflection from a corrugated surface. *J. Appl. Phys.* **27,** 138–148 (1956).

LA GRONE, A. H., STRAITON, A. W. and SMITH, H. W.: Synthesis of radio signals on overwater paths. Elec. Engrg. Res. Lab., Univ. of Texas, Report No. 71, 30 April (1954).

LA GRONE, A. H., STRAITON, A. W. and SMITH, H. W.: Synthesis of radio signals on overwater paths. *Trans. I.R.E.* **AP-3,** 48 (1955).

LA GRONE, A. H. and TOLBERT, C. W.: Reflection studies of millimeter and centimeter radio waves for Gulf of Mexico paths. Elec. Engrg. Res. Lab., Univ. of Texas, Report No. 64. Oct. 31, 1952.

LEADABRAND, R. L. *et al.*: Radio scattering from the surface of the moon. *Proc. I.R.E.* **48,** 932–933 (1960).

LEADABRAND, R. L.; DYCE, R. B., FREDRIKSEN, A., PRENSNELL, R. I. and SCHLOBOHM, J. C.: Evidence that the moon is a rough scatterer at radio frequencies. *J. Geophys. Res.* **65,** 3071–3078 (1960).

LEPORSKI, A. V. (Лепорский, A. B.): Scattering of sound waves by sinusoidal and saw-tooth surfaces (in Russian). *Akust. Zhurn.* **2,** 177–181 (1956).

LEWIS, E. A. and CASEY, J. P.: Electromagnetic reflection and transmission by gratings of resistive wires. *J. Appl. Phys.* **23,** 605–608 (1952).

LYSANOV, YU. P. (Лысанов, Ю. П.): On the problem of scattering of electromagnetic waves from a rough surface (in Russian). *Dokl. Akad. Nauk SSSR* **87,** 719–722 (1952).

LYSANOV, YU. P. (Лысанов,Ю. П.): An approximate solution of the problem of scattering of sound waves from an irregular surface (in Russian). *Akust. Zhurn.* **2,** 182–187 (1956).

MACDONALD, F. G.: The correlation of radar sea clutter on vertical and horizontal polarization with wave height and slope. *Conv. Rec. I.R.E.* **4,** 29–32 (1956).

MACKUNE, W. J. and SMITH, H. W.: Comparison of power spectrum estimates of overwater microwave radio signal and associated water waves. E.E.R.L., Univ. of Texas, Report No. 68, May 6, 1953.

MACLUSKY G.S.R. and DAVIES H.: T.R.E. Rep. No. T 1933 (1945).

MAGNUS, W.: On the scattering effect of a rough plane surface. Research Report EM-40 Inst. Math. Sc., New York University (1952).

MAGNUSKI, H. and KOCH, T. F.: Passive repeater bends microwave beam. *Electronics* **26,** 134 (1953).

MANDRAZHI, V. P. and BURNUS, V. V. (Мандражи, В. П., Бурнус, В. В.): Propagation of radio waves in irregular terrain at 300 Mc (in Russian). *Techn. Televid.* No. 30, 50–61 (1959).

MARKOV, A. V. (Марков, A. B.): Distribution of brightness over the lunar disk at full moon (in Russian). *Astronomicheskii Zh. USSR* **25,** 172–179 (1948).

MCGAVIN, R. E. and MALONEY, L. J.: Study at 1.046 Mc/s of the reflection coefficient of irregular terrain at grazing angles. *J. Res. NBS* **63D,** 235–248 (1959).

MCGINN, J. W. JR. and PIKE, E. W.: A study of sea clutter spectra. *Statistical Methods in Radio Wave Propagation.* Pergamon Press, London, 1960, 49–92.

MCPETRIE, J. S.: The reflection coefficient of the earth's surface for radio waves, *Journ. I.E.E.* **82,** 214 (1938).

MCPETRIE, J. S. and FORD, L. H.: Some experiments of the propagation of 9.2 cm wavelength, especially on the effect of obstacles. *Journ. I.E.E.*, Part III A, **93,** 531 (1946).

MEECHAM, W. C.: Variation method for the calculation of the distribution of energy reflected from a periodic surface. *J. Appl. Phys.* **27,** 361–367 (1956a).

MEECHAM, W. C.: A method for the calculation of the distribution of energy reflected from a periodic surface. *Trans. I.R.E.* **AP-4,** 581 (1956b).

MEECHAM, W. C.: Fourier transform method for the treatment of the problem of the reflection of radiation from irregular surfaces. *J. Acoust. Soc. Amer.* **28,** 370 (1956c).

MEECHAM, W. C. and PETERS, C. W.: Reflection of plane-polarized electromagnetic radiation form an echelette diffraction grating. *J. Appl. Phys.* **28,** 216 (1957).

MEGAW, E. C. S.: Experimental studies of the propagation of very short radio waves. *J. Inst. Elect. Engrs*, Part 3A, **93,** 79 (1946).

MEN', A. V., BRAUDE, S. YA. and GORBACH, V. I. (Мень А. В., Врауде С. Я., Горбач В. И.): Fluctuations of the phase front in the propagation of decimetre waves over the surface of the sea (in Russian). *Dokl. Akad. Nauk. SSSR* **125,** 1019–1022 (1959a).

MEN', A. V., BRAUDE, S. YA. and GORBACH, V. I. (Мень А. В., Врауде С. Я., Горбач В. И.): Experimental investigation of the phase fluctuations of centimetre radio waves over the surface of the sea (in Russian). *Izv. Vyssh. Zav. Radiofiz.* **2,** 848–857 (1959b).

MEN', A. V., BRAUDE, S. YA. and GORBACH, V. I. (Мень А. В., Врауде С. Я., Горбач В. И.): The effect of the boundary on the fluctuations of radio waves propagated in an inhomogeneous medium (in Ukrainian). *Dopovidi Akad. Nauk Ukr. RSR*, No. 7, 740–744 (1957).

MILES, J. W.: On nonspecular reflection at a rough surface. *J. Acoust. Soc. Am.* **26,** 191–199 (1954).

MILLAR, J. Z. and BYAM, L. A.: A microwave propagation test. *Proc. I.R.E.* **38,** 619 (1950).

MILLAR, R. F.: Scattering by a grating, I, II. *Canad. J. Phys.* **39,** 81–103 (1961).

MILLINGTON, G.: The reflection coefficient of a linearly graded layer. *Marconi Rev.* **12,** 140 (1949).

MISME, P.: Evanouissements d'interférences causés par des discontinuités frontales. *L'Onde Electrique* **36,** 43 (1956).

MISME, P.: The correlation between the electric field at a great distance and a new radio-meteorological parameter. *Trans. I.R.E.* AP, 289–292 (1958).

MOFENSEN, J.: Radar echoes from the moon. *Electronics* **19,** 92–98 (1946).

MONTIETH, J. L. and SZEICZ, G.: The radiation balance of bare soil and vegetation. *Quart. J. Roy. Meteorol. Soc.* **87,** 159 (1961).

MOORE, R. K.: Resolution of vertical incidence radar return into random and specular components. Univ. of New Mexico, Albuquerque Engrg. Exp. Stat., Tech. Report E.E.T, July 1951.

MOORE, R. K. and WILLIAMS, C. S.: Radar terrain return at near vertical incidence. *Proc. I.R.E.* **45,** 228–238 (1957).

MYAKISHEV, B. YA. (Мякишев, Б. Я.): *Trudy MAI* (Moscow Aviation Institute) No. 98, Oboronizdat, Moscow, 1958 (in Russian).

MURCRAY, D. G., BROOKS, J. N., SIBLE, N. J. and WESTDAL, C.: Infrared background radiation as measured from 85,000 ft. Special report No.1, Contract A.F. 33 (616) — 5199. Denver Res. Inst., Univ. of Denver.

NORTON, K. A.: Transmission loss of space wave propagation over irregular terrain. *Trans. I.R.E.* AP-3, joint U.R.S.I.,-I.R.E. Meeting, 152–166 (1952).

NORTON, K. A. and OMBERG, A. C.: Maximum range of a radar set. *Proc. I.R.E.* **35,** 4 (1947).

NORTON, K. A., SCHULKIN, M. and KIRBY, R. S.: Ground wave propagation over irregular terrain above 50 Mc. Ref. C. Rep. of the Ad Hoc Committee of the F.C.C. for the evaluation of the radio propagation factors concerning the television and frequency modulation broadcasting services in the frequency range between 50 and 250 Mc. 6 June 1949.

NORTON, K. A., VOGLER, L. E., MANSFIELD, W. V. and SHORT, P. J.: The probability distribution of the amplitude of a constant vector plus a Rayleigh-distributed vector. *Proc. I.R.E.* **43,** 1354–1361 (1955).

ONOUE, M., NISHIKORI, K. and NENOHI, M.,: Microwave propagation over the sea beyond the line of sight. *J. Rad. Res. Lab.* (*Tokyo*) **4,** 395–406 (1957).

ORNSTEIN, L. S. and VAN DER BURG, A.: Reflectivity of corrugated surfaces. *Physica* **4,** 1181 (1937).

OXEHUFWUD, A.: Tests conducted over highly reflective terrain at 4000, 6000 and 11,000 Mc. *Trans. Amer. I.E.E.* I, **78,** 265–270 (1959).

PARKER, J. G.: Reflection of plane sound waves from an irregular surface. *J. Acoust. Soc. Amer.* **28,** 627 (1956).

PETERSON, D. W. and EPSTEIN, J.: A method of predicting the coverage of a television station. *R.C.A. Review* **17,** 571 (1956).

PETTENGILL, G. H.: Measurement of lunar reflectivy using the Millstone Radar. *Proc. I.R.E.* **48,** 933–934 (1960).

POLESCHINSKI, E. R.: Propagation conditions for radio-telephone links in mountainous country. *Brown-Boveri Rev.* **46,** 564–572 (1959).

PRICE, R., GREEN, P. E. JR., GOBLICK, I. J. JR., KINGSTON, R. H., KRAFT, L. T., PETTENGILL, G. H., SILVER, R. and SMITH, W. B.: Radar echoes from Venus. *Science* **129,** 751–753; 20 March, 1959.

PRIMICH, R. K.: Some electromagnetic transmission and reflection properties of a strip grating. *Trans. I.R.E.* **AP-5,** 176 (1957).

PROUD, J.: Reflection of sound from a surface of saw-tooth. *J. Appl. Phys.* **28,** 1298 (1957).

PROUD, J. M., JR., BEYER, R. T. and TAMARKIN, P.: Reflection of sound from randomly rough surfaces. *J. Appl. Phys.* **31,** 543–553 (1960).

RAYLEIGH, LORD: *The Theory of Sound.* 3rd ed. MacMillan, London 1896.

RAYLEIGH, LORD: On the light dispersed from fine lines ruled upon reflecting surfaces or transmitted by very narrow slits. *Phil. Mag.* **14,** 350–359 (1907).

REDHEFFER, R.: The dependence of reflection on the incidence angle. *Trans. I.R.E.* **MTT-7,** 423–429 (1959).

RICE S. O.: Mathematical analysis of random noise. *Bell Syst. T.J.*, No. 3, 1944 and No. 1, 1945.

RICE, S. O.: Reflection of electromagnetic waves from slightly rough surfaces. *Comm. Pure Appl. Math.* **4,** 351–378 (1951).

RICHTER, R., BESSIS, J. and CATELLA, P.: Application au radar du phénomène de réflexion des ondes électromagnétiques par le sol. *L'Onde Electrique* **40,** 392–410 (1960).

RIVET, P.: Essais de diversité et étude de l'effet de focalisation sur les liaisons longues en visibilité. *L'Onde Electrique* **36,** 24 (1956).

RUBINSTEIN, P. J. and FISHBACK, W. T.: Preliminary measurements of 10 cm reflection coefficient of land and sea at small grazing angles R. L. Report No. 478, 11 Dec. 1943.

SAXTON, J. A.: Propagation à travers la troposphère. *L'Onde Electrique* **40,** 505–514 (1960).

SAXTON, J. A. and HARDEN, B. N.: Ground wave propagation at UHF. *Nature* **173,** 573–574 (1954).

SAXTON, J. A. and LANE, J. A.: Electrical properties of sea water. Reflection and attenuation characteristics at VHF. *Wirel. Eng.* **29,** 269–275 (1952).

SCHOOLEY, A. H.: A simple optical method for measuring the statistical distribution of water surface slopes. *J. Opt. Soc. Amer.* **44,** 37–40 (1954).

SCHOOLEY, A. H.: Radar reflections from sea waves. *Tele-Tech.* **14,** 70–71 (1955).

SCHOOLEY, A. H.: Some limiting cases of radar sea clutter noise. *Proc. I.R.E.* **44,** 1043 (1956).

SCHOUTEN, J. P. and DE HOOP A. T.: Sur la reflexion d'une onde électromagnétique plane par une surface rugueuse parfaitement conductrice. *Ann. Télécomm.* **12,** 211–214 (1957).

SEMENOV, A. A., KVAVADZE, D. K., NAZAROVA, L. G. and ZVYAGINTSEVA, I. I. (Семенов, А. А., Квавадзе, Д. К., Назарова, Л. Г., Звягинцева, И. И.): Investigation of the reflecting properties of some systems with a periodic structure (in Russian). *Vyest. Mosk. Univers., Ser. Matem., Astron., Fiz.*, No. 1, 107–114 (1958).

SENIOR, T. B. A.: The scattering of electromagnetic waves by a corrugated sheet. *Canad. J. Phys.* **37,** 787–797 (1959). Cf. also important correction. *Canad. J. Phys.* **37,** 1572 (1959).

SENIOR, T. B. A.: Radar reflection characteristics of the moon. Paris Symposium ou Radioastronomy. Stanford Univ. Press (1959).

SENIOR, T. B. A. and SIEGEL, K. M.: A theory of radar scattering by the moon. *J. Res. NBS* **64D,** 217–229 (1960).

SHERWOOD, E. M.: S. Band measurements of reflection coefficients for various types of earth. Sperry gyroscope Co. Report No. 5220–129, 19 Oct. 1943.

SHERWOOD, E. M.: Reflection coefficients of irregular terrain at 10 cm. *Proc. I.R.E.* **43** 877–878 (1955).

SCHERWOOD, E. M. and GINZTON, E. L.: Reflection coefficient of an irregular terrain at 10 cm. *Proc. I.R.E.* **43,** 877 (1955).

SHULMAN, L. A. (Шульман, Л. А.): The reflection of electromagnetic waves from a semi-infinite periodic-layer structure. *Uch. Zap. Tadzh. Univ.* **10,** 103–109 (1957).

SMITH-ROSE, R. L.: A preliminary investigation of radio transmission conditions over land and sea on cm wavelengths. *Journ. I.E.E.* Pt. IIIA **93,** 98–100 (1946).

SMYTH J. B. and ANDERSON L. J.: L'importance du processus de réflexion dans les communications au delà de l'horizon. *Onde électr.* **37,** 485–490 (1957).

SOFAER, E.: Phase-coherent back-scatter of radio waves at the surface of the sea. *Proc. I.E.E.*, Pt. B, **105,** 383–394 (1958).

SOLLFREY, W. and SCHMOYS, J.: Gratings of circular cylinders. Research Report EM-41, Inst. Math. Sci., New York University (1952).

SPETNER, L. M.: A statistical model for forward scattering of waves of a rough surface. *Trans. I.R.E.* **AP-6,** 88–94 (1958).

SPETNER, L. M. and KATZ, I.: Two statistical models for radar terrain return. *Trans. I.R.E.* **AP-8,** 242–246 (1960).

SPIZZICHINO, M. A.: La reflexion des ondes électromagnétiques par une surface irregulière. Research Report No. 549 T, Centre National d'Etudes des Télécommunications, 4.11.1959 (1959).

STANLEY, R. C.: Some problems in modelling water waves relating to the back-scattering of microwave radio-elergy. Texas University Report No. 98 (1958).

STANLEY, R. C.: Some problems in modelling water waves for the study of back-scattering of microwave radio elergy. *Deutsche Hydrogr. Z.* **11,** 255–258 (1958).

STOKER, J. J.: Water waves. *The Mathematical Theory with Applications* Interscience Publishers New-York-London 1957.

STRAITON, A.: Reflection of centimeter waves from ground and water. *Trans. I.R.E.* **AP-4,** August 1955.

STRAITON, A. W., GERHARDT, J. P., LA GRONE, A. H. and TOBERT, C. W.: Reflection of centimeter and millimeter radio waves from the surface of a small lake. Elec. Engrg. Res. Lab., Univ. of Texas. Report No. 63, 15 May 1952.

STRAITON, A. W. and TOLBERT, C. W.: Measurement and analysis of instantaneous radio height-gain curves at 8.6 mm over rough surfaces. *Trans. I.R.E.* **AP-4,** 346–351 (1956).

STRAITON, A. W. and TOLBERT, C. W.: Moon reflection studies with bistatic radar at 3000 mc/S. *Comm. and Electr.* **50,** 436–460 (1960). *Elec. Engng.* **79,** 503–506 (1960).

SUZUKI, M.: Equivalent reflection coefficients in Gc/s band propagation over snow covered terrain. *J. Inst. El. Comm. Eng. Japan* **42,** 490–496 (1959).

SWARM, H. M., GHOSE, R. N. and KEITEL, G. H.: Tropospheric propagation in horizontally stratified media over rough terrain. *Proc. I.E.E.*, Pt. II, **100,** 77–84 (1953).

TAI, C. T.: Reflection and refraction of a plane electromagnetic wave at a periodical surface. Harvard Techn. Rep. No. 28, Cruft Lab., January 1948.

THOMPSON, L. E.: Microwave propagation experiments. *Proc. I.R.E.* **36,** 671 (1948).

THUE, M.: Etude de la propagation d'ondes hertziennes de 8cm entre le Mont Pilat et le Mont Ventoux. Etude CNET No. 343T (1955).

TOMLINSON, H. T. and STRAITON, A. W.: Analysis of 3-cm radio height-gain curves taken over rough terrain. *Trans. I.R.E.* **AP-7,** 405–413 (1959).

TREXLER, J. H.: Lunar radio echoes. *Proc. I.R.E.* **46,** 286–292 (1958).

TROLESE, L. G.: Characteristics of tropospheric scattered fields. *Proc. I.R.E.* **43,** 1300–1305 (1955).

TWERSKY, V.: On the non-specular reflection of plane waves of sound. *A. Acoust. Soc. Amer.* **22,** 539–546 (1950).

TWERSKY, V.: On the non-specular reflection of sound from planes with absorbent bosses. *J. Acoust. Soc. Amer.* **23,** 336–338 (1951a).

TWERSKY, V.: On the nonspecular reflection of electromagnetic waves. *J. Appl. Phys.* **22,** 825–835 (1951b).

TWERSKY, V.: Multiple scattering of radiation by an arbitrary planer configuration of parallel cylinders and by two parallel cylinders. *J. Appl. Phys.* **23,** 407–414 (1952).

TWERSKY, V.: Reflection coefficients for certain rough surfaces. *J. Appl. Phys.* **24,** 569–660 (1953).

TWERSKY, V.: Certain transmission and reflection theorems. *J. Appl. Phys.* **25,** 859–862 (1954).

TWERSKY, V.: On the scattering of waves by an infinite grating. *Trans. I.R.E.* **AP-4,** 330 (1956a).

TWERSKY, V.: Scattering theorems for bounded periodic structures. *J. Appl. Phys.* **27,** 1118–1122 (1956).

TWERSKY, V.: On the scattering and reflection of electromagnetic waves by rough surfaces. *Trans. I.R.E.* **AP-5,** 81–90 (1957a).

TWERSKY, V.: On scattering and reflection of sound by rough surfaces. *J. Acoust. Soc. Amer* **29,** 209–225 (1957b).

TWERSKY, V.: Calcul des coefficients de réflexion et des sections droites différentielles de diffraction pour les surfaces irrégulières. *Ann. Télécomm.* **12,** 214–216 (1957c).

TWERSKY, V.: Scattering by quasi-periodic and quasi-random distributions. *Trans. I.R.E.* **AP-7,** 307–319 (1959).

TWERSKY, V.: On multiple scattering of waves. *J. Res. NBS* **64D,** 715–730 (1960).

VAYNSHTEYN, L. A. (Вайнштейн, Л. А.): Diffraction of electromagnetic waves by a grating of parallel conducting strips (in Russian). *Zhurn. Tekh. Fiz.* **25,** 847–852 (1955).

VECCHIACCHI, F.: Prove di propagazione con la frequenza di 1000 MHz su vari percosi. *Alta Frequenza* **25,** 100 (1956).

VICTOR, W. K., STEVENS, R. and GOLOMB, S. W.: Radar exploration of Venus: Goldstone Observatory Report for March–May 1961. Jet propulsion Laboratory. California Institute of Technology, Pasadena, Calif. Technical Report No. 32–132 (August 1961).

VOGE, J.: Réflexion spéculaire et réflexion diffuse sur des feuillets atmosphériques. (Part IV of DU CASTEL, MISME, SPIZZICHINO and VOGE [1960]) *Ann. Télécomm.* **15,** 38–47 (1960).

VOGE, J.: Réflexion et diffusion troposphérique des ondes radioélectriques. (Part V of DU CASTEL, MISME, SPIZZICHINO and VOGE [1960]) *Ann. Télécomm.* **15,** 107–121 (1960).

VOGE, J.: Note sur le role des feuillets à surface irrégulière dans la formation des couches E sporadique. *Ann. Télécomm.* **16,** 62–63 (1961).

VVEDENSKY, B. A. and ARENBERG, A. G. (Введенский, В. А., Аренберг, А. Г.): *Propagation of Ultrashort Waves* (in Russian). Sovyetskoe Radio. Moscow 1948.

WAIT, J. R.: Reflection at arbitrary incidence from a parallel wire grid. *Appl. Sci. Res. B* **4,** 393 (1955).

WAIT, J. R.: On the theory of reflection from a wire grid parallel to an interface between homogeneous media. *Appl. Sci. Res. B.* **7,** 355–360 (Part II) (1958).

WAIT, J. R.: The calculation of the field in a homogeneous conductor with a wavy interface. *Proc. I.R.E.* **47,** (1959a).

WAIT, J. R. Guiding of electromagnetic waves by uniformly rough surfaces. *Trans. I.R.E.* **AP-7,** 154–162 (Part I), 163–168 (Part II) (1959b).

WAIT, J. R.: *Electromagnetic Waves in Stratified Media.* Pergamon Press, 1962.

WAIT, J. R. and CONDA, A. M.: Pattern of an antenna on a curved lossy surface. *Trans. I.R.E.* **AP-6.** 348–359 (1958).

WATERMAN, A. T.: A rapid beam swinging experiment in transhorizon propagation. *Trans. I.R.E.* **AP-6,** 338–340 (1958).

WEBER, J.: Scattering of electromagnetic waves by wires and plates. *Conv. Rec. I.R.E.* Pt.I, 31–38 (1954).

WEINSTEIN, L. A., cf. VAYNSHTEYN, L. A.

WHEELER, B. F. and MATHWICH, H. R.: Use of distribution curves in evaluating microwave path clearance. *Proc. I.R.E., Trans. on Instrumentation* **31,** (1955).

WILTSE, I. C., SCHLESINGER, S. P. and JOHNSON C.M.: Back-scattering characteristics of the sea in the region from 10 to 50 kMc. *Proc. I.R.E.* **45,** 220–228, 244–246 (1957).

YAMPOL'SKI, V. G. (Ямпольский, В. Г.): Reflection of a plane wave from a wire grid at normal polarization (in Russian). *Radiotekhn.* **11,** No. 11, 33–37 (1956).

YAPPLEE, B. S., BRUTON, R. H., CRAIG, K. J. and ROMAN, N. G.: Radar echoes from the moon at the wavelength of 10 cm. *Proc. I.R.E.* **46,** 293–297 (1958).

YAPPLEE, B. S., ROMAN, N. G., CRAIG, K. J. and SCANLAN, T. F.: A lunar radar study at 10 cm wavelength. Paris Symposium on radio-astronomy. Stanford Univ. Press (1959).

ZUHRT, H.: Die Summenhäufigkeitskurven der exzentrischen Rayleigh-Verteilung und ihre Anwendung auf Ausbreitungsmessungen. *Arch. el Übertr.* **11,** 478–484 (1957).

AUTHOR INDEX

SUBJECT INDEX

www.ingramcontent.com/pod-product-compliance
Lightning Source LLC
Chambersburg PA
CBHW051255200926
51612CB00001B/51
* 9 7 8 0 8 9 0 0 6 2 3 8 8 *